Contemporary Precalculus

through applications

functions, data analysis and matrices

Contemporary Precalculus

through applications

functions, data analysis and matrices

Department of Mathematics and Computer Science
The North Carolina School of Science and Mathematics

Gloria B. Barrett
Kevin G. Bartkovich
Helen L. Compton
Steve Davis
Dorothy Doyle
John A. Goebel
Lawrence D. Gould
Julie L. Graves
Jo Ann Lutz
Daniel J. Teague

Janson Publications, Inc
Dedham Massachusetts

The North Carolina School of Science and Mathematics is an affiliate member of the University of North Carolina System.

Acknowledgment: Portions of this text were previously published in the series New Topics for Secondary School Mathematics: *Data Analysis*, *Geometric Probability*, and *Matrices*, © 1988, National Council of Teachers of Mathematics, and appear here with permission of the publisher.

Library of Congress Cataloging -in-Publication Data

Contemporary precalculus through applications: functions, data analysis, and matrices/Department of Mathematics and Computer Science, the North Carolina School of Science and Mathematics, Gloria B. Barrett ...[et al.].
p. cm.
Includes index.
ISBN 0-939765-54-3
1. Functions. 2. Matrices. 3. Mathematical statistics.
I. Barrett, Gloria B. 1948- . II. North Carolina School of Science and Mathematics. Dept. of Mathematics and Computer Science.
QA331.3.C67 1991
515--dc20 91-17557
CIP
AC

Revised, 1992

Printed in the United States of America.

9 8 7 6 5 4 3 2 2 3 4 5 6 7 8 9

Design: Malcolm Grear Designers

Typesetting: Rosenlaui Publishing Services and Verbatim, Inc.

Illustration: LM Graphics

To Henry Pollak for his inspiration and support
and to the students of NCSSM who learned along with us.

CONTENTS

PREFACE

Contemporary Precalculus through Applications provides students with an applications-oriented, investigative mathematics curriculum in which they use technology to solve problems and to enhance their understanding of mathematics. The topics presented lay a foundation to support future course work in mathematics including calculus, finite mathematics, discrete mathematics, and statistics. The topics also provide an introduction to the mathematics used in engineering, the physical and life sciences, business, finance, and computer science. Whenever possible, new material is presented in the context of real-world applications. Students are active learners who generate ideas for both the development of problem statements and for the solution of problems. Students learn to use a variety of techniques to solve problems that are investigated from a number of perspectives as they proceed through the course. Problems are presented in the context of real-world applications, and, consequently, the interpretation of solutions is given strong emphasis.

The goals of *Contemporary Precalculus through Applications* parallel those of the National Council of Teachers of Mathematics *Curriculum and Evaluation Standards*, a document developed simultaneously with this textbook. A primary goal of the authors is to foster the development of mathematical power in students. Other student needs addressed include: exposure to real-world applications of mathematics in a wide variety of disciplines so that students can learn to value mathematics; preparation for future course work in mathematics; development of the self-confidence necessary to undertake further study; and opportunities to use modern technology to enhance understanding and to solve problems.

The fabric of *Contemporary Precalculus through Applications* is woven of six spiraling themes treated with increasing depth and breadth at each exposure as follows.

Mathematical Modeling

The use of mathematics to model a wide variety of phenomena is central to the course. The modeling approach is used in analyzing sets of data; introducing and applying the various elementary functions; and applying matrices to various problem-solving situations. Additionally, the modeling approach provides motivation for student study. Further, problem situations are often presented in such a way that the student must supply the mathematical framework required in the solution process.

Computers and Calculators as Tools

Technology has lessened the need for extensive computation using pencil and paper techniques. Now students can focus on mathematical concepts and structure while calculators or computers carry out the computations. In this text, calculators are used in evaluating expressions, in applying numerical algorithms, and as an important aid in making conjectures.

The graphing calculator and the computer are used for calculations related to functions and matrices and also for quick and accurate graphing. Empirical models are central to the course, and the graphing calculator and the computer are used to develop such models from actual data (which students sometimes gather themselves) through re-expression and curve fitting. Throughout, the design of the material is based on the assumption that appropriate technology is available to students. This might include a single microcomputer available for demonstrations by the teacher, graphing calculators for individual students, or a computer lab for class and individual use.

Applications of Functions

The overall goal of the study of elementary functions is to illustrate how functions serve as bridges between mathematics and the situations they model. The study of specific functions is motivated by the need for tools to build empirical models and to approximate trends in data. The importance of compositions and inverses of functions is heightened by the frequency of their use in building mathematical models. The focus on understanding the behavior of functions leads to an emphasis on graphing, using hand-drawn sketches and a computer or graphing calculator. The geometry of the functions is often used to enhance understanding of the algebra.

Data Analysis

The principal goal of data analysis is to deduce information from data. A conceptual treatment of resistant and least squares techniques of curve fitting is given here. Students are reminded that observation and measurement in the real world result in values for variables, rather than formulas for functions. Techniques of data analysis allow students to uncover what, if any, functional relationship exists between the relevant variables. Re-expression of data by means of elementary functions is helpful in extracting information and gaining insight from data. Through discussions of residuals and causation versus association, students learn that predictions based on techniques of data analysis have limits to their accuracy and reliability.

Discrete Phenomena

The traditional precalculus-calculus course is based primarily on the mathematics of continuous functions. In contrast, however, discrete techniques are required for the mathematical analysis of many phenomena. Discrete mathematics topics included in this text include finance, population growth, economic production, and Markov chains.

Numerical Algorithms

The pervasiveness of fast and inexpensive calculating tools requires that students be able to use numerical methods to solve problems for which no exact method exists or for which the exact method is beyond the ability of the student. Examples of such situations include determining extreme values of a function, identifying the zeros of a function, and finding the points of intersection of polynomial and transcendental functions. Such problems, which illustrate applications of elementary mathematics, can be solved with the assistance of a computer and appropriate software or a graphing calculator.

Contemporary Precalculus through Applications is designed to encourage students to approach mathematics in new and innovative ways. A conscious effort is made to combine several of the major themes in examples and problems, to ask familiar questions in new contexts, and to apply new concepts to familiar questions. Each theme that spirals through the course has an effect on the instructional approach, but none more than the use of the computer and graphing calculator as tools. Having a computer in front of the classroom or regularly using a graphing calculator during discussions enables students to ask and answer "What if ...?" questions, to make conjectures, to check their guesses and analyses, and to work with real data. All these activities are invaluable contributions to the learning process.

The models that are developed in the course come from many diverse areas. Naturally, models from the sciences are included, but, in addition, models from banking and finance, anthropology, economics, sociology, sports, and environmental issues also appear. The applications vary in complexity and depth and are often revisited as new techniques are learned. For example, characteristics of the simple exponential function are investigated and applied. Later, data analysis is used to model the phenomenon of population growth with the exponential function. Finally, the Leslie matrix is used to explore age-specific population growth, with the computer assisting in the consideration of long run expectation.

The course is not designed for a head-down march through the syllabus. The authors expect and have worked to create opportunities for discussions in class, additional questions, and reflection. For some problems there is no single "best" answer. The course has been constructed with the philosophy that the quality of learning is more important than completing a syllabus.

Acknowledgments

The curriculum and ideas of this textbook were developed with funding from the Carnegie Corporation of New York and the National Science Foundation. The endorsement and support of the Carnegie Corporation of New York through Alden Dunham encouraged us to develop a different precalculus course and to ask hard questions about what concepts are essential to the mathematical development of students. The National Science Foundation support gave us the opportunity to make revisions based on teaching experience and interaction with teachers across the United States. Any opinions, finding, and conclusions or recommendations expressed through this project are those of the authors and do not necessarily reflect the views of the Carnegie Corporation of New York or the National Science Foundation.

Three chapters of this book were published previously by the National Council of Teachers of Mathematics. NCTM has given permission for those chapters to be included in this textbook.

The years spent working on this project have confirmed the excitement of learning — by teachers and by students. A remarkable number of people have contributed. The students of the North Carolina School of Science and Mathematics from 1984 to 1991 played a significant role in the development of this book. A number of colleagues at NCSSM gave us support and help throughout the project; however, very special thanks go to Billie Bean, Nicole Holbrook, John Kolena, Mary Malinauskus, John Parker, Bonnie Ramey, Donita Robinson, and Dotte Williams. Their contributions included frequent feedback and ideas based on their classroom teaching experiences, proofreading and providing answers. Doug Shackelford, a 1985 graduate of NCSSM, created the software that enabled us to implement these ideas and continued to support that software until the NCTM publications were complete. A grant from IBM Corporation provided the equipment our students used and funded the beginning of the software development.

A number of teachers from across the United States became valued friends and colleagues through our work in this project. Landy Godbold from the Westminster Schools in Atlanta, Georgia, is an extraordinary and forward-looking teacher and was willing to gamble on this new curriculum in its earliest forms. We benefited from his creative ideas and findings throughout the project. David Bannard of The Collegiate Schools in Richmond, Virginia, Tom Seidenberg of Phillips Exeter Academy in Exeter, New Hampshire, Rick Jennings of Eisenhower High School in Yakima, Washington, Stan Mathes of Niskayuna High School in Schenectady, New York, Jon Choate of The Groton School in Groton, Massachusetts, Wally Green of Jordan High School in Durham, North Carolina and Jim Ebert of Durham Academy in Durham, North Carolina, were all very helpful in suggesting improvements and giving us constructive feedback. Teachers who took part in NCSSM summer math workshops, the Secondary School Mathematics and Computer Conferences at Phillips Exeter Academy, the Woodrow Wilson Institutes, and especially the NSF teachers were all essential to the development of the book. We learned a lot together.

Dr. Mary Kiely, of Stanford University and formerly with the Carnegie Corporation of New York, was a wonderfully supportive grant officer. Her enthusiasm and confidence in the project gave us the courage to pursue new ideas in secondary mathematics.

Zedra Williams and Cindy Colimon worked diligently in the production of the text and daily supported us in our work. Corlise Ferrell's technical expertise in the preparation of the manuscript was invaluable.

A very special thanks to the publisher, Barbara Janson, whose patience and expertise have brought this textbook to fruition. We particularly appreciate the freedom she gave us to produce a text that we believe supports the current needs in precalculus mathematics.

Our families — Rita, Doc, Jo-Ann, Judd, Chris, Emily, Scott, Bill, Kate, Ted, Lillian, and Nancy — deserve our greatest thanks for giving us the time and the support first to dream about and then to complete this project.

We are sad that Lawrence Gould, our colleague and co-author, did not see the completion of this textbook. Throughout the development of the course and in our daily work, Lawrence offered us the wisdom that experience brings. He passed away in May of 1990.

CASIO SCIENTIFIC CALCULATOR fx-7000GA
SHIFT
Prog
MODE
Graph
Range
ENG
TEXAS INSTRUMENTS TI-81
Y=
RANGE
ZOOM
TRACE
GRAPH
2nd
INS
DEL
ALPHA
XIT
MODE
MATRX
VARS
CLEAR
SIN
Exercises

Contemporary Precalculus

through applications

functions, data analysis and matrices

ARMSPAN
HEIGHT
158.7cm
170.3cm

CHAPTER 1

Data Analysis One

1

Some Thoughts about Models and Mathematics

When children think about models, they are generally considering some kind of a toy, perhaps an airplane or a dinosaur. When scientists and mathematicians think about models, they are generally considering a model as a tool, even though they may be thinking about the same airplane or dinosaur. Scientists and mathematicians use models to help them study and understand the physical world. People in all walks of life use models to help them solve problems; problems in this course will involve models used by bankers, city planners, anthropologists, geologists, economists, automobile makers, and many, many others.

So just what is a model? Models are representations of phenomena. In order to be useful, a model must share important characteristics with the phenomenon it represents, and it must also be simpler than what it represents. A model usually differs significantly from what it represents, but these differences are offset by the advantage that comes from simplifying the phenomenon. A good example is a road map, which models the streets and highways in a particular area. Clearly, a map has a lot in common with the actual streets and highways — it shows how roads are oriented and where they intersect. A road map simplifies the situation; it ignores stoplights, steep ascents, and back alleys and instead focuses on major thoroughfares. Such a map is very useful for traveling from one city to another, but is not much good for finding the quickest route to the shopping mall or the best street for skateboarding. Road maps, and most other models, are useful precisely because they ignore some extraneous information and thereby allow you to see other information more clearly.

Another fairly common model is an EKG, which models the electrical activity of the heart. The EKG is an excellent model when used to determine the heart rate or to find which regions of the heart may be damaged after a heart attack. It is a very poor model for determining the volume of blood flowing through the heart or the condition of the valves. Different models emphasize different aspects of a phenomenon; the choice of what model to use depends on what aspect is under investigation.

The ability to predict is the ultimate test for a model. A good model allows us to make accurate predictions about what will occur under certain conditions. If what actually occurs is very different from our prediction, then the model is of little use.

Scientists and mathematicians often need to update or revise models as more is learned about the phenomenon under study. Sometimes a model needs to be completely discarded and replaced with a new one. The pre-Columbian model of the flat world was discarded, and Ptolemy's geocentric model of the universe was revised by Copernicus and then overthrown by Galileo.

Even though Isaac Newton's models for the actions of a gravitational field have been replaced by Einstein's relativistic model, we still use Newtonian physics under everyday conditions because it is easier and because it gives reasonably accurate results. The aspects of Einsteinian mechanics that are ignored are largely irrelevant in most everyday applications, so the Newtonian model is still a good one.

As we move through the course, we will encounter phenomena that we want to know more about. Our task will be to find a mathematical expression or formula or picture that mimics the phenomenon we are interested in. This model must accurately represent the aspects of the phenomenon that we care about, but it may be very different from the phenomenon in other ways. To be able to find a model to represent a problem, we need to have a large toolkit of mathematical information and techniques at our disposal. The fundamental concepts learned in Algebra 1, Geometry, and Algebra 2 are all a part of our toolkit. We will also use the calculator and computer as tools to construct and analyze models for the phenomena we study. Probably the most important tools necessary for model making are an inquisitive mind and a determined spirit.

Often we will not stop after we have developed one model but will form two or three to get a better view of the subject. For example, suppose a rock is thrown into the air. How can its path be modeled? It would be informative if we had a picture of the flight. We can use a graph as a model for this phenomenon (see Figure 1).

We could also describe the flight by an equation. If the height above the ground is called h and the horizontal distance away from the thrower is called d, then we can represent the flight by the formula

$$\text{Model 2:} \quad h = -\frac{2}{5}d^2 + \frac{4}{5}d + 1.$$

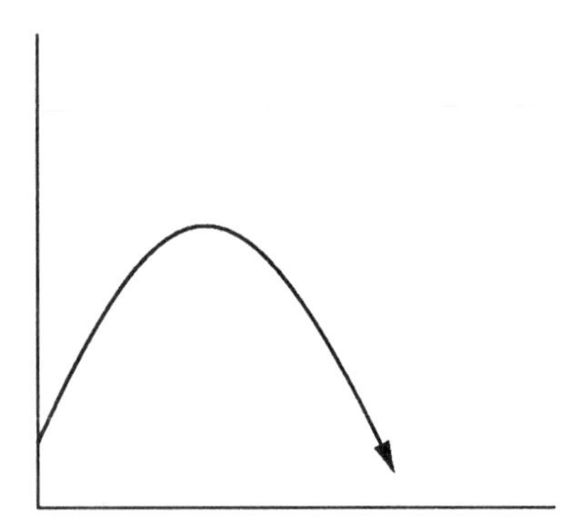

Figure 1 Model 1

Notice that these models do not give complete information about the problem. We cannot tell from the models what type of rock was thrown, who threw it, or why. These aspects of the problem are not relevant — all we really care about is distance and time. Both models give this information. The strength of any model is in what it emphasizes and what it ignores.

2

Analysis of Sets of Ordered Pairs

We begin our course with graphical models of data. These models provide information so that we can answer questions like the following:

— are high school graduation rates related to state spending on education?

— what will be the winning speed for the New York Marathon in the year 2000?

— is there a relationship between the amount of time a student spends studying for a test and the grade on the test?

Though the questions posed above are different in many respects, each requires the collection,

Table 2 Public School Spending per Pupil and High School Graduation Rates

	Spending	Graduation Rate		Spending	Graduation Rate
Alaska	$8,842	67.1%	California	$3,751	65.8%
New York	$6,299	62.7%	Iowa	$3,740	86.5%
Wyoming	$6,229	74.3%	Maine	$3,650	78.6%
New Jersey	$6,120	77.3%	Texas	$3,584	63.2%
Connecticut	$5,532	80.4%	New Mexico	$3,537	71.9%
Dist. of Columbia	$5,349	54.8%	North Carolina	$3,473	70.3%
Massachusetts	$4,856	76.3%	Nebraska	$3,437	86.9%
Delaware	$4,776	69.9%	New Hampshire	$3,386	75.2%
Pennsylvania	$4,752	77.2%	Indiana	$3,379	76.4%
Wisconsin	$4,701	84.0%	Missouri	$3,345	76.1%
Maryland	$4,659	77.7%	Louisiana	$3,237	54.7%
Rhode Island	$4,574	67.6%	North Dakota	$3,209	86.1%
Vermont	$4,459	83.4%	South Dakota	$3,190	85.1%
Hawaii	$4,372	73.8%	Georgia	$3,167	62.6%
Minnesota	$4,241	90.6%	Kentucky	$3,107	68.2%
Oregon	$4,236	72.7%	South Carolina	$3,005	62.4%
Kansas	$4,137	81.4%	West Virginia	$2,959	72.8%
Colorado	$4,129	72.2%	Tennessee	$2,842	64.1%
Montana	$4,070	82.9%	Arkansas	$2,795	75.7%
Florida	$4,056	61.2%	Arizona	$2,784	64.5%
Illinois	$3,980	74.0%	Oklahoma	$2,701	71.1%
Michigan	$3,954	71.9%	Alabama	$2,610	63.0%
Virginia	$3,809	73.7%	Idaho	$2,555	76.7%
Washington	$3,808	74.9%	Mississippi	$2,534	61.8%
Ohio	$3,769	76.1%	Utah	$2,455	75.9%
Nevada	$3,768	63.9%			

organization, and interpretation of data. To answer each question, we need to analyze the relationship between several variables. We will limit our attention to paired measurements, that is, the *analysis of two variables.* Sometimes one variable actually depends on the other; for example, we expect that blood pressure in adults of the same height in some way depends on weight and that crop yield depends on amount of rainfall. Other times there is a relationship between the variables, but it is not one of cause and effect or dependence. For example, we can show that there is a relationship between points scored and personal fouls committed by college basketball players, but we would certainly not consider one of these variables to be dependent on the other. Moreover, sometimes there is no relationship at all between the two variables; for example, we do not expect there to be a relationship between the distance a student lives from school and his or her height.

To determine whether there is a relationship between two variables, we must analyze data consisting of ordered pairs. Sometimes these data are gathered from a well-designed, carefully controlled scientific experiment. Other times we want to analyze data that exist in the world around us. The

September 7, 1987, *U.S. News and World Report* claims that "spending heavily on teachers doesn't always yield a bumper crop of graduates." The comment is followed by the data provided in Table 2. Study this list to determine whether you agree with the statement made by the magazine.

How do we get information out of the list of numbers in Table 2? Did you actually read all of it, or did you skip to this paragraph? Presented as just a table of numbers, the data are difficult to interpret. In analyzing data, we search for information hidden in the numbers. It should be obvious that we need to have some way of organizing and simplifying the data so that we can see its essential characteristics without getting lost in a jumble of numbers. Then we can decide for ourselves whether or not graduation rate is related to educational spending.

3
Scatter Plots

One step in analyzing the relationship between two variables is to make a *scatter plot.* A scatter plot is simply a graph in a rectangular coordinate system of all ordered pairs of data. Scatter plots display data so that we can see the general relationship between two variables. A computer-generated scatter plot of the fifty-one data points listed in Table 2 is shown in Figure 3. Each observation in the data set is represented by one point in the plane. State spending is plotted on the horizontal axis, and graduation rate is plotted on the vertical axis. For example, the ordered pair representing Montana is ($4070, 82.9%). When making a scatter plot, it really does not matter which variable is plotted on which axis. If we suspect that one variable depends on the other, however, we usually plot the dependent variable on the vertical axis and the independent variable on the horizontal axis.

Study the scatter plot in Figure 3. Do you agree with the *U.S. News and World Report* claim? Does there appear to be any relationship between state spending and graduation rate? Are you surprised by the point that lies far to the right of the others? Which display do you find easier to interpret, the table or the graph?

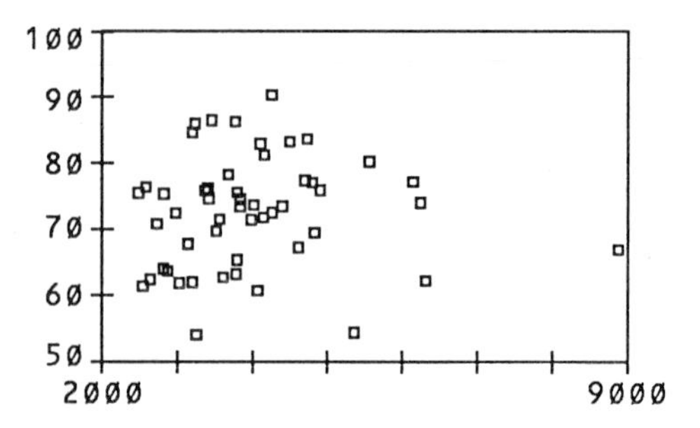

Figure 3 Graduation Rate versus Public School Spending

A scatter plot is an effective tool for analyzing data. Special characteristics of data that may be unnoticed in a table are more obvious from a graph. If there is some relationship between the variables, a pattern or trend is usually apparent in the scatter plot.

Example 1

Basketball Statistics

The data in Table 4, provided by the Sports Information Offices at Duke University and the University of North Carolina, Chapel Hill, summarize the total number of rebounds, assists, personal fouls committed, and points scored by individual basketball players during the 1984–85 basketball season.

Do you think there is a relationship between the number of personal fouls that a basketball player commits in a season of play and the number of points he scores? The relationship (or lack thereof) should be more obvious if we plot the data. On the computer-generated scatter plot shown in Figure 5, page 8, personal fouls are plotted on the horizontal axis and points scored are on the vertical axis.

Looking at the scatter plot should convince us that there is a relationship between these variables. The points tend to slope upward to the right, so we

Table 4 Basketball Statistics

Player	Rebounds	Assists	Total Fouls Committed	Total Points Scored
Duke University				
Alarie	158	49	78	492
Amaker	69	184	55	253
Anderson	20	0	12	15
Bilas	186	14	96	312
Bryan	8	2	6	12
Dawkins	141	154	64	582
Henderson	98	51	69	317
King	50	22	34	51
Meagher	130	46	85	241
Nessley	18	3	15	16
Strickland	36	15	37	108
Williams	29	2	15	47
University of North Carolina, Chapel Hill				
Brust	4	3	2	2
Daugherty	349	77	112	623
Daye	3	2	1	6
Hale	119	168	102	349
Hunter	36	32	34	120
Martin	199	44	103	347
Morris	5	0	0	8
Peterson	65	57	31	198
Popson	87	19	66	211
Roper	1	0	2	0
K. Smith	92	235	54	444
R. Smith	19	7	18	65
Wolf	158	58	79	274

observe that the players who commit the most personal fouls also score the most points. When both variables increase together, we say there is a *positive association* between them. What is the general shape of the plot? There does not appear to be any obvious curvature; rather, the points seem to be increasing steadily. How strong is the relationship between these variables? To answer this question, you might want to consider how well you would be able to predict the points scored by a player commiting 25 personal fouls. The scatter plot shows quite a bit of spread in the data. Though the variables are related, we would have to consider the relationship loose, or weak, in the sense that knowing a value of one variable does not give us confidence in predicting a value for the other variable. Are there any clusters of points on the graph? If so, where are they? Can you offer a reasonable explanation for the clusters? Are there any points that appear to stand out from the rest, that is, points that do not seem consistent with the other observations? If so, such points deserve special attention in our analysis. Sometimes we observe points that follow the general pattern of the data but are far removed from other points. Other times there are points that are inconsistent with the general trend. Such points may indicate errors in measurement or in plotting that need to be corrected, or they may indicate the presence of some factor that deserves special attention. Whatever the cause, we should look for, and attempt to explain, odd points, called *outliers*, that do not appear to fit the general pattern of the scatter plot.

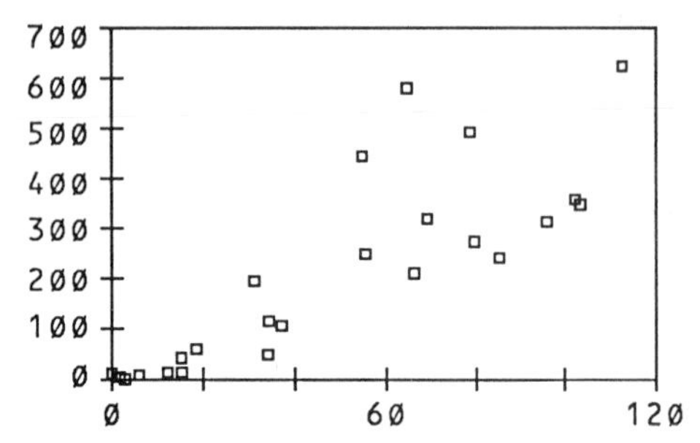

Figure 5 Points Scored versus Personal Fouls

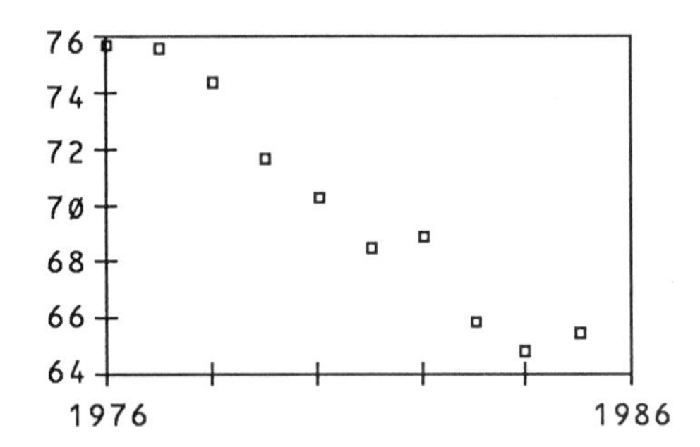

Percent of People Retiring Who Are Age 64 or Older versus Year

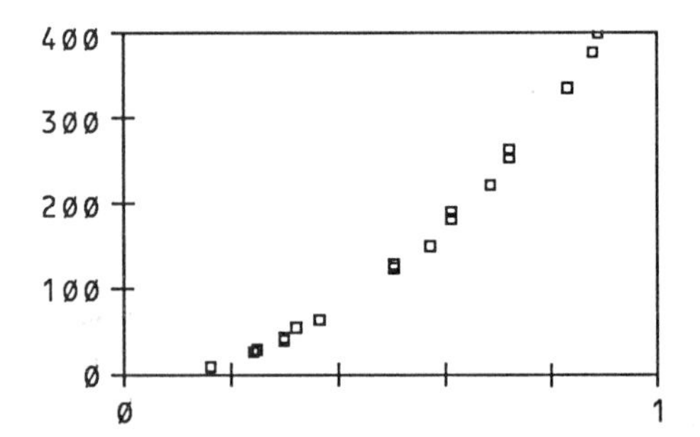

Distance Fallen for Object in Free-Fall versus Time

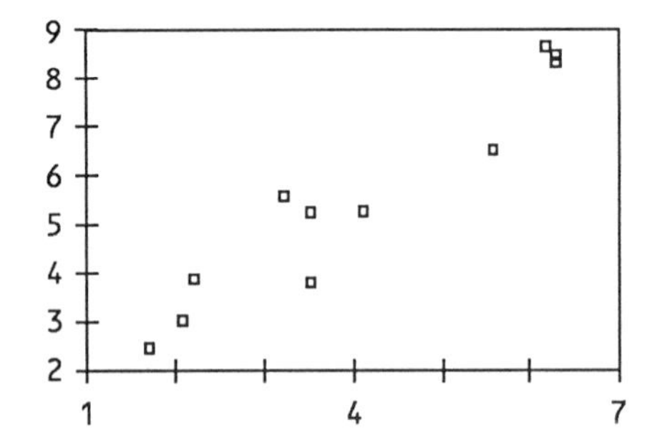

Unemployment Rates for Eleven Counties: January 1986 versus December 1985

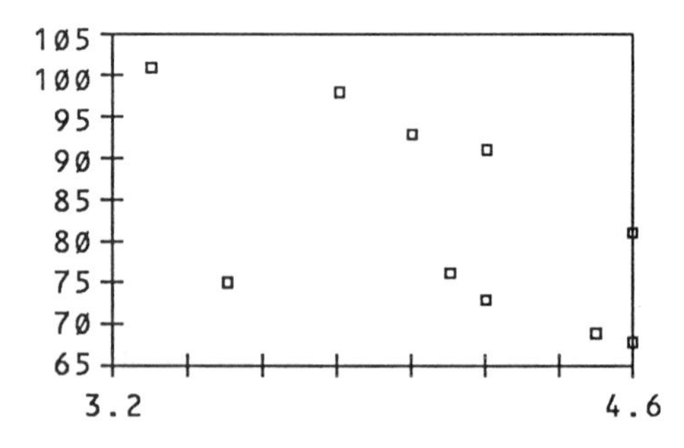

Ten NFL Football Teams, 1979 Season: Number of Punts versus Average Yards per Running Play

Figure 6 Scatter Plots for Class Practice

One final point should be considered regarding this example. We mentioned earlier that sometimes when there is a relationship between variables we can consider one variable independent and the other dependent. How do you feel about the variables in this example? Does a change in one variable cause the other to change, or is this a situation where they simply change together as the result of some underlying factor? Considered in another way, would you expect the coach to encourage his players to commit lots of fouls in order to be sure of scoring many points? This does not seem to be a relationship of cause and effect. What underlying factors could be influencing both of our variables? ■

Class Practice

1. Comment on the important characteristics of the scatter plots provided in Figure 6. Consider the shape of the relationship (linear or curved), the strength of the relationship (tight or loose), whether the relationship is positive (as in Example 1) or *negative* (one variable decreases when the other variable increases), and whether there are gaps, clusters, or outliers apparent in the data.

When a relationship is suggested by a scatter plot, we usually want to describe it mathematically. We describe the relationship by finding an equation that summarizes the way the two variables are related; such an equation is another example of a mathematical model. When we discussed mathematical models at the beginning of the chapter, we pointed out that a good model simplifies the phenomenon it represents and gives us the ability to predict. If we can find the equation of a curve that closely "fits" a scatter plot, we can focus on the important characteristics of the relationship between the variables without the clutter of a scatter plot. We can also use this equation to predict the values of one variable for specific values of the other variable. Sometimes we use the model to *interpolate*, or estimate values among observed values; sometimes we use the model to *extrapolate*, or predict values outside the region of observations.

Exercise Set 3

1. Make a scatter plot to analyze the relationship between rebounds and assists for the basketball data provided in Table 4. Use the graph to comment on the important characteristics of the relationship.
2. A student, who was helping at a yard sale, watched as a customer looked through the pile of jeans for sale. The man would bend his hand back at the wrist, bend his arm at the elbow, and then wrap the waist of the pants around his forearm. The man explained that his waist was the same size as his forearm, so he never needed to try slacks on for size. The student was studying data analysis in his mathematics class, so he decided to gather some data and model the relationship between the forearm circumference and waist size. The scatter plot in Figure 7, with forearm circumference measured in inches on the horizontal axis, is a graph of his data set. Since the points appear to increase steadily, a relationship between these two variables is apparently linear. Find an equation of the form $y = mx + b$ to fit the points. (You might want to experiment by moving a clear ruler or dark thread through the scatter plot until you find the line that seems to fit the points best.) After you find the equation of the line you feel is best, comment on the criterion you used to determine this line. Compare your line with those found by other students.

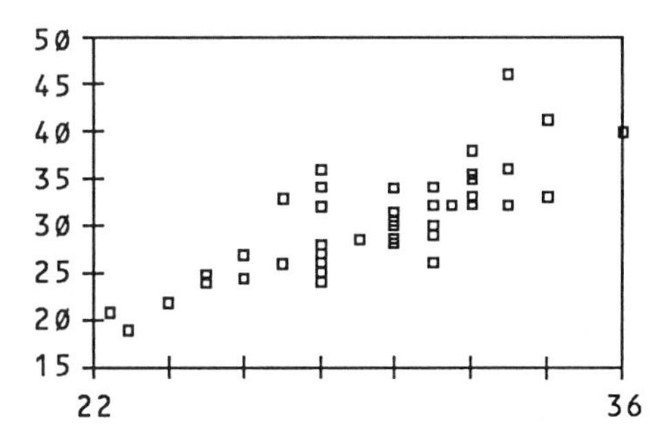

Figure 7 Waist Measure versus Forearm Circumference

3. Refer to Exercise 2.

a. In terms of the relationship between forearm and waist, what information is provided by the slope in the equation you found in Exercise 2? What information is provided by the y-intercept?

b. Use the model to predict your waist size. How accurate is the prediction? What message or advice would you give to a person whose data point lies above the line? What message or advice would you give to a person whose data point lies below the line?

4

The Median-Median Line

The preceding exercises illustrate that freehand methods of curve fitting, though helpful, may vary from person to person and to some extent allow personal bias to enter the data analysis process. We are accustomed in mathematics to objectivity and exactness, so we become concerned when different people fit reasonable but different lines to the same set of data. Which line is best? The concept of "best" is a defined one, and we will discuss it at length in "Data Analysis Two." For now, however, we will be satisfied with a standard, repeatable procedure that produces a good, reasonable line, called the *median-median line.* This line is also referred to as a *median fit line* or a *resistant line.*

As before, our first step is to look at a scatter plot of our data to be sure that the relationship between the variables is linear. We certainly do not want to blindly fit a line to a data set that is obviously nonlinear. (Later we will discuss how to transform data sets that appear nonlinear.) A look at the scatter plot will also make us aware of outliers in our data set. As we mentioned previously, such points may indicate errors in measurement, errors in data entry if using the computer, or errors in graphing when working without a computer. Errors should be corrected before proceeding. If the points are correct, it is helpful, if possible, to gather additional data in the vicinity of each outlier to determine whether it is an isolated extreme observation. If the extreme point is an isolated one, the median-median technique will provide a line that fits the central part of the data set and is rather insensitive to outliers that are inconsistent with the general trend.

To explain the procedure for finding a median-median line, we use data from an article in *Journal of Environmental Health,* May–June 1965, Volume 27, Number 6, pages 883–897. Robert Fadeley, author of the article, explains that the Hanford, Washington, Atomic Energy Plant has been a plutonium production facility since the Second World War. Some of the wastes have been stored in pits in the same area. Radioactive waste has been seeping into the Columbia River since that time, and eight Oregon counties and the city of Portland have been exposed to radioactive contamination. Table 8 lists the number of cancer deaths per 100,000 residents for Portland and these counties. Also provided is an index of exposure that measures the proximity of the residents to the contamination. The index is formulated on the assumption that county or city exposure is directly proportional to river frontage and inversely proportional both to the distance from the Hanford, Washington, site and to the square of the county's (or city's) average depth away from the river. We

Table 8 Rate of Cancer Deaths versus Exposure Index

County/City	Index	Deaths
Umatilla	2.5	147
Morrow	2.6	130
Gilliam	3.4	130
Sherman	1.3	114
Wasco	1.6	138
Hood River	3.8	162
Portland	11.6	208
Columbia	6.4	178
Clatsop	8.3	210

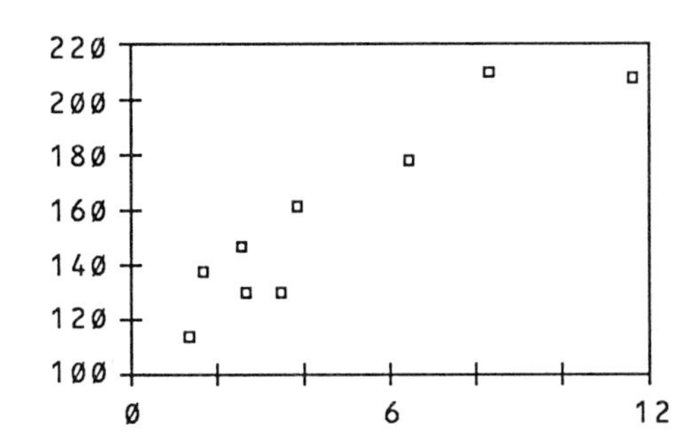

Figure 8 Rate of Cancer Deaths versus Exposure Index

are interested in a model to describe the relationship between the index of exposure and the rate of cancer deaths. The scatter plot provided has the exposure index on the horizontal axis.

The scatter plot indicates a positive linear relationship between these two variables. To fit a median-median line to these points, we divide them into three groups. The grouping is based on the x-values of the points. To do this graphically, we create a left-most group consisting of points with the smallest x-values, a middle group, and a right-most group. If the number of points is not divisible by three, extra points should be assigned symmetrically. If there is only one extra point, it should be placed in the middle group; if there are two extra points, place one in each outer group. Even if it makes equal allocation impossible, points with the same x-values must always be placed in the same group. In the example above, our groups consist of the following observations: left-most group contains (1.3, 114), (1.6, 138), and (2.5, 147); middle group contains (2.6, 130), (3.4, 130), and (3.8, 162); right-most group contains (6.4, 178), (8.3, 210), and (11.6, 208). Now consider each group of observations separately and order the values of each variable. Data pairings should be ignored at this stage. For example, in the right-most group, the ordered x-values are 6.4, 8.3, and 11.6; the ordered y-values are 178, 208, and 210. We now create a *summary point* for this portion of data by using the median x-value, 8.3, and the median y-value, 208, and combining them to create the ordered pair (8.3, 208). We repeat this process to obtain (1.6, 138) and (3.4, 130) as summary points for the other two groups. Notice that two of the summary points are actual data points, but the third one is not. These three summary points are marked on the scatter plot in Figure 9.

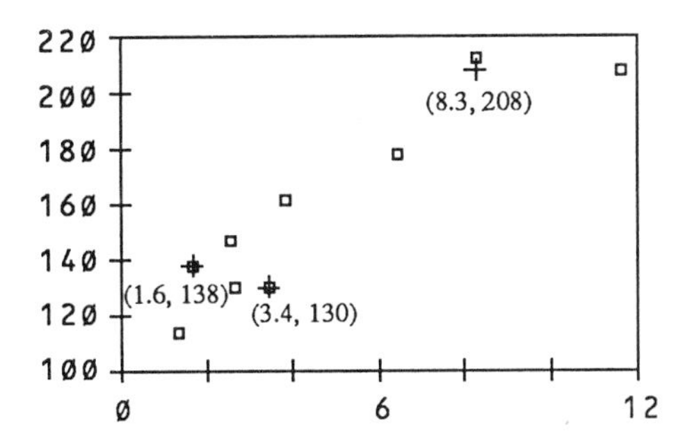

Figure 9 Summary Points for Cancer Data

We now use the summary points to determine the median-median line. We will use the summary points from the two outer groups to determine the slope and all three summary points to determine the y-intercept. To understand graphically what is being done, place a clear ruler on the scatter plot to connect the two outer summary points. Now move the ruler one-third of the way toward the middle summary point, being sure to keep it parallel to its original position. If you now trace the ruler, the line you draw is the median-median line. By moving the ruler one-third of the way toward the middle point, we give each summary point equal weight in determining the y-intercept (see Figure 10).

To find the equation of the median-median line, first use the two outer summary points to calculate the slope,

$$m = \frac{208 - 138}{8.3 - 1.6} = 10.4,$$

and find the equation of the line connecting them:

$$y - 138 = 10.4(x - 1.6),$$

or

$$y = 10.4x + 121.3.$$

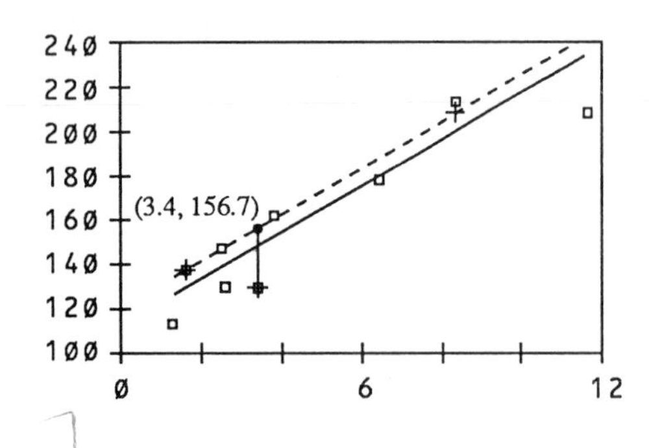

Figure 10 Median-Median Line for Cancer Data

This is the line we need to slide. It is the dashed line on the scatter plot in Figure 10. To determine how far to slide it, we hold the x-value of the middle summary point constant and compare the y-values of the point on the line and the middle summary point. On this line, what value of y is paired with the x-value of the middle summary point? That is, what value of y does this equation produce when $x = 3.4$? We substitute into the equation above to get

$$y = (10.4)(3.4) + 121.3 = 156.7,$$

so the point (3.4, 156.7) lies on the line connecting the two outer summary points. To slide the line one-third of the way toward the summary point (3.4, 130), we subtract the y-values, $130 - 156.7 = -26.7$. To calculate the directed distance from the line to the summary point, take one-third of this amount, which is $-26.7/3 = -8.9$, and move the line *down* 8.9 units, since the directed distance is negative. Our median-median line, then, will pass through (3.4, 156.7−8.9), or (3.4, 147.8), and have slope 10.4 as calculated above. The equation is

$$y - 147.8 = 10.4(x - 3.4),$$

or

$$y = 10.4x + 112.4.$$

It is the solid line drawn on the scatter plot in Figure 10.

We now have a model that describes the relationship between the index of exposure to radioactive contamination and the rate of cancer deaths in the region near Hanford, Washington. What information does the algebraic equation provide? First consider the y-intercept in the equation. Its value is 112.4; algebraically, it is the y-value associated with an x-value of zero. What is its significance in terms of the model? It is the cancer death rate we predict when the index of exposure is zero, that is, when there is no radioactive contamination. Does it seem reasonable for there to be approximately 112 cancer deaths per 100,000 residents in an area without exposure to radioactive contamination? Now consider what values the variables can assume. Algebraically, we can extend the line over all real number values of both variables. Many real numbers, however, are not reasonable values for the exposure index defined

Summary of Procedure for Finding Median-Median Line

- Separate the data into three groups of equal size (or as close to equal as possible) according to values of the horizontal coordinate.
- Find the summary point for each group based on the median x-value and the median y-value.
- Find the equation of the line through the summary points of the outer groups. We will call this line L.
- Slide L one-third of the way to the middle summary point.
  1. Find the y-coordinate of the point on L with the same x-coordinate as the middle summary point.
  2. Find the vertical distance between the middle summary point and the line by subtracting y-values.
  3. Find the coordinates of the point P one-third of the way from the line L to the middle summary point.
- Find the equation of the line through the point P that is parallel to line L.

for this analysis. There is no way to have a negative index, and on the basis of the indices in the data set, we would question how large an index could reasonably be. Finally, consider the slope. Its value is 10.4; algebraically, it is the amount that y-values change for unit changes in x-values. In terms of the model, the slope indicates that for each unit increase in the exposure index, the cancer deaths increase by 10.4 per 100,000 residents. Do you expect this to be true for all possible values of the exposure index?

Before proceeding, review the summary procedure on page 12 for finding the median-median line through a set of points.

Table 11 Winning Speeds for Indianapolis 500

Year	Speed (mph)	Year	Speed (mph)
1961	139.1	1971	157.7
1962	140.3	1972	163.5
1963	143.1	1973	159.0
1964	147.4	1974	158.6
1965	151.4	1975	149.2
1966	144.3	1976	148.7
1967	151.2	1977	161.3
1968	152.9	1978	161.4
1969	156.9	1979	158.9
1970	155.7	1980	142.9

Source: *The World Almanac*

Exercise Set 4

1. Using the basketball statistics provided in Table 4, find the median-median line to describe the relationship between personal fouls committed and points scored. Draw the line through a scatter plot of the data. Use the equation of the median-median line to calculate y-values for several x-values given in the table. How do these computed y-values compare to the observed ones? Circle those data points that seem to be most distant from the line.
2. Explain why it is important to examine a scatter plot of data before calculating the equation of a median-median line.
3. The Indianapolis 500 auto race is held each year. Table 11 provides winning speeds in miles per hour for the years 1961 through 1980. Make a scatter plot of Indianapolis 500 winning speeds for the period 1961 through 1974. Plot the year on the horizontal axis and the speed on the vertical axis. Find the equation of the median-median line through the points. Interpret the slope. Draw the line through the scatter plot. Do you feel that this line is a good model? Explain.
4. Use the median-median line from Exercise 3 to predict winning speeds for the years 1975 through 1980. Compare each prediction with the actual winning speed provided in Table 11. Do you feel that this line is a good model for these years? Explain.
5. Use the data provided in Table 12 to make a scatter plot for records for running the mile. Find the median-median line to fit the data, and interpret the slope. Predict what the record will be in the year 2000. Also use the line to predict when the record will be three minutes. According to your model, when will the record be one minute? Do you feel that this model is a good one? Explain.
6. Re-examine the scatter plot used in Exercise 5. Does it seem that you could fit the data better with two equations, one for early records and a different equation for more recent records? Separate the data into two parts as you feel appropriate. Fit a median-median line to each data set and record the equations. You now have two models for this data set. Compare this model with the one you found in Exercise 5. Do they differ appreciably? Which one do you feel is best? Explain.
7. Make a scatter plot of the data provided in Table 13. Describe the shape of the graph.

Calculate the square root of y-values and graph ordered pairs of the form $(x, \sqrt{y})$. Describe the shape of this graph. Now calculate the square of x-values and graph ordered pairs of the form (x^2, y). Describe the graph.

Table 12 Record Times for Running a Mile

Year	Time (sec)	Year	Time (sec)
1880	263.2	1945	241.4
1882	261.4	1954	239.4
1882	259.4	1954	238.0
1884	258.4	1957	237.2
1894	258.2	1958	234.5
1895	257.0	1962	234.4
1895	255.6	1964	234.1
1911	255.4	1965	233.6
1913	254.6	1966	231.3
1915	252.6	1967	231.1
1923	250.4	1975	231.0
1931	249.2	1975	229.4
1933	247.6	1979	229.0
1934	246.8	1980	228.8
1937	246.4	1981	228.5
1942	246.2	1981	228.4
1942	244.6	1981	227.3
1943	242.6	1985	226.3
1944	241.6		

Source: *The World Almanac*

Table 13 Data for Exercise 7

x	0.8	1.5	3.2	2.6	1.9
y	0.7	2.1	10.5	6.8	3.5
x	2.4	3.5	0.6	2.1	
y	5.9	12.4	0.3	4.3	

5

How Good Is Our Fit?

Whenever we fit a curve or a line to a data set, we should examine that curve in relation to our scatter plot to determine how well it models the relationship between the variables. Most of the points on the scatter plot will not fall exactly on the fitted curve. The question we are interested in is "How close to the curve are the data points?" Since we often fit curves in order to predict y-values, we usually study the vertical distance between each observed point and the fitted curve. This difference in y-values is called a *deviation*, or a *residual.* (You were asked to examine some residuals in Exercise 1 of the previous section dealing with points and fouls in basketball.)

The graph in Figure 14 shows the median-median line we previously fitted in the plutonium/cancer example. Recall that the horizontal axis gives values of the exposure index and the vertical axis gives cancer deaths per 100,000 residents. The brackets show the residuals or vertical deviations of the data points from the fitted line.

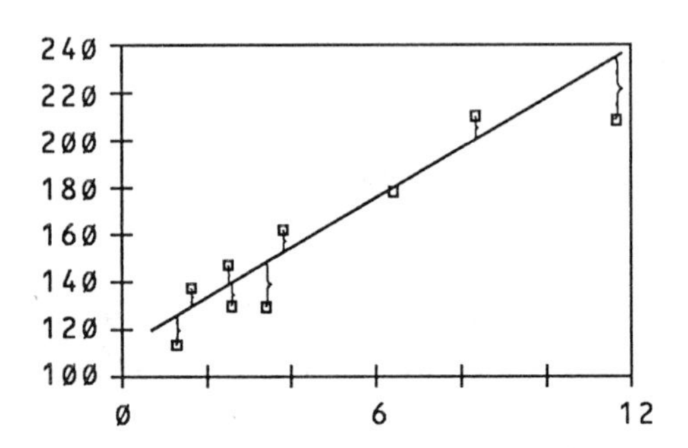

Figure 14 Residuals for Median-Median Line

If we denote observed data points as (x_i, y_i) and the corresponding points on the curve we fit as $(x_i, \hat{y}_i)$, the residual is simply $y_i - \hat{y}_i$. Table 15 lists the residuals associated with the graph in Figure 14.

Examination of the residuals allows us to get a better sense of how well our curve fits the data

Table 15 Residuals for Median-Median Line

x_i	y_i (data)	$\hat{y}_i$ (fit)	**residual** = **data** − **fit** $r_i = y_i - \hat{y}_i$
1.3	114	126	−12
1.6	138	129	9
2.5	147	138	9
2.6	130	139	−9
3.4	130	148	−18
3.8	162	152	10
6.4	178	179	−1
8.3	210	199	11
11.6	208	233	−25

and sometimes alerts us to characteristics of the data that we may not have previously noticed. We usually look first at the size of the residuals. Residuals that are small relative to observed y-values provide evidence of a good fit. Obviously, many large residuals would cause us to question how good our model is, because the predictions based on our model are not accurate. If there are many large residuals, we may need to choose a different mathematical function to model the relationship. One or two large residuals, however, may draw our attention to outliers that were not previously detected. As discussed previously, we should attempt to determine why these points do not fit the general pattern of the data. If they result from errors in measurement, the data points should be corrected or excluded from the analysis. If, on the other hand, all points are correct, a large residual may provide interesting and useful information. If the nonconformity can be explained, we often gain additional information about factors influencing the relationship we are trying to model. In our example, it is clear that our line is closer to some of the points than to others, but the only residual that seems to really stand out is the final one. Looking back at the graph in Figure 14, we may wonder if the points tend to level off as the index becomes large. We really cannot answer this question without more data points, but this observation would probably make us feel less confident when we use our model to extrapolate.

The second feature of the residuals that we need to examine is whether they follow any trend or pattern as the x-values vary. Are residuals in the middle all positive or all negative and at the ends all opposite in sign? Do the residuals at one end of the graph seem to be larger in magnitude than they do at the other end? If we observe a pattern, it might indicate that we chose the wrong curve to fit the data. In our example, there are both positive and negative residuals scattered throughout the graph, so we have no problem of this type. Except for our final point, the residuals are of similar magnitude at both ends of the graph. If there is a problem in this area, however, we must recognize that the fit is less accurate at one end and be careful when using the model to predict in that region.

Class Practice

1. The four sets of residuals in Table 16 were obtained by a student who was trying to fit linear equations to various sets of data. What conclusions can you draw from these residuals regarding his success? Be precise and specific; include a reason for each claim.

Often it is easier to analyze the residuals when they are paired with corresponding x-values and studied as a new data set $\{(x_i, r_i)\}$. A scatter plot of this data set, often called a *residual plot*, lets us check the size and pattern of the residuals. If our model provides a good fit for the original data, the residual plot should show points scattered randomly within a horizontal band about the horizontal axis. In Figure 17 we have graphed the ordered pairs (x_i, r_i) created from the plutonium/cancer example. This plot shows that the residuals are reasonably small in magnitude and do not follow a trend or pattern; therefore, our linear model is a good one.

After doing a median-median fit, we will examine the residuals to determine how good our fit is. We can fit a median-median line to any data set; the technique will work regardless of whether

Table 16 Residual Sets for Class Practice

Data Set A

x_i	1	2	3	4	5	6	7	8	9	10	11	12	13	14	15
r_i	−.77	1.14	−.69	.23	−.08	.83	.75	.35	.18	.75	−.15	.42	−.88	− .64	.26

Data Set B

x_i	1	2	3	4	5	6	7	8	9	10	11	12	13	14	15
r_i	−.88	−.37	−.12	−.01	.00	−.12	.42	1.03	.49	.11	.86	−.07	−.56	−.07	−.16

Data Set C

x_i	1	2	3	4	5	6	7	8	9	10	11	12	13	14	15
r_i	−.37	−.48	−.62	−.15	−1.11	−.15	1.31	−.09	.71	.19	.47	.02	.02	.58	1.28

Data Set D

x_i	1	2	3	4	5	6	7	8	9	10	11	12	13	14	15
r_i	−.16	.04	.15	−.53	.47	.10	−.65	.57	.31	−1.09	−.74	−.09	10.02	.25	−.08

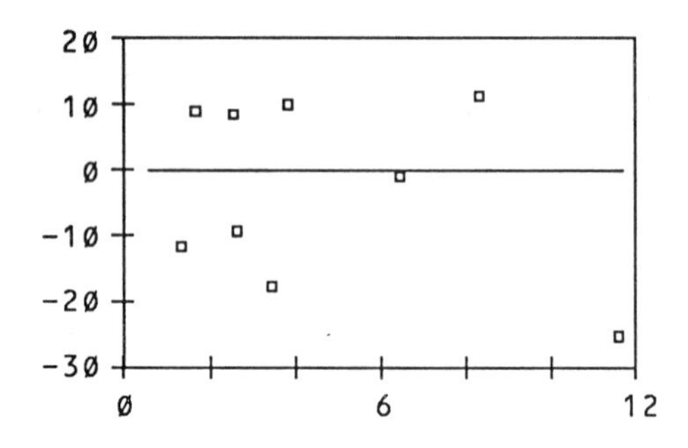

Figure 17 Plot of Residuals for Cancer Data

or not using it is appropriate. That is why it is very important to look at the scatter plot of data to decide whether a linear model is reasonable. If we fit a straight line to data when the true relationship is not linear, the residual plot will usually show a pattern. Because the scale on a residual plot is usually much smaller than on the original scatter plot, curvature is more easily noticed on the residual plot. Using both the scatter plot and the residual plot in the curve-fitting process will help ensure a good fit.

Exercise Set 5

1. Refer to Exercise 3 on page 10 in which you found a median-median line to fit data on the Indianapolis 500 for the years 1961 through 1974. Use the equation of this line to calculate winning speeds for the years 1961 through 1974. List these speeds as an additional column of Table 11. Then calculate the residuals and list them in another column. Make a residual plot and discuss whether the median-median line is a good model.
2. Make a scatter plot using all of the Indianapolis 500 data provided in Table 11. Plot the year on the horizontal axis and the speed on the vertical axis. Does this scatter plot indicate a linear relationship throughout the period 1961 through 1980? Fit a median-median line to this data set and draw the line through the scatter plot. Using the graph to estimate the size of the residuals, make a freehand residual plot. Is a better model needed to fit this data set? Explain.

6

Introduction to Re-expression

Now that we have an objective procedure for fitting lines to data, that process can be expanded to find equations for curves through scatter plots with obvious curvature. The technique illustrated in the following example avoids fitting curves directly to data sets. Instead, we re-express our data to make it linear and then fit a median-median line through the transformed data.

Students in a physics class are studying free-fall to determine the relationship between the distance an object has fallen and the amount of time since release. They collect the data in Figure 18 and use a computer or calculator to generate the scatter plot. Time is graphed on the horizontal axis, and the distance fallen by the object is graphed on the vertical axis.

The students examine the scatter plot and feel that the points lie roughly along a positively sloped line. They then find the median-median line through the data set. The equation of the line they fit is $d = 520.10t - 113.64$; this line is superimposed on the scatter plot in Figure 19.

Several students observe that points to the far left and far right on the scatter plot are above the line, while points in the middle lie below the line. When they graph the residual plot, which is shown in Figure 20, it becomes even more obvious that the residuals have a very definite pattern. The residuals in the middle are all negative, and the residuals at both ends are all positive. As we mentioned previously, such a pattern often indicates that the curve we fit to the data is not appropriate. The students look once again at the scatter plot of their original data set. Now they notice that the points seem to curve upward rather than to rise steadily, and they realize that a linear model is not appropriate for this data set.

How can we find the equation of a curve that better fits this data set? One way is to transform, or *re-express*, the data in such a way that it becomes linear. Then we can fit a median-median line to the transformed data and, if satisfied with the fit, "undo" the transformation and change the variables back to their original state. How can we "straighten" the scatter plot in Figure 18? Graphically what we need to do is pull down the data points with large d-values. We can do this by taking the square root of each d-value. The scatter plot in Figure 21 shows time on the horizontal axis and the square root of the distance measurements on the vertical axis.

These points seem to lie along a line, so we can proceed to fit a median-median line to the transformed data. Doing so yields an equation

Time (sec)	Distance (cm)	Time (sec)	Distance (cm)
.16	12.1	.57	150.2
.24	29.8	.61	182.2
.25	32.7	.61	189.4
.30	42.8	.68	220.4
.30	44.2	.72	254.0
.32	55.8	.72	261.0
.36	63.5	.83	334.6
.36	65.1	.88	375.5
.50	124.6	.89	399.1
.50	129.7		

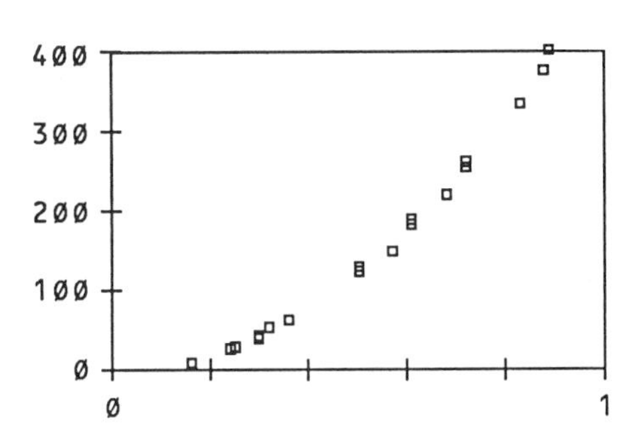

Figure 18 Distance Fallen in Free Fall versus Time

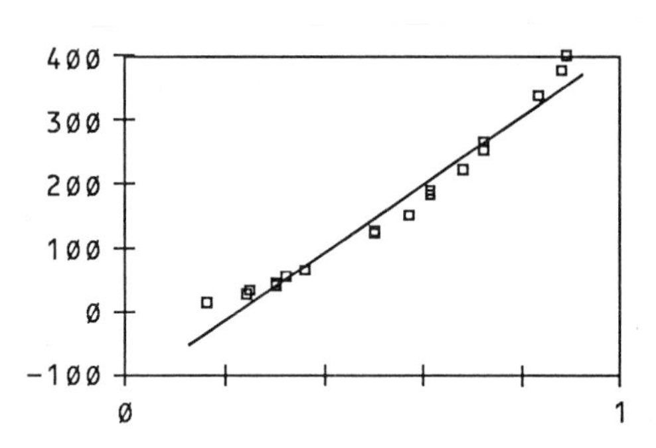

Figure 19 Median-Median Line for Free-Fall Data

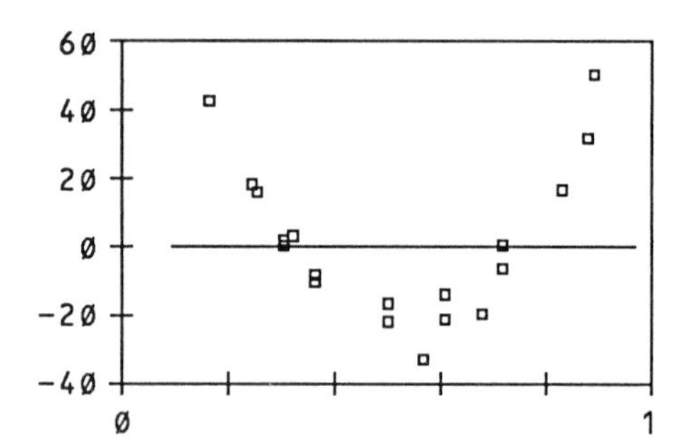

Figure 20 Plot of Residuals for Free-Fall Data

Time (sec)	$\sqrt{\text{Distance}}$ (cm)	Time (sec)	$\sqrt{\text{Distance}}$ (cm)
0.16	3.48	0.57	12.26
0.24	5.46	0.61	13.50
0.25	5.71	0.61	13.76
0.30	6.54	0.68	14.84
0.30	6.65	0.72	15.94
0.32	7.47	0.72	16.15
0.36	7.97	0.83	18.29
0.36	8.07	0.88	19.38
0.50	11.16	0.89	19.98
0.50	11.39		

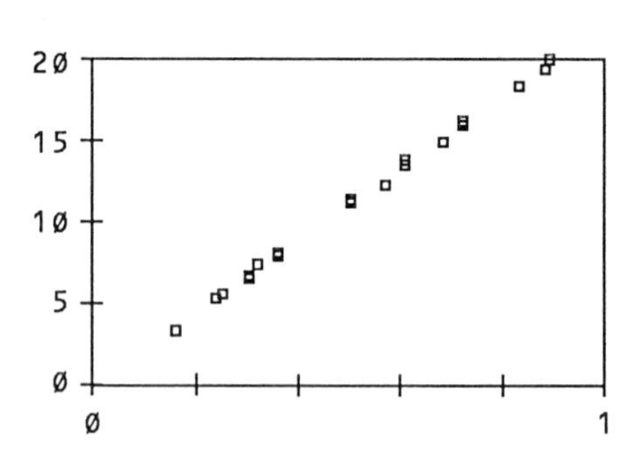

Figure 21 Square Root of Distance Fallen versus Time

through points of the form $(t, \sqrt{d})$. In this example we get

$$\sqrt{d} = 22.19t + 0.12.$$

This equation relates the square root of the distance that an object falls to the time since it was released. The linear fit and corresponding residual plot are shown in Figure 22.

Since the residuals are scattered randomly about the horizontal axis, we feel even more confident that we have successfully straightened the original curvature and proceed to transform the variables back to the original ones. We really want an equation that shows how the actual distance fallen, not the square root, is related to time. So we need to square both sides of the equation above. Doing so yields the equation

$$d = 492.40t^2 + 5.33t + 0.01.$$

How does the graph of this equation fit the original scatter plot? To answer this question we can graph the parabola on the original scatter plot and then examine a residual plot. These plots are provided in Figure 23.

The residuals indicate a satisfactory fit. They are randomly scattered with some positive and

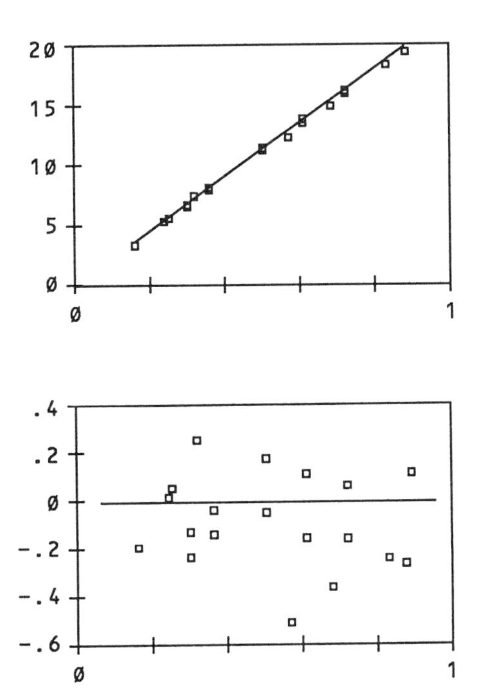

Figure 22 Linear Fit and Residual Plot for Re-expressed Free-Fall Data

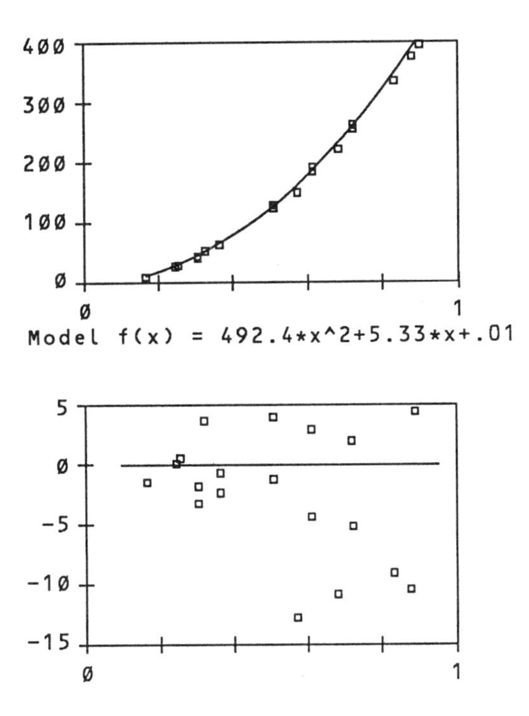

Figure 23 Quadratic Model and Residual Plot for Free-Fall Data

some negative values appearing throughout the plot. Though the size of the residuals seems to increase as t-values increase, this is often the case when the relationship is quadratic. Since this pattern was not observed in the residual plot from the linear fit to re-expressed data, the relative size of residuals is probably consistent throughout. We would want to be aware of this trend in using the model to extrapolate, however.

As we attempt to fit models to data, we use both the shape of a scatter plot and our knowledge of mathematical functions to help us decide how to re-express the data. Re-expressions may involve powers, roots, reciprocals, combinations of these, or other functions yet to be studied. Examples requiring the application of new functions to linearize data will be presented in later chapters.

Example 1

College Cost

A high school junior who is beginning to research colleges wants to determine how much it will cost for her to attend a private college for the four academic years beginning in August 1988. She finds data in the Fall 1985 issue of *The College Board News* and hopes to fit this data with a model that will allow her to predict college costs for the next few years. As a first step, she examines the scatter plot that is shown with the data in Figure 24.

Notice that the years have been entered simply as 75, 76, and so forth, and are plotted on the horizontal axis. The graph follows the general curvature of our previous example, so the appropriate transformation seems to be to take the square root of the y-values in order to linearize the data. The scatter plot of ordered pairs representing year and square root of private college cost is provided in Figure 25.

Academic Year	Total Annual Cost
1975	$4,205
1976	4,460
1977	4,680
1978	4,960
1979	5,510
1980	6,060
1981	6,845
1982	7,600
1983	8,435
1984	9,000
1985	9,659

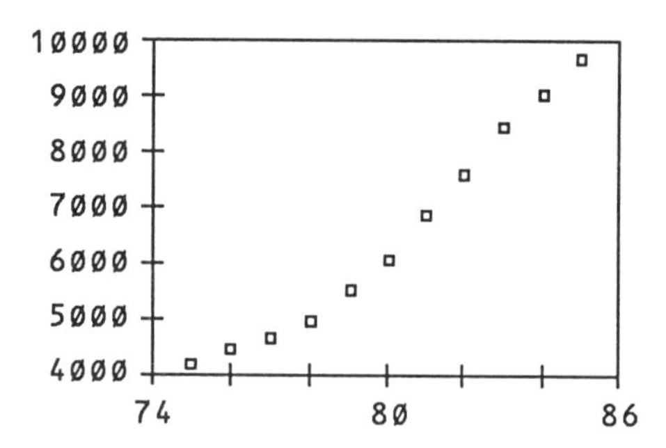

Figure 24 Cost at Four-Year Private Colleges versus Year

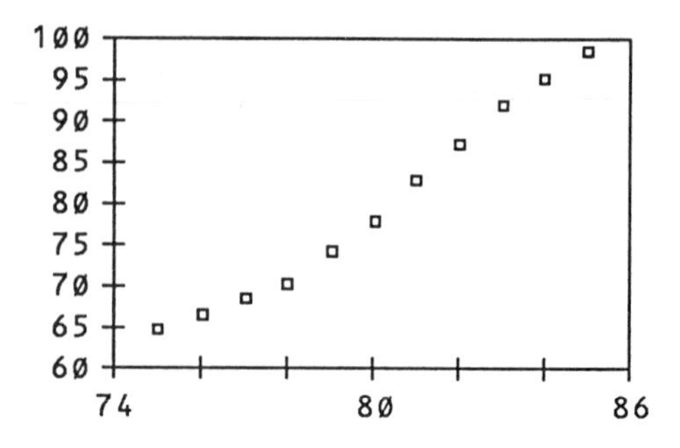

Figure 25 Square Root of College Cost versus Year

Does this scatter plot seem linear? It is certainly straighter than the original one, though some of the curvature is still present. Fitting a median-median line to the transformed data set, we obtain the equation

$$\sqrt{y} = 3.68x - 214.78.$$

Examination of the linear fit and corresponding residual plot in Figure 26 indicates even more strongly that there is curvature that has not been accounted for by this model. Squaring both sides yields the equation

$$y = 13.54x^2 - 1580.78x + 46130.45.$$

To assess how well the quadratic model fits the original data set, we can examine the graph of

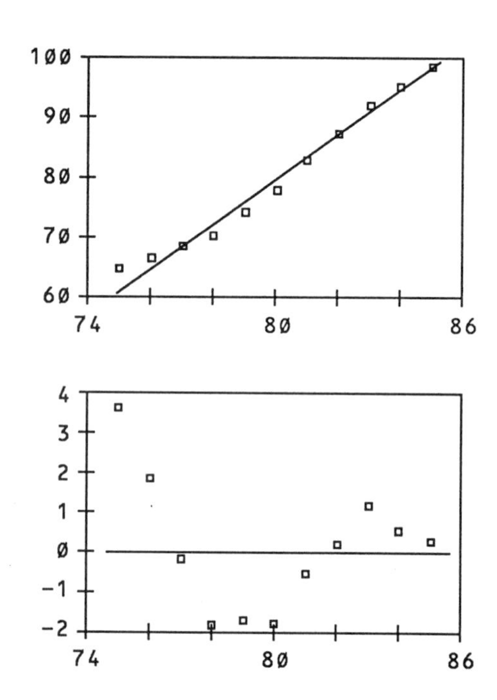

Figure 26 Linear Fit and Residual Plot for Re-expressed College Cost Data

the model superimposed on the original scatter plot and the corresponding residual plot (see Figure 27).

An examination of this residual plot convinces us that our fit is not a good one. There is an obvious pattern in this graph, and we therefore would not feel confident using this model to predict over

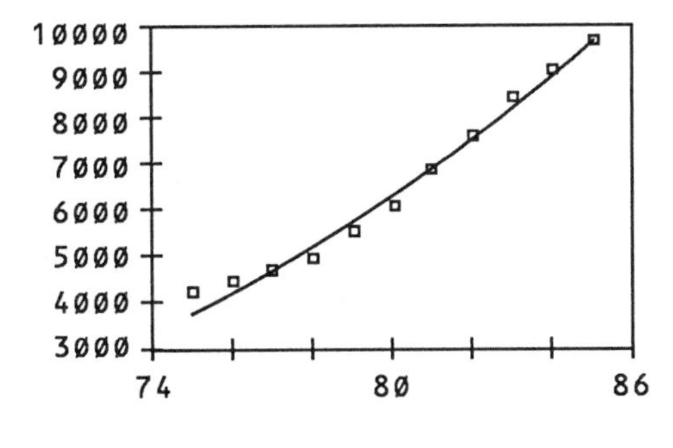

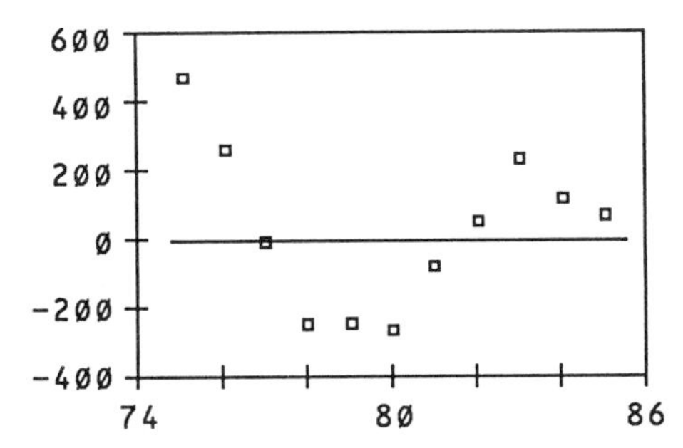

Figure 27 Quadratic Model and Residual Plot for College Cost Data

the next six years. The student who is trying to anticipate future college costs must find a better method of re-expressing the data in order to obtain a good model. ■

The re-expression process used in the previous examples is a helpful tool in analyzing data. Often a great deal of trial and error is required as we search for the mathematical relationship between the variables. A thorough knowledge of functions, special properties of functions, and inverses will aid in this stage of the analysis. In subsequent units we will study new functions and thus expand the models available. We will also study another method of finding linear models for data. Throughout the course, models will be an important part of all that we study, since they are one of the most important tools of mathematics.

We must keep in mind that data sets contain measurement errors, so models based on data only approximate the relationship between variables and are not expected to fit perfectly. When we are able to find a good fit, we can feel confident that we have adequately described the relationship. We cannot assume, however, that a change in one variable necessarily causes a change in the other. Nor can we assume that the relationship will be the same outside the domain of observed values. We must be careful, therefore, when using models to predict beyond the values for which we have observations.

Exercise Set 6

1. We linearized the free-fall data provided in Table 18 by taking the square root of the d-values. Given that the functional relationship between distance fallen and time is quadratic, consider other ways that you could re-express this data in order to make the data set linear.

   a. Square the values of time and examine a scatter plot of the ordered pairs (t^2, d). Find the equation of the median-median line through these points.

   b. Divide the distance values by corresponding time values to create ordered pairs $(t, d/t)$. Find the equation of the median-median line through these points.

   c. Compare the three different models we have found for this data set. Which model do you feel is best?

2. Suppose an object slides without friction down an inclined plane in a laboratory. Sparks are generated on paper tape every tenth of a second to mark the position of the object. The distance between sparks can be measured to generate a data set that relates position of the object with time since release. A data set is provided in Table 28.

   a. Plot time on the horizontal axis and position on the vertical axis to determine how position and time appear to be related.

   b. i. Square the values of time, and find the equation of the median-median line through the points (t^2, s).

Table 28 Inclined Plane Data

Time (sec)	Position (cm)	Time (sec)	Position (cm)
0	0.0	1.0	12.5
.1	0.2	1.1	15.0
.2	0.6	1.2	17.7
.3	1.3	1.3	20.6
.4	2.2	1.4	23.8
.5	3.3	1.5	27.2
.6	4.7	1.6	30.8
.7	6.3	1.7	34.8
.8	8.1	1.8	38.9
.9	10.2	1.9	43.1

ii. Make a residual plot for the linear fit. Are you satisfied that your transformation linearized the data set?

iii. Study the model superimposed on the original scatter plot and the residual plot to determine the goodness of your fit.

c. i. Take the square root of position values, and find the equation of the median-median line through the points $(t, \sqrt{s})$.

ii. Make a residual plot for the linear fit. Are you satisfied that your transformation linearized the data set?

iii. Solve the equation of the median-median line for s.

iv. Study the model superimposed on the original scatter plot and the residual plot to determine the goodness of your fit.

d. Which model provides a better fit for the original data? Explain.

e. If the initial velocity of the object is zero, the theoretical relationship between position and time is $s = 0.5at^2$. If the initial velocity of the object is not zero, the theoretical relationship is $s = 0.5at^2 + v_0 t$ where v_0 represents the initial velocity.

i. Based on your analysis, do you feel that the initial velocity of the object is zero?

ii. Explain why the choice of re-expression seems more important in this example than in the example involving free-fall.

3. Using the data set in the previous exercise, now consider the relationship between the velocity of the object and time since release. We can obtain velocity data by calculating change in position divided by change in time,

$$v_i = \frac{d_{i+1} - d_{i-1}}{t_{i+1} - t_{i-1}}.$$

a. Make a table showing time since release and velocity, (t_i, v_i).

b. Plot time on the horizontal axis and velocity on the vertical axis to determine how velocity and time appear to be related.

c. Fit a median-median line to the data set.

d. Theoretically, the relationship between velocity and time is $v = v_0 + at$ where v_0 represents the initial velocity. Compare the values of acceleration and velocity from this model with the values from the model you found in the previous exercise.

4. High school physics classes often perform a pendulum experiment to gather data that can be used to approximate the strength of gravity and confirm the theoretical relationship between the length and period of a pendulum. To perform the experiment a weight is attached to a string that is suspended from a horizontal rod. The distance from the rod to the center of the weight is measured, the string is pulled back five to ten degrees, and a stopwatch is used to time twenty full cycles. This time is divided by 20 to get the average duration of one cycle (the period) associated with the length of the string. Changing the length of the string and repeating the procedure a number of times, we can accumulate a data set. Use the sample data set in Table 29 or perform the experiment and gather your own data to answer the questions that follow.

a. Plot time on the vertical axis and length on the horizontal axis, and examine the scatter plot.

Table 29 Pendulum Data

Length (cm)	Time (sec)	Length (cm)	Time (sec)
6.5	.51	24.4	1.01
11.0	.68	26.6	1.08
13.2	.73	30.5	1.13
15.0	.79	34.3	1.26
18.0	.88	37.6	1.28
23.1	.99	41.5	1.32

b. Fit a median-median line to the data set.

c. Examine the residual plot. Describe what you see. Is a better model needed?

d. In physics we learn that the relationship is actually

$$T = \frac{2\pi}{\sqrt{g}}\sqrt{L},$$

where g is a constant that represents acceleration due to gravity. Given this information, how can you transform the data to make the plot linear? Re-express the data, and examine a scatter plot of the transformed data to be sure your transformation results in a linear data set.

e. When you are satisfied with the transformation, fit a median-median line. Examine the residual plot from the linear fit to confirm that your transformation is satisfactory. Then solve for the dependent variable T.

f. Examine the residual plot for your new model. Are there any indications that the fit is not good?

CHAPTER 2

Functions

1

Functions as Mathematical Models

Recently the owners of a hamburger chain commissioned a weather consultant to study the relationship between sales of their hamburgers and weather conditions. They wanted to know if there was any relationship between the temperature and daily hamburger sales or between rainfall and hamburger sales. If such relationships were found to exist, they could help the owners predict and anticipate sales patterns; this information would help them make better decisions about ordering inventory and hiring workers.

This example, and others in the preceding chapter, illustrate the need to know about relationships in order to make accurate predictions. We are all familiar with scientific formulas that express the relationship between variables; often these formulas are used to make predictions. But it is not only scientists who use knowledge about relationships this way. In order for each of us to make reasonable decisions, we must know certain relationships and use them to make informed choices. For instance, we know that stopping distance is a function of our driving speed, and we use this information to decide on a safe following distance as we drive.

In relationships like those described above we usually assume that one of the quantities in the relationship determines the value of the other. When this is true, the quantity that depends on the other is called the *dependent* variable; the other quantity is called the *independent* variable. For example, a safe following distance depends on driving speed, so driving speed is the independent variable and following distance is the dependent variable. If a relationship does exist between rainfall and hamburger sales, the rainfall influences the sales but the sales do not influence the rainfall, so rainfall is the independent variable and the dependent variable is sales.

There are several ways to display or describe the relationship between two quantities or variables. One way is to list all of the ordered pairs of values that occur, but it is very difficult to see the characteristics of the relationship in this form. Another way is to give a formula or equation which relates the two variables. For some relationships formulas are already known that express how the variables are related. For other relationships, data can be gathered and a mathematical model could be determined using techniques of data analysis. A third way to present a relationship is in a table or chart. For example, your doctor probably has a chart which shows ideal weights for various heights. A fourth way to show the relationship between two variables is to draw a graph in a rectangular coordinate system. Such a graph can reveal many of the properties of the relationship and thereby help us to better understand it.

You will spend a fair amount of time in this course studying graphs of the relationship between two variables. You will learn to extract information from graphs that have been drawn for you, and also to make graphs to display information.

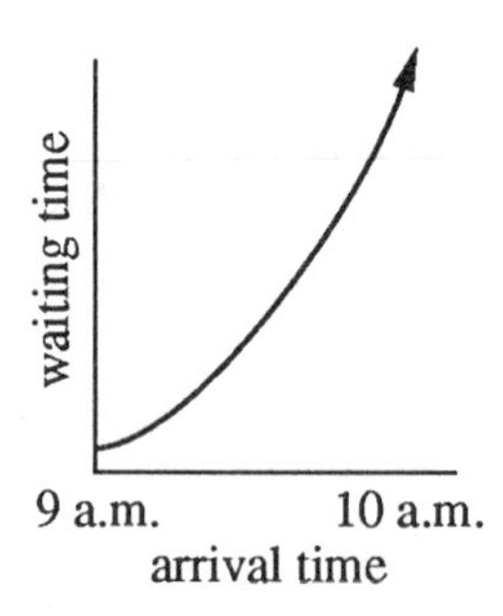

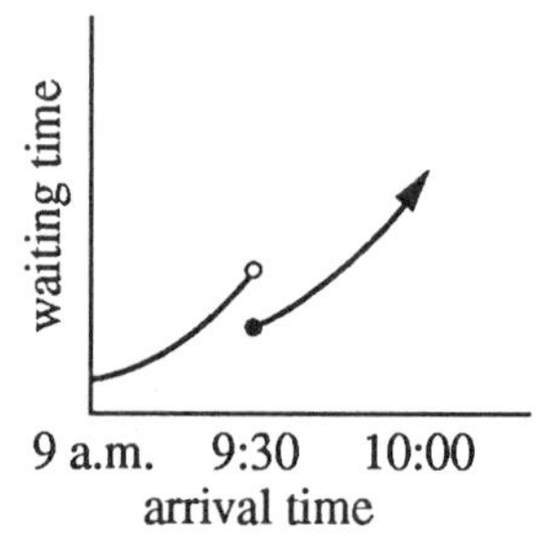

Figure 1 Average Waiting Time versus Arrival Time

Example 1

The ticket booth at a large amusement park opens at 8 a.m. Every morning a long line develops as people arrive to enter the park. The managers of the park are interested in the relationship between when people arrive at the park and how long they spend waiting in line. One busy Saturday in the summer the managers monitor arrival and waiting times between 9 and 10 a.m. The first graph in Figure 1 shows the relationship between arrival time and average waiting time for that morning.

Since the arrival time determines the average waiting time, and not vice versa, the arrival time is the independent variable and is graphed on the horizontal axis. The graph does not convey specific information about how long it takes to purchase tickets, but certain characteristics of the relationship are obvious. It should be clear that arriving later in the hour between 9 and 10 a.m. is associated with a longer waiting time. The graph indicates an *increasing* relationship; as the arrival times increase the average waiting time also increases. The graph is also *continuous* because it does not have any breaks. We do not know the coordinates of any particular points on the graph, nor do we know the precise steepness of the graph. The important characteristics are that the graph is increasing and continuous.

The managers are worried that people might be discouraged from coming to the park if the line at the entrance is too long. To prevent this from happening they open additional ticket booths at 9:30 a.m. on Sunday morning. The second graph in Figure 1 shows how these additional booths change the relationship between arrival times and average waiting times. The break in the graph occurs because the average waiting time is immediately decreased when the new ticket booths are opened. This graph is *discontinuous* at a point. This means that it has a break which requires you to pick up your pencil from the paper as you draw the graph. ■

In any situation in which values of one variable are paired with values of another variable we have what mathematicians call a *relation.* All of the relationships between variables that we have considered so far are examples of relations — a driving speed is paired with a following distance, or an arrival time is paired with a waiting time.

Now let's consider another relation — that between the price you pay for an airplane ticket and the distance you fly. It would seem reasonable that the cost of an airplane ticket should depend on the distance you plan to fly. However, it is entirely possible that two tickets for flights of 400 miles might have very different prices. The relation between price and distance differs from other relations we have considered. Given a value for the independent variable (distance), we cannot confidently determine the value of the dependent variable (price). The price of a ticket is not uniquely determined by the length of the flight; other factors such as the size of the airports, the day of the week, and the date of the ticket purchase all influence the ticket price.

How does the airplane ticket relation differ from the other relations we discussed previously? The difference lies in our ability to use values of the independent variable to predict values of the dependent variable. In many relations, knowing the value of the independent variable guarantees that we can find a unique value for the dependent variable. Whenever this is true, the relation is called a *function.* In the airplane ticket relation we are not guaranteed a unique value for the dependent variable; this relation is not a function.

A function is often defined as *a set of ordered pairs with the property that each first coordinate has a unique second coordinate.* This definition is a formal mathematical one; for our purposes, it is useful to think of a function as a process which maps, or sends, each permissible first coordinate to a unique second coordinate. Thinking about a function as a mapping emphasizes the process of pairing values. In this course it is important to think about functions as being dynamic, as doing something to an x-value to get a y-value.

A variety of notations are used to write about functions. Functions are usually named with a letter; f is a common choice. We can write $f : x \to y$ to show that a value of x is sent to a particular y. To show that f does something to an x-value to get a y-value, we often write $y = f(x)$, which is read "y equals f of x." For example, the notations

$$f : x \to x^2 + 1$$

and

$$f(x) = x^2 + 1$$

and

$$y = x^2 + 1$$

all indicate that a given value of x is paired with a y-value that is obtained by squaring the x-value and then adding 1.

The function f makes ordered pairs of the form $(x, x^2 + 1)$. The *input* is x and the *output* is $x^2 + 1$. In this notation, x is often called the *argument* of the function — it is the variable that the function acts on to produce the second coordinate of each ordered pair. The output of the function is sometimes referred to as the *value of the function.*

The argument of a function does not have to be x. For instance, consider

$$f(a) = a^2 + 1,$$

$$f(-x) = (-x)^2 + 1,$$

$$\text{and } f(w + 7) = (w + 7)^2 + 1.$$

In these examples, the arguments of the function are a, $-x$, and $w + 7$, respectively.

Exercise Set 1

1. For each relationship described below, identify the two quantities that vary and decide which should be represented by an independent variable and which by a dependent variable. Sketch a reasonable graph to show the relationship between the two variables. You should only be concerned with the basic shape of the graph, not with particular points. Write a sentence or two to justify the shape and behavior of your graph.

a. The amount of money earned for a part-time job and the number of hours worked.

b. The number of people absent from school each day of the school year.

c. The temperature of an ice-cold drink left in a warm room for a period of time.

d. The amount of daylight each day of the year.

e. The water level on the supports of a pier at an ocean beach on a calm day.

f. The population of the U.S. according to each census since 1790.

g. The height of a baseball after being hit into the air.

Table 3 Income Tax Table

Taxable Income	Income Tax
over \$0 but not over \$18,550	15% of income
over \$18,550 but not over \$44,900	\$2,782.50 + 28% of income over \$18,550
over \$44,900	\$10,160.50 + 33% of income over \$44,900

h. The distance between the ceiling and the tip of the minute hand of a clock hung on the wall.

i. The height of an individual as he or she ages.

j. The size of a person's vocabulary from birth onwards.

k. The number of bacteria in a culture over a period of time.

2. Sketch a graph of the relationship presented in each table. Identify dependent and independent variables. Discuss any conclusions you can make based on the shape of your graph.

a. Refer to the postal rates given in Table 2.

b. Refer to the tax rates given in Table 3.

3. Let $f(x) = 3x - 1$.

a. Find $f(4)$.

b. Find x such that $f(x) = 4$.

c. Write an expression for $f(2x)$, $2f(x)$, $f(x+2)$ and $f(x)+2$.

4. Let $g(x) = x^2 - x$.

a. Find $g(7)$.

b. Find x such that $g(x) = 6$.

c. Write an expression for $g(\frac{1}{x})$, $\frac{1}{g(x)}$, $g(x-1)$, and $g(x) - 1$.

5. If $f(x) = x^2 + x$, write and simplify an expression for each of the following.

a. $f(x+1)$

b. $2f(x) - 3$

c. $f(0.5x)$

d. $f(1/x)$

e. $f(|x|)$

f. $f(x+h)$

6. Let $f(x) = x^2$ and let $g(x) = \frac{1}{x}$. Express each of the following functions as a variation of f or g. For instance, $y = (x+1)^2$ can be expressed as $y = f(x+1)$.

a. $y = (x-4)^2$

b. $y = \frac{3}{x}$

c. $y = x^2 + 5$

d. $y = \frac{1}{x-1}$

e. $y = 2 + 7x^2$

f. $y = \frac{1}{5x}$

Table 2 Third Class Postage Rates

Weight	Rate
not more than 1 oz.	\$0.25
greater than 1 and not more than 2 ozs.	0.45
greater than 2 and not more than 3 ozs.	0.65
greater than 3 and not more than 4 ozs.	0.85
greater than 4 and not more than 6 ozs.	1.00
greater than 6 and not more than 8 ozs.	1.10
greater than 8 and not more than 10 ozs.	1.20
greater than 10 and not more than 12 ozs.	1.30
greater than 12 and not more than 14 ozs.	1.40
greater than 14 and not more than 16 ozs.	1.50

2

Graphing Functions — Developing A Toolkit

One of the goals of this course is to develop your ability to picture relationships between variables. This ability includes graphing functions. A good function-grapher must be able to recognize what functions are best left to a computer or calculator to graph and which are more efficiently tackled with paper and pencil. He or she must also develop knowledge and techniques to facilitate paper-and-pencil graphing. In the next section we will discuss using a computer and a calculator as graphing tools. This section will focus on paper-and-pencil graphing.

Most of the graphs you make by hand do not require pinpoint accuracy. Rather, your graphs should display basic characteristics of the function and of the relationship between the variables. With an eye toward developing graphing skills, we will discuss a collection of eight functions that make up our *toolkit.* These functions are important for their intrinsic mathematical content and as a set of tools for mathematicians, scientists, economists, and anyone else who creates mathematical models to solve problems. This toolkit is not sufficient to tackle every problem, but it gives us some common ground that we will use as we expand our knowledge of functions. A thorough understanding of these toolkit functions is essential, and will make the future study of functions and mathematical modeling more meaningful. As you read about each of the eight toolkit functions you should plot a few points to help you understand the graph.

Constant Function

One of the simplest functions is one in which all first coordinates are paired with the same second coordinate; such a function is called a constant function. For instance, $f(x) = 4$ consists of ordered

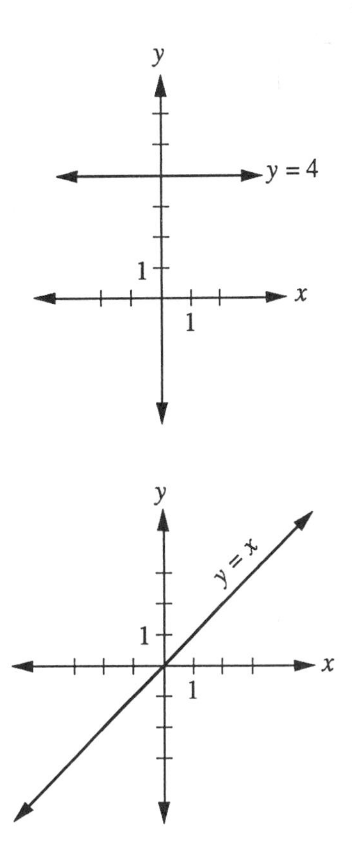

Figure 4 Constant and Linear Functions

pairs all with second coordinate 4. The graph is the horizontal line $y = 4$ (see Figure 4).

Linear Function

Another simple linear function is $f(x) = x$. It is often called the *identity function*; the first and second coordinates in each ordered pair are identical. The graph of this function is the diagonal line shown in Figure 4. Notice that this graph is everywhere increasing and continuous. The general linear function is $f(x) = mx + b$; its graph is a variation of the line $y = x$ that has slope m and y-intercept b. You should be familiar with these variations from study in previous mathematics courses and from your work with the median-median line in Chapter 1.

Quadratic Function

The toolkit quadratic function is $f(x) = x^2$. You can better understand the shape of the graph of $f(x) = x^2$ by thinking about the ordered pairs (x, x^2) that belong to the function. Whenever x is greater than 1, $x^2 > x$ so the graph is above the line $y = x$. When x is between 0 and 1, $x^2 < x$ and the graph is below that of $y = x$. The shape of the graph of $f(x) = x^2$, shown in Figure 5, is called a *parabola.* Notice that it is decreasing for negative x-values and increasing for positive x-values. The graph has a *turning point* at (0, 0), which is called the *vertex* of the parabola.

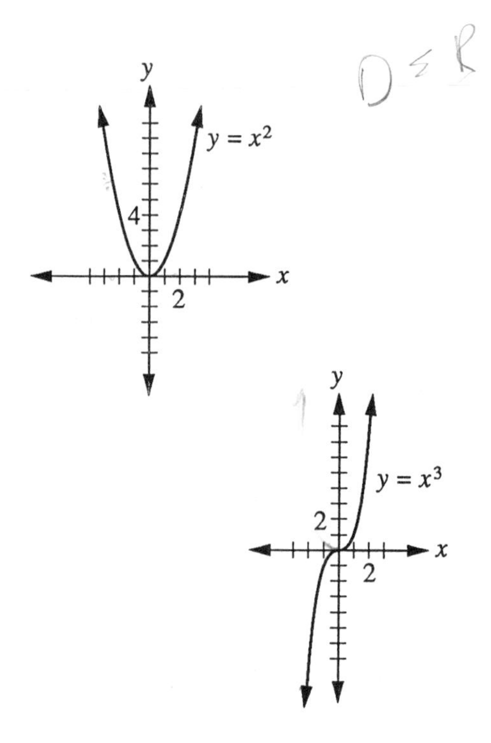

Figure 5 Quadratic and Cubic Functions

Cubic Function

The toolkit cubic function is $f(x) = x^3$; the graph is shown in Figure 5. This graph is increasing for all x-values; it is steeper than the graph of $f(x) = x^2$ for $x > 1$. Notice that the graph lies entirely in the first and third quadrants; in ordered pairs (x, x^3), either both coordinates are positive or both are negative.

Square Root Function

Let us denote the square root function by $f(x) = \sqrt{x}$. Since the symbol $\sqrt{x}$ represents only the *principal* (positive) *square root* of x, each x-value is paired with a unique y-value. (Think about what your calculator does when you compute the square root of 9; the calculator gives a display of 3, not ± 3.) The graph of $f(x) = \sqrt{x}$ is shown in Figure 6. Notice that it is increasing, but the rate of increase is very slow. Another important feature of this graph is the downward curvature. Mathematicians call this curvature *concave down*; in contrast, the shape of the parabola $y = x^2$ is *concave up.*

We can compare the steepness of $f(x) = \sqrt{x}$ with that of linear, quadratic, and cubic functions by determining for each function the x-value that is paired with a y-value of 64. For $f(x) = x^3$, the ordered pair is (4, 64). For $f(x) = x^2$, the ordered pair is (8, 64). For $f(x) = x$, the ordered pair is (64, 64). For $f(x) = \sqrt{x}$, the ordered pair is (4096, 64). Notice that the cubic function attains the value of 64 when $x = 4$, whereas the square root function doesn't attain this value until $x = 4096$.

It is also interesting to study the steepness of $f(x) = \sqrt{x}$ in another way. When $x > 1$, $\sqrt{x} < x$, so the graph is below the line $y = x$. In effect, taking the square root pulls y-values down. On the other hand, when $0 < x < 1$, $\sqrt{x} > x$ and the graph is above the line $y = x$. For these x-values, taking the square root pushes the y-values up.

Absolute Value Function

Recall that the symbol $|x|$ is defined as follows:

$$|x| = x \text{ if } x \geq 0 \text{ and } |x| = -x \text{ if } x < 0.$$

For positive x-values and zero, $|x|$ is equal to x; for negative x-values $|x|$ is equal to the opposite

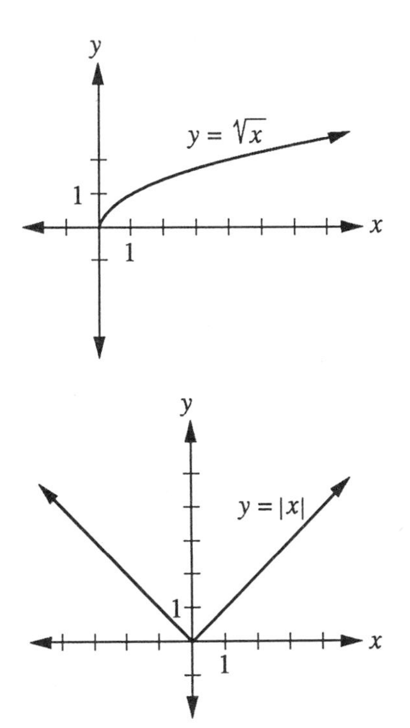

Figure 6 Square Root and Absolute Value Functions

of x. The absolute value function $f(x) = |x|$ is actually an example of a *piecewise-defined function.* This means that the function is defined by different formulas on different intervals. The function $f(x) = |x|$ is the same as the piecewise-defined function

$$f(x) = \begin{cases} x & \text{if } x \geq 0 \\ -x & \text{if } x < 0. \end{cases}$$

The graph of this function is shown in Figure 6. Notice that f pairs positive x-values with themselves, and pairs negative x-values with their opposites. For positive x-values the graph of f is identical to the line $y = x$; for negative x-values the graph is identical to the line $y = -x$.

Reciprocal Function

The graph of the reciprocal function $f(x) = \frac{1}{x}$ is shown in Figure 7, page 32; this shape is called a *hyperbola.* Notice that the graph is discontinuous at $x = 0$ and is decreasing on $x < 0$ and on $x > 0$. The function f produces an output by taking the reciprocal of its input; since taking the reciprocal does not cause a sign change, the graph lies entirely in the first and third quadrants. In the first quadrant where x-values are positive, the reciprocals of small numbers are big numbers and the reciprocals of big numbers are small numbers. As x-values get larger, y-values get smaller and approach zero, so the graph gets closer and closer to the x-axis. We say that the graph is *asymptotic* to the x-axis. Similarly, x's close to zero are paired with large y's, and the graph also has the y-axis as an *asymptote.* Similar analysis for negative values of x explains the behavior of the graph in the third quadrant.

Sine Function

The sine function defined by the equation $f(x) = \sin x$ is graphed in Figure 7, page 32. The graph of the sine function oscillates in a regular pattern between a maximum y-value of 1 and a minimum y-value of -1. The function has a value of 0 at $x = 0$. It increases to its maximum of 1 at $x = \frac{\pi}{2}$, returns to 0 at $x = \pi$, decreases to its minimum of -1 at $x = \frac{3\pi}{2}$, and returns to 0 at $x = 2\pi$. Then it repeats the process indefinitely. If you look at the section of the sine function between -2π and 0, or between 2π and 4π, you notice that it has the same shape as the section between 0 and 2π. Functions with such repetition of a basic shape are called *periodic* functions. Since the portion being repeated is 2π units long, the *period* is said to be 2π. That is, the sine function completes one cycle every 2π units. The values $\frac{\pi}{2}$, 2π, and so on may seem unusual to you. Since π is related to circles, you may guess that the sine function is somehow related to circles. This is, in fact, true, but we will wait until a later chapter for a thorough discussion of the relationship. For now, it is important to know the basic shape of the sine function and coordinates of intercepts and turning points. You can verify that $\sin \frac{\pi}{2} = 1$, $\sin \pi = 0$, and so forth by using your calculator in radian mode. Radian mode is what your calculator uses to evaluate the sine of a real number.

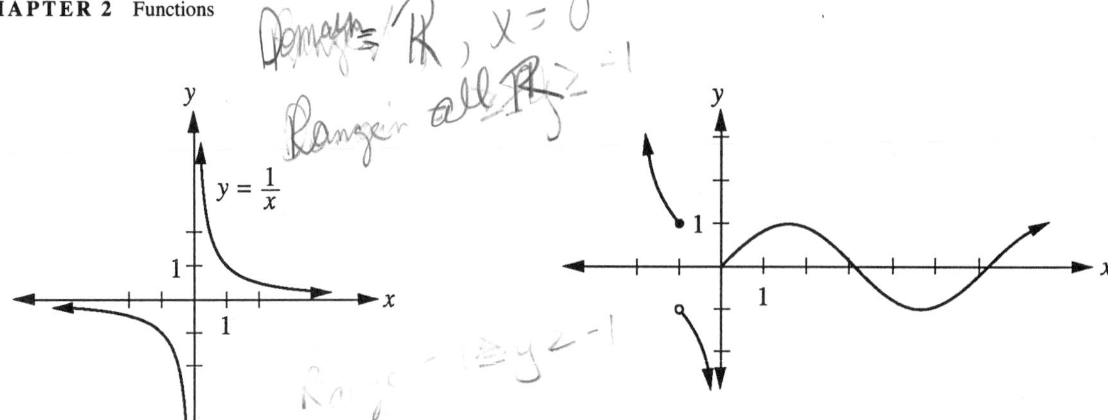

Figure 8 Graph of Piecewise-Defined Function

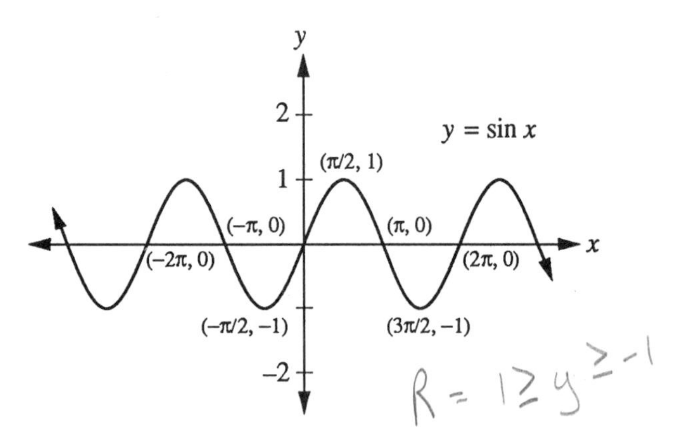

Figure 7 Reciprocal and Sine Functions

Example 1

Sketch a graph of the following piecewise-defined function.

$$f(x) = \begin{cases} x^2, & \text{if } x \le -1 \\ \frac{1}{x}, & \text{if } -1 < x < 0 \\ \sin x, & \text{if } x \ge 0 \end{cases}$$

Solution The graph is shown in Figure 8. For x-values less than or equal to -1, the function is defined by the equation $y = x^2$, and the graph consists of a section of the toolkit parabola. For x-values between -1 and 0, the function is defined by the equation $y = \frac{1}{x}$ and the graph consists of a section of the toolkit hyperbola. Note the open circle at $(-1, -1)$; this point is not on the graph of the function. For x-values greater than or equal to 0, the function is defined by the equation $y = \sin x$. ■

Exercise Set 2

1. Sketch graphs of $y = x^4, y = x^5, y = x^6$, and $y = x^7$. Make a generalization about the graph of $y = x^n$ for n an even integer and for n an odd integer.
2. Sketch a graph of each piecewise-defined function.

   a. $f(x) = \begin{cases} x, & \text{if } x \ge -2 \\ |x|, & \text{if } x < -2 \end{cases}$

   b. $k(x) = \begin{cases} x^3, & \text{if } x < 0 \\ \frac{1}{x}, & \text{if } x > 0 \end{cases}$

   c. $g(x) = \begin{cases} x+2, & \text{if } x \le 0 \\ 2, & \text{if } 0 < x < 4 \\ \sqrt{x}, & \text{if } x > 4 \end{cases}$

   d. $f(x) = \begin{cases} 1, & \text{if } |x| < 1 \\ \\ x^2, & \text{if } x \ge 1 \end{cases}$
3. Suppose $f(x)$ is a linear function and you know that $f(0) = 4$ and that $f(x) - f(x+2) = -3$ for all values of x. Write an expression for $f(x)$.
4. Let $p(x) = \sqrt{x}$, $q(x) = \frac{1}{x}$, and $r(x) = x^3$. Express each of the following as a variation of p, q, or r. For instance, $y = \sqrt{x+1} + 2$ can be expressed as $y = p(x+1) + 2$.

a. $y = 1 + \dfrac{1}{x}$
b. $y = 8x^3$
c. $y = \dfrac{1}{2x - 2}$
d. $y = \sqrt{6 - x}$
e. $y = 3 - \sqrt{x}$
f. $y = (2x)^3$

5. Write one or two sentences to compare the graphs of $y = x$, $y = x^2$, and $y = x^3$ on the interval $0 \le x \le 1$.

6. Write one or two sentences to compare the graphs of $y = x$ and $y = \sqrt{x}$ on the interval $0 \le x \le 1$.

3

Graphing Functions on a Computer or Graphing Calculator

Now that you are familiar with the simple functions that make up the toolkit, we will begin to explore more complex and interesting functions. In this section we will concentrate on using the computer or graphing calculator to graph functions. You should bear in mind that when either of these tools makes a graph it is just plotting discrete points. Computers and graphing calculators can plot lots of points quickly, which makes them efficient tools for such tasks. However, the ease with which we can create these graphs is deceptive, since these tools can plot points but are unable to make judgments about these points. The power of the computer and graphing calculator must be combined with our knowledge and common sense; if not, we may be misled by information they give us.

No matter what computer graphing software or graphing calculator you will be using, you will need to choose the interval of x-values over which you want the function graphed. After you specify this interval, the computer or calculator calculates the coordinates of points and then plots them. With some software and graphing calculators the interval of y-values is automatically chosen to include all possible y-values that result when the function acts on the x-values. Other software and graphing calculators require that the user choose an interval of y-values. In either case, the result is a *viewing window* that shows the graph of a function in one particular area. When the viewing window is large, we see the *global* behavior of the graph, but some of the details about its behavior are usually obscured. When the viewing window is small, we see *local* behavior of the graph; the details are apparent but the big picture is missing.

Example 1

Use a computer or graphing calculator to graph $y = x^3 - x$.

Solution Before you even turn on the computer or graphing calculator, think about how you expect the graph to look. The equation of this function suggests that it is related to the toolkit function $y = x^3$. For large positive x-values, the x^3-term is so much bigger than the x-term that the graph will behave like $y = x^3$ and will thus be increasing. The x^3-term also dominates when x is very negative, so the graph is also increasing in the third quadrant. You can find the x-intercepts of the graph by setting $y = 0$ and solving $x^3 - x = 0$. Factoring the expression $x^3 - x$ as $x(x^2 - 1)$ or $x(x+1)(x-1)$ yields $x = 0$, $x = -1$ and $x = 1$ as x-intercepts. These x-values that produce a y-value of 0 are called *zeros* of the function. The observations we have made about $y = x^3 - x$ are recorded in Figure 9.

Now we can use a computer or calculator to complete the graph. Let's first graph over the x-interval $[-10, 10]$. (This interval contains all x-values that satisfy $-10 \le x \le 10$.) We need to choose an interval of y-values that contains outputs

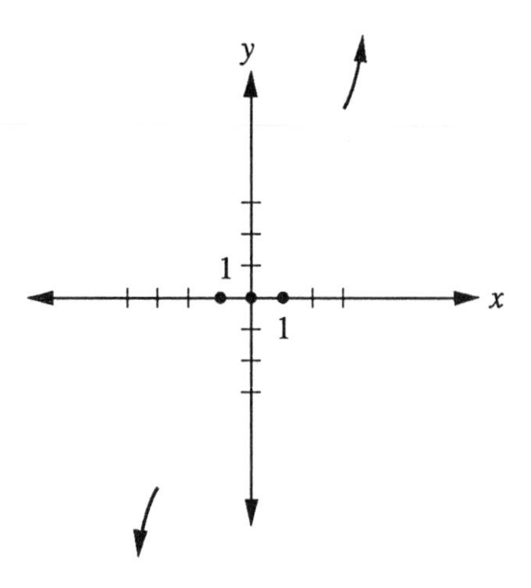

Figure 9 Key Features of Graph $y = x^3 - x$

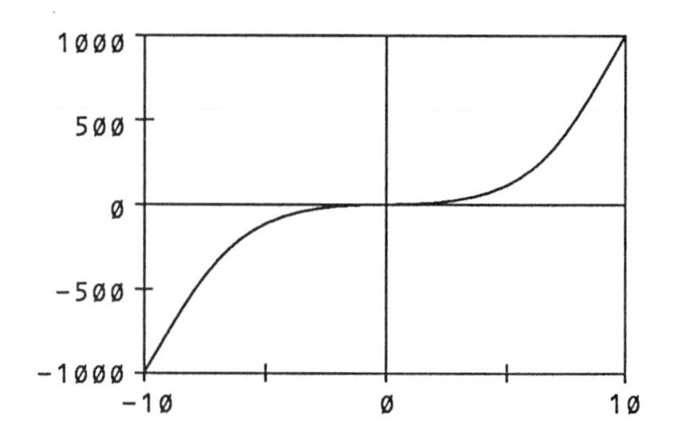

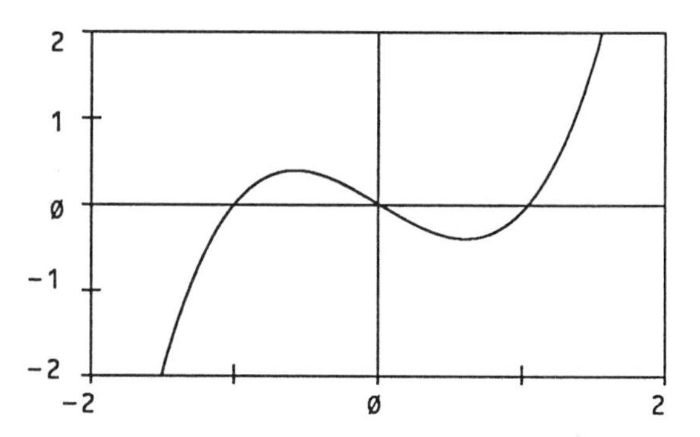

Figure 10 Graphs of $y = x^3 - x$ with Different Viewing Windows

which correspond to the inputs in our x-interval; an appropriate y-interval is $[-1000, 1000]$. The graph in this viewing window is first shown in Figure 10. You should notice that from this perspective the graph of $y = x^3 - x$ is very much like that of $y = x^3$; this is the global behavior of the graph. You cannot distinguish the three x-intercepts that we know the graph should have. This choice of a viewing window has the effect of disguising information about the graph.

We have two options for getting a better picture of the graph's local behavior. We can either use special features of the software or calculator to "zoom in" on our graph, or we can choose a smaller viewing window, say x-interval $[-1.5, 1.5]$ and y-interval $[-2, 2]$. A graph over this smaller viewing window is also shown in Figure 10. Now we can see the x-intercepts, but some of the global behavior of the graph is missing. What do you think you will see if you graph over the x-interval $[-0.5, 0.5]$ and y-interval $[-1, 1]$? Check your prediction.

The two turning points on the graph can be identified by moving the cursor to trace along the graph. The leftmost turning point is located at $(-0.58, 0.385)$, and the rightmost point is at $(0.58, -0.385)$. ■

Example 2

Graph this function using a calculator or computer:

$$f(x) = \frac{x^4 - 1}{x + 2}.$$

Solution Again, it is a good idea to make a few initial observations about what we expect to see in the graph. You should verify that $f(1) = 0$, $f(-1) = 0$, $f(0) = -\frac{1}{2}$, and $f(-2)$ is undefined.

A graph over the x-interval $[-10, 10]$ and y-interval $[-1500, 1000]$ is shown in Figure 11. From this perspective, we see the global behavior, and observe that the graph of f resembles that of $y = x^3$. This is because for large x-values, the x^4 term dominates the numerator, the x-term dominates the denominator, and the ratio of x^4 to x is x^3.

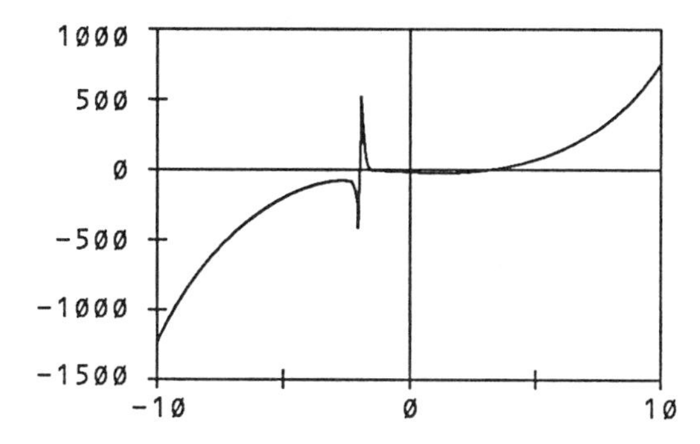

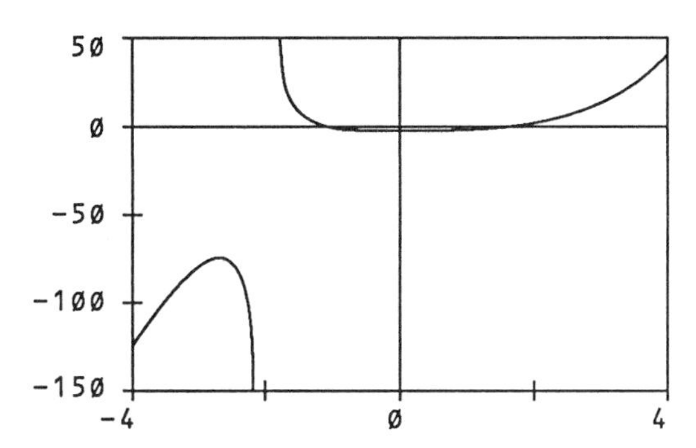

Figure 11 Graphs of $y = \frac{x^4-1}{x+2}$ with Different Viewing Windows

If we choose a smaller window, say x in $[-4, 4]$ and y in $[-150, 50]$, the local behavior of the graph near the origin can be seen. This graph is also shown in Figure 11. Notice that the graph has two x-intercepts as expected, as well as a vertical asymptote at $x = -2$. Use a calculator to verify that $f(-1.9) = 120.321$ and $f(-2.1) = -184.481$; this should help you understand the behavior of the graph near $x = -2$. ■

Bear in mind that computers and calculators generate graphs by plotting discrete points. Most computer software simply calculates coordinates of points, plots the points, and then connects consecutive points. If your software does this, then your graph in Example 2 may include a nearly vertical line near $x = -2$. This could be the line connecting the points $(-2.1, -184.481)$ and $(-1.9, 120.321)$. Graphing software that connects consecutive points may connect points on a graph where there should be a discontinuity.

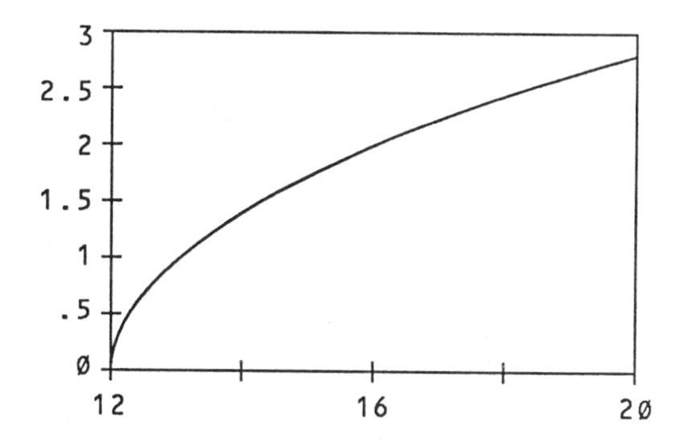

Figure 12 Graph of $y = (x - 12)^{\frac{1}{2}}$

Example 3

Graph $g(x) = \sqrt{x - 12}$.

Solution You should try graphing this function in the x-interval $[-10, 10]$ and y-interval $[-10, 10]$. What happens? You will either get an error message, or no graph will be visible on the display screen. Why has this happened, and what does it mean? For any x-value between -10 and 10, the quantity $x - 12$ is negative, and therefore the square root of $x - 12$ is not a real number. Thus, there are no points on the graph with x-coordinates between -10 and 10.

How can we see what the graph of g looks like? As observed above, $x - 12$ must be a nonnegative number so that its square root will be a real number. Thus, we must look in a viewing window that contains x-values that satisfy $x - 12 \geq 0$, or $x \geq 12$. A graph in the window $[12, 20]$ and $[0, 3]$ is shown in Figure 12. ■

Exercise Set 3

1. Use a computer or a graphing calculator to graph $y = \frac{1}{x+15}$ for each of the following viewing windows. Write a sentence or two describing what aspects of the graph you can see with

each window. Experiment with different windows to get the best view.

a. x-interval $[-5, 5]$ and y-interval $[-2, 2]$

b. x-interval $[-10, 10]$ and y-interval $[-5, 5]$

c. x-interval $[-15, 15]$ and y-interval $[-15, 15]$

d. x-interval $[-20, 20]$ and y-interval $[-20, 20]$

2. Graph $y = \dfrac{x^5 - 1}{x^2 - 1}$ with a graphing calculator or computer. Experiment with different viewing windows. Write one or two sentences to describe the graph.

3. Use a computer or graphing calculator to graph $y = \sqrt{20 - x}$. Experiment with different viewing windows. Write one or two sentences to describe the graph.

4. With a graphing calculator or computer, graph each function f listed below. Compare each graph to the graph of the toolkit function g; specifically tell how you could alter the graph of g to obtain the graph of f.

a. $f(x) = x^2 + 2$ $\quad g(x) = x^2$

b. $f(x) = (x - 1)^3$ $\quad g(x) = x^3$

c. $f(x) = \sqrt{x + 4}$ $\quad g(x) = \sqrt{x}$

d. $f(x) = |\frac{1}{x}|$ $\quad g(x) = \frac{1}{x}$

e. $f(x) = \sqrt{|x|}$ $\quad g(x) = \sqrt{x}$

f. $f(x) = -\sin x$ $\quad g(x) = \sin x$

g. $f(x) = 3\sin x$ $\quad g(x) = \sin x$

4

Domain and Range

Each time we work with a function we are going to be interested in its domain and its range. To understand these terms, refer back to the table of postage rates given in Table 2 on page 28. Weights and postage rates are given in the columns of this table. Since the postage rate depends on the weight, the weight is the independent variable. We can think about a function that consists of ordered pairs (weight, postage rate). The numbers that represent weights make up the *domain* of the function — they are *the first coordinates in the ordered pairs of the function.* The table shows postage rates for weights from 0 to 16 ounces, so the domain consists of all numbers from 0 to 16. The numbers that represent postage rates make up the *range* of the function — they are *the second coordinates in the ordered pairs of the function.* Based on the information in the table, the possible rates are 0.25, 0.45, 0.65, 0.85 and so forth; this set of numbers is the range of the function. Notice that the domain contains infinitely many numbers, whereas the range contains only 10 numbers.

Throughout this book we will use only real numbers for the domain and range of functions. Whenever the domain for a function is not specified, you should assume that the domain is the largest set of real number inputs into the function that produce outputs that are also real numbers. Restrictions on the domain are often necessary in order to avoid division by zero and square roots of negative numbers. Failure to recognize appropriate domain restrictions can result in a blank coordinate system or in error messages. It can also lead to unreasonable conclusions about the phenomenon that the function is modeling.

Example 1

Find the domain of $f(x) = \dfrac{3x - 1}{x^2 - 4}$ and of $g(x) = \sqrt{6 - 7x}$.

Solution The domain of f cannot include x-values that result in a zero in the denominator. Since $x^2 - 4$ is equal to 0 when $x = 2$ and when $x = -2$, the domain consists of all real numbers except 2 and -2.

To ensure real values in the range, the domain of g can include only x-values for which $6 - 7x$ is greater than or equal to 0. Solving the inequality $6 - 7x \geq 0$ gives $x \leq \frac{6}{7}$; therefore the domain of g is $x \leq \frac{6}{7}$. ■

Example 2

Find the domain of $f(x) = \sqrt{x^3 - 4x - 1}$.

Solution The domain of f includes only x-values for which $x^3 - 4x - 1$ is greater than or equal to 0.

There are no good algebraic techniques for solving the inequality $x^3 - 4x - 1 \geq 0$. An alternative to algebraic manipulation is to use graphical analysis. Let's define a new function $h(x) = x^3 - 4x - 1$. We can then use a computer or calculator graph of h to find all x-values for which $h(x) \geq 0$; these are approximately $-1.86 \leq x \leq -0.25$ and $x \geq 2.12$. For x-values in these intervals, $x^3 - 4x - 1$ is greater than or equal to 0. Thus, an approximation for the domain of the original function f is the set of all x such that $-1.86 \leq x \leq -0.25$ or $x \geq 2.12$. ■

Example 3

Imagine that you have a rectangular piece of cardboard that measures 18 inches by 24 inches. You can form an open box by cutting congruent squares from each of the corners of the cardboard and folding up the flaps (see Figure 13). The volume of the resulting open box is a function of the size of the square cut from each corner.

If you cut a square that measures x by x from each corner, then the dimensions of the open box will be x by $18 - 2x$ by $24 - 2x$. The volume of the box is given by the function

$$V(x) = x(18 - 2x)(24 - 2x).$$

What is the domain of this function V?

Solution All real numbers are algebraically legitimate inputs for the function V; you can see from the equation for V that you do not need to be concerned about division by zero or square roots of negative numbers. However, the phenomenon that the function V is modeling imposes some restrictions on the domain. It should be clear that x must be a positive number, and that x cannot exceed 9 inches. Thus, when you consider V as a model the domain is all real numbers between 0 and 9. ■

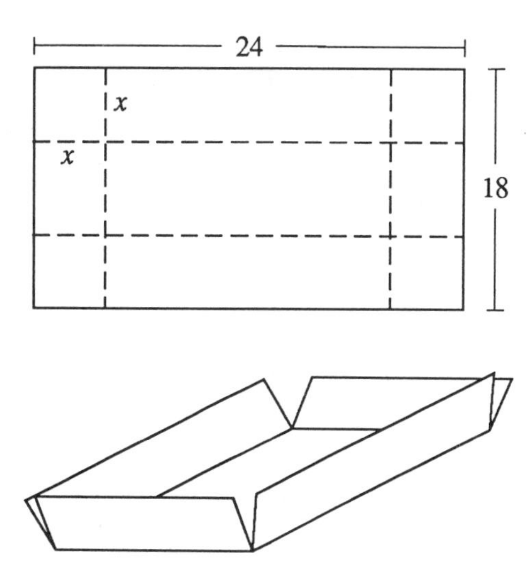

Figure 13 Cuts to Form Box

When we use functions to model phenomena, there will be some restrictions on the domain that arise out of purely algebraic concerns — these are usually apparent from the equation of the function. There will be other restrictions that are implied by the situation we are modeling. If the independent variable represents a number of objects or people, we will not want to include fractions or negative numbers in the domain. In other situations, like the box volume, the physical aspects of the problem impose additional limitations on the domain.

The values in the range of a function are determined by the values in the domain and by how the function acts on these inputs. Finding the range of a function is generally more difficult than finding the domain. One reliable way to determine the range of a function is by looking at its graph. The range consists of all y-values that belong to the function; by looking at a graph you can see what numbers occur as y-coordinates — these numbers make up the range.

Example 4

Identify the range of the function $y = \frac{1}{x} + 1$.

Solution The graph of the function is shown in Figure 14. The graph has the horizontal line $y = 1$ as an asymptote; y-values get close to 1 but are never equal to 1. Every real number except 1 occurs as the y-coordinate of some point on this graph, so the range is all real numbers except 1. ■

Figure 14 Graph of $f(x) = \frac{1}{x} + 1$

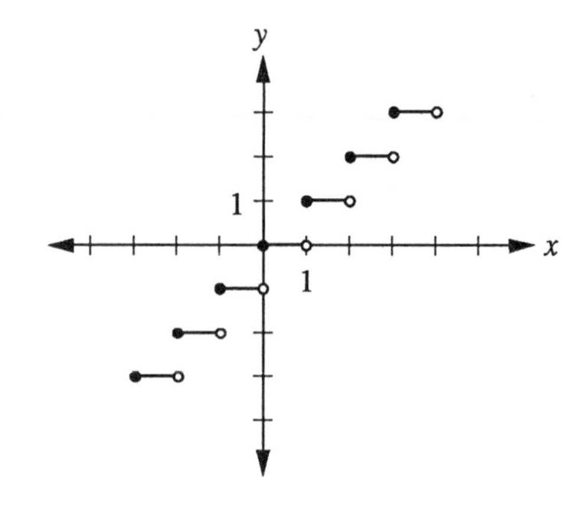

Figure 15 Graph of Function $f(x) = [x]$

Exercise Set 4

1. Identify the domain and the range of each toolkit function.
2. State the domain of each function. You may want to refer to a computer- or calculator-generated graph.
   - **a.** $y = \frac{1}{x-11}$
   - **b.** $y = \frac{x+2}{x^2-3x+2}$
   - **c.** $y = \sqrt{3-x}$
   - **d.** $y = \sqrt{x^2 - 5x - 14}$
   - **e.** $f(x) = \frac{1}{x-3}$
   - **f.** $p(x) = \frac{1}{x^2-5x+6}$
   - **g.** $h(x) = \sqrt{2x-7}$
   - **h.** $f(x) = \sqrt{(x-1)(x+2)}$
   - **i.** $f(t) = \frac{1}{\sqrt{t^2-4}}$
   - **j.** $g(x) = \sqrt{\frac{1-x}{x+1}}$
   - **k.** $y = \sqrt{x^2 - 9}$.
   - **l.** $f(x) = \frac{x^2-4}{x+7}$
   - **m.** $f(x) = \sqrt{\frac{2x-1}{x+1}}$
   - **n.** $y = \frac{1}{\sin x}$
3. State the range of each function. You may want to use a computer- or calculator-generated graph.
   - **a.** $f(x) = x^2 - 1$
   - **b.** $f(x) = x^3$
   - **c.** $f(x) = \sin x + 2$
   - **d.** $f(x) = \sqrt{x-3}$
   - **e.** $f(x) = 0.01x^4$
   - **f.** $f(x) = |x+3|$
   - **g.** $f(x) = |x| + 3$
   - **h.** $f(x) = 3 \sin x$
4. The symbol $[x]$ denotes the greatest integer that is less than or equal to x. For example, $[11.02] = 11$, $[4.9] = 4$, $[2] = 2$, and $[-2.4] = -3$. The function $f(x) = [x]$ is called the *greatest integer function*. For any x-value between 0 and 1, $f(x) = 0$, so over this part of the domain f behaves like a constant function. Similarly, for any x-value between 1 and 2, $f(x) = 1$. Thus, the graph of f consists of a collection of horizontal line segments, or *steps*. The graph is shown in Figure 15. Note that it is discontinuous at every integer value of x.
   - **a.** Explain the open and closed circles at the ends of the steps.
   - **b.** Identify the domain and the range of the function $f(x) = [x]$.
5. Answer the following questions about the function $g(x) = x - [x]$.

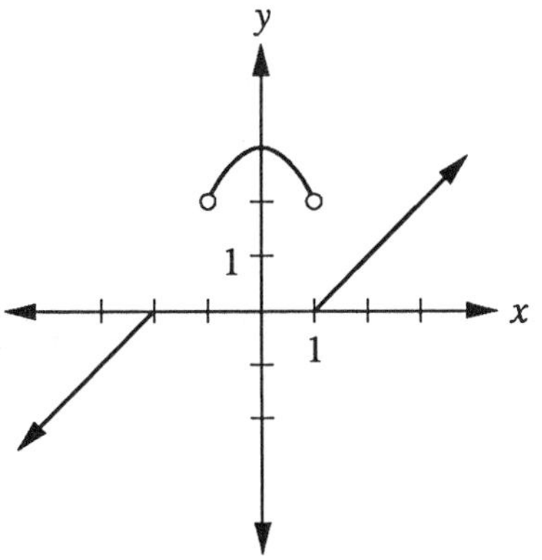

Figure 16 Graph for Exercise 6

a. Evaluate $g(n)$ where n is an integer.

b. Evaluate $g(n + .5)$ where n is an integer.

c. What value(s) of x satisfy the equation $g(x) = 0.75$?

d. Identify the domain and the range of g.

6. State the domain and range of the function graphed in Figure 16.

7. Determine whether or not each of the following values belongs to the range of the function

$$f(x) = \frac{2x - 1}{3x + 4}.$$

a. 0

b. $\frac{-4}{3}$

c. $\frac{1}{2}$

d. $\frac{2}{3}$

8. The two functions f and g defined below appear to be identical. However, they are identical functions only if they have the same domain and the same range. Compare the domains of f and g to help you decide if they really are identical.

$$f(x) = \frac{\sqrt{x+1}}{\sqrt{x-2}} \qquad g(x) = \sqrt{\frac{x+1}{x-2}}$$

9. A delivery service handles only containers whose weight does not exceed 50 lbs. The minimum charge for a delivery is \$5. A charge of 10 cents per pound is added for each full pound in excess of 5 lb. Express the cost c as a function of weight x. Identify the domain and the range of this function.

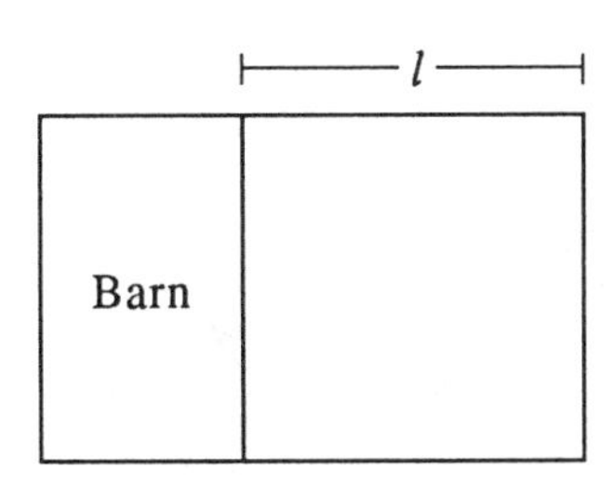

Figure 17 Fencing Three Sides of a Field

10. A piece of fencing 120 ft. long will be used to enclose three sides of a rectangular field. The fourth side of the field will be the wall of a barn. Let l be the length of a side of the field as shown; let A be the area of the field (see Figure 17).

a. Express A as a function of l.

b. What is the domain of A?

c. Use a graph to determine the range of A.

5

Symmetry

The graphs of several toolkit functions exhibit symmetry. For instance, the graph of $y = x^2$ is symmetric about the y-axis. The ability to identify symmetries will help you as you learn to graph other functions that do not belong to the toolkit.

Example 1

Discuss the symmetry of the graph of $f(x) = |x| - 2$ shown in Figure 18.

Solution The graph resembles the V-shape of the toolkit absolute value function. Notice that the expression for $f(-x)$ is identical to that for $f(x)$.

$$f(x) = |x| - 2$$
$$f(-x) = |-x| - 2 = |x| - 2$$

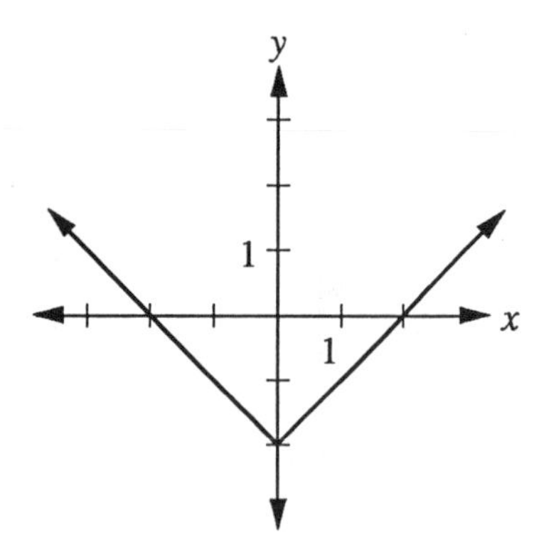

Figure 18 Graph of $f(x) = |x| - 2$

Since $f(-x) = f(x)$, for every point (x, y) on the graph of f there is also a point $(-x, y)$. The fact that these points have the same y-coordinate paired with opposite x-coordinates means that the graph will fit onto itself if it is folded along the y-axis. The graph is *symmetric about the y-axis.* A function that has this type of symmetry is called *even.* ∎

Example 2

Discuss the symmetry of the graph of $f(x) = 5x - x^3$ shown in Figure 19.

Solution Let's first decide whether the function is even by comparing the expressions for $f(x)$ and $f(-x)$.

$$f(x) = 5x - x^3$$

$$f(-x) = 5(-x) - (-x)^3 = -5x + x^3$$

Clearly, $f(-x)$ is not equal to $f(x)$. But notice that $f(-x)$ is equal to the opposite of $f(x)$; that is, $f(-x) = -f(x)$. This shows that for every point (x, y) that is on the graph of f the point $(-x, -y)$ is also on the graph. The graph of f is *symmetric about the point* (0, 0). A function with this type of symmetry is called *odd.* An odd graph will fit onto itself if it is rotated 180 degrees about the origin. ∎

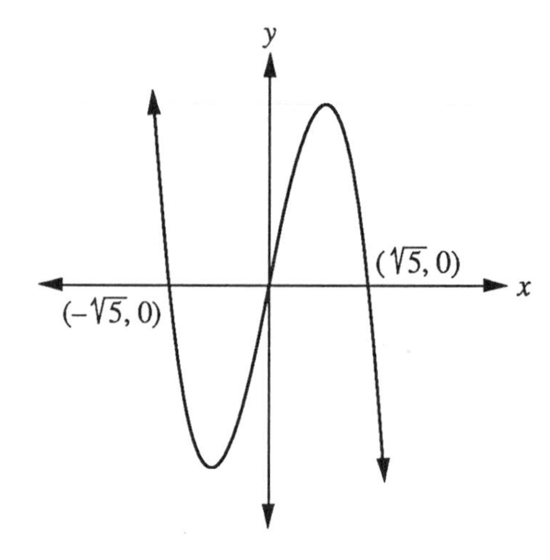

Figure 19 Graph of $f(x) = 5x - x^3$

Example 3

Discuss the symmetry of the graph of $f(x) = 1 + \frac{1}{x-1}$ shown in Figure 20.

Solution The graph of this function illustrates a third type of symmetry. You should verify that this function is neither even nor odd, that its graph contains the point (0, 0), and that the value $x = 1$ is not in the domain. Observe also that as x-values get very large, y-values get closer and closer

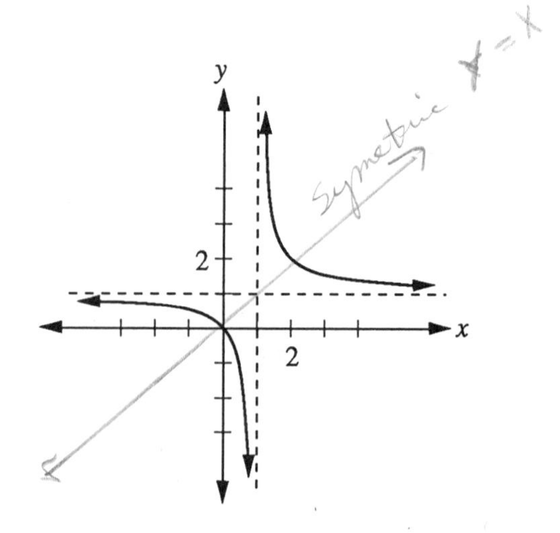

Figure 20 Graph of $f(x) = 1 + \frac{1}{x-1}$

to 1. Similarly, when x-values are just slightly larger than 1, y-values are very large.

Although the function is neither even nor odd, the graph does exhibit some symmetry. For example, the graph is symmetric about the line $y = x$; if you folded the graph on this line it would be superimposed on itself. Do you understand why the graph exhibits this symmetry? To explore this question, consider two points P and Q shown first in Figure 21.They are symmetric about the line $y = x$. Now look at the rectangle that contains P. If P has coordinates (a, b), then this rectangle has height b and width a. Now imagine folding the paper along $y = x$. Point P would now be in the position of Q, and the rectangle would be on its side. Its height is now a and its width b. This means that Q must have coordinates (b, a). Our conclusion is that if a point has coordinates (a, b) then its mirror image in $y = x$ has coordinates (b, a). This fact provides us with a way to test an equation to determine if its graph is symmetric about $y = x$. Whenever interchanging x and y produces an equivalent equation, then that equation's graph is symmetric about $y = x$. You should verify that our original equation $f(x) = 1 + \frac{1}{x-1}$ passes this test. ■

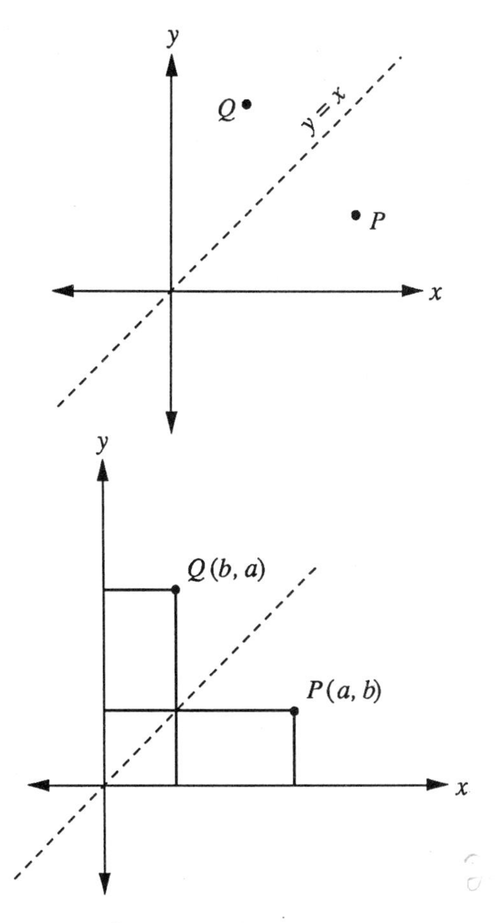

Figure 21 Symmetry about $y = x$

The concept of symmetry has significance in several different situations. Sometimes we look at a graph, observe symmetry, and make an inference about the equation of the graph or about the phenomenon that the graph represents. Whenever we are graphing with paper and pencil, symmetry can serve as a graphing tool. If we test the equation of a function and find that it is symmetric, this will reduce the number of x-values we need to investigate. For instance, if we find that a function is even, we need to investigate points only in the first and fourth quadrants; the graph can then be completed by reflecting across the y-axis.

Exercise Set 5

1. Why are the terms "even" and "odd" used to describe graphs with certain symmetries? **Hint**: Think about exponents.
2. Identify the symmetries of each of the toolkit functions.
3. Can the graph of a function be both even and odd? Explain.
4. Identify the symmetries of each relation. Sketch a graph for each by first plotting a few points, and then use symmetry to complete the graph.

   a. $y = x^2 + 4$
   b. $x^2 + y^2 = 1$
   c. $x^2 = y^2$
   d. $4x^2 - 9y^2 = 36$
   e. $y = x^3 + x$
   f. $f(x) = x^3 - 9x$
   g. $y = 5$

h. $y = \frac{1}{x^2}$

i. $y = x^{-3}$

j. $y = x^2 + 4x + 4$

5. Suppose $f(x) = 3x - 2$. Write an expression for $g(x)$ if the graph of g is obtained from the graph of f by each reflection described.

a. Reflection across the y-axis

b. Reflection across the x-axis

c. Reflection across the line $y = x$

d. Reflection about the line $x = 1$

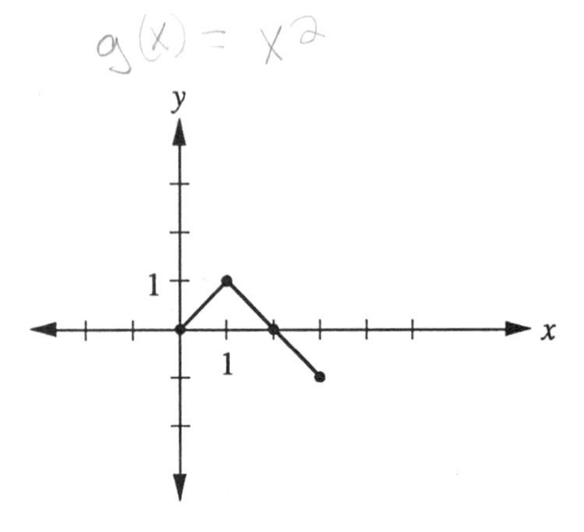

Figure 22 Graph of Function h for Exercise 6

6. Figure 22 shows the graph of a function h for $0 \le x \le 3$.

a. Sketch the graph of h from -3 to 3 if h is an even function.

b. Sketch the graph of h from -3 to 3 if h is an odd function.

6

Transformations of Functions

There are a great many functions that do not belong to the toolkit but nonetheless lend themselves to paper and pencil graphing. In many instances, sketching a graph of such a function by hand actually takes less time than using a computer or graphing calculator.

The examples in this section will provide you with information that will make graph sketching faster and easier. As stated previously, the graphs you sketch do not require pinpoint accuracy, but rather should show characteristics such as intercepts, asymptotes, some turning points, and general shape.

Example 1

Sketch a graph of $y = x^2 - 1$.

Solution The equation of this function suggests that it is related to the toolkit function $y = x^2$; it also shows that the function is even. By thinking about the equation for this function, you should be able to see that for any fixed x-value, the y-coordinate on $y = x^2 - 1$ is 1 less than the y-coordinate on $y = x^2$. This observation suggests that the graph of $y = x^2 - 1$ can be obtained by shifting the toolkit graph $y = x^2$ down 1 unit. The result is the parabola with vertex $(0, -1)$ shown in Figure 23.

Notice that if we define $f(x) = x^2$, then the equation of the shifted parabola is $y = f(x) - 1$. ■

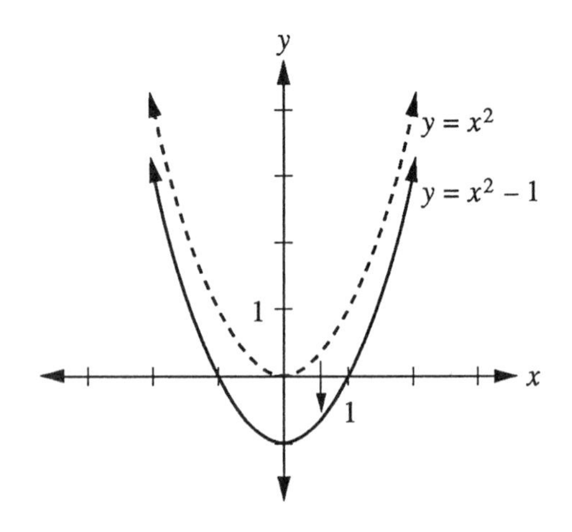

Figure 23 Graph of $y = x^2 - 1$

Example 2

Graph $y = |x| + 1$.

Solution This function is related to the toolkit function $y = |x|$. For any fixed x-value, the y-value on the graph of $y = |x| + 1$ is 1 more than the y-value on the graph of $y = |x|$. This fact is illustrated in the graph in Figure 24.

If we define $f(x) = |x|$, then this variation of the absolute value toolkit function can be written as $y = f(x) + 1$. ■

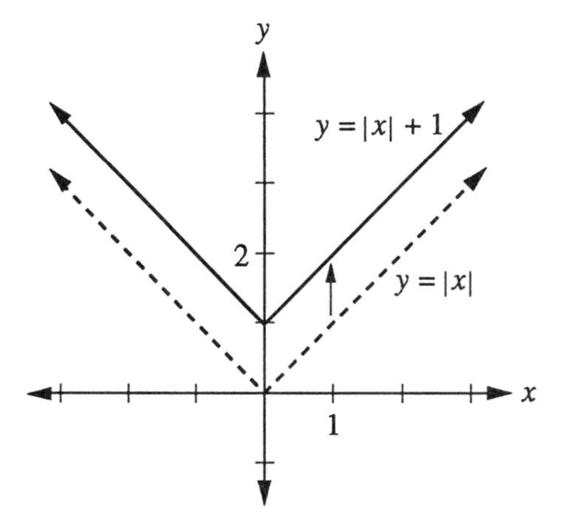

Figure 24 Graph of $y = |x| + 1$

Example 3

Graph $y = \sqrt{x-3}$.

Solution This is another function that appears to be a variation of a toolkit function. Note that the function's argument is $x - 3$; x-values are decreased by 3 before the square root function acts on them to produce an output. The domain of $y = \sqrt{x-3}$ contains x-values that are greater than or equal to 3. This is because the expression under the square root symbol must be zero or greater for the square root to be a real number; $x - 3 \geq 0$ is equivalent to $x \geq 3$. Figure 25 shows the graph of $y = \sqrt{x-3}$ as well as that of $y = \sqrt{x}$. Notice the shape that is characteristic of the square root function from the toolkit.

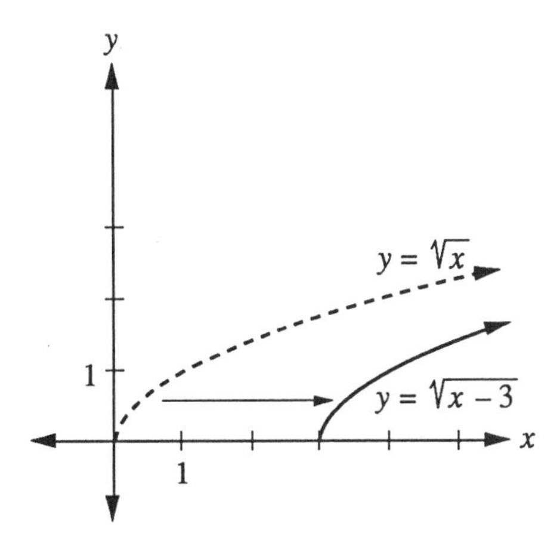

Figure 25 Graph of $y = (x-3)^{\frac{1}{2}}$ and $y = (x)^{\frac{1}{2}}$

If we define $f(x) = \sqrt{x}$, then this variation can be written as $y = f(x-3)$. Writing the equation in this way should help you see that to obtain a particular y-value, the x-value must be 3 more than it was for $y = f(x)$. This accounts for the fact that the graph of $y = \sqrt{x}$ has been shifted 3 units to the right to get the graph of $y = \sqrt{x-3}$. ■

There is an important difference between the notations $y = f(x-k)$ and $y = f(x) - k$, where k is a constant. In the former, the argument of the function is $x - k$, which means that k is subtracted from x before the function acts on its input. In the latter, the argument of the function is x, and k is subtracted after the function acts on its input.

Example 4

Sketch a graph of $y = (x+1)^2$.

Solution This function appears to be related to the quadratic toolkit function. If we define $f(x) = x^2$, then this equation can be written as $y = f(x+1)$; the argument is $x + 1$. To analyze this function, think about what x-value is needed to produce a particular y-value. For the function $f(x) = x^2$, the y-value is 9 when x is 3. However, for the function $f(x+1) = (x+1)^2$, the y-value is 9 when x is 2. In general, to obtain a particular y-value on $y = f(x+1)$ we need an x-value that is 1 less than we needed to obtain the same y-value

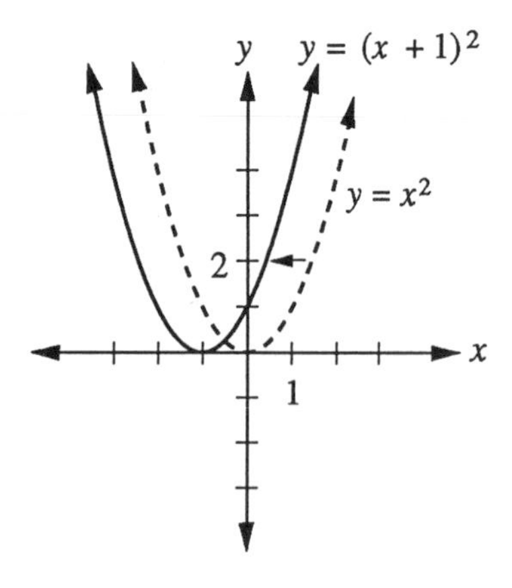

Figure 26 Graph of $y = (x+1)^2$

on $y = f(x)$. The graph of $y = f(x+1)$ will be the toolkit parabola shifted 1 unit to the left (see Figure 26). ∎

We will use the function $f(x) = x^3$ to review what we have observed in the preceding examples. First, consider the function defined by $y = (x-1)^3 = f(x-1)$. For any fixed y-value, the x-value will be 1 more than it was for $y = f(x)$, and the graph will be shifted one unit to the right. This is an example of a *horizontal translation.* Second, consider the function defined by $y = x^3 + 1 = f(x) + 1$. For any fixed x-value the y-value is one more than before, and the graph has been shifted one unit up. This is an example of a *vertical translation.* We can generalize these observations as follows:

The graph of $y = f(x-h)+k$ can be obtained by translating the graph of $y = f(x)$. The translation is h units to the right and k units up if h and k are positive. If h and k are negative the translation is to the left and down.

Example 5

Graph $y = 2\sin x$.

Solution This function is related to the sine function in the toolkit. If we define $f(x) = \sin x$, then the function we want to graph can be written as $y = 2f(x)$. For any fixed x-value, the y-coordinate on the graph of $y = 2f(x)$ will be twice as large as the y-coordinate on $y = f(x)$. Thus, the points $(0,0)$, $(\frac{\pi}{2},1)$, $(\pi,0)$, and $(\frac{3\pi}{2},-1)$ on $y = \sin x$ become $(0,0)$, $(\frac{\pi}{2},2)$, $(\pi,0)$, and $(\frac{3\pi}{2},-2)$ on $y = 2\sin x$. These points and your familiarity with the graph of $f(x) = \sin x$ lead to the graph shown first in Figure 27. You should compare this graph to that of $f(x) = \sin x$ which is also shown in Figure 27. What relationships can you see between the domains and the ranges of the two functions? ∎

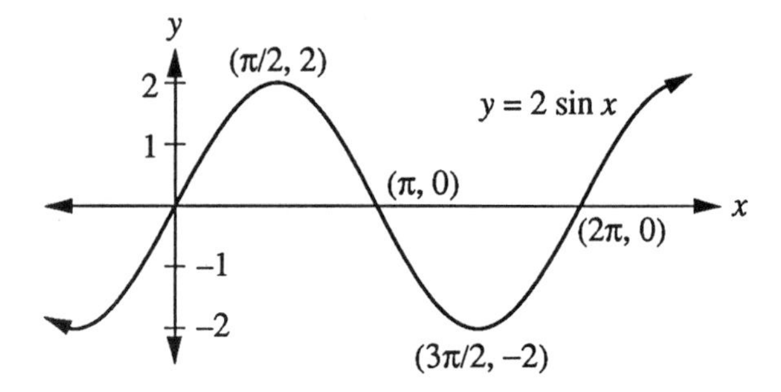

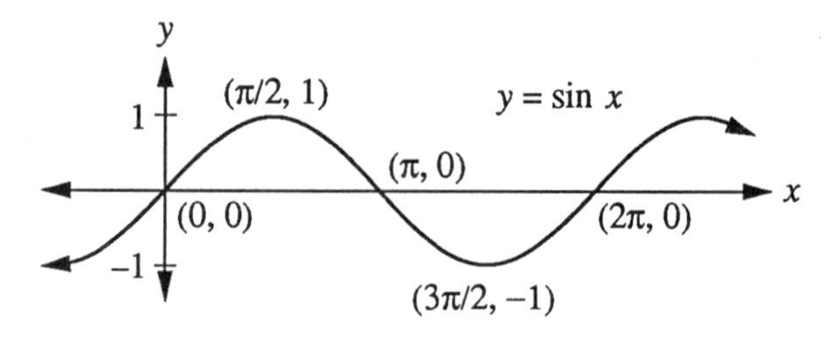

Figure 27 Graphs of $y = 2\sin x$ and $y = \sin x$

Example 6

Sketch a graph of $y = \sin(2x)$.

Solution This is another function that is related to the sine function in the toolkit. Note that the argument of this function is $2x$; x-values are doubled before the function acts on them to produce an output. The effect on the graph of this doubling will become clear as we examine several y-values and determine the corresponding x-values. You should verify the following facts about $y = \sin(2x)$ for special points between 0 and 2π.

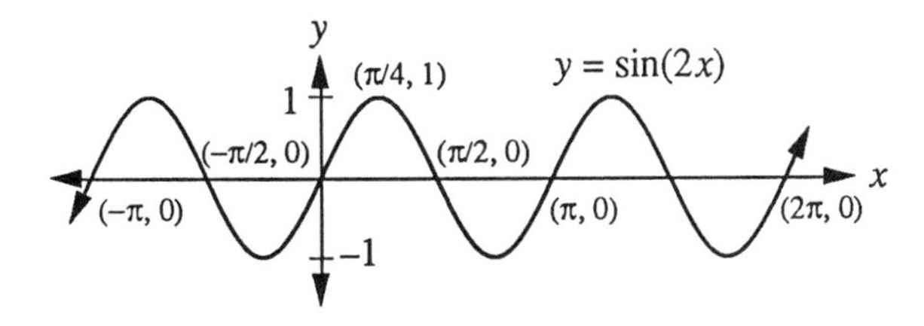

Figure 28 Graph of $y = \sin(2x)$

— if $y = 0$, then $2x = 0$, and $x = 0 \longrightarrow (0, 0)$

— if $y = 1$, then $2x = \frac{\pi}{2}$, and $x = \frac{\pi}{4} \longrightarrow (\frac{\pi}{4}, 1)$

— if $y = 0$, then $2x = \pi$, and $x = \frac{\pi}{2} \longrightarrow (\frac{\pi}{2}, 0)$

— if $y = -1$, then $2x = \frac{3\pi}{2}$, and $x = \frac{3\pi}{4} \longrightarrow (\frac{3\pi}{4}, -1)$

— if $y = 0$, then $2x = 2\pi$, and $x = \pi \longrightarrow (\pi, 0)$

Continuing this type of analysis results in the graph of the function shown in Figure 28. Compare this graph to that of the toolkit function $f(x) = \sin x$; notice that the period of $y = \sin(2x)$ is π, which is half as long as the period of $y = \sin x$. ■

It should be clear that the graphs of $y = 2\sin x$ and $y = \sin(2x)$ are both closely related to the graph of $f(x) = \sin x$, but the 2 in each equation has a different effect. The 2 in $y = 2\sin x$ doubles y-values and causes a *vertical stretching* about the x-axis by a factor of 2. The 2 in $y = \sin(2x)$ halves x-values and causes a *horizontal compression* about the y-axis by a factor of $\frac{1}{2}$.

Class Practice

1. Graph $y = \frac{1}{2}\sin x$.
2. Graph $y = \sin(\frac{1}{2}x)$.

You have now seen several different ways that constants and coefficients in the equation of a function can affect the graph. These effects are summarized in the table that follows.

Graphing Transformations of Functions

$y = f(x) + c$	graph is translated c units vertically
$y = f(x - c)$	graph is translated c units horizontally
$y = cf(x)$	graph is stretched or compressed vertically by a factor of c
$y = f(cx)$	graph is stretched or compressed horizontally by a factor of c

Notice that changes in the argument of the function alter the graph horizontally by influencing the domain, whereas changes outside the argument alter the graph vertically by influencing the range.

Exercise Set 6

1. Graph each function. Describe how the graph is related to that of a toolkit function. Label coordinates of two or three points.
   a. $y = |x| + 5$
   b. $y = x^3 - 2$
   c. $y = \frac{1}{x+2}$
   d. $y = (x - 4)^3$
   e. $y = |x + 3| - 3$
   f. $y = \sin(x + \pi)$
   g. $y = \sqrt{9x}$
   h. $y = \frac{1}{5}x^3 + 1$
   i. $y = 4 + \frac{2}{x}$
2. How can the graph of the toolkit function $g(x) = \sqrt{x}$ be transformed to give the graphs of $y = \sqrt{4x} = g(4x)$ and $y = 2\sqrt{x} = 2g(x)$? Sketch both graphs and write a sentence or two describing how compressions and stretches are related for this particular function.

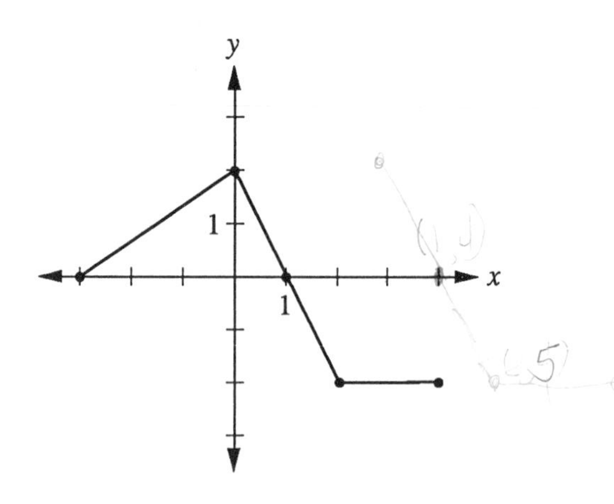

Figure 29 Graph of g for Exercise 3

3. Figure 29 shows the graph of a function g. The domain is $-3 \le x \le 4$ and the range is $-2 \le y \le 2$. Make a graph of each variation of g and state the domain and range for each.
   a. $y = g(x) - 3$
   b. $y = \frac{1}{3}g(x)$
   c. $y = g(\frac{1}{3}x)$
   d. $y = g(x+3)$
4. Graph each piecewise-defined function.
   a. $y = \begin{cases} (x-1)^2 & \text{if } x \ge 0 \\ x-2 & \text{if } x < 0 \end{cases}$
   b. $y = \begin{cases} \sqrt{\frac{1}{2}x} & \text{if } x > 0 \\ \sin(3x) & \text{if } x \le 0 \end{cases}$
   c. $y = \begin{cases} |x+4| & \text{if } x \ge 2 \\ \sqrt{x+7} & \text{if } -3 < x < 2 \\ \frac{1}{x} & \text{if } x \le -3 \end{cases}$

7

Combinations of Transformations

Each of the generalizations about transformations listed in the summary table of the previous section is true for both positive and negative c-values. However, stretches and compressions occur in a slightly different way when c is negative, as the next two examples will illustrate.

Example 1

Graph $f(x) = -\sqrt{x}$ and $g(x) = \sqrt{-x}$.

Solution Each of these graphs could be gotten by plotting points or by using the computer or calculator to create a graph. However, both functions are closely related to the toolkit function $y = \sqrt{x}$ and are easily graphed by hand with the benefit of knowledge about transformations.

Let's work on $f(x) = -\sqrt{x}$ first. For each fixed x-value, the y-value on the graph of f will be the opposite of the y-value on the graph of the toolkit function. For every point (x, y) on the graph of $y = \sqrt{x}$ there is a point $(x, -y)$ on the graph of f. Thus, the graph of the toolkit function can be reflected about the x-axis to produce the graph of $f(x) = -\sqrt{x}$. The graph of f is shown in Figure 30.

Now consider $g(x) = \sqrt{-x}$. Since the domain of the square-root function is the nonnegative real numbers, the x-values in the domain of g must satisfy the inequality $-x \ge 0$. This means that the domain is $x \le 0$. Another way to look at this is to see that for each fixed y-value, the x-value on the graph of g will be the opposite of the x-value on the graph of the toolkit function. In either case, for every point (x, y) on the graph of $y = \sqrt{x}$ there is a point $(-x, y)$ on the graph of g. This means that the graph of the toolkit function can be reflected about the y-axis to produce the graph of $g(x) = \sqrt{-x}$. The graph of g is shown in Figure 30. You should verify the coordinates of a few points

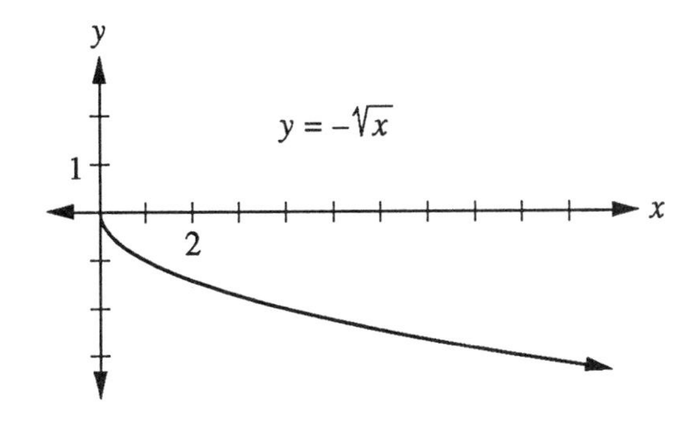

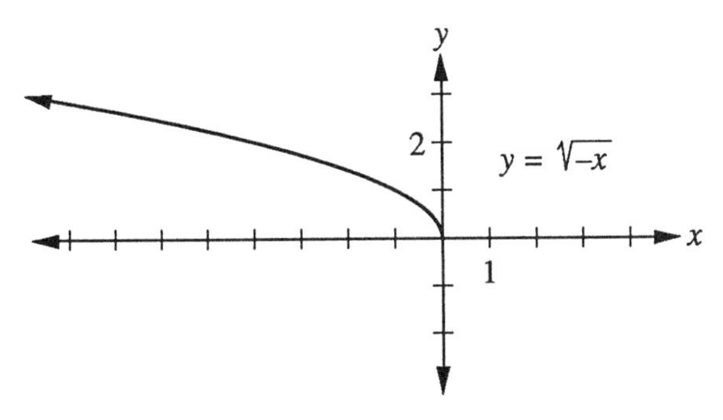

Figure 30 Graphs of $f(x) = -(x)^{\frac{1}{2}}$ and $g(x) = (-x)^{\frac{1}{2}}$

on the graphs of f and g to help you understand why their graphs look the way they do. ∎

Example 2

Make a graph of $h(x) = \sqrt{-2x}$.

Solution The function h combines a horizontal compression with a reflection about the y-axis. Notice that the x-values in the domain of h must satisfy the inequality $-2x \geq 0$, so the domain is $x \leq 0$. A bit of point plotting yields the second graph shown in Figure 31.

How is this graph related to that of the toolkit function $y = \sqrt{x}$ shown first in Figure 31? Since the constant -2 is in the argument of h, we can best analyze the relationship between the graphs by thinking about what x-value is needed to produce a particular y-value. On the toolkit function, $y = 1$ when $x = 1$. On h, $y = 1$ when $-2x = 1$ or $x = -\frac{1}{2}$. Similarly, the point $(9, 3)$ on the toolkit

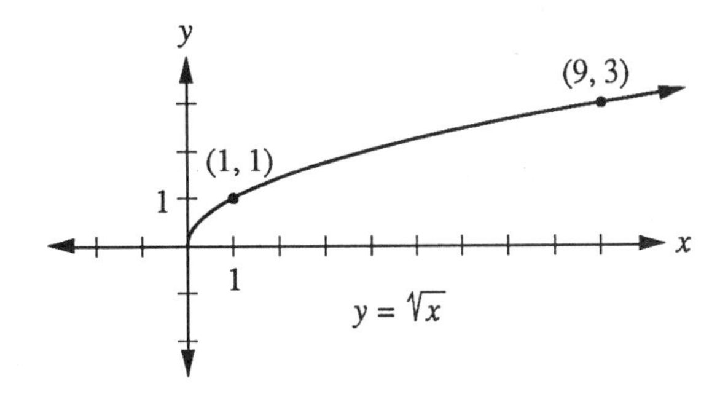

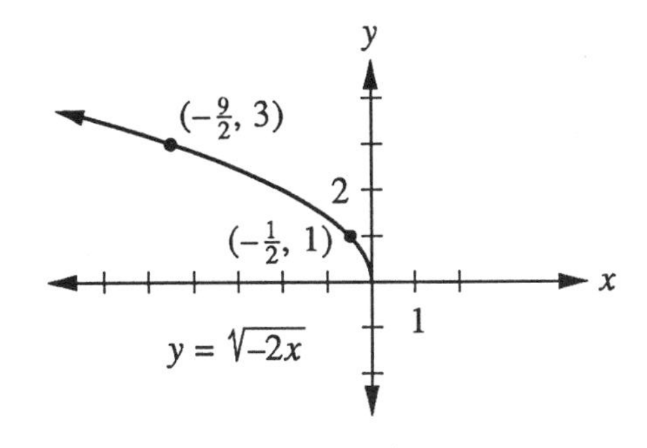

Figure 31 Graph of $y = (x)^{\frac{1}{2}}$ and $h(x) = (-2x)^{\frac{1}{2}}$

function has become $(-\frac{9}{2}, 3)$ on h. These observations suggest that the graph of h can be obtained by reflecting the toolkit graph about the y-axis and compressing it horizontally by a factor of $\frac{1}{2}$. ∎

In general, a negative value for c in $y = cf(x)$ or $y = f(cx)$ has an effect similar to one of those observed in Examples 1 and 2. The absolute value of c will determine the compression or stretch factor, and the negative will result in a reflection about either the x- or y-axis.

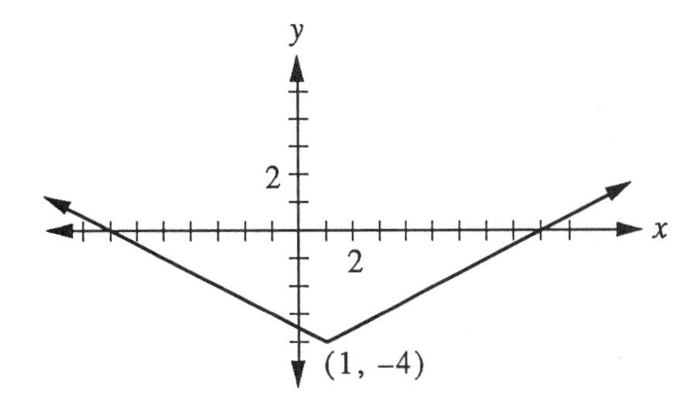

Figure 32 Graph of $y = \frac{1}{2}|x - 1| - 4$

More complicated functions often exhibit a combination of the four types of transformations. For instance, the graph of $y = \frac{1}{2}|x - 1| - 4$ will incorporate three different transformations of the graph of the toolkit function $g(x) = |x|$. Since $y = \frac{1}{2}|x-1|-4$ can be written as $y = \frac{1}{2}g(x-1)-4$, the absolute value V-shape should be compressed vertically by a factor of $\frac{1}{2}$, translated 1 unit to the right, and translated 4 units down (see Figure 32).

Class Practice

1. We have already seen that the graph of $y = \frac{1}{2}|x - 1| - 4$ exhibits three transformations of the toolkit function $y = |x|$.
   a. What would the graph look like if you first shifted 1 unit to the right, then shifted 4 units down, and finally compressed vertically by a factor of $\frac{1}{2}$?
   b. What would the graph look like if you first compressed vertically by a factor of $\frac{1}{2}$, and then shifted 4 units down and 1 unit to the right?
   c. What can you conclude about the order of transformations?

Example 3

Refer to the function f whose graph is shown in Figure 33. Make graphs of $y = f(\frac{1}{2}x)$ and of $y = f(\frac{1}{2}x + 1)$.

Solution Notice that the domain of f is $-3 \le x \le 5$. The point $(-3, 0)$ on the graph indicates that $f(-3) = 0$; similarly $f(-1) = 2, f(1) = 0$, and so on.

What does the graph of $y = f(\frac{1}{2}x)$ look like? You should be able to predict that the $\frac{1}{2}$ in the argument of the function causes a horizontal stretch by a factor of 2. Therefore, the domain of $y = f(\frac{1}{2}x)$ is $-6 \le x \le 10$. Another way to reach this conclusion is to reason that the argument a of the function $y = f(a)$ must satisfy the inequality

$$-3 \le a \le 5.$$

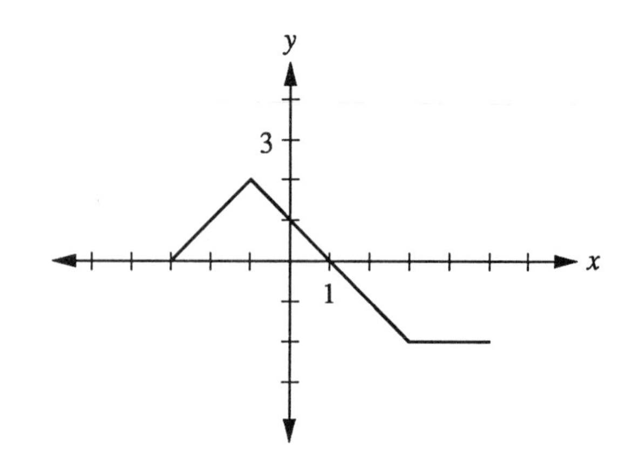

Figure 33 Graph of $y = f(x)$ for Example 3

Therefore, the argument $\frac{1}{2}x$ of the function $y = f(\frac{1}{2}x)$ must satisfy the inequality

$$-3 \le \tfrac{1}{2}x \le 5$$

which is equivalent to

$$-6 \le x \le 10.$$

Points on the graph of $y = f(\frac{1}{2}x)$ can be determined by choosing a particular y-value and then asking yourself what x-value will produce that y-value as output. For instance, what happens to the point $(-1, 2)$ on the graph of f? On $y = f(\frac{1}{2}x)$ the x-value must be twice as big to produce the same y-value, so the new point is $(-2, 2)$. Similarly, the graph of $y = f(\frac{1}{2}x)$ contains the points $(-6, 0), (2, 0), (6, -2)$, and $(10, -2)$. The complete graph is shown in Figure 34.

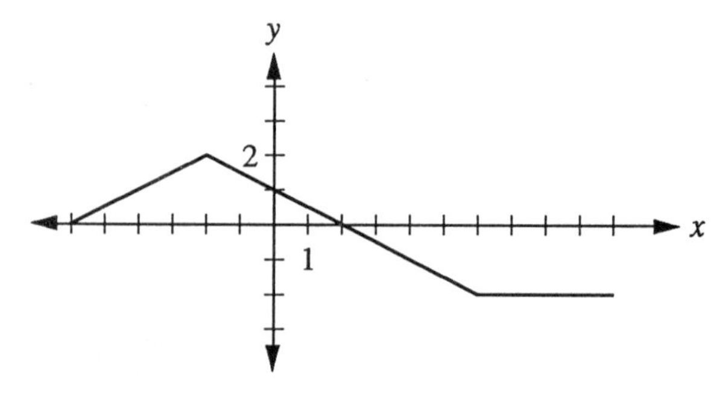

Figure 34 Graph of $y = f(\frac{1}{2}x)$

Can you predict what the graph of the function $y = f(\frac{1}{2}x+1)$ will look like? You may expect to translate the graph of $y = f(\frac{1}{2}x)$ one unit to the left. After we find the domain and plot some points you will see the error in this prediction.

Since the argument of this function is $\frac{1}{2}x+1$, values in the domain must satisfy the inequality

$$-3 \leq \tfrac{1}{2}x + 1 \leq 5$$

which is equivalent to

$$-4 \leq \tfrac{1}{2}x \leq 4$$

or

$$-8 \leq x \leq 8.$$

We can also do some point plotting and determine what happens to the point $(-1, 2)$ on the graph of f. What x-value will produce an output of 2? Since $f(-1) = 2$, the argument $\frac{1}{2}x + 1$ must be equal to -1. Solving for x gives $x = -4$, so the graph of $y = f(\frac{1}{2}x + 1)$ contains the point $(-4, 2)$. Recall that the point $(-1, 2)$ on $f(x)$ became $(-2, 2)$ on $f(\frac{1}{2}x)$; now we also know that this point becomes $(-4, 2)$ on $f(\frac{1}{2}x + 1)$. The complete graph of $y = f(\frac{1}{2}x + 1)$ over the domain $-8 \leq x \leq 8$ is shown in Figure 35.

Comparing this graph to that of $y = f(\frac{1}{2}x)$ in Figure 34 seems to indicate that the graph of $y = f(\frac{1}{2}x)$ is shifted 2 units to the left to obtain the graph of $y = f(\frac{1}{2}x + 1)$. Why is the shift two units instead of one unit? One way to answer this question is to reason that one unit on the graph of $y = f(x)$ becomes like two units on the graph of $y = f(\frac{1}{2}x)$ because of the horizontal stretch.

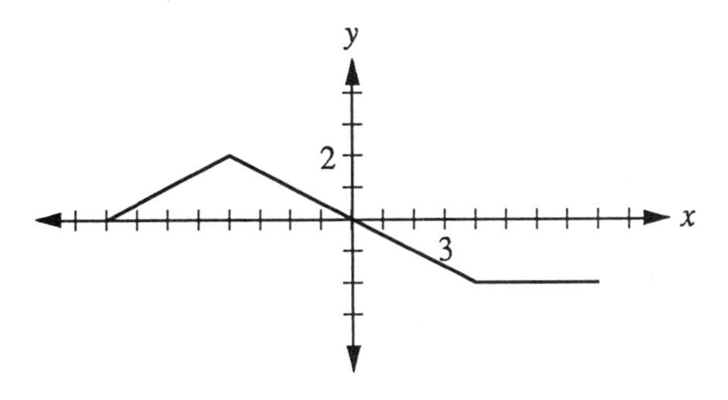

Figure 35 Graph of $y = f(\frac{1}{2}x + 1)$

You can help yourself remember this slightly surprising result in the future by writing the equation of this function as $y = f(\frac{1}{2}(x+2))$. This form indicates that the graph of $y = f(x)$ has been stretched horizontally by a factor of 2 and then translated 2 units to the left. ■

Sometimes it is difficult to recognize the transformations that are involved in the graph of a function. The next example reviews an algebraic technique to help you recognize transformations that are camouflaged.

Example 4

Graph $y = 2x^2 + 12x + 19$.

Solution This equation contains an x^2-term, an x-term, and a constant term. It is difficult to tell what toolkit function has been transformed. Fortunately, the equation can be rewritten using an algebraic technique called *completing the square.* It is assumed that you learned this technique in a previous course; this example is included to refresh your memory and to illustrate the value of completing the square in graphing quadratic functions.

$$y = 2x^2 + 12x + 19$$
$$y = 2(x^2 + 6x) + 19$$
$$y = 2(x^2 + 6x + \ ?\ ) + 19$$
$$y = 2(x^2 + 6x + 9) + 19 - 18$$

The "square was completed" by adding 9 to make a perfect square trinomial inside the parentheses. Writing 9 inside the parentheses has the effect of adding 18 to the right side of the equation, so 18 is subtracted to maintain equality. Simplifying the final equation above gives

$$y = 2(x + 3)^2 + 1.$$

With the equation written in this form, you should be able to recognize that the toolkit function $f(x) = x^2$ has been transformed. Specifically, our function can be expressed as $y = 2f(x + 3) + 1$, so the toolkit parabola has been stretched vertically by a factor of 2, shifted 3 units to the left, and shifted 1 unit up. The graph is shown in Figure 36. ■

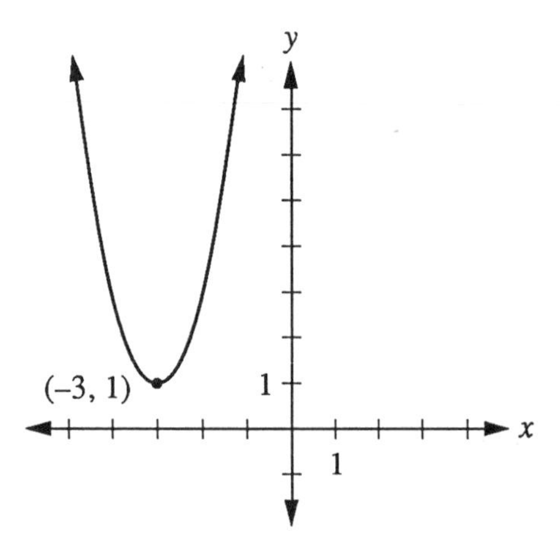

Figure 36 Graph of $y = 2x^2 + 12x + 19$

Exercise Set 7

1. The graph of a piecewise-defined function $y = f(x)$ is shown in Figure 37. For each variation of f, sketch a graph and identify the domain and the range.
   a. $y = f(x+3)$
   b. $y = 2f(x) + 3$
   c. $y = f(-2x)$
   d. $y = -2f(x)$
   e. $y = f(-x+1)$
   f. $y = \frac{1}{2}f(x) - 4$
   g. $y = f(2x-2)$

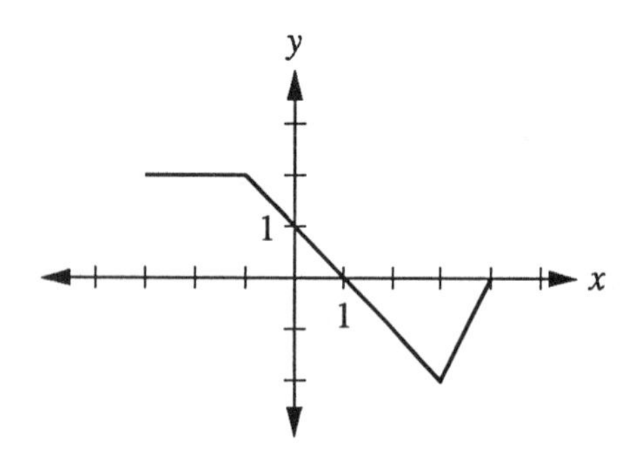

Figure 37 Graph for Exercise 1

2. Identify the related toolkit function and sketch a graph of each function.
   a. $y = -2\sqrt{x}$
   b. $y = 3\sin(2x)$
   c. $y = 3(x+1)^2 - 2$
   d. $y = -\sin(-x)$
   e. $y = |3x - 6|$
   f. $f(x) = \sqrt{x-5}$
   g. $f(x) = \sqrt{5-x}$
   h. $f(x) = 2(x+3)^3 - 1$
   i. $f(x) = -\frac{1}{5}x^2$
   j. $f(x) = 4 - |x|$
   k. $f(x) = -\sin(4x)$
   l. $f(x) = \sin(x - \frac{\pi}{2})$
   m. $f(x) = [4x]$ (greatest integer function)
   n. $f(x) = 2\sin(\frac{1}{2}x)$
   o. $f(x) = (5x)^2$
   p. $f(x) = \frac{1}{2x+4}$
   q. $f(x) = -2\sqrt{x-4} + 3$
3. Rewrite each equation in a form that allows you to recognize it as a transformation of a toolkit function. Then sketch a graph.
   a. $y = x^2 - 2x - 24$
   b. $y = 4x^2 + 28x + 53$
   c. $y = -2x^2 + 6x - 3$
   d. $y = \frac{1}{2}x^2 + 4x + 7$
4. Sketch a graph of each function. Use transformation techniques whenever they are helpful.
   a. $y = \frac{3}{x-2}$
   b. $y = -3|x+5| - 2$
   c. $y = 3\sqrt{x+2}$
   d. $y = -2\sin(x+\pi)$
   e. $y = \frac{1}{2}(x-1)^3 - 2$
   f. $y = x^2 + 7x + 12$
   g. $y = 2\sqrt{x+1}$
   h. $y = \frac{4}{x-3}$
   i. $y = \frac{-2}{x+1}$
   j. $y = [x] - 2$ (greatest integer function)

k. $y = \sin(2x - \pi)$

l. $y = \sin(3x + 2\pi)$

m. $y = -\frac{1}{3}x^3 + 6$

n. $y = -\sqrt{2 - x}$

o. $y = -2|x - 4|$

p. $y = \begin{cases} \sqrt{x-2} & \text{if } x \geq 3 \\ 2x^2 & \text{if } 1 \leq x < 3 \\ \frac{4}{x} & \text{if } x < 0 \end{cases}$

8 *Investigating Functions Using a Computer or Graphing Calculator*

In this section we will concentrate on using the computer and calculator as tools to investigate functions. We will use these tools to study many different attributes of functions, including the basic shape of the graph, where the graph is increasing and where it is decreasing, the domain, the range, the zeros, the turning points, and the asymptotes.

Example 1

Box Volume

Recall the problem we studied (Example 3 in section 4, page 37) about creating an open box by cutting squares from the corners of a piece of cardboard with dimensions 18 inches by 24 inches. The function

$$V(x) = x(18 - 2x)(24 - 2x) = 4x^3 - 84x^2 + 432x$$

expresses the volume of the box as a function of the size of the squares cut from the corners. When we discussed this function previously, we left an important question unanswered: what is the range of this function? If we can graph the function we will be able to "see" the range—the graph will show what y-values occur as second coordinates of ordered pairs.

Solution Figure 38 shows a computer-generated graph of V over the interval $[0, 9]$. Recall that this interval corresponds to the domain of the function V considered as a model. The graph shows that for x-values between 0 and 9 the y-values increase, reach a maximum, and then decrease. The presence of this turning point indicates that there is a maximum box volume that we can create. The x-coordinate of this turning point tells us what size square to cut from the corners in order to create the box with the greatest possible volume, and the y-coordinate tells us just what that maximum volume is.

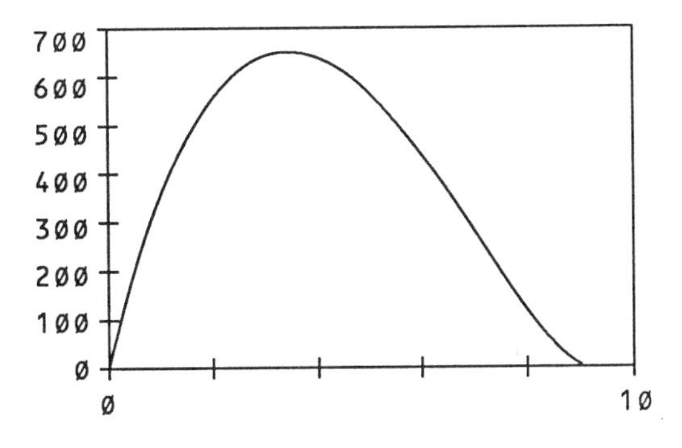

Figure 38 Graph of $V(x) = x(18 - 2x)(24 - 2x)$

The coordinates of the turning point may be approximated by using the cursor to trace along the curve. According to one graphing calculator, the y-coordinate of the turning point is around 654.8, which occurs when $x \approx 3.3$. This means that the largest possible box has volume of approximately 655 cubic inches, and this volume is achieved when a square with side about $3\frac{1}{3}$ inches is cut from each corner. Our graph of V also shows that the range of V contains all real numbers between 0 and 654.8. ∎

Example 2

Seismic Reflection

Geologists and other scientists are interested in the nature of rock formations beneath the surface of the earth. They can determine the depth and composition of layers under the surface without resorting to drilling that is costly, time consuming, and ecologically destructive. Instead of drilling, they use sound waves and mathematical models!

One procedure they use is called seismic reflection; it is based on the fact that sound travels at different rates in different media. For example, sound travels at 330 meters per second in air, 1450 meters per second in water, and about 5000 meters per second in granite. In general, the more compact a medium, the faster sound will travel through it. Figure 39 shows a layer of earth of uniform thickness on top of a layer of sandstone. A seismologist will set up a noise source at point A and a receiver at point B. Sound waves will radiate from A in all directions; some will travel to B by traveling through the top layer of earth, along the interface between the two layers, and then back up to B. There is one path from A to B that requires the least amount of time; sound waves that follow this path will be the first to reach the receiver at B. We will try to find this path that takes the least time. Notice that the path of least time is not necessarily the same as the path of least distance; this is because the elapsed time depends on both the distance travelled and the speed.

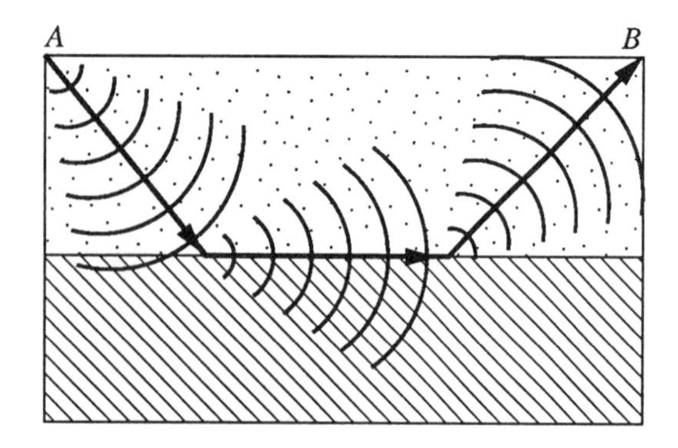

Figure 39 Earth and Sandstone

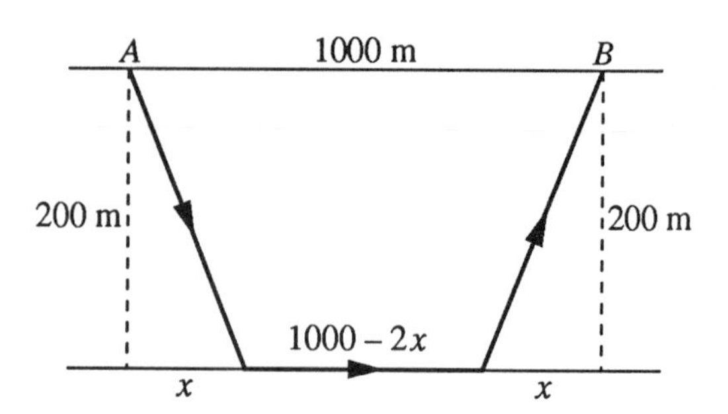

Figure 40 Path from A to B

Assume that sound travels through the upper layer of earth at 150 meters per second and through the sandstone layer along the interface at 1500 meters per second, that A and B are 1000 meters apart, and that the top layer of earth is 200 meters deep. We will also assume that the first sound wave to reach B begins traveling along the interface x meters from the point directly below A (see Figure 40). Under these assumptions, we can write a function to show the relationship between x and the elapsed time. We will need to use the Pythagorean Theorem and the fact that elapsed time is equal to distance divided by velocity.

The least-time path through the earth and along the interface consists of three sections. Each diagonal section of the least time path has length $\sqrt{40000+x^2}$, so the elapsed time for these two sections is

$$2\frac{\sqrt{40000+x^2}}{150}.$$

The horizontal section of this path has length $1000-2x$, so the elapsed time for this section is

$$\frac{1000-2x}{1500}.$$

Thus, the total elapsed time $T(x)$ depends on x in the following way:

$$T(x) = 2\frac{\sqrt{40000+x^2}}{150} + \frac{1000-2x}{1500}.$$

This is quite a complicated function. It should be clear from the context of the problem that the domain for this function contains x-values from

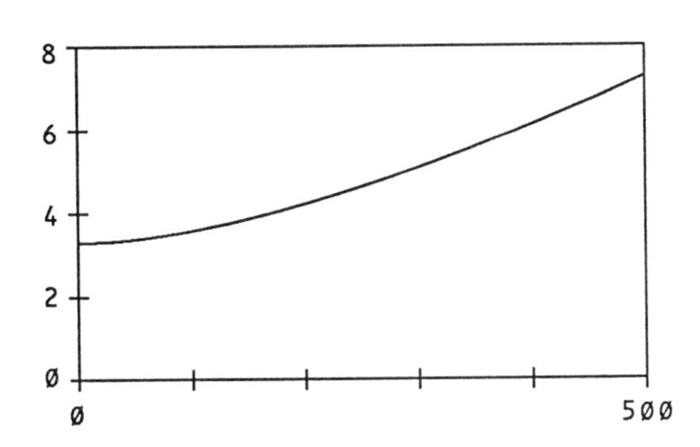

Figure 41 Graph of T from Example 2

0 to 500. Apart from this, however, you probably have very little intuition about what its graph may look like. This function is a prime candidate for computer or calculator assisted graphing. Figure 41 shows a computer generated graph of $T(x)$ for $0 \le x \le 500$. Notice how little information about the function this graph conveys.

We can get more information about T by having a computer or calculator generate a *table of values* of T; this is simply a listing of ordered pairs that belong to the function. Studying the values on the left in Table 42 shows that when $x = 0$, $T = 3.333$; then as x-values increase T-values decrease for awhile. Somewhere between $x = 10$ and $x = 30$ the T-values reach a minimum, and then begin to increase. We can determine the location of this turning point more precisely by making another table on the interval $[10, 30]$; we find that an approximation for the turning point is $x = 20$.

Table 42 Tables of Values for T

x	T	x	T
0	3.33333	10	3.32333
10	3.32333	12	3.32213
20	3.31997	14	3.32119
30	3.32317	16	3.32052
40	3.33281	18	3.32011
50	3.34874	20	3.31997
60	3.37075	22	3.32008
70	3.39862	24	3.32046
80	3.43209	26	3.32111
90	3.47089	28	3.32201
100	3.51476	30	3.32317

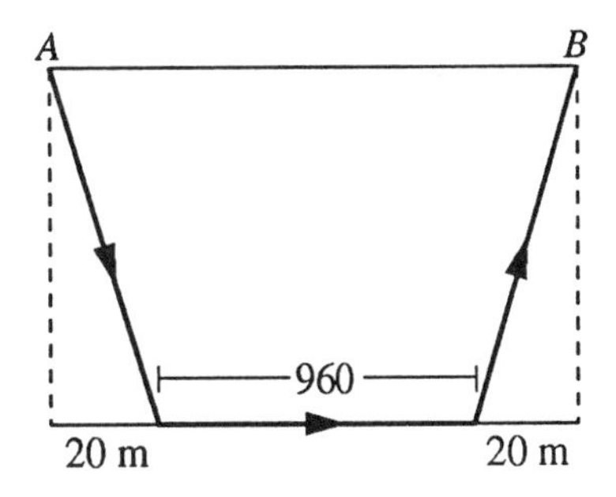

Figure 43 Least-time Path

In conclusion, we know that the least time path reaches the interface between the two layers about 20 meters from a point directly below A (see Figure 43). It takes the sound waves about 3.3 seconds to travel the entire path from A to B. Later in this course we will consider other problems involving seismic reflection. At that time we will study how this model is combined with techniques of data analysis to help geologists determine both the depth and composition of rock formations beneath the surface of the earth. ■

Example 3

Firecracker

When a firecracker is shot upward into the sky, its path resembles a parabola. The path of a particular bottle rocket is modeled by the function

$$h(t) = -16t^2 + 117t + 5$$

where t represents the number of seconds that have elapsed and $h(t)$ represents the height of the rocket in feet. According to this model, how long will the bottle rocket stay in the air?

Solution We can get a sense of what the firecracker's path looks like by having a computer or calculator generate a graph of h. The graph shows that the height increases, reaches a maximum, and then decreases (see Figure 44).

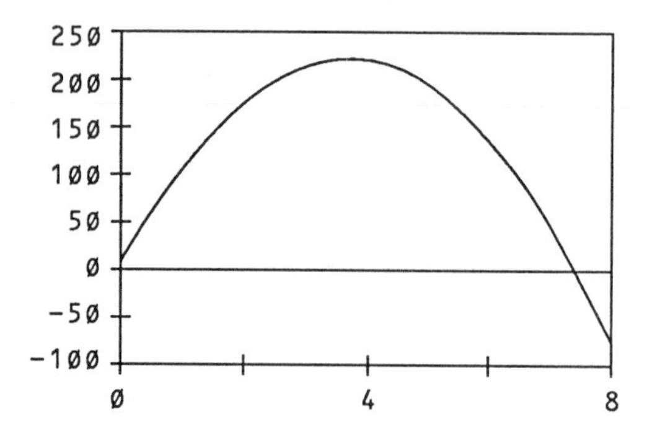

Figure 44 Graph of h from Example 3

The firecracker will stay in the air until its height is zero, so we need to find the value of t for which $h(t)$ is equal to zero. Based on the graph, it is clear that this value is between $t = 6$ and $t = 8$. The graph shows that $h(6)$ is positive and $h(8)$ is negative, so a zero of h must be located between 6 and 8. We can get a more accurate answer by inspecting a table of values for h over the interval $[6, 8]$ (see Table 45).

Table 45 shows that h has a zero between 7.3 and 7.4. You could list more values in this interval to increase the accuracy of your estimate, or you could use the zoom and trace features of a computer or calculator graph to locate a zero of h at $t = 7.35$, accurate to the nearest hundredth. This means that the firecracker will stay in the air for about 7.35 seconds. ■

Table 45 Values for $h(t) = -16t^2 + 117t + 5$

t	$h(t)$	t	$h(t)$
6.0	131.00		
6.1	123.34	7.1	29.14
6.2	115.36	7.2	17.96
6.3	107.06	7.3	6.46
6.4	98.44	7.4	−5.36
6.5	89.50	7.5	−17.50
6.6	80.24	7.6	−29.96
6.7	70.66	7.7	−42.74
6.8	60.76	7.8	−55.84
6.9	50.54	7.9	−69.26
7.0	40.00	8.0	−83.00

Exercise Set 8

1. Refer to the firecracker example.

a. Identify the domain and range of the function used to model the firecracker's flight.

b. According to the model, how high will the firecracker go?

2. Graph the function $f(x) = x^3 - 3x + 1$ on a computer or calculator. Be sure your graph displays the significant features of the function. Identify the turning points and the zeros (accurate to hundredths). Describe what happens to the graph as x increases without bound and as x decreases without bound. What symmetry do you observe in the graph?

3. Predict how the graphs of the functions $g(x) = x^3 - 3x$ and $h(x) = -x^3 + 3x - 1$ compare to that of f in the preceding exercise. Check your predictions by graphing g and h with a computer or calculator.

4. Use a computer or graphing calculator to help you find the range of each function.

a. $f(x) = \frac{1}{x^2+1}$

b. $y = \sqrt{x^2 + 4x + 5}$

c. $f(x) = \frac{x^2}{x^2-1}$

d. $y = \frac{x-3}{x+2}$

e. $g(x) = \frac{1}{x^2+4}$

f. $y = \frac{1}{x^2-4}$

5. Use a computer or graphing calculator to help you find the zeros of each function accurate to hundredths.

a. $f(x) = 2x^2 - 1$

b. $y = x^3 + 2x^2 - 12x$

c. $f(x) = \frac{x-3}{x+2}$

d. $y = x^3 - 2x - 1$

e. $y = 2\sin(2x - 1)$

f. $g(x) = \sin(2x) - 0.8$

6. Use graphs and tables to investigate these functions. Identify zeros and turning points.

a. $f(x) = x^3 + x^2 - 12x$

b. $f(x) = \frac{4x^4}{1+x^6}$

c. $f(x) = \frac{x^2+x-1}{x-1}$

d. $f(x) = \frac{1}{x^2+1}$

e. $f(x) = \frac{1}{x^2-1}$

f. $f(x) = \sin(x-1) + 1$

g. $f(x) = \sqrt{x^2 + 4x - 5}$

7. Use a computer or graphing calculator to graph $f(x) = \sin x - 0.6$.

a. How many zeros does f have?

b. Use the graph of f to identify all zeros between 0 and 2π.

8. The amount of power that can be generated by an undershot waterwheel in a stream depends on several factors, including the velocity of the waterwheel and the velocity of the stream. When the velocity of the stream is 4.1 feet per second, the power P generated is a function of the wheel's velocity v:

$$P = 1600v(4.1 - v)^2.$$

Use a computer or calculator to graph this function and find the maximum power that can be generated.

9. Imagine that a rectangular coordinate system is superimposed over the area where a submarine is traveling. In this system, a sonar buoy is located at the point with coordinates $(1, -1.5)$ and the submarine's path is along the graph of the function $y = x^2$. (This means that the submarine is located at a point with coordinates (x, x^2).) Write a function to express the distance between the submarine and the buoy in terms of the submarine's x-coordinate. Where must the submarine be located so that its distance from the sonar buoy is minimized?

10. An open box is constructed by cutting congruent squares from the corners of a 30 inch by 20 inch piece of aluminum. What are the dimensions of the largest box that can be constructed?

11. The owners of a theme park know that an average of 50,000 people visit the park each day. They are presently charging $15.00 for an admission ticket. Each time in the past that they have raised the admission price, an average of 2,500 fewer people have come to the park for each $1.00 increase in ticket price. Using x to represent the number of $1.00 price increases, write a function to express the relationship between x and the revenue from ticket sales. What ticket price will maximize the revenue from ticket sales?

12. A museum hires a graphics company to produce a poster for their upcoming exhibit. The graphics company charges $1,000 for production and design work and an additional $2.00 for each poster printed. The museum decides to sell the posters for $7.50 each. Write a function to represent the profit the museum realizes when it sells x posters. How many posters should they sell to maximize their profit?

13. Suppose a customer wants to borrow $10,000 and that the annual interest rate is 12%. This rate is equivalent to 1% per month. The monthly payment (P) is related to the duration of the loan in months (n) by the formula

$$P = \frac{10,000(0.01)(1.01)^n}{(1.01)^n - 1}.$$

a. Determine the monthly payment for loans of various duration: one month, one year, ten years, and thirty years ($n = 1, n = 12, n = 120, n = 360$).

b. Identify the domain and the range of the function.

c. The value $P = 100$ is not in the range of the function. Explain why this makes sense. What would happen to the outstanding balance on the loan if the monthly payment were $100?

14. Refer to the preceding exercise about loan repayment. Write a function to express the total amount the customer repays for the loan

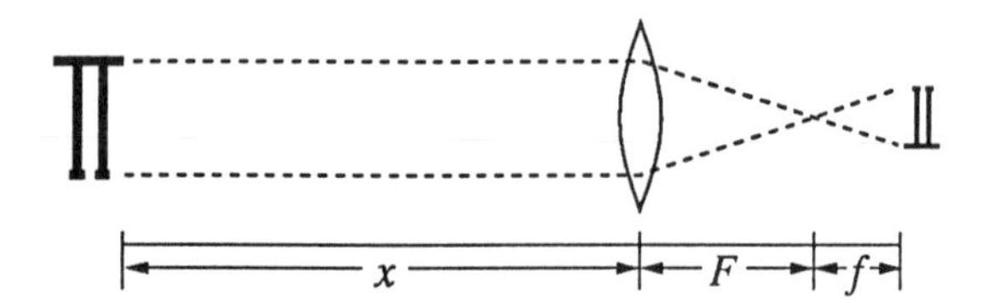

Figure 46 Camera Diagram

in terms of the duration of the loan. Use the computer to investigate how the total amount repaid varies as the duration of the loan increases from 1 year to 30 years. What implications does this have for borrowers and lenders?

15. Focusing a camera involves moving the lens so that it is an appropriate distance away from the film. When a camera is focused on a far away object, the lens is at the infinity setting. The distance f that the lens must move from the infinity setting depends on two things: (1) the distance x from the object to the lens, and (2) the focal length F of the lens (see Figure 46). In particular,

$$f = \frac{F^2}{x - F}.$$

Since the focal length of any lens is a constant, f is a function of x. Use the computer to investigate the relationship between f and x for a 50 mm lens ($F = 50$) between $x = 55$ mm and $x = 500$ mm. Compare the way f varies for close objects (small x-values) with the way it varies for distant objects (large x-values). Can you explain why close-up photographs are difficult to focus?

9
Solving Inequalities

The need to solve inequalities arises frequently in our work with functions. In an earlier section of this chapter we needed to solve inequalities to determine the domains of various functions. Many applications of functions also require that we solve inequalities. In this section we will develop algebraic and computer techniques to solve inequalities.

To find exact solutions of inequalities, we must be able to find the x-values that make the expression zero or undefined. These two types of values are important because they are the only places where most functions that we work with can change sign from positive to negative. For instance, if $f(x) = 5 - 2x$, then $x = \frac{5}{2}$ is a zero of the function. Note that x-values greater than $\frac{5}{2}$ result in negative values for f, and x-values less than $\frac{5}{2}$ produce positive values for f; therefore f changes sign at this zero. The function $g(x) = 1/(2x - 5)$ has the same important value, $x = \frac{5}{2}$. The values of g are positive for $x > \frac{5}{2}$ and negative for $x < \frac{5}{2}$, and g is undefined when $x = \frac{5}{2}$.

Example 1

Find the domain of the function

$$y = \sqrt{x^2 - x - 12}.$$

Solution Use the expression under the square root to define a new function: $f(x) = x^2 - x - 12$. To find the domain we will need to find where $f(x)$ is greater than or equal to zero. Because f can easily be factored, we can accomplish this task without a computer. Since

$$f(x) = x^2 - x - 12 = (x - 4)(x + 3),$$

the factors of f are $(x - 4)$ and $(x + 3)$, and the zeros of f are $x = 4$ and $x = -3$. The sign of f is determined by the signs of its factors; $f(x)$ is greater than zero when both factors are positive and also when both factors are negative. Figure 47 shows a useful way to display the signs of the factors on a number line. We see that both factors are positive for x-values greater than 4 and both are negative for x-values less than -3. Therefore, the domain of $y = \sqrt{x^2 - x - 12}$ is $x \geq 4$ or $x \leq -3$. ■

The technique presented in the previous example depends on finding zeros and deciding whether

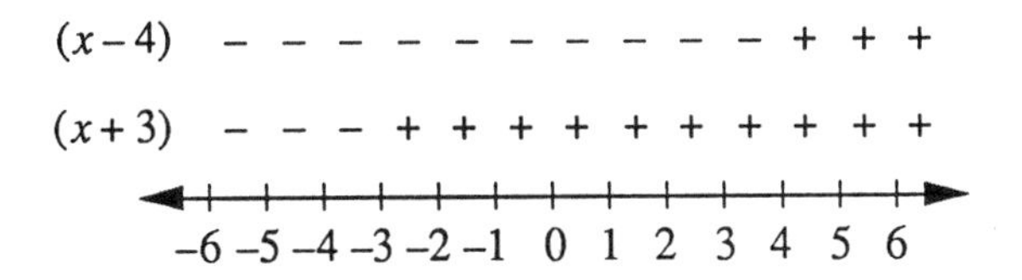

Figure 47 Signs of Factors on a Number Line

or not the function changes sign at these x-values. For instance, the function $g(x) = (x-5)^2$ has a zero at $x = 5$, but it does not change sign here; in fact $g(x)$ is nonnegative for all values of x. Note that this technique requires that the inequality be written so that an expression is compared to zero; zero is the unique number that separates positive and negative real numbers.

Class Practice

1. Use algebraic methods to solve each inequality.
   a. $x^2 + x + 3 > 5$
   b. $x^3 - 2x^2 + x \leq 0$
   c. $\frac{x(x-2)^2}{(x+1)^3} \geq 0$

Not every inequality can be solved algebraically. A significant number of the mathematical problems encountered in daily life involve inequalities which are difficult to solve without the aid of a calculator or computer.

Example 2

Mortgage Payments

A real estate agent has a client who needs to borrow \$65,000 to purchase a house. The client wants a 30-year variable rate mortgage, with a monthly mortgage payment of no more than \$500. What annual interest rates will result in an affordable monthly payment?

Solution The formula

$$P = \frac{Ar(1+r)^n}{(1+r)^n - 1}$$

expresses the monthly payment P in terms of A, the amount borrowed, r, the monthly interest rate, and n, the duration of the loan in months.

In this problem, $A = \$65{,}000$ and $n = 360$ months; we need to find the values of r for which P will be less than or equal to \$500. That is, we need to solve the following inequality for r:

$$\frac{65,000r(1+r)^{360}}{(1+r)^{360} - 1} \leq 500.$$

The technique used in Example 1 is not helpful in solving this inequality. Fortunately, a computer or calculator can be of assistance, but we first need to formulate the problem in a way that allows us to take advantage of these tools. First define the function

$$P(r) = \frac{65,000r(1+r)^{360}}{(1+r)^{360} - 1}.$$

Then have a computer or calculator list values of this function. Inspecting the table will show when function values fall below \$500. What r-values should we include in the table? Mortgage interest rates are usually quoted as annual rates that are generally between 6% and 18%. Since r represents a monthly interest rate, annual rates are divided by 12. Therefore, our table should include r-values between .005 and .015 (see Table 48).

The table shows that r-values of .007 and smaller result in P-values that are less than 500. A second table over a smaller interval would show shows that P is less than 500 when r is less than

Table 48 Table of Values for Monthly Mortgage Payments

r	P	r	P
.005	389.71	.011	729.21
.006	441.21	.012	790.79
.007	495.19	.013	853.16
.008	551.30	.014	916.14
.009	609.21	.015	979.61
.010	668.60		

0.00709. This means that the solution to the inequality

$$\frac{65,000r(1+r)^{360}}{(1+r)^{360}-1} \leq 500$$

is $r \leq 0.00709$. An r-value of 0.00709 corresponds to an annual interest rate of about 8.5%. Thus, the client will have a monthly payment below \$500 whenever the annual interest rate on the variable mortgage is less than or equal to 8.5%. ■

Example 3

Find the solution set of the inequality

$$x^3 > 3x + 1.$$

Solution There are no straightforward algebraic techniques that allow us to solve this inequality. However, if we think of this problem in terms of functions, we can use a computer or graphing calculator to assist us. Define $f(x) = x^3$ and $g(x) = 3x + 1$; solving our inequality is equivalent to finding where $f(x) > g(x)$. In terms of a graph, we need to determine where the y-coordinate on the graph of f is greater than the y-coordinate on the graph of g. Figure 49 shows f and g graphed in the same coordinate system; the x-values for which $f(x) > g(x)$ are the solutions of the inequality. You can use the trace feature of the computer or calculator to find the x-coordinates of the points where the graphs of f and g intersect. These x-values are approximately -1.53, -0.35, and 1.88. This information, together with the graph in Figure 49, implies that the solution set of the inequality is approximately $-1.53 < x < -0.35$ or $x > 1.88$. ■

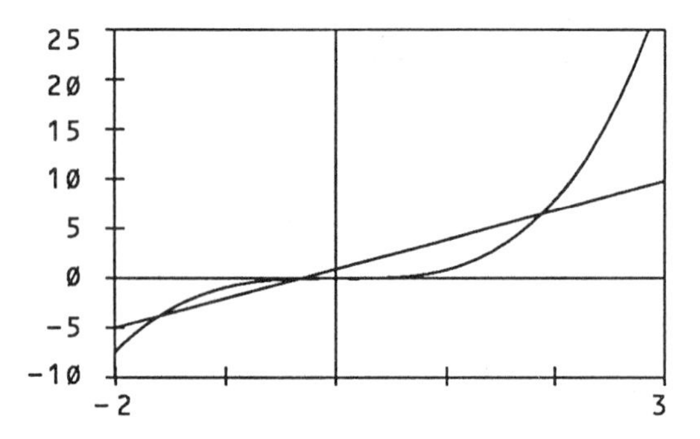

Figure 49 Graphs of $f(x) = x^3$ and $g(x) = 3x + 1$

Example 4

Solve the inequality

$$\frac{1}{x-1} \geq -2.$$

Solution You could multiply both sides of the inequality by $(x-1)$ if you consider separately the cases $(x-1) > 0$ and $(x-1) < 0$.

This inequality also can be solved algebraically by first writing it as

$$\frac{1}{x-1} + 2 \geq 0.$$

Combining the two terms on the left-hand side yields

$$\frac{2x-1}{x-1} \geq 0.$$

We can now find the zeros of $2x - 1$ and $x - 1$ and use the number line technique of Example 1 to find that the solutions are $x \leq \frac{1}{2}$ or $x > 1$.

Instead of using algebraic manipulations, we could approach the inequality in terms of a function. If we define $f(x) = \frac{1}{x-1}$, then the inequality $\frac{1}{x-1} \geq -2$ is equivalent to $f(x) \geq -2$. Thus, we simply need to determine when the function values of f are greater than or equal to -2. We can do this with either a table of values or a graph. Figure 50 shows graphs of f and of the line $y = -2$; the trace feature can be used to find where the hyperbola and the line intersect. Since the solution set consists of all values of x for which $f(x)$ is above the line $y = -2$, we find that the solution set of the inequality is $x \leq \frac{1}{2}$ or $x > 1$. ■

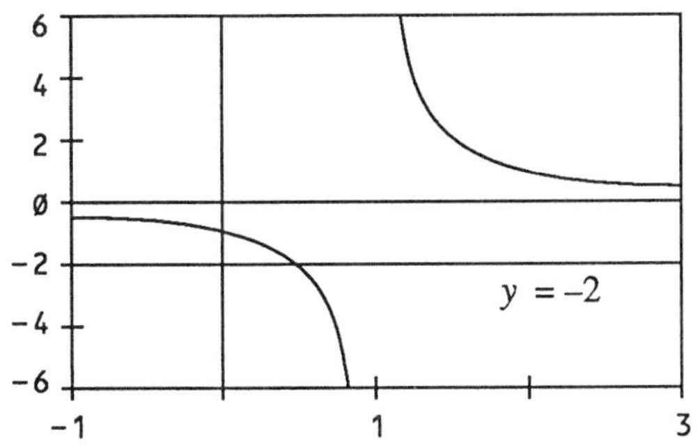

Figure 50 Graphs of $f(x) = \frac{1}{x-1}$ and $y = -2$

Exercise Set 9

1. Refer to Example 2. At what annual interest rates will a 30-year loan for $80,000 require a monthly payment of less than $750?
2. The inequality $x^3 > 3x + 1$ can be rewritten as $x^3 - 3x - 1 > 0$. Describe how you could use the graph of $f(x) = x^3 - 3x - 1$ to solve the inequality.
3. Solve each inequality. Find exact solutions if possible. You may want to use some computer or calculator graphing.
   - **a.** $x^2 - x - 6 > 0$
   - **b.** $x^3 - 3x > x^2 + 7$
   - **c.** $\frac{2}{x-1} < x + 2$
   - **d.** $\sin x \leq 0$
   - **e.** $\sin 2x > 0$
   - **f.** $x - \frac{1}{x} \geq x^2$
   - **g.** $x^4 - 5x^2 < 36$
   - **h.** $|2x - 5| \leq 9$
   - **i.** $|5x - 7| < 3x + 2$
   - **j.** $\frac{1}{x^3 - x} \geq 0$
   - **k.** $x^3 + 2x^2 - 8x \leq 0$
   - **l.** $\frac{1}{x} + \frac{1}{x-1} \leq 0$
   - **m.** $\frac{x-1}{x+2} < \frac{x+4}{x-3}$
   - **n.** $x^2 + 2x \leq 3x + 20$
   - **o.** $\sqrt{x+2} > x + 1$
   - **p.** $2 \sin x > \sin x$
   - **q.** $2x - 3 \leq \frac{x}{x-3}$
   - **r.** $x^2 - 10 < \frac{x-2}{|x|-3}$
   - **s.** $|x^2 - 4| > 1$
   - **t.** $\frac{|x+1|}{|x-2|} < 1$
   - **u.** $3x + 2 < x - 1 < 2x + 1$
4. Refer to problem 9 in section 8 about the submarine and the sonar buoy. Where can the submarine be located so that its distance from the buoy will be less than 4 units?
5. The formula

$$d = .05s^2 + s$$

expresses an approximate relationship between the distance d required to stop a car and the speed s of the car (where d is measured in feet and s in miles per hour). At what speeds does it take more than 100 feet to stop?

6. Suppose you invest p dollars each month in an account that pays r percent interest per month. If you do this for n months, the balance T in the account is given by

$$T = \frac{p(1+r)((1+r)^n - 1)}{r}.$$

A student invests $25 each month in an account that pays 9% annual interest. How long will it take until the balance in the account exceeds $1000?

10

Composition

Suppose that the owners of a small manufacturing firm have determined that the labor force required to produce x units of their product is a function of x; in particular, the number of workers needed is given by $f(x) = 6\sqrt{x}$. The demand for their product in 1988 is 40,000 units and is increasing at the rate of 10,000 units per year. Thus, the number of units demanded is given by the function $g(t) = 40,000 + 10,000t$ where t is the number of years after 1988.

Since the size of the labor force is a function of the demand, and the demand in turn is a function of time, it should be clear that the size of the labor force is a function of time. This is true because any value for t is paired with a unique value for $g(t)$, a value for $g(t)$ corresponds to a value for x, and a value for x is paired with a unique value for $f(x)$. The flow of values through g, and then through f, is shown in Figure 51.

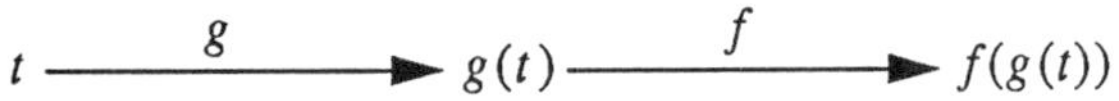

$$t \xrightarrow{\quad g \quad} g(t) \xrightarrow{\quad f \quad} f(g(t))$$

Figure 51 Flow of Composition

What the owners really want to be able to do is use t-values to directly determine the required number of workers. Mathematically speaking, they want a function that expresses the relationship between time and the size of the labor force. Clearly this third function, which we will call h, is closely related to the functions f and g discussed previously. If t represents the number of years after 1988, then $g(t)$ is the number of units demanded, and $f(g(t))$ is the size of the required labor force. The symbol $f(g(t))$, read "f of g of t" indicates that a value of t is input into the function g, the output is $g(t)$, and this output is then the input for the function f. This is exactly what we require of the function h; the input to h should be the value of t, and the output from h should be the size of the labor force.

$$\begin{aligned} h(t) &= f(g(t)) = f(40000 + 10000t) \\ &= 6\sqrt{40000 + 10000t} \end{aligned}$$

This function h is called the *composition* of the functions f and g. The symbol $f \circ g$ is also used to denote the composition of f and g; it can be read "f composed with g." The notation $f \circ g(t)$ is equivalent to $f(g(t))$; g is the *inner* function and acts on t, whereas f is the *outer* function and acts on the output of g, namely $g(t)$.

Example 1

Let $f(x) = x^2$ and $g(x) = 2x + 1$. Find $f(g(2))$, $g(f(2))$, an expression for the composition $f \circ g$, and an expression for the composition $g \circ f$.

Solution

$$\begin{aligned} f(g(2)) &= f(5) = 25 \\ g(f(2)) &= g(4) = 9 \\ f \circ g(x) &= f(g(x)) = f(2x + 1) \\ &= (2x + 1)^2 = 4x^2 + 4x + 1 \\ g \circ f(x) &= g(f(x)) = g(x^2) \\ &= 2(x^2) + 1 = 2x^2 + 1 \end{aligned}$$

■

Whenever two functions are composed, there are important relationships between their domains and ranges. In the composition $g \circ f$, values in the range of f serve as the domain of g. Therefore, the range of f and the domain of g must have a nonempty intersection. Also, the domain of the function $g \circ f$ depends on the relationship between the functions that are composed. This dependence is illustrated in the following examples. When studying these examples, bear in mind that a real number x is in the domain of the composition $g \circ f$ only if it satisfies two conditions. First, x must be in the domain of the inner function f. Second, x must produce an output $f(x)$ that is in the domain of the outer function g.

Example 2

Let $f(x) = \frac{1}{x}$ and $g(x) = \frac{1}{x-2} + 1$. Find an expression for $g \circ f$ and identify the domain of the composition.

Solution It is useful to begin by examining the domain and range of f and g.

Domain of f: $x \neq 0$ Range of f: $f(x) \neq 0$
Domain of g: $x \neq 2$ Range of g: $g(x) \neq 1$

Now we can find an expression for $g \circ f$:

$$g(f(x)) = g\left(\frac{1}{x}\right) = \frac{1}{\frac{1}{x} - 2} + 1$$

$$g(f(x)) = \frac{1}{\frac{1-2x}{x}} + 1 = \frac{x}{1-2x} + \frac{1-2x}{1-2x} = \frac{1-x}{1-2x}.$$

What is the domain of this composition? A number can be in the domain of $g \circ f$ only if it is in the domain of the inner function f. This means that $x = 0$ is not in the domain of the composition. The expression for $g \circ f$ makes it clear that $x = \frac{1}{2}$ is also not in the domain. Even though $x = \frac{1}{2}$ is in the domain of f, $f(\frac{1}{2}) = 2$ and 2 is not in the domain of the outer function g.

Thus,

$$g \circ f = \frac{1-x}{1-2x}, \qquad x \neq 0, x \neq \frac{1}{2}. \qquad ■$$

Note that the resulting function $h(x) = \dfrac{1-x}{1-2x}$, $x \neq 1/2$, has a larger domain than the composition $g \circ f$.

Example 3

Let $f(x) = \sqrt{x}$ and let $g(x) = x^2 - 2$. Find $f \circ g$ and $g \circ f$ and the domain of each composition.

Solution We begin by identifying the domain and range of f and g.

Domain of f: $x \geq 0$	Range of f: $f(x) \geq 0$
Domain of g: all real nos.	Range of g: $g(x) \geq -2$

Now find an expression for $f \circ g$.

$$f(g(x)) = f(x^2 - 2) = \sqrt{x^2 - 2}$$

The expression for $f \circ g$ makes it clear that its domain must be restricted to values of x which satisfy $x^2 - 2 \geq 0$. This is equivalent to $x \leq -\sqrt{2}$ or $x \geq \sqrt{2}$. Since all of these values are acceptable inputs to the inner function g, and also produce outputs that are in the domain of f, the domain of $f \circ g$ is $x \leq -\sqrt{2}$ or $x \geq \sqrt{2}$.

Now consider $g \circ f$ and its domain.

$$g(f(x)) = g(\sqrt{x}) = (\sqrt{x})^2 - 2 = x - 2$$

The expression for $g \circ f$ gives the impression that this function's domain consists of all real numbers. However, a number x is in the domain of $g \circ f$ only if it is in the domain of f and its output $f(x)$ is in the domain of g. Therefore, since the domain of f is $x \geq 0$, the domain of $g \circ f$ is also $x \geq 0$. We should write

$$g \circ f(x) = x - 2, \qquad x \geq 0$$

to indicate that the function has domain restrictions. ■

The preceding example brings out two important points about domains of compositions. First, note that while the domain of the function $y = x - 2$ consists of all real numbers, the domain of the composition $g(f(x)) = x - 2$ consists only of nonnegative real numbers. Second, it would be misleading to write that $g(f(x)) = x - 2$ without explicitly stating the domain of this function; the domain would be assumed to be all real numbers unless stated otherwise.

Composition as a Graphing Tool

The fact that composition can serve as an aid to graphing functions may surprise you. The utility of composition for graphing does not involve combining two functions to produce a third function. Rather, it involves *decomposing* a complicated function into two simpler functions.

For instance, the function $h(x) = |x^2 - 9|$ can be expressed as $f \circ g$ when f and g are defined as $f(x) = |x|$ and $g(x) = x^2 - 9$. Can you decompose the function $k(x) = (\frac{1}{x})^3 + 1$? This function starts with an input x, takes the reciprocal of x, then cubes that reciprocal and adds one. Two simpler functions can be composed to do the same thing. If we define $f(x) = \frac{1}{x}$ and $g(x) = x^3 + 1$, then the function $g \circ f$ is equal to k. Both $g \circ f$ and k produce ordered pairs by taking the reciprocal of the input, cubing that reciprocal, and then adding one.

Class Practice

1. Express each of the following functions as a composition of $f(x) = x^3, g(x) = \sqrt{x}, h(x) = 3x$, and $j(x) = x - 2$.

   a. $y = \sqrt{3x - 2}$

   b. $y = 3(x - 2)$

c. $y = (x-2)^3$

d. $y = 3\sqrt{x-2}$

Example 4

Graph the function $h(x) = |3\sin x + 2|$.

Solution If we express h as $f \circ g$ with $f(x) = |x|$ and $g(x) = 3\sin x + 2$, then we will be able to take advantage of the fact that f is a simple toolkit function and g is a transformation of a toolkit function. Figure 52 illustrates the "flow" of the composition $f \circ g$; we start with an input x, take the sine, triple the sine, add two, and then take the absolute value.

$$x \xrightarrow{g(x) = 3\sin x + 2} 3\sin x + 2 \xrightarrow{f(x) = |x|} |3\sin x + 2|$$

Figure 52 f Composed with g

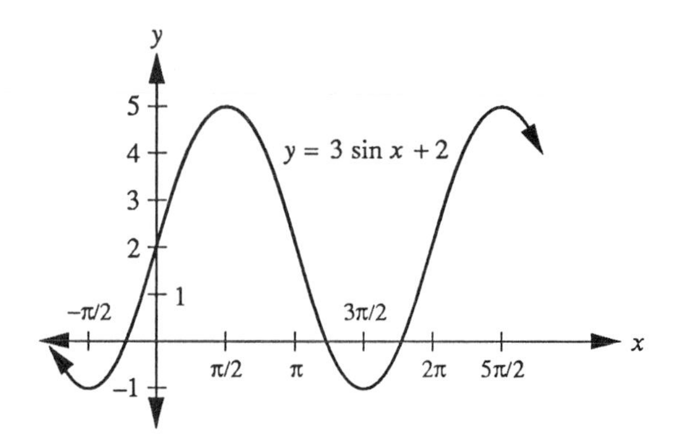

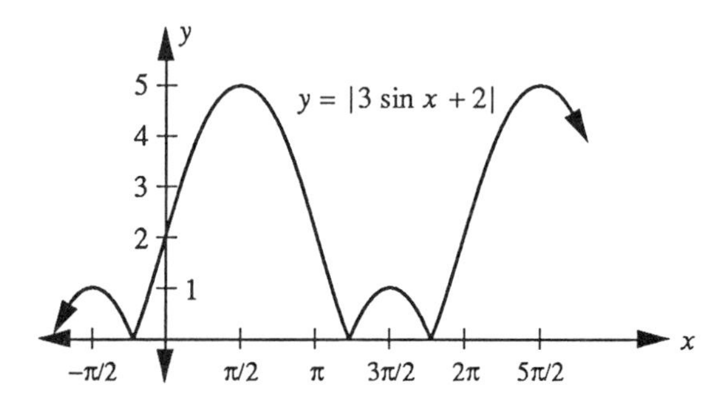

Figure 53 Graphs of $g(x) = 3\sin x + 2$ and $h(x) = |3\sin x + 2|$

We already know that the graph of g is a sine wave that is stretched vertically by a factor of 3 and shifted up 2 units. Its maximum y-value is 5, and its minimum is -1. The graph of g is shown in Figure 53. We want to determine what the graph of $|g|$ looks like. To do this, we need to think about the effect of taking the absolute value of the y-coordinates on the graph of g. Since $|y| = y$ whenever $y \geq 0$, the y-coordinates on the graph of $|g|$ will be the same as those on g whenever those y-coordinates are nonnegative. In terms of the graph, this means that the graph of $|g|$ is identical to the graph of g wherever that graph is on or above the x-axis. Similarly, $|y| = -y$ whenever $y < 0$, so the y-coordinates on the graph of $|g|$ will be the opposite of those on g whenever these y-coordinates are negative. This means that any portion of the graph of g that lies below the x-axis should be reflected about the x-axis to produce the graph of $|g|$. The complete graph of $h = f \circ g = |g|$ is also shown in Figure 53. How does the range of $h(x) = |3\sin x + 2|$ differ from the range of $g(x) = 3\sin x + 2$? ■

Example 5

Graph $k(x) = (|x| - 3)^2 - 1$.

Solution Although this function looks rather complicated, it is actually the composition of two simple functions. We can express k as $t \circ r$ where $t(x) = (x-3)^2 - 1$ and $r(x) = |x|$. This decomposition will serve as an aid to graphing because we are already familiar with the graphs of t and r. The graph of t is a transformation of the toolkit parabola (see Figure 54) and r is the simple absolute value function. Notice that this composition differs from that in Example 4; here the absolute value function is the inner function in the composition. The effect of having $r(x) = |x|$ as the inner function is that k is an even function—it is easy to verify that $k(-x) = k(x)$.

Since $|x| = x$ when $x \geq 0$, the graph of k is identical to the graph of t on the right side of the

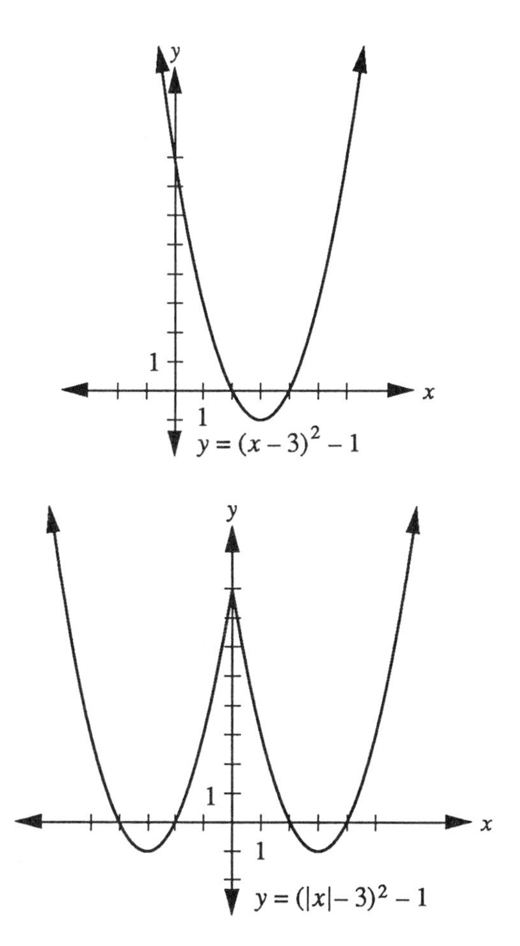

Figure 54 Graphs of $t(x) = (x-3)^2 - 1$ and $k(x) = (|x| - 3)^2 - 1$

y-axis. Recall that k is an even function, so its graph will be symmetric about the y-axis. This means that we can finish the graph of k by reflecting the right half about the y-axis. The complete graph of k is also shown in Figure 54. ∎

The functions we have graphed in the previous two examples have both involved composition with $y = |x|$. The nature of absolute value makes these functions well suited to the type of analysis we have used in the examples; we have graphed without the computer and without plotting points. Instead, we have analyzed the effect of taking the absolute value of a number and used this as a graphing guide. In the next two examples we will use a similar approach and will supplement this analysis with some point plotting.

Example 6

Graph $r(x) = \sqrt{2x^2 - 12x + 10}$.

Solution This function will look much simpler if we decompose it; $r = g \circ f$ where $f(x) = 2x^2 - 12x + 10$ and $g(x) = \sqrt{x}$. We already know what the graph of f will look like—it is a transformation of $y = x^2$. Completing the square on the equation of f gives $f(x) = 2(x-3)^2 - 8$, which means that the graph of f is a parabola that has been stretched vertically by a factor of 2 and whose vertex is at $(3, -8)$. Solving the quadratic equation $0 = 2x^2 - 12x + 10$ shows that the zeros of f are $x = 1$ and $x = 5$; a graph of f is shown in Figure 55. The function r that we want to graph can be expressed as $\sqrt{f}$, so we need to think about the effect of taking the square root of the y-values on the graph of f. First of all, it is possible to take the square root of these y-values only when they are nonnegative. The graph of f in Figure 55 shows that the y-values are greater than or equal to zero when $x \le 1$ or when $x \ge 5$. Therefore, the domain of r is $x \le 1$ or $x \ge 5$. Second, whenever the y-values on the graph of f are greater than 1, taking their square root decreases their size. (Recall that $\sqrt{a} < a$ for all $a > 1$.) This means that the graph of $\sqrt{f}$ will rise less steeply than that of f. Clearly, whenever the y-value on the graph of f is equal to 1, then the y-value on the graph of $\sqrt{f}$ will remain 1. Also, whenever the y-values on the graph of f are positive and less than 1, the y-values on the graph of $\sqrt{f}$ will be greater than the y-values on f (because $\sqrt{a} > a$ when $0 < a < 1$). These observations make an important contribution to our understanding of the behavior of r. We can verify these observations by plotting a few points; the graph of r is shown in Figure 55. Compare the steepness of the graph of r with that of f. Since f increases at a quadratic rate and $r(x) = \sqrt{f(x)}$, it should make sense that r increases at a rate that is approximately linear. ∎

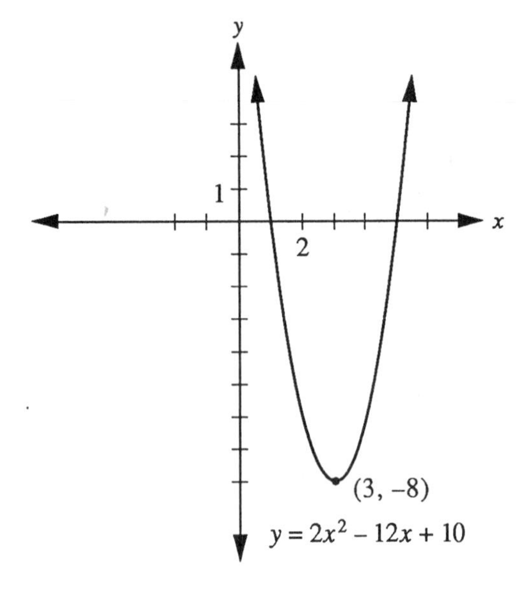

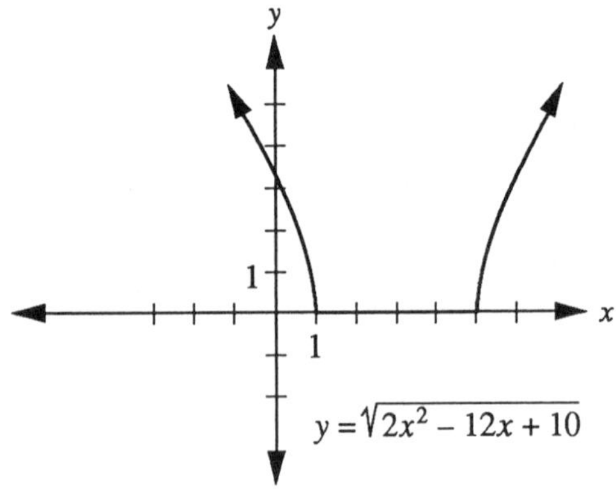

Figure 55 Graphs of $f(x) = 2x^2 - 12x + 10$ and $r(x) = (2x^2 - 12x + 10)^{\frac{1}{2}}$

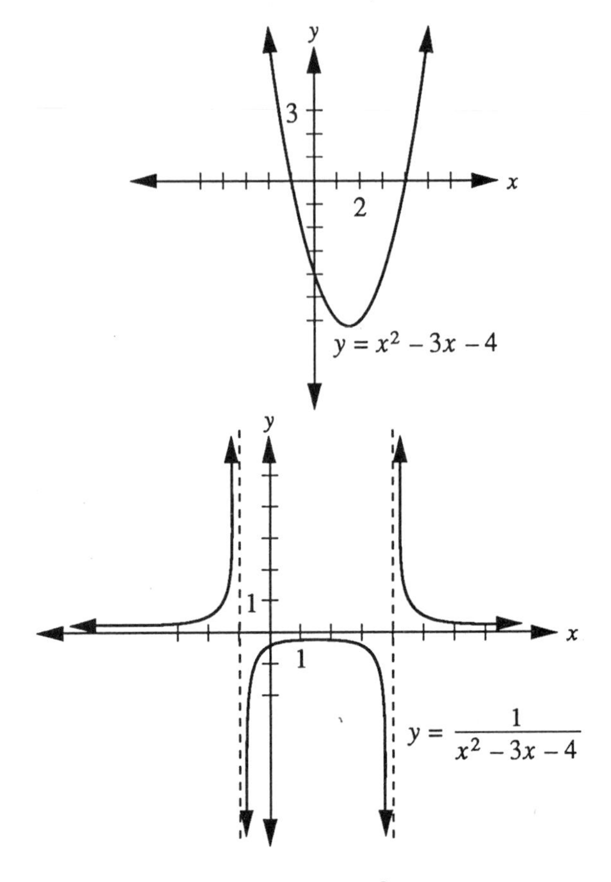

Figure 56 Graphs of $h(x) = x^2 - 3x - 4$ and $f(x) = \dfrac{1}{x^2 - 3x - 4}$

Example 7

Graph $f(x) = \dfrac{1}{x^2 - 3x - 4}$.

Solution This function can be decomposed as $g \circ h$ where $g(x) = \frac{1}{x}$ and $h(x) = x^2 - 3x - 4$. The graph of h is a parabola with vertex $(\frac{3}{2}, -\frac{25}{4})$ and zeros at $x = -1$ and $x = 4$; h is graphed in Figure 56. Note that f can be expressed as $\frac{1}{h}$; to explore the graph of f we need to think about the effect of taking the reciprocal of the y-values on the graph of h. Clearly, we can take the reciprocal of only the nonzero y-values, so $x = -1$ and $x = 4$ are not in the domain of f.

What other observations can we make about the reciprocals of the y-values on the graph of h? Taking the reciprocal of a number does not change its sign, so $\frac{1}{h}$ has positive y-values wherever h has positive values and negative y-values wherever h has negative values. Since the reciprocal of a large number is a small number, $\frac{1}{h}$ will have y-values close to zero where h has y-values with large absolute value. Also, $\frac{1}{h}$ will have y-values with large absolute value wherever h has y-values close to zero. Finally, wherever $h(x)$ is equal to 1 or -1, $\frac{1}{h(x)}$ has the same value as $h(x)$. These observations can be supplemented with some point plotting to produce the graph of f in Figure 56.

Before proceeding, take a minute to study and compare the graphs of h and f shown in Figure 56. Notice how the observations we made previously about taking the reciprocal are manifested in these graphs. The graph of f has three asymptotes. The vertical lines $x = -1$ and $x = 4$ correspond to the fact that h has zeros at these x-values; the horizontal line $y = 0$ corresponds to the fact that h increases without bound. ■

Example 8

Based on the graph of $y = f(x)$ in Figure 57, sketch a graph of $y = \dfrac{1}{f(x)}$.

Solution This example obviously differs from others we have already worked on — we do not have a formula for f. We will not be able to do any point plotting and will instead need to rely solely on an analysis of tendencies in the graph of f.

Let's first describe the behavior of f:

$$\text{As } x \longrightarrow \infty, \quad f(x) \longrightarrow \infty, \text{ read,}$$

"as x increases without bound, $f(x)$ increases without bound."

$$\text{As } x \longrightarrow 1^+, \quad f(x) \longrightarrow 0^+, \text{ read,}$$

Figure 57 Graph of $y = f(x)$ for Example 8

"as x approaches 1 from the right, $f(x)$ approaches zero from above."

$$\text{As } x \longrightarrow 1^-, \quad f(x) \longrightarrow 0^-, \text{ read,}$$

"as x approaches one from the left, $f(x)$ approaches zero from below."

$$\text{As } x \longrightarrow 0^+, \quad f(x) \longrightarrow -\infty, \text{ read,}$$

"as x approaches zero from the right, $f(x)$ decreases without bound."

To analyze the behavior of $y = \frac{1}{f(x)}$ we will need to refer back to the observations about reciprocals that we made in Example 7.

— The reciprocal of a number far from zero is close to zero.

— The reciprocal of a number close to zero is large in magnitude.

— The reciprocal of 1 is 1 and of -1 is -1.

— The reciprocal of a positive number is positive and of a negative number is negative.

Applying these observations to the problem of graphing $y = \frac{1}{f(x)}$ leads to an expanded table that shows tendencies of x, $f(x)$, and $\frac{1}{f(x)}$:

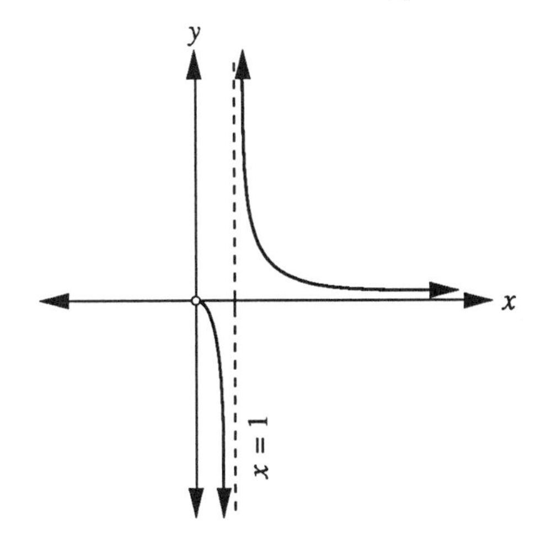

Figure 58 Graph of $y = \frac{1}{f(x)}$ for Example 8

as $x \longrightarrow \infty$, $f(x) \longrightarrow \infty$ so $\frac{1}{f(x)} \longrightarrow 0^+$

as $x \longrightarrow 1^+$, $f(x) \longrightarrow 0^+$ so $\frac{1}{f(x)} \longrightarrow \infty$

as $x \longrightarrow 1^-$, $f(x) \longrightarrow 0^-$ so $\frac{1}{f(x)} \longrightarrow -\infty$

as $x \longrightarrow 0$, $f(x) \longrightarrow -\infty$ so $\frac{1}{f(x)} \longrightarrow 0^-$

To graph $y = \frac{1}{f(x)}$ we need information about ordered pairs $(x, \frac{1}{f(x)})$, so we will use the first and last columns of the above table. The graph of $y = \frac{1}{f(x)}$ is shown in Figure 58. Notice that there is little scale on either the x- or y-axis; the graph does not show specific points, but does show general tendencies. ■

Exercise Set 10

1. Identify the domain and the range of the function f in Example 8.
2. Write and simplify an expression for $f \circ g$ and $g \circ f$. State the domain of each composition.
   a. $f(x) = \frac{1}{x-1}$ and $g(x) = \frac{1}{x+1}$
   b. $f(x) = x^2 - 5$ and $g(x) = \sqrt{4x-5}$
   c. $f(x) = \sqrt{x-5}$ and $g(x) = \frac{1}{x^2}$
   d. $f(x) = x^2 + 1$ and $g(x) = x - 4$
   e. $f(x) = |x+1|$ and $g(x) = \sqrt{x-1}$
   f. $f(x) = \frac{1}{x-2}$ and $g(x) = x + 3$
   g. $f(x) = \frac{x}{x-1}$ and $g(x) = \sqrt{x+2}$
   h. $f(x) = x^3 - 3x - 1$ and $g(x) = \frac{1}{x}$ (**Hint**: Use the computer or a calculator to help you determine the domains.)
   i. $f(x) = \sqrt{x}$ and $g(x) = \sin x$
3. Use the idea of decomposition to help you sketch a graph of each function. Check your graphs using a computer or graphing calculator.
   a. $y = |(x-4)^3|$
   b. $y = \frac{1}{1-x^2}$
   c. $y = \frac{1}{\sin x}$
   d. $y = |x|^3$
   e. $y = \frac{1}{x^2+1}$
   f. $y = |2 + \frac{1}{x}|$
   g. $y = |2 + \frac{1}{x+2}|$
   h. $y = \sqrt{|x-4|}$
   i. $y = \sqrt{|x-4|+1}$
   j. $y = \sqrt{\sin x}$
   k. $y = |x^3 - 2|$
   l. $y = -|\frac{1}{2}x^3 + 1|$
   m. $y = \frac{1}{x^2-3x+2}$
   n. $y = ||x| - 4|$
   o. $y = |3\sin x - 2|$
   p. $y = (|x| - 2)^2 - 1$
   q. $y = \frac{1}{|x|-2} - 1$
   r. $y = \sqrt{\frac{1}{2}x^2 - 4x + 8}$
4. The graph of $y = g(x)$ is shown in Figure 59. Sketch a graph of each of the following.
   a. $y = \sqrt{g(x)}$
   b. $y = |g(x)|$
   c. $y = 1/g(x)$
   d. $y = g(|x|)$
   e. $y = g(-x)$
5. You already know how to graph $y = \frac{1}{x-5}$ as a transformation of $y = \frac{1}{x}$. Describe how to arrive at the same graph by considering the composition $f \circ g$ where $f(x) = \frac{1}{x}$ and $g(x) = x - 5$.

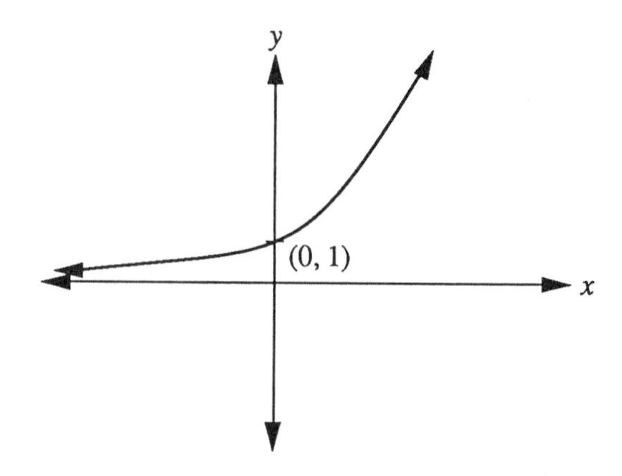

Figure 59 Graph of $y = g(x)$ for Exercise 4

6. You already know how to graph $y = |x| + 3$ as a transformation of $y = |x|$. Describe how to arrive at the same graph by considering the composition $f \circ g$ where $f(x) = x + 3$ and $g(x) = |x|$.

7. You already know how to graph $y = |x + 3|$ as a transformation of $y = |x|$. Describe how to arrive at the same graph by considering the composition $p \circ q$ where $p(x) = |x|$ and $q(x) = x + 3$.

8. You already know how to graph $y = \sqrt{x + 1}$ as a transformation $y = \sqrt{x}$. Describe how to arrive at the same graph by considering the composition of $f(x) = \sqrt{x}$ and $t(x) = x + 1$.

9. Suppose you were asked to graph the function $g(x) = \sqrt{7 - \frac{3}{4}x}$. Would you rather consider this as a transformation of the toolkit function $y = \sqrt{x}$ or as a composition of $f(x) = 7 - \frac{3}{4}x$ and $h(x) = \sqrt{x}$? Explain the reason for your choice.

10. Graph $y = \sqrt{\frac{1}{4}x^2 - 2x + 4}$. Identify the domain. Write a sentence or two to explain why the graph looks the way it does.

11. Suppose $p(x) = x^2 + 1$ and $q(x) = [x]$, the greatest integer function. Evaluate each of the following.

a. $p(q(3)), p(q(3.3)), p(q(3.5))$ and $p(q(3.9))$

b. $q(p(3)), q(p(3.3)), q(p(3.5))$ and $q(p(3.9))$

12. Sketch a graph of $y = [x^2 + 1]$ over the domain $-2 \le x \le 2$.

13. Sketch a graph of $y = [x]^2 + 1$ over the domain $-2 \le x \le 2$.

14. Sketch a graph of $y = [\sin x]$.

15. Suppose $g(x) = \frac{x+3}{2}$. Evaluate $g(g(1))$ and $g(g(g(1)))$.

16. Use the idea of decomposition to help you sketch a graph of the function $y = \sqrt{25 - x^2}$. Identify the domain and the range.

17. Find a function f which satisfies the following conditions, and then evaluate $f(f(f(f(f(f(f(f(f(6)))))))))$.

$$f(x) = \begin{cases} f(x-2), & \text{if } x > 5 \\ x^2 - 3, & \text{if } x = 5 \\ x^2 + 1, & \text{if } x < 5 \end{cases}$$

18. Find a function f which satisfies the following:

$$f(x) = \begin{cases} x - 10, & \text{if } x > 100 \\ f(f(x + 11)), & \text{if } x \le 100 \end{cases}$$

The domain of this function consists of the set of integers. You may be surprised to discover what the range is. Evaluate each of the following and then identify the range.

a. $f(102)$

b. $f(80)$

c. $f(97)$

d. $f(-10)$

e. $f(301)$

11
Inverses

A car salesperson advises potential customers that the monthly payment on a new car depends on four factors: the price of the car, the size of the down payment, the interest rate, and the term of the loan. These quantities are related by the formula

$$p = \frac{(P - d)r(1 + r)^n}{(1 + r)^n - 1}$$

where p is the monthly payment, P is the price of the car, d is the down payment, r is the monthly interest rate, and n is the number of months required to pay back the loan. In most instances the interest rate and the duration of the loan are determined by the seller, and the down payment is agreed upon by the buyer and seller. Since these three quantities do not vary, the monthly payment and the price of the car are the only true variables in the formula. The price of the car determines the monthly payment, so price is usually thought of as the independent variable and the monthly payment is a function of the price.

Suppose a customer who has \$2400 for a down payment wants to buy a \$12,000 car. The annual interest rate is 9% (which is equivalent to 0.75% per month), and the term of the loan is 48 months. Using the formula given above, the salesperson determines that the monthly payment will be \$238.90. Thus, the ordered pair (12000, 238.90) belongs to the function that contains ordered pairs of the form (price of car, monthly payment).

As often happens, this customer cannot afford a monthly payment of \$238.90. The buyer's income and other expenses limit the monthly payment to \$200. The buyer will have to shop around to find an affordable car. In this situation, the monthly payment is no longer the independent variable. Now the affordable price depends on the affordable monthly payment, and the roles of independent and dependent variables have been reversed. In effect, there is now a new function involved in this problem—it consists of ordered pairs of the form (monthly payment, price of car). This second function has a special relationship to the first function that consisted of ordered pairs (price of car, monthly payment). The two functions are said to be *inverses* of each other. The symbol f^{-1} is used to represent the inverse of the function f.

Since f and f^{-1} have reversed ordered pairs and have interchanged the role of independent and dependent variables, the domain of f is the same as the range of f^{-1} and the range of f is the same as the domain of f^{-1}. If $f(a) = b$, then $f^{-1}(b) = a$.

Before proceeding, let's find out how expensive a car the customer can buy with a monthly payment of \$200. Solving the equation

$$200 = \frac{(P - 2400)(.0075)(1.0075^{48})}{1.0075^{48} - 1}$$

gives $P = 10436.96$, so the customer can afford a car that costs about \$10,437.

Not every function has an inverse that is a function. For instance, the function $f(x) = x^2$ contains ordered pairs $(2, 4)$ and $(-2, 4)$. The inverse f^{-1} must contain ordered pairs $(4, 2)$ and $(4, -2)$. However, since 4 is paired with both 2 and -2 this relation is not a function. What condition must a function f satisfy in order for its inverse to be a function? To avoid the problem encountered above with $f(x) = x^2$, each y-value must come from only one x-value; in other words, two different x-values cannot be paired with the same y-value. A function that has this property is called *one-to-one.* Every one-to-one function has an inverse that is a function. If the domain of $f(x) = x^2$ is restricted to $x \geq 0$, then f is a one-to-one function and $f^{-1}(x) = \sqrt{x}$.

A calculator is a useful tool for exploring the concept of inverse functions. Suppose you were doing some calculation and the number 37.958127 was on your calculator display. If you were to accidently press the square root key, the new display would be 6.1610167. How could you retrieve your original display? It should be clear that pushing the x^2 key will achieve this result, since squaring a number "undoes" the effect of taking its square root. This sequence of events can be understood in terms of inverses. The square root key acts as a function that pairs the input 37.958127 with the output 6.1610167. The number 6.1610167 was then the input to a second function that was the inverse of the first; this function squared its input, so 6.1610167 was paired with 37.958127. This "undoing" operation can be described mathematically with function composition. If f and g undo each other, then

$$g(f(x)) = x$$

for all x's in the domain of f, and

$$f(g(x)) = x$$

for all x's in the domain of g, and f and g are *inverse functions.*

Recall that the function $y = x$ is often called the *identity* function and consists of ordered pairs of the form (x, x). A function and its inverse compose to the identity function because $f : a \longrightarrow b$ and $f^{-1} : b \longrightarrow a$; therefore the composition $f^{-1} \circ f$ pairs a with a and $f \circ f^{-1}$ pairs b with b.

The graphs of a function and of its inverse have a special relationship. For each point (a, b) on the graph of f there is a point (b, a) on the graph of f^{-1}. Recall from the earlier discussion of symmetry that the points (a, b) and (b, a) are mirror images in the line $y = x$. Thus, when f and f^{-1} are graphed on the same coordinate axes, the two graphs viewed together will appear symmetric about $y = x$. In fact, the graph of f can be reflected about $y = x$ to obtain the graph of f^{-1}.

Example 1

Graph the inverse of $g(x) = x^3$.

Solution A graph of g is shown in Figure 60. It is one-to-one, so its inverse is a function. A graph of the inverse can be obtained by reflecting the graph of g about the line $y = x$; g^{-1} is also graphed in Figure 60. ■

The function g in Example 1 contains ordered pairs of the form (x, x^3), so g^{-1} contains pairs of the form (x^3, x) in which the first coordinate is the cube of the second coordinate. The form of the reversed ordered pairs can also be expressed as $(a, \sqrt[3]{a})$; this suggests that $g^{-1}(x) = \sqrt[3]{x}$.

Example 2

Graph the inverse of $h(x) = \sqrt{x-1}$. Identify the domain and range of h and of h^{-1}.

Solution The graph of h is a horizontal translation of the toolkit square root function; the graph of h^{-1} is obtained by reflecting over the line $y = x$. The graphs of h and h^{-1} are shown in Figure 61.

The domain and range of h^{-1} can be read from the graph. Notice that the domain of h is the same as the range of h^{-1} and vice versa.

Domain of h: $x \geq 1$	Domain of h^{-1}: $x \geq 0$
Range of h: $y \geq 0$	Range of h^{-1}: $y \geq 1$

The graph of h^{-1} in Figure 61 should suggest to you that $h^{-1}(x) = x^2 + 1, x \geq 0$. Think about why this is reasonable in light of the equation of h. Since h subtracts 1 from its input and then takes the square root, h^{-1} squares its input and then adds 1. ■

Figure 62 shows the graphs of the function $f(x) = x^2 - 3$ and of the inverse of f. These graphs make it clear that the inverse of f is not a function; one x-value is paired with two different y-values. This occurs because f is not one-to-one. When this happens we frequently want to restrict the domain of f in such a way as to make f a one-to-one function. This restriction can be made in a variety of ways. We will discuss two ways that retain as much of the graph of f as possible.

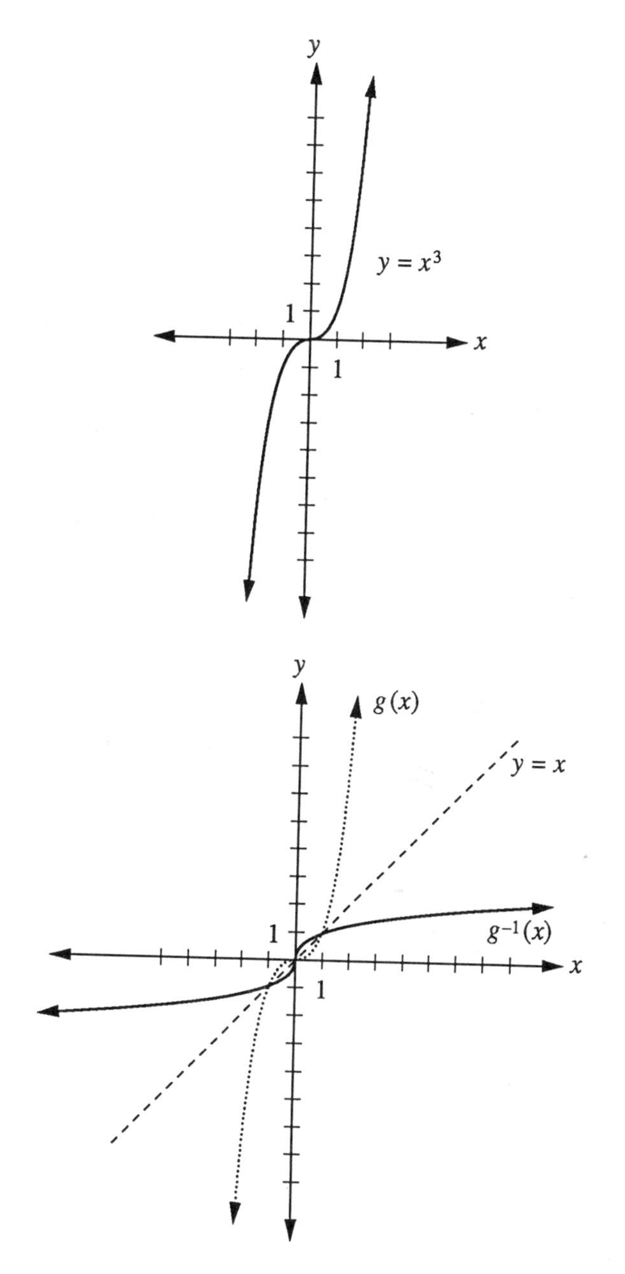

Figure 60 Graphs of $g(x) = x^3$ and $g^{-1}(x)$

If we choose $x \geq 0$ as the domain of f, then the graph of f will be the right half of a parabola and the graph of f^{-1} will be the upper half of a parabola. This choice is illustrated first in Figure 63. You should be able to recognize from the

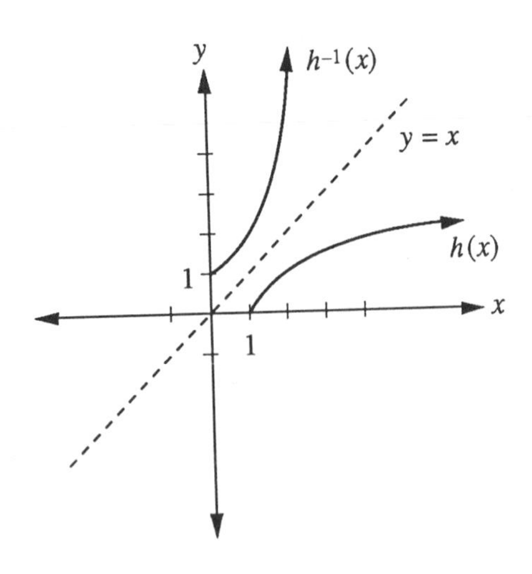

Figure 61 Graphs of $h(x) = (x-1)^{\frac{1}{2}}$ and $h^{-1}(x)$

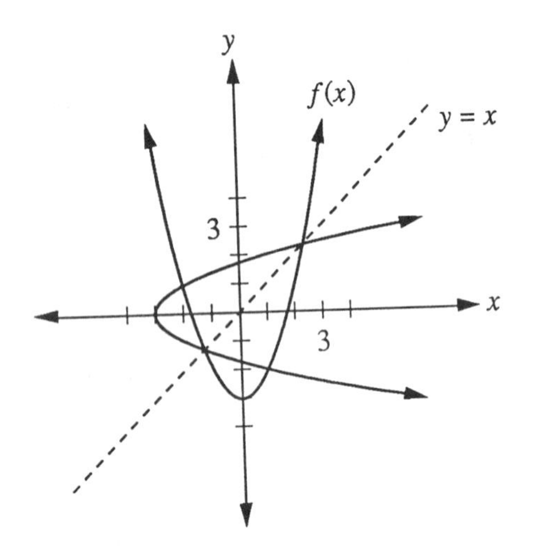

Figure 62 Graphs of $f(x) = x^2 - 3$ and its reflection across $y = x$

graph that $f^{-1}(x) = \sqrt{x+3}$. The domain and range of f and f^{-1} follow.

Domain of f: $x \geq 0$ Domain of f^{-1}: $x \geq -3$
Range of f: $y \geq -3$ Range of f^{-1}: $y \geq 0$

On the other hand, we can also choose $x \leq 0$ as the domain of f. In this case, the graph of f will be the left half of a parabola, and the graph

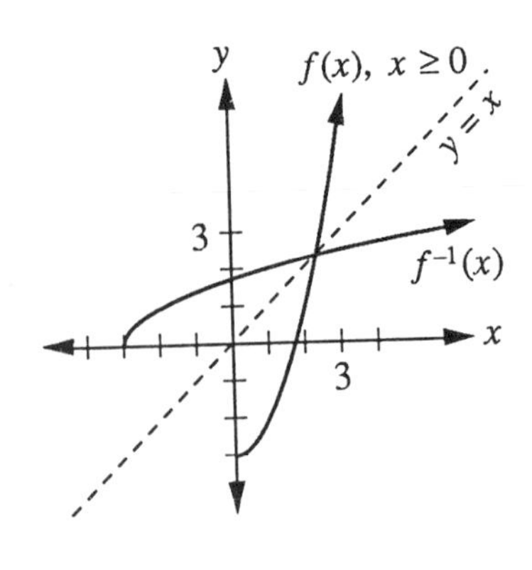

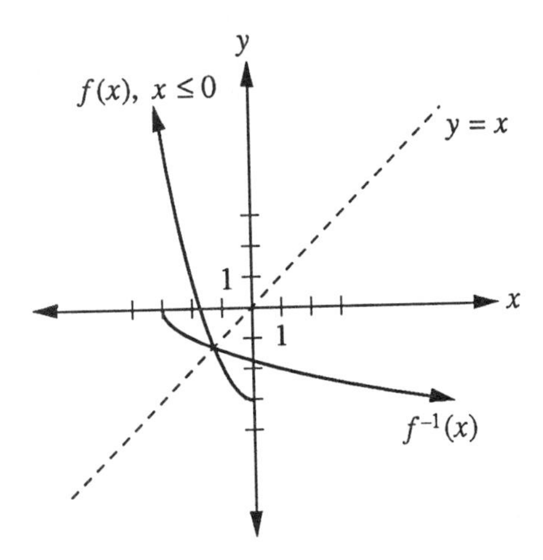

Figure 63 Restricting the Domain to Create a One-to-One Function

of f^{-1} will be the lower half of a parabola; see the other graph in Figure 63. With this domain restriction for f, the expression for the inverse is $f^{-1}(x) = -\sqrt{x+3}$. You should be able to identify the domain and range of f and f^{-1} based on this domain restriction.

The preceding discussion should make it clear why it is not correct to say that $f(x) = x^2 - 3$ and $g(x) = \sqrt{x+3}$ are inverse functions unless we specifically mention restrictions on the domain of f. It is true that $f(x) = x^2 - 3$, $x \geq 0$, and $g(x) = \sqrt{x+3}$ are inverses.

Note that

$$\begin{aligned} f(g(x)) &= f(\sqrt{x+3}) = (\sqrt{x+3})^2 - 3 \\ &= x + 3 - 3 = x, \qquad x \geq -3 \end{aligned}$$

and

$$g(f(x)) = g(x^2 - 3) = \sqrt{(x^2 - 3) + 3}$$
$$= \sqrt{x^2} = x, \qquad x \geq 0.$$

This confirms that with an appropriate restriction on the domain of f, $g(f(x)) = x$ and $f(g(x)) = x$.

Example 3

If $f(x) = \frac{1}{2}x^3 - 4$, write an expression for $f^{-1}(x)$.

Solution The expression $f(x) = \frac{1}{2}x^3 - 4$ defines a cubic function that is one-to-one and therefore has an inverse that is a function. We can analyze f as follows:

independent variable → cube → times $\frac{1}{2}$
→ minus 4 = dependent variable

We know that f^{-1} must start with the dependent variable and end up with the independent variable. Therefore f^{-1} must do the following:

dependent variable → add 4 → times 2
→ cube root = independent variable

The above analysis of f^{-1} suggests that the expression for f^{-1} is $f^{-1}(x) = \sqrt[3]{2(x+4)}$. This procedure can be used to find the inverse of a function f only if f is one-to-one and the independent variable occurs exactly once in the expression for f.

The work to find an expression for f^{-1} can be organized as follows. Given the function $f(x) = \frac{1}{2}x^3 - 4$, we can write

$$y = \tfrac{1}{2}x^3 - 4.$$

Now solve for x:

$$y + 4 = \tfrac{1}{2}x^3$$
$$2(y+4) = x^3$$
$$\sqrt[3]{2(y+4)} = x$$

The final equation above gives all the information we need to write the expression for f^{-1}; it shows that f^{-1} adds 4 to its input, then multiplies by 2, and then takes the cube root. Since we usually name the input x, we write

$$f^{-1}(x) = \sqrt[3]{2(x+4)}.$$ ■

Example 4

If $g(x) = x^2 - 4x + 5$, write a formula for g^{-1}.

Solution Completing the square on g gives $g(x) = (x-2)^2 + 1$, so the graph of g is a parabola with vertex at (2, 1). This graph is shown first in Figure 64; it is clearly not one-to-one, so we need to restrict the domain if g^{-1} is to be a function. We can choose $x \geq 2$ as the domain to make g one-to-one and retain as much of its graph as possible.

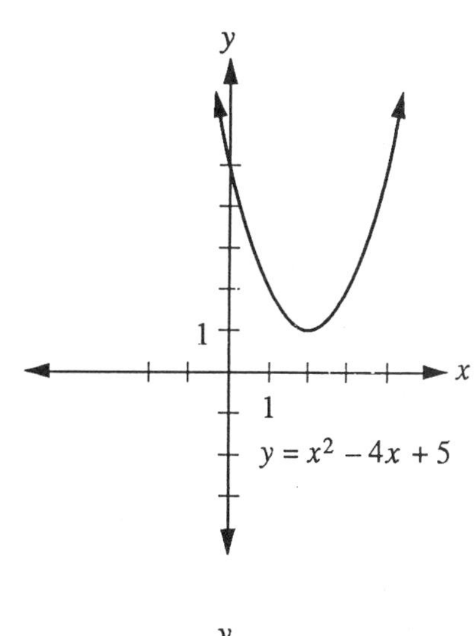

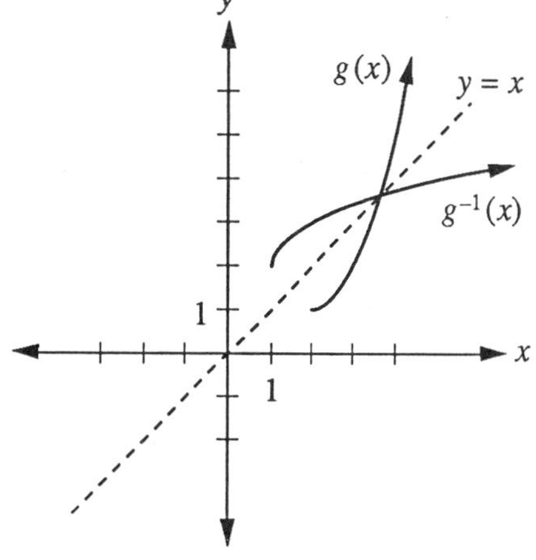

Figure 64 Graphs of $g(x) = x^2 - 4x + 5$ and related functions

Now let's find the formula for g^{-1}.

$$\begin{aligned} y &= (x-2)^2+1 \\ y-1 &= (x-2)^2 \\ \pm\sqrt{y-1} &= (x-2) \end{aligned}$$

Either $(x-2)$ is equal to $+\sqrt{y-1}$ or it is equal to $-\sqrt{y-1}$. Which should we choose? Consideration of the domain of g can guide us. The domain of g is $x \geq 2$; therefore $x-2 \geq 0$. Since $x-2$ is 0 or positive, it cannot be equal to a negative square root. This implies that $x-2$ must be equal to the positive square root of $y-1$.

$$\begin{aligned} \sqrt{y-1} &= x-2 \\ 2+\sqrt{y-1} &= x \end{aligned}$$

Finally, we know that

$$g^{-1}(x) = 2+\sqrt{x-1}.$$

This is the inverse when the domain of g is restricted to $x \geq 2$. What would be the formula for g^{-1} if the domain of g were restricted to $x \leq 2$? The graphs of g with a restricted domain and of g^{-1} are shown next in Figure 64. Can you understand why these graphs look the way they do? Identify the domains and ranges of g and g^{-1}. ■

Example 5

If $g(x) = (x-2)^2+1$, $x \geq 2$, we know that $g^{-1}(x) = 2+\sqrt{x-1}$, $x \geq 1$. Write a formula for the composition $g^{-1} \circ g$.

Solution

$$\begin{aligned} g^{-1}(g(x)) &= g^{-1}((x-2)^2+1) \\ &= 2+\sqrt{(x-2)^2+1-1} \\ &= 2+\sqrt{(x-2)^2} \\ &= 2+(x-2) \\ &= x \end{aligned}$$

This shows that $g^{-1}(g(x)) = x$; this confirms an important property of inverse functions. Note that the domain of this composition is $x \geq 2$. ■

Example 6

Find the coordinates of the point(s) where the graph of $g(x) = x^2-4x+5$, $x \geq 2$, intersects the graph of its inverse.

Solution Based on the work in Example 4, when the domain of g is restricted to $x \geq 2$ the formula for g^{-1} is

$$g^{-1}(x) = 2+\sqrt{x-1}.$$

To find where the graphs of g and g^{-1} intersect, we could solve the equation

$$x^2-4x+5 = 2+\sqrt{x-1}.$$

There is no straightforward algebraic way to solve this equation. However, we can benefit from an important characteristic of functions and their inverses to simplify the solution process. Recall that if the point (a,b) is on the graph of a function, then (b,a) is on the graph of the inverse. For this example a point that lies on the graph of both g and g^{-1} must be of the form (a,a), and therefore lies on the line $y = x$. This means that finding where the graphs of g and g^{-1} intersect is equivalent to finding where g or g^{-1} intersects the line $y = x$. In this case, we will solve $x^2-4x+5 = x$ to find the x-coordinate of the required point(s) of intersection.

$$\begin{aligned} x^2-4x+5 &= x \\ x^2-5x+5 &= 0 \\ x &= \frac{5\pm\sqrt{5}}{2} \\ x \approx 3.62 &\text{ or } x \approx 1.38 \end{aligned}$$

Only one of these x-values, $x \approx 3.62$, is contained in the domain of g. Therefore, the point of intersection of g and g^{-1} is approximately $(3.62, 3.62)$. ■

Exercise Set 11

1. Determine a formula for $f^{-1}(x)$.
   a. $f(x) = (x-1)^3+2$
   b. $f(x) = \frac{1}{x+1}$

c. $f(x) = \sqrt{5-x}$

d. $f(x) = 5 - 2x$

e. $f(x) = \sqrt{9-x^2}, x \geq 0$

f. $f(x) = \frac{x}{x-1}$

g. $f(x) = 6x - x^2, x \leq 3$

h. $f(x) = -\sqrt{x+1}$

2. For each function in Exercise 1, graph f and f^{-1} in the same coordinate system.

3. For each function in Exercise 1, identify the domain of f and of f^{-1}.

4. Find where the graphs of g and g^{-1} intersect.

a. $g(x) = 4x - 1$

b. $g(x) = -\sqrt{4-x}$

5. Find an equation for h^{-1}. Based on your answer, what symmetry must be present in the graph of h?

a. $h(x) = (1-x^3)^{\frac{1}{3}}$

b. $h(x) = \sqrt{1-x^2}, x \geq 0$

c. $h(x) = \frac{3x-1}{2x-3}$

6. What happens when you try to find the inverse of the function $f(x) = \frac{x^2}{x-3}$? What does this tell you about the graph of f?

7. Find the inverse of each function. When necessary, give a restriction on the domain of f so that f^{-1} will be a function. A sketch of f and f^{-1} will be helpful.

a. $f(x) = x(x-2)$

b. $f(x) = |x-1| + 1$

c. $f(x) = \sqrt{1-x^2}$

8. Let $f(x) = \sin x$.

a. What happens when you try to sketch a graph of the inverse of f. Have you graphed a function?

b. Suggest a domain restriction so that the inverse of f will be a function.

9. Let $p(x) = -\sqrt{x+3}$.

a. Find a formula for $p^{-1}(x)$.

b. Graph p and p^{-1} on the same coordinate axes.

c. Identify the domain of p and of p^{-1}.

d. For what values of x is it true that $p(p^{-1}(x)) = x$?

e. For what values of x is it true that $p^{-1}(p(x)) = x$?

f. Graph $p \circ p^{-1}$. Be sure to use the appropriate domain.

g. Graph $p^{-1} \circ p$. Be sure to use the appropriate domain.

10. Write a formula for the inverse of g. A sketch of the graph will be helpful.

a. $g(x) = \begin{cases} x-2, & x \leq 2 \\ 2x-4, & x > 2 \end{cases}$

b. $g(x) = \begin{cases} -(x-1)^2 + 3, & x \leq 1 \\ \frac{1}{2}x + \frac{5}{2}, & 1 < x < 7 \\ \sqrt{x-7} + 6, & x \geq 7 \end{cases}$

CHAPTER 3

Polynomials, Rational Functions, and Algorithms

1 *Introduction to Polynomials*

Polynomial functions are important both for mathematical modeling and for the mathematical concepts these functions illustrate. One of the goals of this section is to help you develop an intuitive feel for the graphs of polynomial functions. This will help you to understand how these functions behave and will enable you to check the reasonableness of graphs that are generated by computers or calculators.

The building blocks of polynomials are *power functions*; these are functions of the form $f(x) = x^n$ where n is a nonnegative integer. The graphs of power functions fall into two categories. When n is even, $f(x) = x^n$ is an even function; its graph is symmetric about the y-axis and looks something like a parabola. When n is odd, $f(x) = x^n$ is an odd function; its graph is symmetric about the origin and has a shape similar to that of $y = x^3$. These general shapes are illustrated in Figure 1.

A single power function, perhaps multiplied by a coefficient, is called a *monomial.* The following are all examples of monomials: $3x^2$, $-7x$, and 9. When two or more of these monomials are added in a function, the sum is a *polynomial.* For example, the function $f(x) = x^6 - 11x^4 + 30x^2 - 5x - 14$

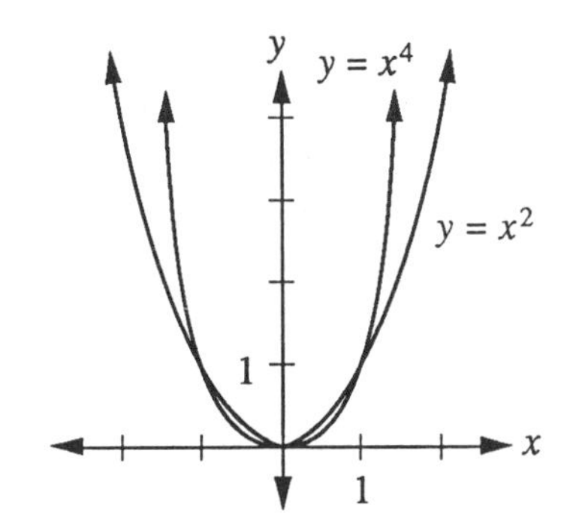

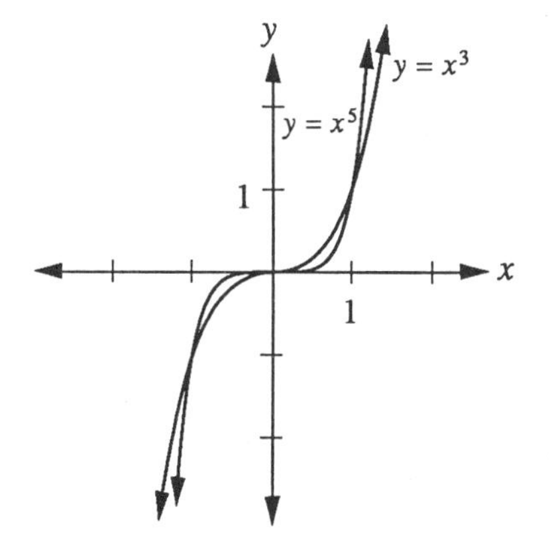

Figure 1 Graphs of Power Functions

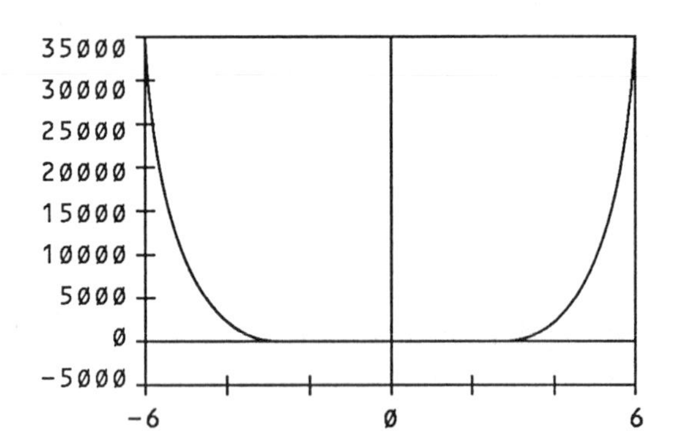

Figure 2 Graph of $f(x) = x^6 - 11x^4 + 30x^2 - 5x - 14$ over $[-6, 6]$

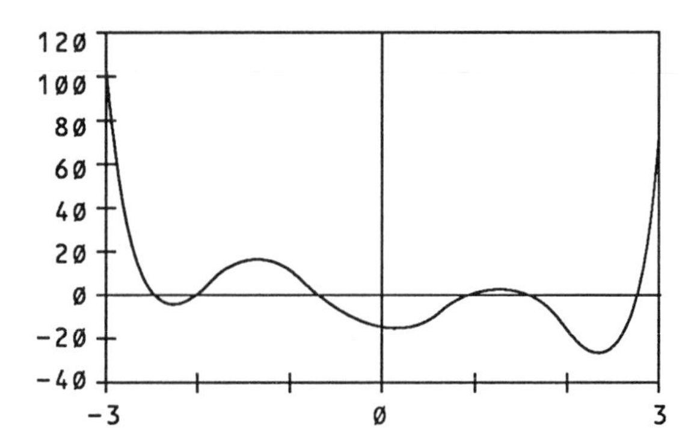

Figure 3 Graph of $f(x) = x^6 - 11x^4 + 30x^2 - 5x - 14$ over $[-3, 3]$

is the sum of the monomials, x^6, $-11x^4$, $30x^2$, $-5x$, and -14. Thus $f(x)$ is a polynomial function. Figure 2 shows a computer-generated graph of this function over $[-6, 6]$. You can use it to assess the global behavior of this sixth degree polynomial. Notice that the graph looks much like those shown in Figure 1 for power functions of even degree. This is true, because for large x-values the x^6-term dominates the other terms in the polynomial. Therefore,

$$\text{as } x \longrightarrow \infty, y \longrightarrow \infty$$

and

$$\text{as } x \longrightarrow -\infty, y \longrightarrow \infty.$$

In Figure 3 we show a smaller portion of the graph. Now you can see that the graph of this polynomial is actually quite different from that of a power function. As shown in the graph, this function has six zeros and five turning points.

The constant, linear, quadratic, and cubic functions are all simple polynomial functions that belong to the toolkit. As you know, it is relatively easy to graph linear and quadratic functions. Every linear function has an equation of the form $y = mx + b$ and is a transformation of the toolkit function $f(x) = x$. Every quadratic function has an equation of the form $y = ax^2 + bx + c$. By completing the square, any quadratic function can be written in the form $y = a(x - h)^2 + k$, and is therefore a transformation of the function $g(x) = x^2$. The general form for the equation of a cubic function is $y = ax^3 + bx^2 + cx + d$. It is not always possible to write such an equation in the form $y = a(x - h)^3 + k$, so we cannot graph every cubic function as a transformation of the toolkit function $y = x^3$. Other graphing techniques must be used.

Example 1

Graph $f(x) = x^3 - x^2$.

Solution The function $f(x)$ cannot be rewritten as a transformation of the toolkit function $g(x) = x^3$. When it is not possible to analyze a polynomial function in terms of transformations, it may be useful to use a technique called *addition of ordinates*. If f can be written as the sum of two simpler functions, then f can be graphed by adding the y-values of the simpler functions for each x-value. In this example $f(x)$ is the sum of the two functions $g(x) = x^3$ and $h(x) = -x^2$.

On the graph of $g(x) = x^3$, the vertical distance of any point from the x-axis is equal to x^3. On the graph of $h(x) = -x^2$, the vertical distance is $-x^2$. On the graph of $f(x)$, the vertical distance is $x^3 + (-x^2)$, the sum of the two vertical distances of $g(x)$ and $h(x)$. We can determine the shape of the graph of $f(x)$ by analyzing where the sum of these distances is positive, where it is negative, and where it is zero. Graphs of $g(x) = x^3$

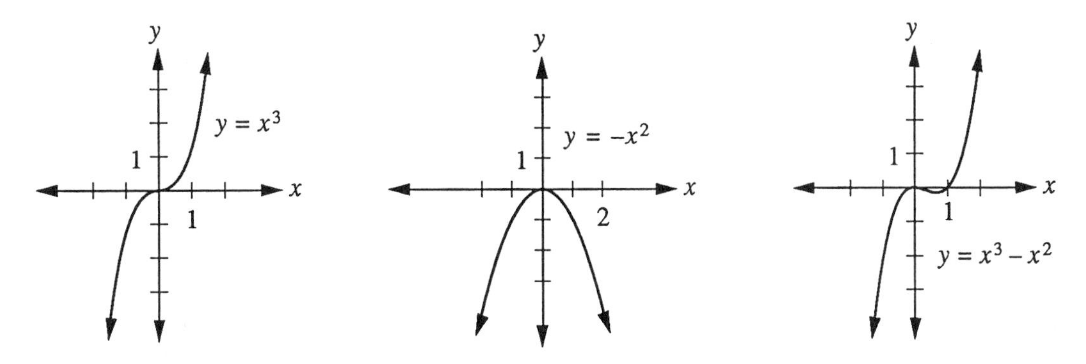

Figure 4 Graphs of $g(x) = x^3$, $h(x) = -x^2$, and $f(x) = x^3 - x^2$

and $h(x) = -x^2$ are shown in Figure 4. The zeros of $f(x)$ will occur when $g(x) + h(x) = 0$. At $x = 0$ both $g(x)$ and $h(x)$ equal 0; therefore the sum of the vertical distances is zero. When $x = 1$, the vertical distance on $g(x) = x^3$ is 1 and on $h(x) = -x^2$ is -1; the sum of the vertical distances is 0. Therefore, $f(0) = 0$ and $f(1) = 0$.

The sum of the ordinates is negative for all x-values less than zero. The graphs in Figure 4 show that the sum of the ordinates is also negative when x is between 0 and 1; this is because the positive vertical distance of $g(x) = x^3$ is smaller than the negative vertical distance of $-x^2$. When $x > 1$, the positive vertical distance from $g(x) = x^3$ more than offsets the negative vertical distance from $h(x) = -x^2$, so the graph of $f(x) = x^3 - x^2$ is positive. Also notice that for large values of x, the x^3-term dominates the expression for $f(x)$. Thus, as $x \to \infty$, $f(x) \to \infty$. The complete graph of $f(x)$ is shown on the right in Figure 4. ■

The task of graphing polynomials of degree three or higher is often quite challenging. Sometimes such polynomials can be factored, sometimes they can be expressed as transformations of simpler functions, and sometimes they lend themselves to the use of addition of ordinates. Often, however, they are best left to a computer or graphing calculator. It is worthwhile to know a little about the general shape of these graphs so that you can judge whether computer- and calculator-generated graphs are reasonable.

Class Practice

Graph the following functions with a calculator or computer over the given x-interval and again over the x-interval $[-50, 50]$. You may need to experiment to find the appropriate y-interval. Notice the global and local behavior of the function, the number and location of the zeros, and points where the function turns.

1. $f(x) = x^4 - 3x^2 + 2x$ on $[-2.5, 2.5]$
2. $f(x) = x^5 - 6x^4 + 8x^3 + 6x^2 - 8x + 1$ on $[-1.5, 3.5]$
3. $f(x) = -x^3 + 4x^2 - x - 6$ on $[-2, 4]$
4. $f(x) = x^4 - 3x^2 + 2x + 5$ on $[-2, 2]$
5. $f(x) = x^6 - 3x^3 + x^2 - 7$ on $[-2, 2]$

Zeros and Factors of Polynomials

When you graph a polynomial or use a polynomial as a model, it is often important to know the location of its zeros. There are two situations in which zeros can be found by using techniques of algebra. First, if a polynomial can be written in the form $y = a(x-h)^n + k$ (that is, as a transformation of a power function), then you can solve the equation

$$0 = a(x - h)^n + k$$

to determine zeros. Second, if a polynomial can be

written as a product of linear factors, you can solve the equation

$$0 = a(x - r_1)(x - r_2)\cdots(x - r_n)$$

to determine zeros.

Example 2

Find the zeros of the function $f(x) = 2x^3 - 6x^2 + 6x - 18 = 2(x-1)^3 - 16$.

Solution We need to find the values of x for which $f(x) = 0$.

$$\begin{aligned} 0 &= 2(x-1)^3 - 16 \\ 16 &= 2(x-1)^3 \\ 8 &= (x-1)^3 \\ 2 &= (x-1) \\ 3 &= x \end{aligned}$$

Thus $x = 3$ is a zero of this function. Thinking of $f(x)$ as a transformation of the toolkit function $y = x^3$ should make it clear why this function has exactly one zero. ■

Example 3

Find the zeros of the function $g(x) = x^3 + 4x^2 - 11x + 6 = (x+6)(x-1)(x-1)$.

Solution We need to solve the equation

$$0 = (x+6)(x-1)(x-1).$$

The product will be zero whenever a factor is zero; this occurs when x is either -6 or 1. Note that each factor of $g(x)$ corresponds to a zero of the function. Because the factor $(x-1)$ occurs twice, the zero $x = 1$ is said to have *multiplicity* two. The factor $(x+6)$ occurs only once, and the zero $x = -6$ has multiplicity one. ■

Example 3 suggests an important relationship between the factors and the zeros of a polynomial. This relationship, sometimes called the *Factor Theorem*, states that a polynomial $f(x)$ has a factor $(x-a)$ if and only if $f(a) = 0$. A polynomial of degree n has at most n distinct factors of the form $(x-a)$; therefore, it has at most n distinct zeros.

We know that if $f(a) = 0$ then $(x-a)$ is a factor of $f(x)$. What exactly does it mean to say that $(x-a)$ is a factor of $f(x)$? It means that $f(x)$ can be evenly divided by $(x-a)$. The result of the division is a quotient $q(x)$ and a remainder zero, so we can write $f(x) = q(x)\cdot(x-a)$. Similarly, if $(x-a)$ is not a factor of $f(x)$, then the result of dividing by $(x-a)$ is a quotient $q(x)$ and a nonzero remainder r, so we can write $f(x) = q(x)\cdot(x-a)+r$. This last equation gives us important information about what happens when $(x-a)$ is not a factor of $f(x)$. We can use the equation

$$f(x) = q(x)\cdot(x-a) + r$$

to evaluate $f(a)$:

$$f(a) = q(a)\cdot(a-a) + r.$$

It does not matter that we do not know the value of $q(a)$ because it is multiplied by $(a-a)$, which is zero. Thus, $f(a) = r$, which means that $f(a)$ is equal to the remainder when $f(x)$ is divided by $(x-a)$. This fact is often referred to as the *Remainder Theorem.* Notice that if $f(a)$ happens to be zero, then $f(x)$ divided by $(x-a)$ results in a zero remainder.

Example 4

Find the zeros and the factors of $j(x) = x^3 - 7x^2 + 13x - 3$.

Solution Since $j(x)$ is a cubic polynomial, $j(x)$ has at most three linear factors. The coefficient of the leading term of $j(x)$ is 1 and the constant term is -3, so four possible candidates for those factors are $(x-1)$, $(x+1)$, $(x-3)$, and $(x+3)$. The Factor Theorem says that if $(x-a)$ is a factor of a function $f(x)$ then $f(a) = 0$. Using this idea, we can evaluate $j(1)$, $j(-1)$, $j(3)$, and $j(-3)$ to see which if any equals 0. In fact, $j(3) = 0$, so we know that $(x-3)$ is a factor of $j(x)$.

Using long division, we find that $j(x) = (x-3)\cdot(x^2 - 4x + 1)$. Applying the quadratic formula to $x^2 - 4x + 1 = 0$ yields the zeros $\frac{4+\sqrt{12}}{2}$ and $\frac{4-\sqrt{12}}{2}$. Therefore, the zeros of $j(x)$ are 3, $2 + \sqrt{3}$, and $2 - \sqrt{3}$, and $j(x)$ can be factored as

$$j(x) = (x - 3) \times (x - (2 + \sqrt{3})) \times (x - (2 - \sqrt{3})).$$ ■

Example 5

Find the zeros and the factors of $m(x) = x^3 - 7x^2 + 10x - 3$.

Solution Notice that $m(x)$ is a cubic polynomial in which only one term is different from $j(x)$ in Example 4. Again, four possible candidates for factors are $(x - 1)$, $(x + 1)$, $(x - 3)$, and $(x + 3)$, but evaluating $m(1), m(-1), m(3)$, and $m(-3)$ shows that none of these equals zero. If we use a computer or calculator to graph $m(x)$ over the interval $[-1, 6]$, we see that m does indeed have three real zeros. We can use the trace and zoom features to estimate that these zeros are located at approximately 0.412, 1.406, and 5.182. Since we do not have exact values for the zeros of m, we must write

$$m(x) \approx (x - 0.412) \times (x - 1.406) \times (x - 5.182)$$

to communicate that the factors are also inexact. ■

Example 6

Use a computer or graphing calculator to graph $f(x) = x(x - 1)(2x - 3)^2(2x + 1)^3$.

Solution A graph of $f(x)$ is shown in Figure 5; since $f(x)$ has degree seven, its graph has the characteristic shape of a polynomial of odd degree. The graph has four distinct zeros. At the zeros of odd multiplicity the function changes sign (from positive to negative or vice versa) and the graph crosses the x-axis. At the zero of even multiplicity the function does not change sign, so the graph touches the x-axis but does not cross it. This zero is a turning point on the graph. ■

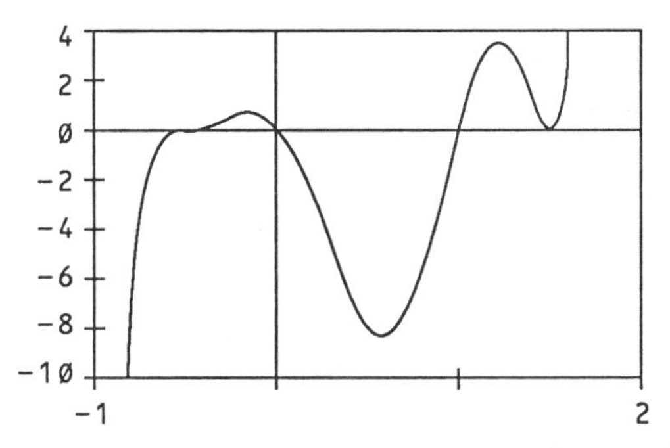

Figure 5 Graph of $f(x) = x(x-1)(2x-3)^2(2x+1)^3$

Example 7

How many times does the graph of

$$g(x) = x(x - 1)(2x - 3)^2(2x + 1)^3 + 10$$

cross the x-axis?

Solution The graph of $g(x)$ has the same shape as the graph of $f(x)$ in the preceding example, but the graph has been vertically shifted up 10 units. Since the polynomial is of degree seven, you might expect seven factors and therefore seven zeros. However, the vertical shift moves the curve away from the x-axis and changes the zeros of the function. A computer- or calculator-generated graph will confirm that the graph of $g(x)$ has only one real zero. Some of the zeros of this polynomial are not in the real number system, but are complex numbers. Within our study of polynomial functions we will work with only the real zeros of polynomials, that is, the zeros visible on the x-axis in the real number plane. Whenever we discuss a zero of a polynomial, we are referring to an x-intercept of the graph. ■

Example 8

If $f(x) = 2x^4 + x^3 - 5x^2 + 2x - 11$, evaluate $f(3.217)$.

Solution This example clearly calls for the use of a calculator. To evaluate $f(3.217)$, we can raise 3.217 to the fourth power, multiply the result by 2, raise 3.217 to the third power, add this to the previous result, and so forth. Fortunately, there is a way to organize these calculations more efficiently; this alternative is based on writing the equation for f in the *nested form* shown below. In each subsequent equation x has been factored out of as many terms as possible.

$$\begin{aligned} f(x) &= 2x^4 + x^3 - 5x^2 + 2x - 11 \\ &= x(2x^3 + x^2 - 5x + 2) - 11 \\ &= x(x(2x^2 + x - 5) + 2) - 11 \\ &= x(x(x(2x + 1) - 5) + 2) - 11 \end{aligned}$$

To evaluate $f(3.217)$ we start from the innermost set of parentheses and work outward, substituting 3.217 for each x: 3.217 times 2, plus 1, times 3.217, minus 5, times 3.217, plus 2, times 3.217, minus 11. We find that $f(3.217) \approx 191.19$. This nested form allows you to evaluate $f(3.217)$ with a very straightforward sequence of calculator key strokes involving only multiplication, addition, and subtraction. On some calculators, equals must be used after addition or subtraction. By storing the value of x, in this case $x = 3.217$, in the memory of your calculator, the number of keystrokes is reduced. ■

Exercise Set 1

1. Use addition of ordinates to sketch the graph of $f(x) = x^4 + x^3$. What are the zeros of the function?

2. Sketch the graphs of the following polynomial functions. Show all intercepts.

   a. $f(x) = x^3 - 2$

   b. $f(x) = (x - 2)^2(x + 2)$

   c. $f(x) = (x - 1)^3$

   d. $f(x) = (x - 1)^3(x + 2)^2$

   e. $f(x) = (x - 2)^3 + 2$

   f. $f(x) = x^4 + 1$

   g. $f(x) = (x + 2)^4$

   h. $f(x) = (x + 2)^4(x - 3)^3$

   i. $f(x) = 2(x - 1)^4 - 3$

3. Write a statement to describe the general shape of a 17th degree polynomial.

4. Write a statement to describe the general shape of a 44th degree polynomial.

5. Describe the difference between the range of an nth degree polynomial when n is even and the range when n is odd.

6. The function $p(v) = kAv(V - v)^2$ expresses the relationship between the power generated by an undershot waterwheel and the velocity at which the wheel is turning. In the formula, k is a constant of proportionality, A is the area of the waterwheel in contact with the water, V is the velocity of the water, and v is the velocity of the wheel. For a particular waterwheel, k and A will be constants. Let V equal 4, 8, 10, and 20 feet per second. In each case, what is a reasonable domain of the function and what are the zeros?

7. For each of the following polynomial functions, show that the given x-values are zeros of the function. Write the polynomial in factored form.

   a. $f(x) = x^3 - 6x^2 - x + 30$, $x = -2, 3, 5$

   b. $f(x) = x^4 - 2x^3 - 13x^2 + 14x + 24$, $x = -3, -1, 2, 4$

   c. $f(x) = x^3 - 2.8x^2 - 5.39x + 10.29$, $x = -2.1, 1.4, 3.5$

   d. $f(x) = 4x^3 - 28x^2 + 57x - 36$, $x = 4, \frac{3}{2}$

8. Evaluate each polynomial using its nested form.

   a. $f(4.438)$, where $f(x) = x^3 - 4x^2 - 6x + 18$

   b. $f(-3.28)$, where $f(x) = 2x^3 + 3x^2 - 10x + 3$

   c. $f(1.114)$, where $f(x) = x^4 - 3x^2 - 7x - 13$

2

Writing the Expression for a Polynomial

If we know the coordinates of some points, we can find an equation of a polynomial passing through these points. However, without additional information, it is usually not possible to find a unique polynomial for a particular set of points. For example, the zeros of a polynomial do not uniquely determine that polynomial. By considering different vertical stretches and compressions we can find an infinite number of polynomials that pass through the same zeros. In this section we will explore some methods for finding expressions for polynomials.

Example 1

Find an expression for a polynomial with zeros at -2 and 5. (Assume that the polynomial has no other zeros.)

Solution We know that each zero is associated with a factor of the polynomial, so this polynomial has factors $(x+2)$ and $(x-5)$. Therefore, the polynomial could be $f(x) = (x+2)(x-5)$. Is this the only polynomial that will have these zeros? What about $f(x) = 3(x+2)(x-5)$? We can generate a whole family of functions $f(x) = k(x+2)(x-5)$ that are all different but all have zeros at -2 and 5. Several such functions are graphed in Figure 6.

We also need to know the multiplicity of each zero. Many different polynomials of at least second degree can be written that have zeros at $x = -2$ and $x = 5$. If the zero $x = -2$ has multiplicity two and $x = 5$ has multiplicity one, then the function will have the form $f(x) = k(x+2)^2(x-5)$. Some of these polynomials are shown in Figure 7. ∎

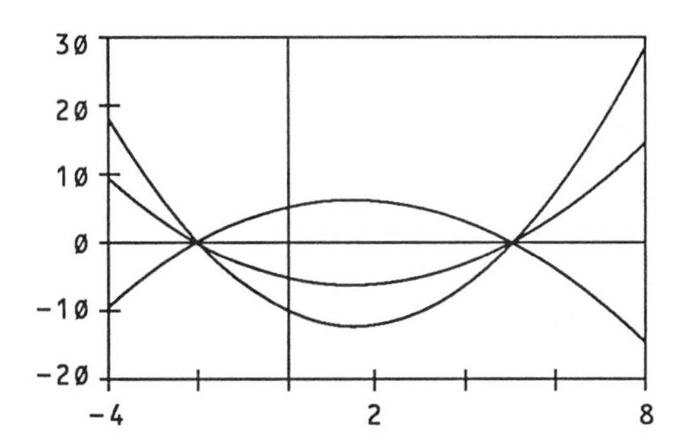

Figure 6 The Family of Curves
$f(x) = k(x+2)(x-5)$

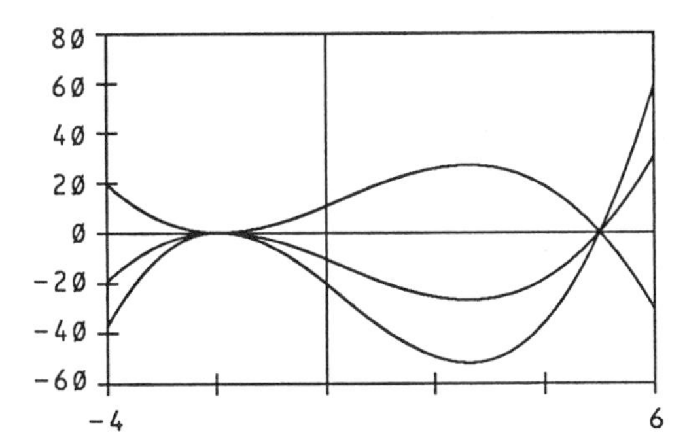

Figure 7 The Family of Curves
$f(x) = k(x+2)^2(x-5)$

Example 2

Write an expression for the polynomial function of least degree that passes through $(-2, 0)$, $(5, 0)$, and $(8, 60)$.

Solution The expression for this polynomial can be uniquely determined since we know one point that is not a zero and we are asked for the polynomial of least degree. We need a formula of the form $f(x) = k(x+2)(x-5)$ that passes through $(8, 60)$. Now $f(8) = 60$, so $60 = k(8+2)(8-5)$ or

$60 = 30k$ and $k = 2$. Thus $f(x) = 2(x + 2)(x - 5)$ is the unique polynomial of degree two that passes through the two given zeros and the third point. ■

Example 3

Write an expression for the polynomial function of least degree that passes through $(-2, 5)$, $(0, 4)$, $(2, 7)$, and $(3, 5)$.

Solution Since none of the four given points is a zero, we don't know any of the factors of the polynomial. What degree polynomial will go through these four points? We need to recognize that if we are trying to fit a polynomial of least degree to n points, the degree of the polynomial will be at most $n - 1$. For instance, two points determine a line $(n = 2)$ and three noncollinear points determine a unique parabola $(n = 3)$. The scatterplot in Figure 8 shows that the four points are not collinear and don't lie on a parabola. Thus, a cubic is the polynomial of least degree that will pass through the points.

A polynomial function of degree three has the form $y = ax^3 + bx^2 + cx + d$. Four ordered pairs (x, y) are required to determine the four unknown coefficients a, b, c, and d. By substituting each of the four values for x and y into the equation $y = ax^3 + bx^2 + cx + d$, we find the four equations:

$$\begin{aligned} 5 &= -8a + 4b - 2c + d \\ 4 &= \phantom{-8a + 4b - 2c + {}} d \\ 7 &= 8a + 4b + 2c + d \\ 5 &= 27a + 9b + 3c + d \end{aligned}$$

The solution to this system of equations yields $a = -\frac{1}{3}$, $b = \frac{1}{2}$, $c = \frac{11}{6}$, and $d = 4$. Therefore, a polynomial function which passes through all these points is $f(x) = -\frac{1}{3}x^3 + \frac{1}{2}x^2 + \frac{11}{6}x + 4$. The graph in Figure 8 shows $f(x)$ with the four given points. ■

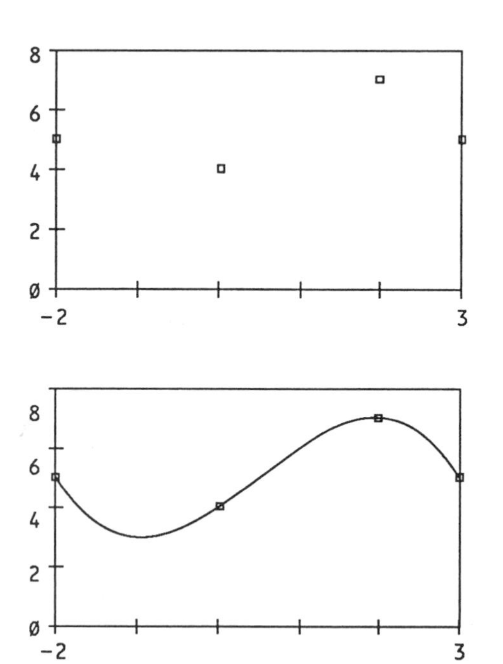

Figure 8 Points $(-2, 5), (0, 4), (2, 7)$, and $(3, 5)$ and Graph of $f(x) = -\frac{1}{3}x^3 + \frac{1}{2}x^2 + \frac{11}{6}x + 4$

Suppose we are conducting an experiment and we gather the following data: $(0, 0)$, $(1, 0.4)$, $(2, 1.1)$, $(3, 2.7)$, $(4, 4.7)$, and $(5, 7.6)$. The polynomial that fits all six of these points is $f(x) = 0.0175x^5 - 0.2208x^4 + 0.9875x^3 - 1.52917x^2 + 1.145x$. However, when we plot these points we see that they lie very near a parabola (see Figure 9). The question that arises is whether we want a polynomial of degree five that passes exactly through all of the points or whether a polynomial of lower degree (or some other less complicated curve) that

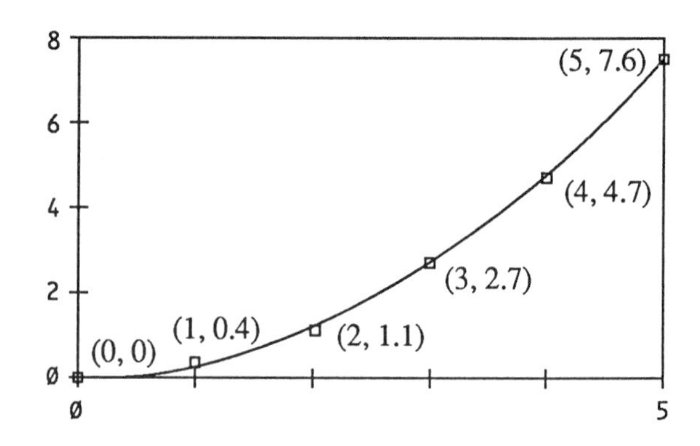

Figure 9 Graph of $g(x) = 0.3x^2$ and $(0, 0)$, $(1, 0.4)$, $(2, 1.1)$, $(3, 2.7)$, $(4, 4.7)$, and $(5, 7.6)$

passes near the points might be a better model for the experimental data.

In fact, the function $g(x) = 0.3x^2$ passes through two of the points and is very close to the others as shown in Figure 9. When you consider the error that was probably involved in gathering the data, it doesn't make much sense to make the mathematics precise when the data is not. For most practical purposes, we would much rather use the simpler equation. If we find later that our data were very accurate, and we need the more precise, higher degree polynomial, we can find it. In a later chapter we will look more closely at techniques for fitting curves to data so that we have this choice.

Exercise Set 2

1. Explain why a polynomial of degree 3 is the polynomial of least degree that goes through exactly three distinct zeros.
2. Given the zeros -1, 3, and 0 of a polynomial, find four different polynomials that pass through these zeros. Do not simplify the equations. Sketch a graph of each.
3. Find one expression for a polynomial function of least degree that passes through the points (2, 0), (-3, 0), (1, 0), (0,-24), and (-4, 0).
4. Find two fourth degree polynomial functions that pass through the points (-4, 0), (-1, 0), (0, 5), and (5, 0). Sketch the graph of each.

3
Turning Points

In an earlier chapter we located turning points in order to solve problems that asked for a maximum or minimum value. Several techniques were used to find the coordinates of these points. Tables of values or the trace feature of your calculator or graphing software can be used to estimate the coordinates of the turning points of a function.

Example 1

For the polynomial function $g(x) = x^4 - 3x^3 - 8x^2 + 12x + 16$, find the coordinates of the turning points.

Solution Figure 10 shows the graph of the function $g(x) = x^4 - 3x^3 - 8x^2 + 12x + 16$; it shows three turning points. By making a list of values of $g(x)$ or by tracing the graph, we find a minimum point located at approximately $(-1.557, -4.877)$, a maximum point at approximately $(0.601, 19.802)$, and another minimum point at approximately $(3.206, -20.967)$. The point $(0.601, 19.802)$ is not an *absolute maximum* since it does not contain the largest y-value of the function. But 19.802 is the greatest y-value in a region around the point $(0.601, 19.802)$; this point is called a *relative maximum* because its y-value is a maximum relative to the points near it on the graph. Based on the graph, we can conclude that the point $(-1.557, -4.877)$ is a *relative minimum.* The point $(3.206, -20.967)$ is called an *absolute minimum,* since it has the smallest y-value of the function over its entire domain. ■

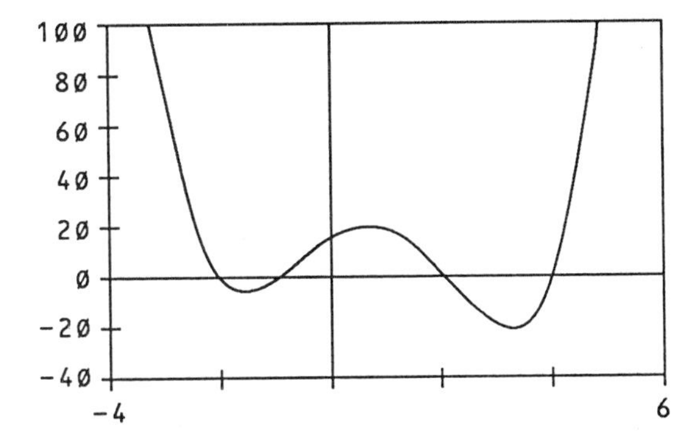

Figure 10 Graph of $g(x) = x^4 - 3x^3 - 8x^2 + 12x + 16$

Example 2

A reinforced box is made by cutting congruent squares from the four corners of a rectangular piece of cardboard that measures 24 inches by 48 inches. The flaps are folded up as shown in Figure 11. What are the dimensions of the box with the greatest volume?

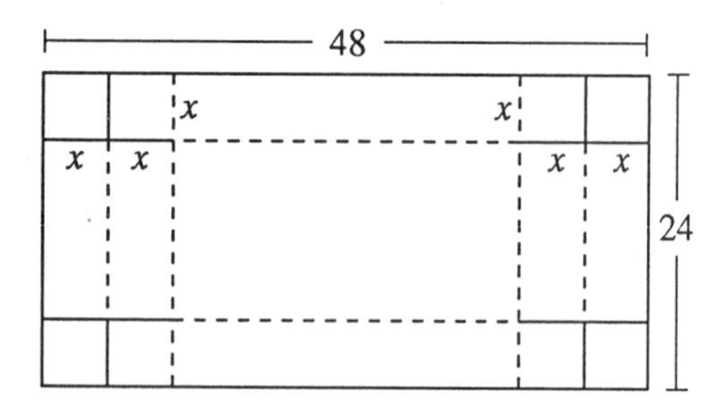

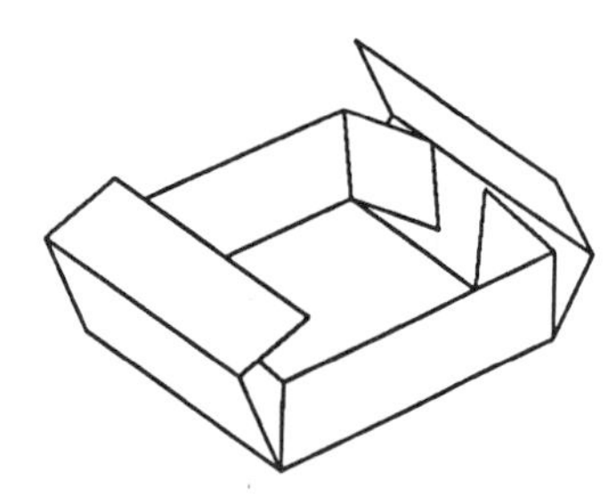

Figure 11 Reinforced Rectangular Box

Solution We need to develop a function that expresses the volume in terms of the length of the side of the squares cut from each corner. If squares x inches on each side are cut from each corner, then the dimensions of the box are x, $48-4x$, and $24-2x$. The volume is thus given by the function

$$V(x) = x(48-4x)(24-2x).$$

When this function is used to model the volume of the box, its domain must be restricted to x-values between 0 and 12. A graph of $V(x)$ over this domain is shown in Figure 12; notice that the volume increases, reaches a maximum, and then decreases. If we make a table of values for $V(x)$ or trace a computer or calculator graph, we see that $V(x)$ reaches a maximum when $x \approx 4$. Thus, the dimensions of the box with the greatest volume are approximately 32 inches, 16 inches, and 4 inches. The maximum volume is about 2048 cubic inches. ■

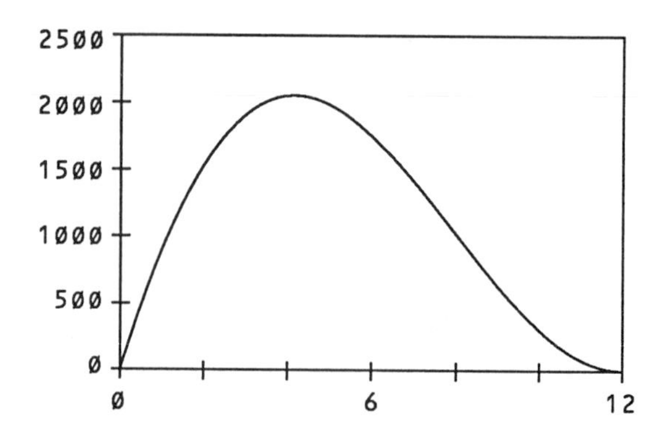

Figure 12 Graph of $V(x) = x(48-4x)(24-2x)$

Example 3

The owner of a small business estimates that the profit from producing x items is given by the function

$$P(x) = 0.003x^3 - 1.5x^2 + 200x - 1000.$$

This function is based on current production levels that cannot exceed 350 items due to limited space and resources. What production plan should the owner follow to maximize profit?

Solution A graph of the profit function $P(x)$ is shown in Figure 13. A relative maximum and a relative minimum occur between 0 and 350, and the values of $P(x)$ increase after the relative minimum. The greatest P-value in the interval $[0, 350]$ occurs at the endpoint of the interval when $x = 350$ and $P = \$13,875$. The graph shows that the owner reaches a relative maximum profit of approximately \$7040 at $x = 92$. When production exceeds 92 items, profit drops, but then climbs again.

The owner's best strategy is to produce 350 items. Although the profit function appears to increase without bound, the function was derived based on production of no more than 350 items, so it may not be accurate beyond that production level. ■

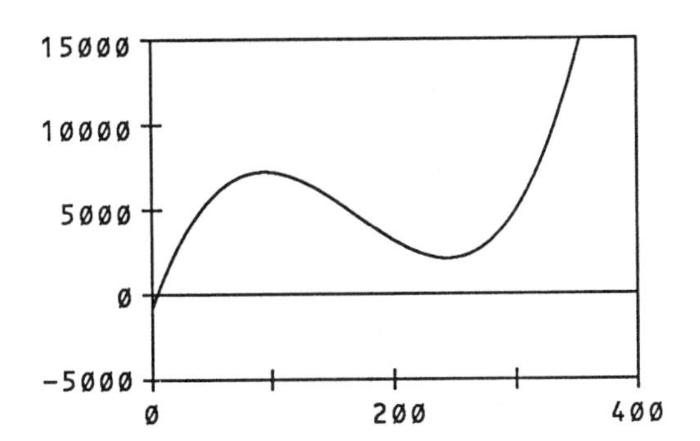

Figure 13 Graph of $P(x) = 0.003x^3 - 1.5x^2 + 200x - 1000$

Exercise Set 3

1. A box with a lid can be constructed as shown in Figure 14. Write a function for the volume of such a box if the original piece of cardboard measures 24 inches by 60 inches. Find the dimensions of the box with the greatest volume that can be constructed from this piece of cardboard.

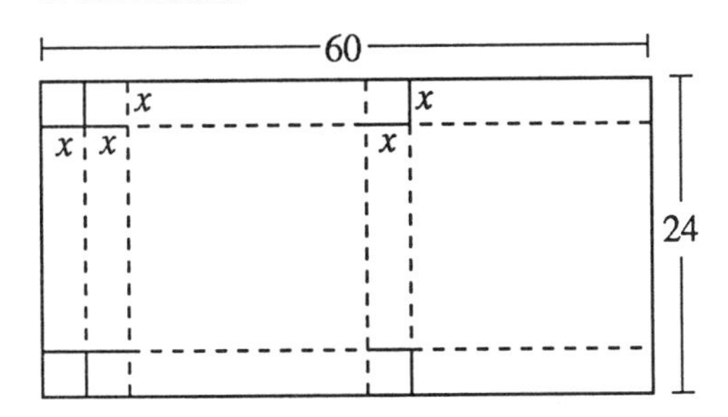

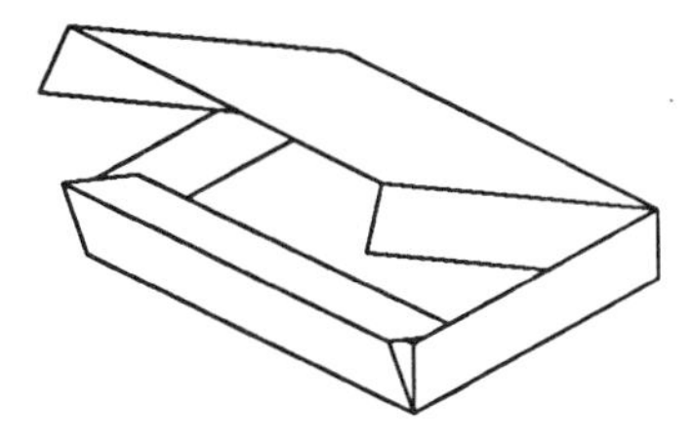

Figure 14 Box with Lid

2. A particle is traveling at a velocity of $v(t) = t^3 - 3t^2 + 5t$, where t represents time in seconds. During the first 10 seconds, when does the particle reach a maximum velocity? When does the velocity reach a minimum?

3. A charter airline company charges \$300 per roundtrip ticket to London if it can sell tickets for all 95 seats on the plane. For each empty seat, the price of a ticket increases by \$3.50. Write a function to show how the company's income depends on the number of empty seats. What is the maximum possible income?

4. The power p generated by a particular waterwheel is given by $p = 0.5v(V - v)^2$ where V represents the velocity of the stream and v the velocity of the waterwheel. Let V equal 4, 8, 10, and 20 feet per second and find the waterwheel velocity at which the power is a maximum for each value of V.

5. At what points on the graph of $f(x) = x^3 - 9x - 5$ are there relative maximum and minimum points?

6. The senior class at Ickum High School is planning to raffle off a one week vacation at the beach as a way of making money for class activities. A member of the Parent's Club is willing to donate a condominium for the week. The seniors want to know at what price they should sell the tickets for the raffle. The higher the price, the fewer tickets they will sell, but the more they will make on each ticket. A survey was sent to all 1100 parents at the school asking what price they would be willing to pay for a ticket. The results are given in Table 15. These data mean that if the tickets sell for \$7.50, the senior class can expect to sell around 390, but if they sell for \$15.00, they can expect to sell close to 150 tickets. The following questions are concerned with a linear model of the relationship between the price of tickets and the number sold.

   a. Which variable in the model is the independent variable? Which variable is the dependent variable?

Table 15 Raffle Data

Price	Number
\$2.50	508
5.00	420
7.50	389
10.00	293
12.50	245
15.00	152
17.50	90
20.00	63

b. Find an equation for the median-median line that fits the data.

c. What does the slope of the median-median line represent?

d. What does your model suggest is the lowest price at which no one will buy a ticket?

e. If the tickets are priced at \$8.25 each, how many tickets does your model predict that the senior class will sell? How much money will they make?

f. Which will make more money from the raffle, selling tickets at \$16.00 or at \$6.00?

g. The profit for the raffle is the product of the number of tickets sold and the price at which the tickets are sold. If the median-median line for the data is represented by a function $f(x)$, and x is the price of a ticket, then the profit function P is represented by $P(x) = xf(x)$. Sketch a graph of P, and find the price of a ticket that gives the senior class its maximum profit.

4

Finding Zeros of Polynomials Using a Calculator or Computer

When we search for zeros of a polynomial function by examining a graph or a table of values, we use the same techniques that were discussed in the preceding chapter. A zero is located by finding where the values of a polynomial change sign.

Example 1

Suppose an object is moving so that its velocity in meters per second at time t seconds is given by

$$v(t) = t^3 - t^2 - 32t + 40, \qquad t \geq 0.$$

What is the object's initial velocity? To the nearest thousandth of a second, find when the object will come to rest.

Solution Since we begin measuring the elapsed time at $t = 0$, the initial velocity is given by $v(0) = 40$. The object has a velocity of 40 meters per second at 0 seconds.

The object will be at rest when $v(t)$ is equal to 0, so we need to find the zeros of the function $v(t)$, where $t \geq 0$. This polynomial, like most that arise as models for realistic problems, is difficult to factor and therefore doesn't lend itself to algebraic techniques for finding zeros. We have other tools that are much more efficient to use in these situations; primary among these are the computer and the calculator. The principal concept used to locate zeros is that the number zero separates the positive numbers from the negative numbers. To find a zero of a polynomial function, we simply look for values of the independent variable near which the values of the function change sign. A computer-generated graph of the velocity function is shown in Figure 16.

The graph shows that two zeros of $v(t)$ exist for $t \geq 0$. We can look at the graph of $v(t)$ over

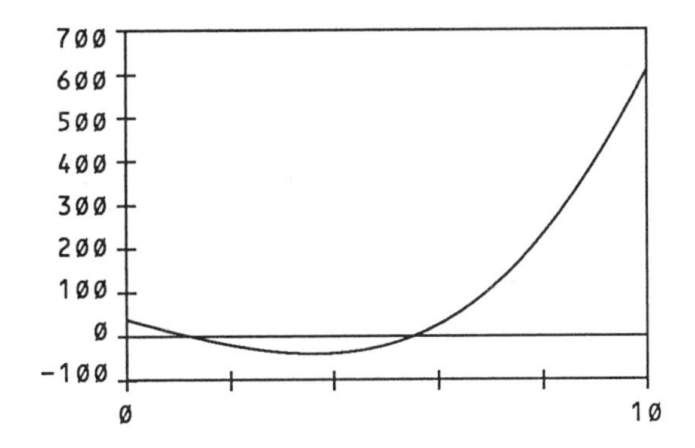

Figure 16 Graph of $v(t) = t^3 - t^2 - 32t + 40, t \geq 0$

Table 17 Values of $v(t) = t^3 - t^2 - 32t + 40$

t	$v(t)$	t	$v(t)$
1.260	0.0928	1.2630	0.00353
1.261	0.0630	1.2631	0.000555
1.262	0.0333	1.2632	−0.00242
1.263	0.00353	1.2633	−0.00539
1.264	−0.0262	1.2634	−0.00837
1.265	−0.0559	1.2635	−0.0113
1.266	−0.0857	1.2636	−0.0143
1.267	−0.115	1.2637	−0.0173
1.268	−0.145	1.2638	−0.0203
1.269	−0.175	1.2639	−0.0232
1.270	−0.205	1.2640	−0.0262

a smaller interval to see that one zero is in the interval $[1, 2]$ and the other is in $[5, 6]$. To find an approximation for the zero in the interval $[1, 2]$, we continue to find smaller intervals that contain the zero. A spreadsheet or software that generates tables of values makes this a simple task. Table 17 displays two lists of function values. We find that a zero is $t \approx 1.263$. Using the same technique we can estimate that $t \approx 5.497$ is also a zero. The object will be at rest approximately 1.263 seconds and 5.497 seconds after we begin to measure the velocity. ∎

Using a table of values to locate zeros can be slow and somewhat cumbersome. There are several other methods for locating zeros that are faster and more elegant. We will introduce one of these methods now. In a later section, we will study this technique in more detail and add another method for finding zeros.

The Bisection Method for Locating Zeros

The *bisection method* for locating zeros is closely related to the process we use to locate zeros with a table of values. A continuous function has a zero in any x-interval in which the y-values change from positive to negative (or vice versa). The search for a zero is narrowed by selecting x-intervals that are smaller and smaller but still contain y-values that change sign in the interval. This idea is formalized with the bisection method as shown in the following example.

Example 2

Find one zero of $f(x) = x^2 - 2$.

Solution We select a simple function so that we can focus on the steps involved in the bisection method and not be distracted by difficult computations.

Since f is continuous with $f(1) = -1$ and $f(2) = 2$, a zero of the function is located in the interval $[1, 2]$, or $1 \leq x \leq 2$. A reasonable guess is that the zero is at the midpoint of the interval, which is $x = 1.5$. Since $f(1.5) = 0.25$, we have not found a zero. However, since $f(1)$ is negative and $f(1.5)$ is positive, we now know that a zero is contained in the interval $[1, 1.5]$. The size of the interval known to contain the zero has been bisected, or cut in half. A check of the midpoint of this new interval produces $f(1.25) = -0.4375$. Since the new y-value is negative, there must be a zero in the interval $[1.25, 1.5]$. Again, the size of the interval known to contain a zero has been bisected. A check of the new midpoint shows

$f(1.375) = -0.109375$. Now we know that a zero is located in the interval $[1.375, 1.5]$. We test the midpoint of the interval $[1.375, 1.5]$ and find that $f(1.4375)$ is positive. Therefore, we know the zero is located in the interval $[1.375, 1.4375]$, so the zero is approximately 1.4. We can continue bisecting until the interval known to contain a zero becomes small enough to provide us with any desired degree of accuracy. ∎

Exercise Set 4

1. For each function listed, use the bisection method to locate at least one zero. Find the zero to the nearest tenth.

a. $f(x) = x^2 - 3x - 10$

b. $f(x) = x^3 - 2x + 7$

c. $f(x) = x^4 - 3x^3 + 1$

2. The bisection method can be used to find some of the zeros of a polynomial, but there may be zeros that cannot be located using bisection. Explain and give an example to illustrate this statement.

3. The graph of $f(x) = 49x^3 + 28x^2 - 115x + 50$ shows a zero for $x < 0$ and an apparent point of tangency to the x-axis near $x = 0.71$. Explain how you would use a computer or calculator to locate the zero near $x = 0.71$.

4. Other types of functions may have $f(a) < 0$ and $f(b) > 0$ without having $f(x) = 0$ for some x-value in the interval $[a, b]$. Explain how this could happen. You may use an example to illustrate.

5
Introduction to Algorithms

In the preceding section we described the bisection method for finding zeros. Similar procedures based on halving a search interval can be used to solve other problems. Suppose, for example, you challenge a fellow student to pick a word in a dictionary and you will guess that word by asking fewer than twenty questions. Seems impossible? If you use the dictionary as a number line is used in guessing zeros of a function, you can keep asking your fellow student whether the word falls in smaller and smaller intervals of the dictionary. By getting the answer to the question, "Does the word come before *lighthouse* in the dictionary?", you have essentially narrowed your search to approximately half of the words in the dictionary. Below we provide clear, precise, and unambiguous instructions for finding a word in a dictionary. Notice that this process parallels in many ways the bisection method for finding zeros.

1. Label the first word in the dictionary FIRST. Label the last word in the dictionary LAST.
2. Open the dictionary to the middle and pick a word, which we will label NEWWORD. If NEWWORD is the word you are looking for, then stop.
3. Determine if the word you are looking for comes before or after NEWWORD in the dictionary. If it comes before NEWWORD, then we need no longer consider the part of the dictionary after NEWWORD, so we relabel the word we have called NEWWORD as LAST. If the word you are looking for comes after NEWWORD, we need no longer consider words before NEWWORD, so we relabel the word we have called NEWWORD as FIRST.
4. Pick a word midway between FIRST and LAST, and label this word NEWWORD. If NEWWORD is the word you are looking for, then stop; otherwise go back to step 3.

The procedure outlined above, like the bisection method for finding zeros, is a *binary search.* Each step in the process cuts in half the number of words still under consideration. For example, in a dictionary with 100,000 words, examining the first new word cuts the number of words still being considered to 50,000. How many words are needed to cut the number of words being considered down to 3,125? In general, after looking at n words, we have cut the number of words being considered down to $\frac{100{,}000}{2^n}$.

How many words might need to be picked in order to complete the binary search process? To determine the maximum number of words needed, we need to know how many times 100,000 can be cut in half. Based on the formula mentioned in the previous paragraph, we can solve the equation

$$1 = \frac{100,000}{2^n},$$

where n represents the number of words chosen. We can simplify this equation to $2^n = 100,000$. How do we solve for n? In the next chapter we will develop algebraic techniques for solving an equation of this type. For now, we can use a calculator and evaluate powers of two to solve this equation using trial and error. We find that $2^{16} = 65,536$ and $2^{17} = 131,072$. This means that at most 17 words need to be examined to find a particular word in a dictionary of 100,000 words.

Definition of Algorithm

The instructions provided above for finding a word in a dictionary comprise an *algorithm*, a precise and unambiguous set of instructions that lead from a starting state to a stopping state. Every algorithm has some standard for determining when a stopping state has been reached.

Often algorithms are described in statements that can be easily converted to computer programming language, since programs also depend on unambiguous statements. According to this definition an algorithm can be decomposed into three parts:

1. starting state,
2. precise instructions for making transitions from one state to another,
3. stopping state and the standard for determining when it is reached.

For our purposes an algorithm must reach a stopping state in a finite amount of time; instructions that cause us to perform some task forever imply that we will never find a solution.

In the sections that follow we will develop two algorithms for finding zeros of a function. The first is a precise description of the bisection process for finding zeros; the second technique is called the secant algorithm. These two algorithms share some basic properties. Each algorithm starts with an initial estimate for a zero of a function. (Often this estimate is based on a graph of the function.) The algorithm then provides a sequence of steps that lead to improvements on the estimate. Each repetition of the steps in the algorithm is called an *iteration* of the algorithm. Ideally, a greater number of iterations leads to a better approximation of the zero. Both algorithms require that we establish a suitable stopping criterion. Since the algorithms are implemented on calculators and computers of limited precision, we cannot simply iterate an arbitrary number of times to approximate a root as closely as we want. Instead, we stop when we attain a specified level of accuracy.

6

Zero-Finding Algorithms

The Bisection Algorithm

You are already familiar with the bisection method for locating zeros of functions. Now we will describe the method more formally as an algorithm. This description will allow you to write your own computer or calculator program to execute the algorithm.

The Bisection Algorithm

---

1. Choose x-values x_1 and x_2 whose function values y_1 and y_2 are opposite in sign. The function must be continuous over $[x_1, x_2]$.
2. Find a new x-value by averaging x_1 and x_2, $x_{new} = \frac{x_1+x_2}{2}$; calculate the corresponding new y-value, y_{new}.

3. If $y_{new} = 0$, then x_{new} is a zero and you can stop the process. If $y_{new} \neq 0$, compare the value of y_{new} with y_1 and y_2. If y_{new} has the same sign as y_1, replace the current (x_1, y_1) with (x_{new}, y_{new}); if y_{new} has the same sign as y_2, replace the current (x_2, y_2) with (x_{new}, y_{new}).
4. Repeat steps 2 and 3 until you obtain the desired degree of accuracy.

Note that steps 2 and 3 are precise instructions that can be easily followed by a person with a calculator. The algorithm is also easy to implement on a computer. It is important to note that each time step 2 is performed, the two x-values being averaged have y-values that are opposite in sign; this was the starting state of the algorithm as well as a property that is maintained by step 3. Each execution of steps 2 and 3 constitutes an iteration. Step 4 is somewhat vague regarding the number of iterations to be performed. We will discuss stopping criteria in more detail in a later section.

We will now use the bisection algorithm to find a zero of $f(x) = x^2 - 2$ correct to the nearest tenth. The list below shows how the bisection algorithm progresses if we start with $x_1 = 1$ and $x_2 = 2$.

$(x_1, y_1) = (1, -1)$ and $(x_2, y_2) = (2, 2)$ yields

$$(x_{new}, y_{new}) = (1.5, 0.25).$$

$(x_1, y_1) = (1, -1)$ and $(x_2, y_2) = (1.5, 0.25)$ yields

$$(x_{new}, y_{new}) = (1.25, -0.4375).$$

$(x_1, y_1) = (1.25, -0.4375)$ and
$(x_2, y_2) = (1.5, 0.25)$ yields

$$(x_{new}, y_{new}) = (1.375, -0.1094).$$

$(x_1, y_1) = (1.375, -0.1094)$ and
$(x_2, y_2) = (1.5, 0.25)$ yields

$$(x_{new}, y_{new}) = (1.4375, 0.0664).$$

$(x_1, y_1) = (1.375, -0.1094)$ and
$(x_2, y_2) = (1.4375, 0.0664)$.

Since any number between $x_1 = 1.375$ and $x_2 = 1.4375$ will round off to 1.4 (correct to the tenths place), we can stop the process and conclude that one zero is $x \approx 1.4$.

Why does the bisection method work? The interval known to contain a zero is made smaller and smaller until it is small enough to satisfy the degree of accuracy specified by the stopping criterion. The method requires that we know an interval in which $f(x)$ changes sign and that $f(x)$ be continuous on that interval. It is the first of these requirements that sometimes proves challenging — before we begin this method we must know an x_1 and an x_2 whose y-values have opposite signs.

The Secant Algorithm

A second algorithm for locating zeros is called the *secant algorithm.* This algorithm does not require that we start with x-values whose y-values differ in sign. We can choose any x-values that do not have the same y-value. The secant algorithm works on the principle that a continuous function can be approximated, at least over a small interval, by a linear function. The x-intercept of the linear function is assumed to be very close to the x-intercept of the function it approximates. For example, we know that the function $f(x) = x^2 - 2$ passes through the points (2,2) and (3,7). Figure 18 shows the graph of $f(x) = x^2 - 2$ and a *secant line* through these points. The x-intercept of the secant line approximates the x-intercept, or zero, of $f(x)$.

Using the x-value associated with this intercept to locate another point on $f(x)$, we can draw another secant line and use its x-intercept as the next approximation for a zero of $f(x)$.

The secant algorithm is described below.

The Secant Algorithm

---

1. Choose x-values x_1 and x_2. The function must be continuous over $[x_1, x_2]$.
2. Calculate the function values y_1 and y_2 associated with x_1 and x_2. If $y_1 = y_2$, repeat step 1 with new choices for x_1 and x_2.

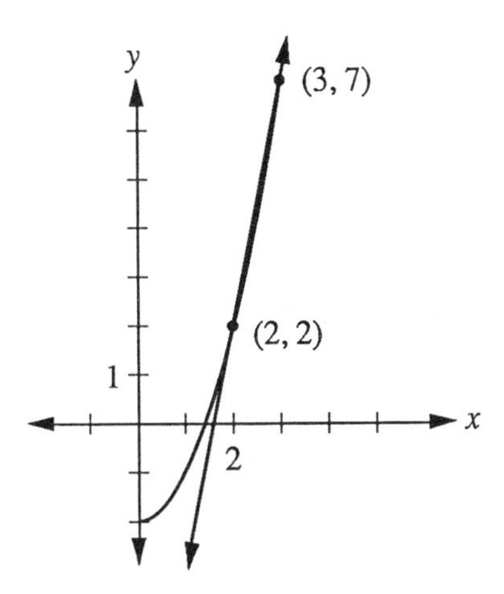

Figure 18 Graph of $f(x) = x^2 - 2$ with Secant Line

3. If $y_1 = 0$ or $y_2 = 0$, then the corresponding x-value is a zero and you can stop the process. If $y_1 \neq 0$ and $y_2 \neq 0$, find the x-intercept of the line through (x_1, y_1) and (x_2, y_2). Call the x-intercept x_{new}.
4. Replace the current x_1 with x_2 and replace the current x_2 with x_{new}.
5. Repeat steps 2, 3, and 4 until you obtain the desired degree of accuracy.

Each x-intercept calculated in step 3 is an approximation of the zero of the function, and each execution of steps 2, 3, and 4 constitutes an iteration of the algorithm. Often, but not always, by the second or third iteration the x-intercept of the secant line is very close to the zero of the function.

Class Practice

1. The graph of $f(x)$ is provided in Figure 19. Using the two points marked as the initial points, sketch the secant lines and show the zero approximations obtained from two iterations of the secant algorithm. Does it matter which of the given points you consider x_1?

In carrying out the calculations involved in step 3 of the secant algorithm, we will repeatedly need to find the x-intercept of a line through

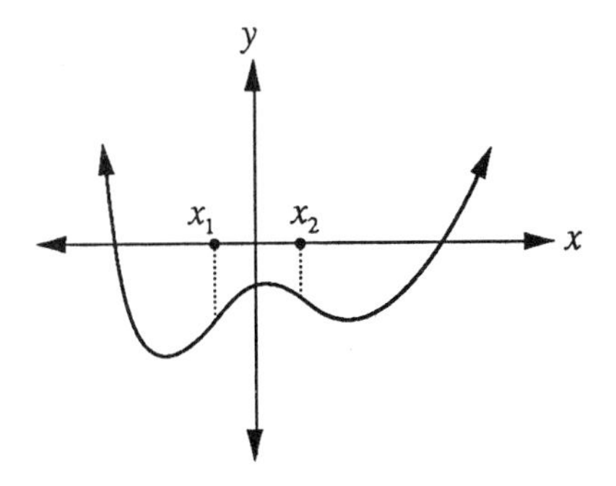

Figure 19 Graph for Class Practice Exercise

two points. Since the slope of the line through (x_1, y_1) and (x_2, y_2) is $\frac{y_1 - y_2}{x_1 - x_2}$, the equation of the line through these two points is

$$y - y_1 = \frac{y_1 - y_2}{x_1 - x_2}(x - x_1).$$

To determine the x-intercept of this line (called x_{new}), we set $y = 0$ and solve for x to obtain

$$x_{new} = \frac{x_2 y_1 - x_1 y_2}{y_1 - y_2}.$$

Using this equation to find the x-intercept in step 3 makes the secant algorithm easy to perform on a calculator or computer.

We will now use the secant algorithm to find a zero of $f(x) = x^2 - 2$. The list below shows how the secant algorithm progresses if we start with $x_1 = 2$ and $x_2 = 3$.

$(x_1, y_1) = (2, 2)$ and $(x_2, y_2) = (3, 7)$ yields

$$x\text{-intercept} = 1.6 = x_{new}.$$

$(x_1, y_1) = (3, 7)$ and $(x_2, y_2) = (1.6, 0.56)$ yields

$$x\text{-intercept} = 1.478 = x_{new}.$$

$(x_1, y_1) = (1.6, 0.56)$ and $(x_2, y_2) = (1.478, 0.184)$ yields

$$x\text{-intercept} = 1.418 = x_{new}.$$

$(x_1, y_1) = (1.478, 0.184)$
and $(x_2, y_2) = (1.418, 0.011)$ yields

$$x\text{-intercept} = 1.414 = x_{new}.$$

At this point we can see that the function $f(x) = x^2 - 2$ has a zero at $x \approx 1.4$.

As we pointed out before, the secant algorithm does not require that the initial x-values have y-values with opposite signs. If, however, we happen to choose x-values whose y-values differ in sign, then the secant algorithm usually finds the zero between them. The secant algorithm often finds zeros with fewer iterations than does the bisection algorithm.

There are several disadvantages to the secant algorithm. One is that we have little control over which zero of the function the algorithm will find. It is possible that repeated different choices for initial x-values will fail to yield the particular zero we desire. A second disadvantage of the secant method occurs only rarely; if two x-values yield the same y-value, then the line through them will not have an x-intercept. This problem is addressed in step 2 of the algorithm by picking different initial x-values.

Exercise Set 6

1. Webster's dictionary has 234,936 words in it. What is the largest number of words that may need to be picked in the process of searching for a word using binary search?

2. For each function listed, use the bisection algorithm to locate at least one zero to the nearest tenth. Keep track of the number of iterations needed to find the zero.

   a. $f(x) = x^2 - 5x - 8$

   b. $f(x) = x^3 - 8x^2 + x - 5$

   c. $f(x) = x^3 - 7x^2 - 6x + 36$

   d. $f(x) = x^4 - 10x^3 + 20x^2 + 10x - 11$

3. Suppose you borrow $4000 from the bank at an annual interest rate of 8%. You agree to pay back the money in four years, in four equal payments. You know that your yearly payment must be more than $1000, since $1000 payments are required just to cover the loan without interest. Follow the steps below to determine how much your yearly payment will be. If you have access to a computer spreadsheet, you may find it very useful for this problem.

   a. Begin by guessing a yearly payment of $1500, a reasonable guess since $1000 is too low and $2000 seems too high. At the end of the first year, your loan balance is $4000 plus interest in the amount of (.08)($4000) = $320 for a total of $4320. If you pay $1500, you enter the second year with a balance of $2820. Verify that the loan balance at the end of the second year is $3045.60. Continue this process to verify that after the third annual payment you have only $169.25 left to pay on the loan.

   b. Since a payment of $1000 is too small and $1500 is too large, guess a payment of $1250 and repeat the procedure above. You should find that a $1250 annual payment is also too high.

   c. Continue your binary search until you find the correct yearly payment to the nearest dollar.

4. Use the secant algorithm to find at least one zero for each polynomial in Exercise 2. Keep track of the number of iterations needed to find the zero.

5. Let $f(x) = x^3 - x^2 - 5x + 5$. Using initial values $x = 2$ and $x = 3$, compare the number of iterations required for the secant algorithm and the bisection algorithm to find the zero in the interval $[2, 3]$ accurate to the nearest hundredth.

6. Approximate $\sqrt{10}$ to the nearest hundredth by finding a zero of $f(x) = x^2 - 10$ with a table of values, with the bisection algorithm, and with the secant algorithm. Which method do you prefer?

7. Use the bisection algorithm to approximate the solution of $x^5 - 7 = 0$ to the nearest thousandth.

8. The polynomial $y = f(x)$ shown in Figure 20 has three zeros marked A, B, and C. Using the initial points x_1 and x_2, sketch the secant lines produced by the secant algorithm. Which zero will be approximated using these initial points?

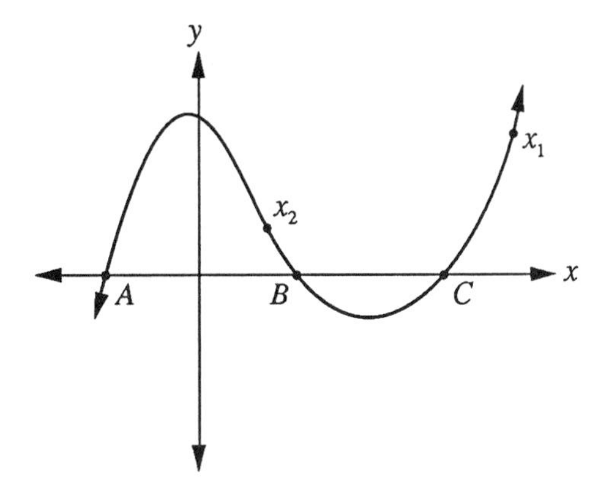

Figure 20 Graph of $y = f(x)$ for Exercise 8

7

Termination of Zero-Finding Algorithms

We say that an algorithm *converges* if it leads to a solution of a problem. Zero-finding algorithms are said to converge if they provide closer and closer approximations to zeros. When we described the bisection and secant algorithms in the previous section, we were vague about stopping criteria. In this section we will examine several stopping criteria for these algorithms. You will see that the stopping criterion we choose often depends on the application that motivates our search for zeros.

One stopping criterion is based on keeping $f(x)$ very close to zero. To implement this criterion, we examine the y-values associated with each iteration. If the y-value is less than some initially specified small number, we stop the process; on the other hand, if the y-value exceeds the specified number, we perform another iteration. Suppose, for example, that $f(x)$ is the measure of the error in an experiment. Since our goal most likely is to keep the error small, a suitable stopping criterion might be to continue the process until the y-value is within 0.0001 of zero. In symbols, we want

$$|y_{new}| < 0.0001 \quad \text{or} \quad |f(x_{new})| < 0.0001.$$

Mathematicians usually write this inequality in a more general form as $|f(x_{new})| < \epsilon$, where ϵ (*epsilon*) is used to represent an arbitrary small positive number.

Another stopping criterion that we often use involves x-values rather than y-values. Specifically, we continue the zero-finding process until successive x-values are very close in value, that is, until $|x_1 - x_2| < \epsilon$. If we know that we want to approximate a zero correct to a certain decimal place, this stopping criterion will often be the one we choose. The value of epsilon in the stopping criterion $|x_1 - x_2| < \epsilon$ is a measure of the *absolute error* in the estimate of the zero.

There are some problems associated with the stopping criterion $|x_1 - x_2| < \epsilon$. Suppose, for example, that x-values represent the distance in miles to different galaxies. We would hardly expect the difference between two such distances to be less than some small epsilon such as 0.05. Instead of using the stopping criterion $|x_1 - x_2| < \epsilon$, an alternative is the following stopping criterion:

$$\frac{|x_1 - x_2|}{1 + |x_2|} < \epsilon.$$

This criterion depends on the concept of *relative error* because the length of the interval between x_1 and x_2 is compared to the size of x_2. The form $\frac{|x_1 - x_2|}{1+|x_2|}$ is useful because it works for both small and large x-values. For small x-values, this criterion behaves like $|x_1 - x_2| < \epsilon$ since $1 + |x_2|$ is approximately 1. For large x-values, $\frac{|x_1 - x_2|}{1+|x_2|}$ has the desired effect of comparing the length of the interval between x-values to the size of x_2. You might question why the denominator in the expression for relative error is $1 + |x_2|$ rather than simply $|x_2|$. One reason for this choice is that it eliminates the potential of division by zero.

In the galaxy example suppose that $x_1 = 2.3 \times 10^{20}$ miles and $x_2 = 2.5 \times 10^{20}$ miles. Then the difference $|x_1 - x_2|$ is 2×10^{19} — a very large number (and certainly not less than a small epsilon). The relative error, on the other hand, is $(2 \times 10^{19})/(1 + 2.5 \times 10^{20}) \approx 0.08$ which is much more likely to satisfy some stopping criterion.

When we want to base our stopping criterion on the difference between two x-values, relative error is usually more useful than absolute error. If the stopping criterion being used is of the form $|f(x)| < \epsilon$, then relative error is not considered.

Round-off Error and Significant Figures

Most of the zero-finding algorithms are cumbersome to implement by hand and require either a calculator or a computer. Both calculators and computers, however, are limited by how many digits they maintain.

Suppose, for example, that we want to evaluate $f(x) = \sqrt{x^4 + 4} - 2$ near $x = 0$. Of course, $f(0) = 0$, but using a computer or calculator, you are very likely to find that $f(x) = 0$ for $-0.001 \leq x \leq 0.001$. However, the function f is increasing for all $x > 0$ and has only one zero at $x = 0$. Computers and calculators report that $f(0.001) = 0$ because their limits of accuracy have been exceeded.

Calculator and computer algorithms will compute values to as many decimal places as the equipment allows. The number of these digits that are significant in the problem under consideration must be determined by the user. The significant figures in a quantity are determined by the number of digits in the quantity that are known for certain. If a measure is recorded as 10.8 liters, the accuracy implied is $\frac{1}{10}$ of a liter, while 10.80 indicates that the accuracy of the measurement is $\frac{1}{100}$. It is important to understand that while the calculator and computer will give many digits, the final solution should have no more significant figures than do the initial values from which it was generated. For example, if a car is driven 154 kilometers on 8.6 liters of gas, the actual distance could be as little as 153.5 kilometers or as much as 154.5 kilometers, while the volume of gas could be anywhere between 8.55 and 8.65 liters. If we compute the number of kilometers per liter, we get $\frac{154}{8.6} = 17.906977$ on a calculator. However, the value could be as high as $\frac{154.5}{8.55} = 18.07018$ or as low as $\frac{153.5}{8.65} = 17.74566$. To try to state the number of kilometers per liter with an accuracy of $\frac{1}{10}$ is misleading. All that can be said for certain is that the car got 18 kilometers per liter. The solution has only two significant figures because the initial information had a factor which had only two significant figures. There is a word of warning, however. Problems of round-off error suggest that as many digits as possible should be used in intermediate steps. In the final solution round your answers to show only significant figures.

Exercise Set 7

1. How many iterations of the bisection algorithm are needed to approximate a root of a function f if it is known that there is a root in the interval $[0, 10]$? (Use absolute error and $\epsilon = 10^{-6}$.)

2. The bisection algorithm is applied to $f(x) = x^3 - 4x + 3$. The steps follow.

$(x_1, y_1) = (-3, -12)$ and
$(x_2, y_2) = (-2, 3)$ yields

$$(x_{new}, y_{new}) = (-2.5, -2.63).$$

$(x_1, y_1) = (-2.5, -2.63)$ and
$(x_2, y_2) = (-2, 3)$ yields

$$(x_{new}, y_{new}) = (-2.25, 0.61).$$

$(x_1, y_1) = (-2.5, -2.63)$ and
$(x_2, y_2) = (-2.25, 0.61)$ yields

$$(x_{new}, y_{new}) = (-2.375, -0.90).$$

$(x_1, y_1) = (-2.375, -0.90)$ and
$(x_2, y_2) = (-2.25, 0.61)$ yields

$$(x_{new}, y_{new}) = (-2.3125, -0.12).$$

Based on this information, what is your estimate of a zero of $f(x)$? Explain the reasons for the degree of accuracy you selected.

3. If there are 127 words in a glossary, how many iterations of the dictionary search algorithm are necessary to find each of the following?

a. A word that occurs as the 16th word?

b. A word that occurs as the 124th word?

c. A word that occurs as the 49th word?

d. A word that is not in the glossary?

8
Graphing Rational Functions

Rational functions are closely related to polynomials. In fact, *rational functions* are defined as functions of the form $f(x) = \frac{p(x)}{q(x)}$ where $p(x)$ and $q(x)$ are both polynomial functions. The zeros of the polynomial in the denominator are values that are excluded from the domain of the rational function. When the rational function is in reduced form, the zeros of the polynomial in the numerator are x-intercepts of the rational function.

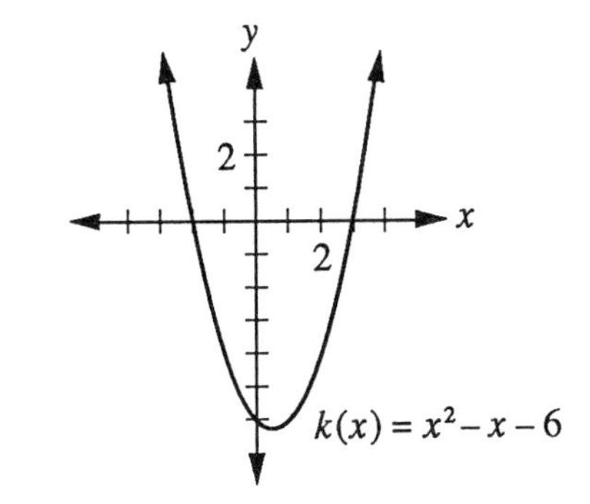

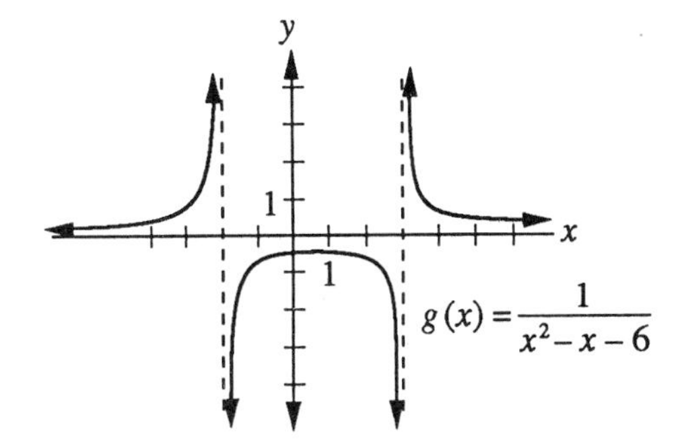

Figure 21 Graphs of $k(x) = x^2 - x - 6$ and $g(x) = \frac{1}{x^2-x-6}$

Example 1

Graph $g(x) = \frac{1}{x^2-x-6}$ using ideas of composition.

Solution Since we are familiar with the graph of $y = x^2 - x - 6$, we can graph this rational function using ideas of composition. Let $g(x) = h(k(x))$, where $k(x) = x^2 - x - 6$ and $h(x) = \frac{1}{x}$.

What is the domain of $g(x)$? Although the domain of the inner function $k(x)$ includes all real numbers, we must eliminate from the domain of g all x-values for which $k(x) = 0$. Factoring $k(x)$ as $(x-3)(x+2)$ shows that $k(3) = 0$ and $k(-2) = 0$, so the domain of $g(x) = \frac{1}{(x-3)(x+2)}$ consists of all real numbers except $x = 3$ and $x = -2$.

To graph g we will first graph the parabola $k(x) = x^2 - x - 6$ (see Figure 21); then we will plot the reciprocal of the y-coordinate of points on this parabola. Completing the square gives $k(x) = (x - \frac{1}{2})^2 - \frac{25}{4}$, so the graph of $k(x)$ has its vertex at $(\frac{1}{2}, -\frac{25}{4})$. This means that the graph of $g(x)$ contains the point $(\frac{1}{2}, -\frac{4}{25})$.

Where $k(x) = -1$, $g(x)$ also equals -1. You can verify that $g(-1.79) \approx -1$ and $g(2.79) \approx -1$. Where $k(x) = 1$, $g(x)$ also equals 1; this occurs at $x \approx -2.19$ and $x \approx 3.19$. Knowledge about the graph of k and about the effect of taking reciprocals provides the information we need to complete the graph:

$$\begin{array}{lll}
\text{as } x \to \infty, & k(x) \to \infty, & \text{so } g(x) \to 0^+ \\
\text{as } x \to -\infty, & k(x) \to \infty, & \text{so } g(x) \to 0^+ \\
\text{as } x \to 3^+, & k(x) \to 0^+, & \text{so } g(x) \to \infty \\
\text{as } x \to 3^-, & k(x) \to 0^-, & \text{so } g(x) \to -\infty \\
\text{as } x \to -2^+, & k(x) \to 0^-, & \text{so } g(x) \to -\infty \\
\text{as } x \to -2^-, & k(x) \to 0^+, & \text{so } g(x) \to \infty.
\end{array}$$

It should be clear from this analysis that the graph of $g(x)$ has vertical asymptotes at $x = 3$ and at $x = -2$, the x-values at which the polynomial in the denominator is equal to zero. Figure 21 shows the graph of $g(x)$. ■

The function graphed in Example 1 displays many of the important characteristics of rational functions. These characteristics suggest the types of questions that you need to ask when graphing rational functions:

— What x-values must be excluded from the domain?

— What happens to function values as x-values approach values that are excluded from the domain?

— What happens to function values as x-values increase without bound or decrease without bound?

We will investigate these questions in more depth in subsequent examples.

Example 2

Graph $f(x) = \frac{x^2-x-12}{x-4}$.

Solution We start with several observations. Notice that the degree of the numerator is greater than the degree of the denominator. We can see that this quotient increases without bound as x-values increase. This may be interpreted as meaning that as x-values increase without bound, x^2-x-12 increases faster than $x-4$. Notice that $f(4)$ is undefined, so $x = 4$ is excluded from the domain of f. However, the graph will not have a vertical asymptote at $x = 4$, because the numerator and denominator both have the factor $(x-4)$. Since $\frac{(x-4)}{(x-4)} = 1$ for all $x \neq 4$, these factors have the effect of cancelling each other for all x-values other than 4. Since

$$f(x) = \frac{x^2 - x - 12}{x - 4} = \frac{(x+3)(x-4)}{x-4},$$

you may be tempted to write that $f(x) = x + 3$. This is not a valid statement, because it completely ignores the fact that $f(4)$ is undefined. If we define $g(x) = x + 3$, then we can say that $f(x) = g(x)$ for all x-values except $x = 4$; $g(4) = 7$ but $f(4)$ is undefined. You can use a calculator or computer to verify that as x approaches 4, $f(x)$ approaches (but never reaches) 7.

The graph of f is shown in Figure 22. An open circle at $(4, 7)$ on the graph indicates the undefined point; 4 is not in the domain but as $x \to 4, f(x) \to 7$. If this function were graphed using a computer or calculator, the graph would not show a "hole" at $(4, 7)$. Only with special attention to domain do you know that this point should be excluded from the graph of $f(x)$. ■

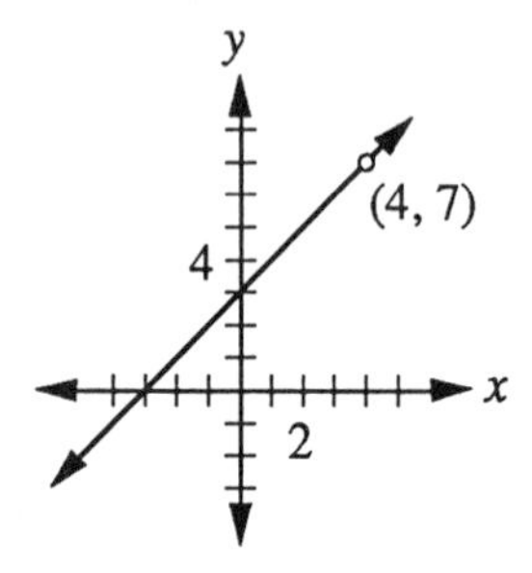

Figure 22 Graph of $f(x) = \frac{x^2-x-12}{x-4}$

Example 2 illustrates that the rational function $\frac{p(x)}{q(x)}$ does not necessarily have a vertical asymptote wherever $q(x) = 0$. The graph will have a vertical asymptote at any x-value for which $q(x) = 0$ and $p(x) \neq 0$. When $q(x) = 0$ and $p(x) = 0$, the graph may have an open circle to denote the undefined point. Reducing the rational function will allow you to analyze its behavior at all points where $q(x) \neq 0$.

Example 3

Identify the asymptotes and graph the function $y = \frac{2x+1}{x-1}$.

Solution The denominator is equal to zero when x is 1, and there will be a vertical asymptote at $x = 1$. To find other asymptotes, we need to investigate what happens to y-values when x gets very far from the origin. Using a computer or calculator to generate a table of values, you should see that when x increases without bound the $+1$ in the numerator and the -1 in the denominator have very little effect on function values. Thus the function approaches $\frac{2x}{x}$, or 2, so the line $y = 2$ is a horizontal asymptote.

Another way of determining the horizontal asymptote is to note the result of doing division:

$$(2x + 1) \div (x - 1) = 2 + \frac{3}{x - 1}.$$

For large values of x, $\frac{3}{x-1}$ becomes very small, so that $2 + \frac{3}{x-1}$ is very nearly 2. The graph of this function can be obtained by transforming the hyperbola $y = \frac{1}{x}$ as follows: a vertical stretch by a factor of 3, a horizontal shift 1 unit to the right, and a vertical shift 2 units up. The vertical shift corresponds to the horizontal asymptote being $y = 2$, and the horizontal shift corresponds to the vertical asymptote being $x = 1$. Figure 23 shows the effect of these vertical and horizontal shifts. ■

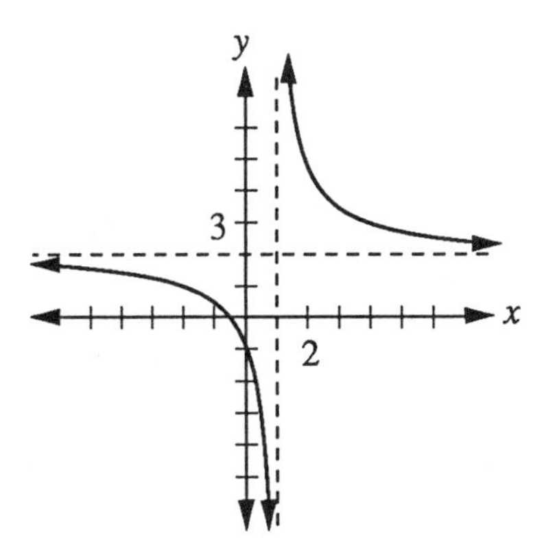

Figure 23 Graph of $y = \frac{2x+1}{x-1} = 2 + \frac{3}{x-1}$

Example 4

Graph the function

$$f(x) = \frac{x^2 - 2x - 3}{2x^2 + 2x - 12}.$$

Solution Factoring the numerator and denominator reveals important information about the function.

$$f(x) = \frac{(x - 3)(x + 1)}{2(x + 3)(x - 2)}$$

Vertical asymptotes occur where the denominator is zero, which happens when $x = -3$ and $x = 2$. We have x-intercepts where the numerator is zero, which happens when $x = 3$ and $x = -1$. We also need to explore the behavior of the function for x-values near -3 and 2, and for x-values far from the origin. Using a calculator or computer to generate a table of values, you should find that as the x-values increase without bound, the y-values move toward $\frac{1}{2}$ but are always slightly smaller than $\frac{1}{2}$. Thus,

$$\text{as } x \longrightarrow \infty, \; f(x) \longrightarrow \tfrac{1}{2}^{-}.$$

Continuing with a table of values, you should find that:

$$\begin{array}{ll} \text{as } x \longrightarrow \infty, & f(x) \longrightarrow \frac{1}{2}^{-} \\ \text{as } x \longrightarrow -\infty, & f(x) \longrightarrow \frac{1}{2}^{+} \\ \text{as } x \longrightarrow 2^{+}, & f(x) \longrightarrow -\infty \\ \text{as } x \longrightarrow 2^{-}, & f(x) \longrightarrow \infty \\ \text{as } x \longrightarrow -3^{+}, & f(x) \longrightarrow -\infty \\ \text{as } x \longrightarrow -3^{-}, & f(x) \longrightarrow \infty. \end{array}$$

These trends, together with the information about intercepts and asymptotes mentioned earlier, are incorporated in the graph in Figure 24. ■

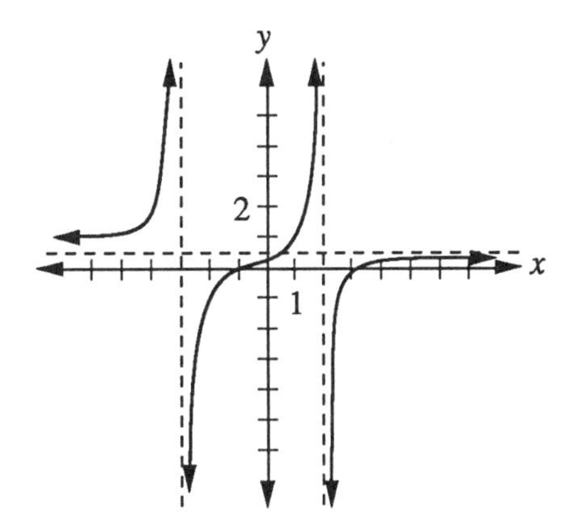

Figure 24 Graph of $f(x) = \frac{x^2-2x-3}{2x^2+2x-12}$.

Examples 3 and 4 both involved rational functions in which the degree of the numerator and the degree of the denominator are the same. In Example 3 the leading coefficients in the numerator and denominator were 2 and 1 respectively, and the horizontal asymptote was $y = \frac{2}{1}$. Similarly, in Example 4 the leading coefficients were 1 and 2 respectively, and the horizontal asymptote was $y = \frac{1}{2}$. In both cases the highest degree terms in the numerator and denominator determine the behavior of the function for x-values far from the origin. We can see the reason for this if we perform a little algebraic manipulation. When we divide both the numerator and denominator of $f(x)$ by x^2 (which is the highest power of x in the denominator) we get

$$f(x) = \frac{x^2 - 2x - 3}{2x^2 + 2x - 12} = \frac{1 - \frac{2}{x} - \frac{3}{x^2}}{2 + \frac{2}{x} - \frac{12}{x^2}}.$$

For very large values of x, the terms with variables in their denominators become insignificant, so function values approach $\frac{1}{2}$ as x moves far from the origin.

Example 5

Analyze the behavior of $y = \frac{x^2-3x+4}{x-2}$ as x gets very far from the origin. Then graph the function.

Solution We can get information about the graph by dividing:

$$y = \frac{x^2 - 3x + 4}{x - 2} = x - 1 + \frac{2}{x - 2}.$$

Notice that the function is the sum of two familiar functions. When x is a large positive number, $\frac{2}{x-2}$ is a very small positive number, so y will be just slightly larger than $x-1$. The larger x is, the closer y is to $x - 1$, or

$$\text{as } x \longrightarrow \infty,\ y \longrightarrow (x-1)^+.$$

Similarly, when x decreases without bound, $\frac{2}{x-2}$ is a negative number very close to zero, so y will be just slightly less than $x - 1$. As x moves far to the left of the origin, y becomes close to $x - 1$ or

$$\text{as } x \longrightarrow -\infty,\ y \longrightarrow (x-1)^-.$$

The graph is asymptotic to $y = x - 1$; this line is called an *oblique* asymptote.

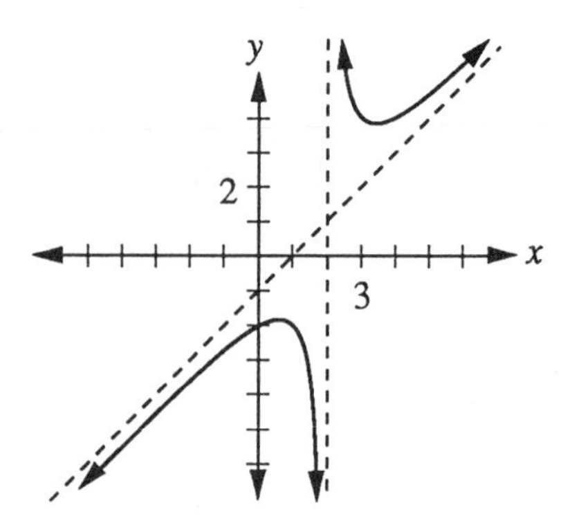

Figure 25 Graph of $f(x) = \frac{x^2-3x+4}{x-2}$

A graph of $y = \frac{x^2-3x+4}{x-2}$ is shown in Figure 25. Notice the two asymptotes, one vertical and one oblique. This graph shows the effect of adding the ordinates of the graphs of p and q, where $p(x) = x - 1$ and $q(x) = \frac{2}{x-2}$. The global behavior of $f(x) = \frac{x^2-3x+4}{x-2}$ is similar to $y = x - 1$. When x is far from the origin, the term $\frac{2}{x-2}$ is insignificant compared to $x - 1$, so the graph resembles that of $p(x) = x - 1$. When x is close to 2, the term $x-1$ is insignificant compared to $\frac{2}{x-2}$, so the graph resembles that of $q(x) = \frac{2}{x-2}$. ■

Once again, as we saw in Example 2, the degree of the numerator is greater than the degree of the denominator, and the quotient increases without bound as x-values increase without bound. In general, in a rational function, when the degree of the numerator is greater than the degree of the denominator, the absolute value of the quotient will increase without bound as x increases without bound.

Example 6

Graph $h(x) = \frac{x-4}{x^2+x-2}$.

Solution We can easily locate the intercepts of this graph; they are $(4, 0)$ and $(0, 2)$. To identify vertical asymptotes it will help to factor the denominator:

$$h(x) = \frac{x - 4}{(x + 2)(x - 1)}.$$

Writing the equation in this form makes it clear that h has asymptotes at $x = -2$ and $x = 1$. How does h behave far from the origin? Since x^2+x-2 increases more quickly than $x-4$, the values of the function will approach zero as x-values increase without bound. Verify the following:

$$\begin{array}{ll} \text{as } x \longrightarrow \infty, & h(x) \longrightarrow 0^+ \\ \text{as } x \longrightarrow 1^+, & h(x) \longrightarrow -\infty \\ \text{as } x \longrightarrow 1^-, & h(x) \longrightarrow \infty \\ \text{as } x \longrightarrow -2^+, & h(x) \longrightarrow \infty \\ \text{as } x \longrightarrow -2^-, & h(x) \longrightarrow -\infty \\ \text{as } x \longrightarrow -\infty, & h(x) \longrightarrow 0^-. \end{array}$$

How does the graph behave between $x = -2$ and $x = 1$? Since the function has no other x-intercepts and no other gaps or holes, it cannot change sign in this interval. We noted earlier that $h(0) = 2$, so the graph must be positive everywhere in $[-2, 1]$. Figure 26 shows a graph of h. Notice the two turning points and the fact that the graph crosses its horizontal asymptote. ■

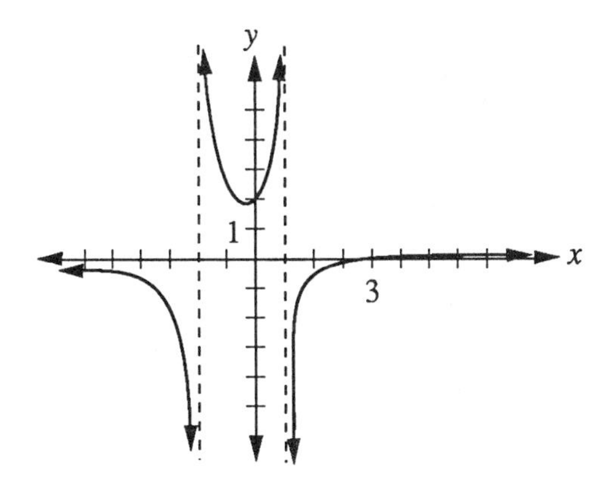

Figure 26 Graph of $h(x) = \frac{x-4}{x^2+x-2}$

Example 7

Graph $k(x) = \frac{x^2-1}{x^2+1}$.

Solution The function $k(x)$ is different from those in other examples; the denominator has no zeros and therefore $k(x)$ has no vertical asymptotes. The degree of the numerator and the degree of the denominator are the same. As values of x increase or decrease without bound, the values of $k(x)$ approach 1 but are always smaller than 1. This tells us that the graph approaches the asymptote $y = 1$ from below. The x-intercepts are at -1 and 1, and the y-intercept is at -1. A sketch of $k(x)$ is shown in Figure 27. ■

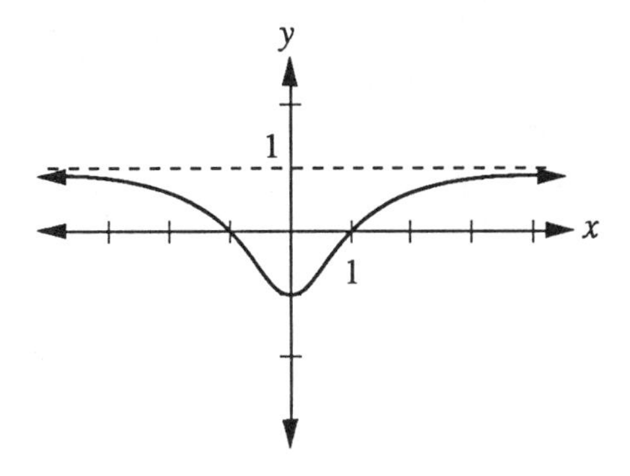

Figure 27 Graph of $k(x) = \frac{x^2-1}{x^2+1}$

You may want to use a graphing calculator or a computer to check the graphs you make while working exercises. When you do so, bear in mind that much of the software available to graph functions plots individual points and then connects adjacent points; the effect is a continuous graph. As you have seen, many rational functions are not continuous, but rather may have asymptotes at points of discontinuity. You will notice that calculator- or computer-generated graphs of rational functions may ignore asymptotes and show connections across x-values that are not in the domain. Another characteristic of calculator- or computer-generated graphs of rational functions is that the scale on the vertical axis may be extremely compressed. This happens because y-coordinates increase or decrease without bound near vertical asymptotes. Whenever possible you should specify maximum and minimum y-values to be plotted on your calculator or computer. In general, it is important to think about the effect of vertical asymptotes when you are interpreting the computer or calculator graph of a rational function.

Exercise Set 8

1. Suppose you want to graph the function $f(x) = \frac{x^3-3x+1}{x^3-x^2-3}$, where there is no simple factorization. Explain what you would do to determine the locations of x-intercepts and vertical asymptotes.
2. If $f(x) = \frac{x^2+1}{x^2-2x+1}$, find the asymptotes of $f(x)$. Investigate whether the graph of $f(x)$ crosses any one of these asymptotes.
3. Use a computer or calculator to graph the following functions over the specified x- and y-intervals. Discuss the global behavior of each function.
   a. $y = \frac{x^2-3x+4}{x-2}$ over $[-1000, 1000]$ and $[-20, 20]$
   b. $y = \frac{x}{x^2-x-6}$ over $[-1000, 1000]$ and $[-20, 20]$
   c. $y = \frac{2x+1}{x-1}$ over $[-1000, 1000]$ and $[-20, 20]$
4. Graph each rational function. Show intercepts, asymptotes, and general shape.
   a. $f(x) = \frac{1}{x+3}$
   b. $f(x) = \frac{1}{x^2-x-12}$
   c. $f(x) = \frac{x-2}{x^2-9}$
   d. $f(x) = \frac{x}{x^2-3x}$
   e. $f(x) = \frac{x^2-2x-8}{x^2-x-6}$
   f. $f(x) = \frac{x+2}{x}$
   g. $f(x) = \frac{x+3}{x^2+2}$
   h. $f(x) = \frac{2x^2+3x-2}{x^2-1}$
   i. $f(x) = \frac{3x-1}{x^2-2x-3}$
   j. $f(x) = \frac{-x^2-2x+5}{x+3}$
   k. $f(x) = \frac{x-1}{(x-2)^2}$
   l. $f(x) = \frac{x^3+x^2-5x}{x^2-1}$
   m. $f(x) = \left|\frac{x+1}{x^2+x-6}\right|$
5. Solve the following equations and inequalities involving rational functions. Computer- or calculator-generated graphs may be helpful.
   a. $2x - 3 = \frac{x}{x-3}$
   b. $\frac{2}{x+2} > \frac{1}{x-3}$
   c. $\frac{1}{x+1} \leq \frac{2x}{3x-1}$
6. Suppose a rational function has vertical asymptotes at $x = -2$ and $x = 3$, has zeros at $x = 1$ and $x = 4$, and has a horizontal asymptote at $y = \frac{1}{2}$.
   a. Sketch the graph of such a function.
   b. Write an equation for such a function.
7. Suppose a rational function approaches the lines $x = 2$ and $y = -x + 3$ asymptotically.
   a. Sketch a graph of this function.
   b. Write an equation for your graph.

9
Applications of Rational Functions

Certain rational functions have special characteristics that are important in applications. One useful relationship that a rational function can describe is one in which the y-values decrease but remain positive as the x-values increase without bound. In a relationship of this type the variables are said to *vary indirectly* or *inversely*. Indirect variation is described by a rational function of the form $f(x) = \frac{k}{x}$.

Example 1

A food packaging company wants to make cylindrical cans that hold 1 liter (1000 cm^3) of their product. What dimensions will enable them to use a minimum amount of material?

Solution In Figure 28 we see that the surface area of a cylindrical can is composed of a rectangular section and two circular sections. To simplify the problem, we will ignore the seams of the can.

The surface area is a function of two variables: the radius and the height of the can. This function

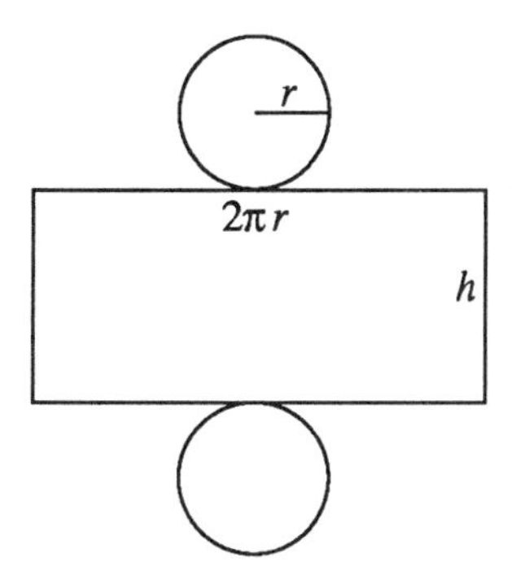

Figure 28 Surface Area of Cylindrical Can

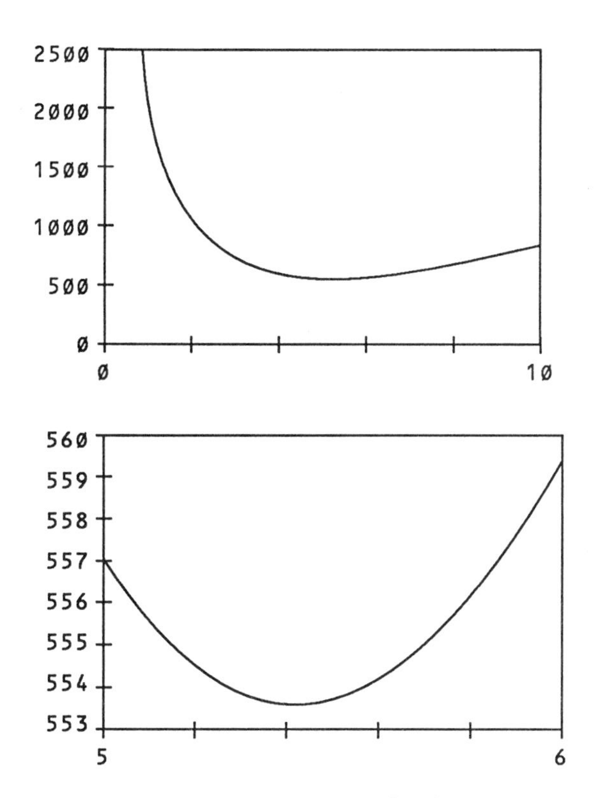

Figure 29 Graphs of $f(r) = 2\pi r^2 + \frac{2000}{r}$

is $f(r,h) = 2\pi r^2 + 2\pi rh$. The notation $f(r,h)$ implies that the function f depends on two variables, in this case r and h. If we could write the surface area as a function of one variable, our work would be much simpler. Since the volume is given by $\pi r^2 h$ and the volume of the can is to be 1000 cm^3, we know that $\pi r^2 h = 1000$, so $h = \frac{1000}{\pi r^2}$. We can now express the surface area as a function of r alone: $f(r) = 2\pi r^2 + \frac{2000}{r}$. Two graphs of this function are shown in Figure 29. The first graph is over the interval $[1, 10]$, and the second graph is over the interval $[5, 6]$.

You will notice a minimum point between 5 and 6. By listing values of this function, or by using the trace feature on the computer or calculator to find the minimum point, we see that when the radius r is approximately 5.42 cm the surface area has a minimum value of 553.58 cm^2. Using $h = \frac{1000}{\pi r^2}$ we can compute the height to be approximately 10.84 cm. ■

Exercise Set 9

1. The weight of an object above the earth's surface is given by the function $w(h) = (\frac{r}{r+h})^2 w_0$, where r is the radius of the earth, w_0 is the weight of the object at sea level, and h is the height of the object above the earth's surface. How much will a 200 pound object weigh when it is 400 miles above the earth's surface? (The earth's radius is approximately 3950 miles.) At what height will the object weigh 100 pounds?

2. In photography, the relationship between the constant focal length of a lens (F), the distance between the lens and the film (v), and the distance between the object and the lens (u) is given by $\frac{1}{u} + \frac{1}{v} = \frac{1}{F}$ (see Figure 30).

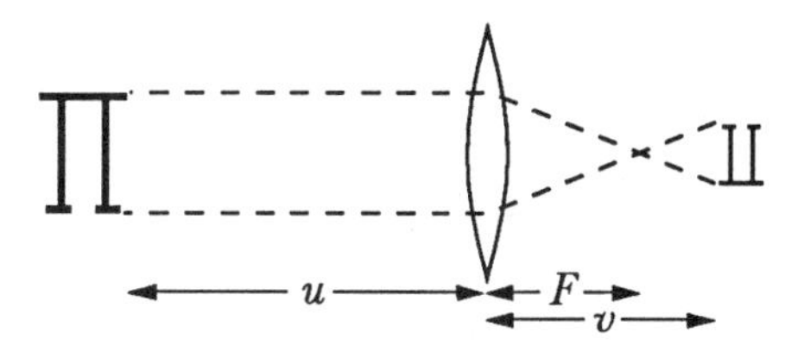

Figure 30 Lens Optics Showing the Relationship between u, v, and F.

a. Suppose the focal length F is 50 mm. Solve for v in terms of u.

b. Graph the function $v = f(u)$ over the interval [50 mm, 50000 mm].

c. "Depth of field" refers to the range of distances in which an object remains in focus. Use the graph to help you explain why the depth of field is broad when you focus on objects that are very far from the lens but is narrow for objects close to the lens.

3. An open box with a square base and rectangular sides is to hold 250 in^3. If sides are double thickness and the bottom is triple thickness, what size box will use the least amount of material?

4. A cylindrical can is to hold 500 cm^3. The material used to make the top and bottom costs 0.012 cents/cm^2, the material used for the sides costs 0.01 cents/cm^2, and the seam joining the top and bottom to the sides costs 0.015 cents/cm. What size can would cost the least to produce?

5. The Durham Tin Can Company minimizes costs by constructing cans from the least possible amount of material. This company supplies many different sizes of cans to packing firms. A designer with the company needs to find the radius that gives the minimum surface area for cylindrical cans with volumes between 100 cc and 1500 cc.

a. Use the technique from Example 1 to find the radius that minimizes surface area for a volume of 100 cc, 300 cc, 500 cc, 700 cc, 900 cc, 1100 cc, 1300 cc, and 1500 cc. Make a scatter plot of the ordered pairs (volume, radius). What type of relationship appears to exist between these variables?

b. Use techniques of data analysis to find a function that expresses the relationship between volume and radius for a can of minimal surface area.

c. Use your function to predict the radius that gives the minimum surface area for a cylindrical can with volume of 2000 cc.

d. What is the ratio of the diameter to the height for the can of minimal surface area that holds a given volume? Do manufacturers generally produce cans with diameter and height in this ratio? Why or why not?

CHAPTER 4

Exponential and Logarithmic Functions

1 *Introduction*

In preceding chapters we have seen how mathematical functions can be used to model real-world phenomena. In this chapter we add the exponential function and its inverse, the logarithmic function, to our toolkit. The exponential function describes many phenomena, including population growth, radioactive decay, compound interest, and the number of iterations required in the bisection algorithm to find zeros with a specified degree of accuracy. We will begin our study of exponential functions by exploring an application in the area of finance.

Example 1

Suppose Maria deposits \$1000 in a credit union savings account that earns 8% interest annually. If the interest is automatically credited to the account at the end of each year and Maria withdraws no money, how much money will be in the account at the end of ten years?

Solution This question asks us to find the *future value* of \$1000 invested for ten years at an annual interest rate of 8%. During the first year, Maria receives 8% of \$1000 in interest. Since this interest is automatically credited to her account, the balance in Maria's account at the end of the first year will be

$$\$1000 + \$1000(0.08) = \$1000(1.08) = \$1080.$$

During the second year, she receives 8% of \$1080 in interest, so her balance at the end of the second year will be

$$\$1080 + \$1080(0.08) = \$1080(1.08) = \$1166.40.$$

To find the balance at the end of year three, we simply multiply the balance at the end of year two by 1.08. A calculator or spreadsheet can be used to continue this process and quickly compute the balances for subsequent years. Table 1 shows balances for ten years. You may be surprised to see that in ten years, Maria's account will more than double in value. Notice also that the increase in balance is higher for the second year than for the first, higher for the third year than for the second, and so forth. As interest accumulates in the account, the balance increases so that the interest earns interest; this is called *compound interest.* ■

We can write a *recursive* function to express mathematically the iterative process that we use to compute successive balances. The word *recursive* refers to the fact that previously obtained values

Table 1 Balance in Maria's Account

Year	1	2	3	4	5
Balance	$1080.00	1166.40	1259.71	1360.48	1469.31
Year	6	7	8	9	10
Balance	$1586.85	1713.79	1850.89	1998.96	2158.87

recur, or occur again, in the process of obtaining new values. In this example, we will let $A(N)$ denote the amount in Maria's account after N years. The initial amount, after zero years, is $A(0) = \$1000$; the amount after one year is $A(1) = \$1080$. To write a recursive function, we want to express $A(N)$ in terms of $A(N-1)$. It should be clear that

$$A(N) = 1.08 \cdot A(N-1).$$

Notice that in the process of computing function values, the output of one computation becomes the input for the next. Using this recursive function, we could easily extend our table to examine even more future values of Maria's $1000 deposit.

To answer Maria's original question about future value in ten years we could use another approach that does not involve the computation of successive balances. Recall that we expressed the balance at the end of the first year as

$$\$1000 + \$1000(0.08) = \$1000(1.08) = \$1080.$$

The balance at the end of the second year can be written as

$$\begin{aligned}\$1080(1.08) &= (\$1000(1.08))(1.08)\\ &= \$1000((1.08)(1.08)) = \$1000(1.08)^2.\end{aligned}$$

At the end of ten years, Maria's account will have a balance of

$$\$1000(1.08)^{10} = \$2158.92.$$

Comparing this future value to the amount in Table 1 we see that the amounts differ by $0.05. In this situation the table value more accurately reflects the balance in Maria's account. At the end of each year the credit union truncates fractional portions of cents. (This truncation is an example of the greatest integer function in action.) This technique was used for all calculations in Table 1, but it was used only once in the computation of $\$1000(1.08)^{10}$.

To generalize this approach for calculating future value we can write

$$A(N) = \$1000(1.08)^N, \text{ where } N = 0, 1, 2, \ldots.$$

This expression for future value allows us to choose a value of N and directly compute the balance without knowing the balance at the end of the preceding year. We call it a *closed form* expression for the function to distinguish it from the recursive expression.

Given any initial deposit A_0 and yearly interest rate r, we can now write two general formulas for the future value of an account after N years. The recursive expression is given by:

$$\begin{aligned}A(0) &= A_0\\ A(N) &= (1+r) \cdot A(N-1).\end{aligned}$$

The closed form expression is given by:

$$A(N) = A_0(1+r)^N.$$

Future value is a function of three variables: the amount initially invested, the yearly interest rate, and the number of years that the interest is allowed to compound. In reality, the amount of the initial investment is a fixed amount, and the money is usually deposited at a fixed interest rate. For fixed values of A_0 and r, future value is a function of N; in the closed form of this function, the variable N is the exponent.

Savings institutions differ in the way they deal with interest when money is withdrawn between compounding dates. Maria's credit union requires that the money remain in the account for the entire year to receive interest. If she withdraws all the money one year and eleven months after her initial deposit, she will receive only $1080. The graph representing how the balance in Maria's account changes over time is a step function and is provided in Figure 2. Notice that the yearly increases in the account balance are not constant; rather, as time increases, the balance increases by larger and larger amounts.

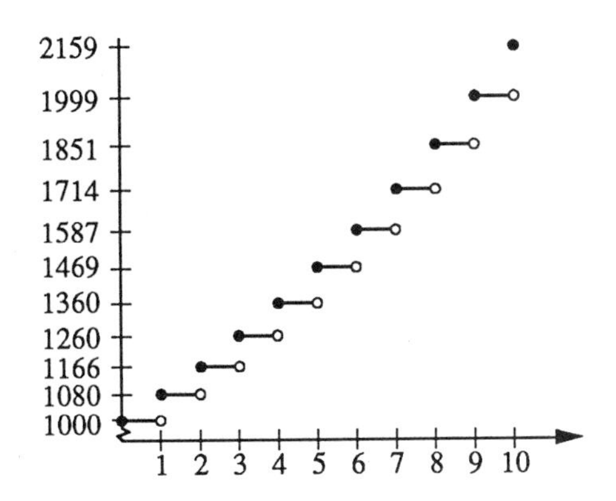

Figure 2 Graph of Balance in Maria's Account over Time

Exercise Set 1

1. Use the information in Table 1 to calculate the interest Maria earns during the tenth year. Compare it to the interest earned during the first and second years.
2. Use Table 1 to estimate how many years it takes to double Maria's $1000 deposit.
3. **a.** Find the future value of $1000 invested for twenty years at an annual interest rate of 8%. Which form of the function did you use to compute this value? Why?

   b. Now find the future value of the $1000 after twenty-one years. Which form of the function is easier to use? Why?
4. Make a computer or calculator graph of the closed form expression for future value when $a_0 = 1000$ and $r = 0.08$. Use your graph to determine how many years it will take to double the amount originally in Maria's account. How many years will it take to quadruple the original amount? (Recall that interest in Maria's account is calculated only once each year. The graph drawn by the computer or calculator is a continuous one, but Maria's balance does not change continuously.)

2 Definition of Exponential Function

Suppose Clovis deposits $1.00 in a credit union account for his little brother. If his account also earns interest at the rate 8% compounded annually, the future value is given by

$$\text{Future Value } = A(N) = 1(1+0.08)^N = 1.08^N, \quad \text{where } N = 0, 1, 2, \ldots.$$

Future values of this account for the first ten years are provided in Table 3. (We have rounded the values and ignored the truncation that takes place when banks compute interest.)

The future value given by $A(N) = 1.08^N$ is an *exponential function* of N with base 1.08. As a model for compound interest, this function is appropriate only for nonnegative integer values of N.

In general, any function of the form

$$f(x) = b^x, \qquad b > 0,\ b \neq 1,$$

is called an exponential function. The base of the

Table 3 Future Value of $1.00 Invested at 8% Compounded Annually

Year N	0	1	2	3	4	5
Future Value $(1.08)^N$	$1.00	1.08	1.17	1.26	1.36	1.47

Year N	6	7	8	9	10
Future Value $(1.08)^N$	$1.59	1.71	1.85	2.00	2.16

exponential function is the constant b; the exponent is a variable.

By plotting several points of the function $f(x) = 2^x$, we can discover the general shape of the graph of the exponential function with base 2. The following points are on the graph of $f(x) = 2^x$:

x	-3	-1	-0.5	0	0.5
2^x	0.125	0.5	0.707	1	1.414

x	1	1.5	2	3	10
2^x	2	2.828	4	8	1024

In the calculation of these values, you may need to review negative and fractional exponents. Recall that $2^{-3} = \frac{1}{2^3} = \frac{1}{8}$ and that $2^{1.5} = 2^{\frac{3}{2}} = \sqrt{2^3}$. Use your calculator to compute the roots. The graph of $y = 2^x$ is shown in Figure 4.

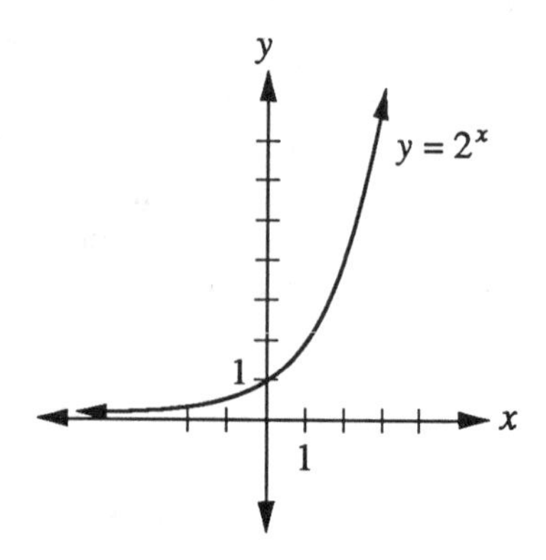

Figure 4 Graph of $f(x) = 2^x$

Now examine the graphs of $g(x) = 3^x$ and $h(x) = (1.5)^x$ shown in Figure 5. The same general shape as $f(x) = 2^x$ appears again. How would you expect the graph of $f(x) = 5^x$ to compare to the ones we have examined?

In each of the above examples, the base was greater than one. Now consider $k(x) = (\frac{1}{2})^x$. Verify the values in the following table and use them to sketch the graph of $y = (\frac{1}{2})^x$. How does this graph compare to the graphs of the exponential functions in Figures 4 and 5?

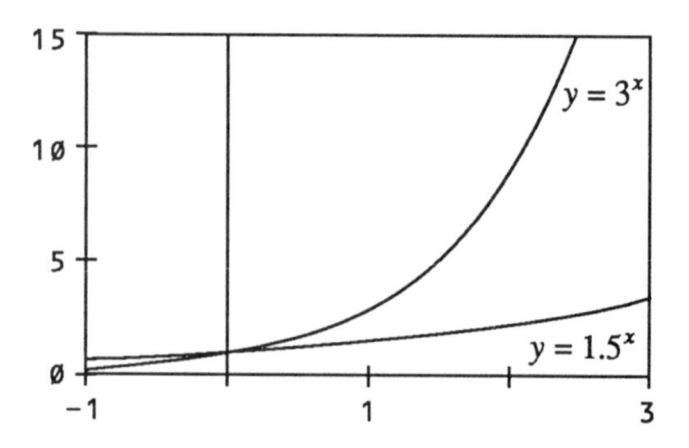

Figure 5 Exponential Functions: $y = 3^x$ and $y = 1.5^x$

x	-3	-1	-0.5	0	0.5
$(\frac{1}{2})^x$	8	2	1.414	1	0.707

x	1	1.5	2	3	10
$(\frac{1}{2})^x$	0.5	0.354	0.25	0.125	0.000977

As you examine the table of values and sketch the graph of $y = (\frac{1}{2})^x$ you probably recognize some of the y-values. Look back at the y-values of $f(x) = 2^x$. The relationship between the functions $f(x) = 2^x$ and $k(x) = (\frac{1}{2})^x$ will become even more obvious if we graph both functions on the same axes. These graphs are provided in Figure 6.

Looking at the graphs together shows that the two curves are symmetric about the y-axis. Some algebra will confirm our suspicion that $k(x)$ is the reflection of $f(x)$ across the y-axis:

$$k(x) = (\tfrac{1}{2})^x = (2^{-1})^x = 2^{-x} = f(-x).$$

Since $k(x) = f(-x)$, we know from our study of transformations that the graph of $k(x) = (\frac{1}{2})^x$ is simply the graph of $f(x) = 2^x$ reflected about the y-axis. In fact, for any $b > 0$, the graph of $y = (\frac{1}{b})^x$ will be a reflection of the graph of $y = b^x$ about the y-axis.

The exponential function of the form $y = b^x$ should be added to our toolkit of functions. Graphs

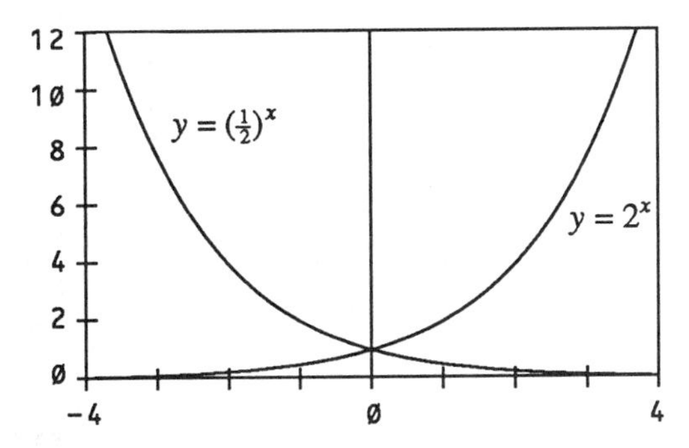

Figure 6 Graphs of $f(x) = 2^x$ and $k(x) = (\frac{1}{2})^x$

of exponential functions of this form share the characteristics that are listed in Table 7. If you look back at the graphs in Figures 4, 5, and 6, you will see these features. Of course, when there are transformations of the exponential function, these characteristics of the graph will also be transformed.

Laws of Exponents

As we continue our study of exponential functions, you will frequently find it necessary to apply the laws of exponents. It is assumed that you have studied these laws in previous math courses. They

Table 7 Characteristics of Exponential Functions of the Form $y = b^x, b > 0, b \neq 1$

1. The point $(0, 1)$ is on every graph.
2. The point $(1, b)$ is on every graph.
3. The x-axis $(y = 0)$ is an asymptote to each curve.
4. The domain of each function is the set of real numbers.
5. The range of each function is the set of real numbers greater than zero.
6. Each function is one-to-one.

are listed below without proof so that you can review them as needed.

For all positive real numbers a and b and all real numbers x and y,

$a^x a^y = a^{x+y}$.

$\frac{a^x}{a^y} = a^{x-y}$.

$(a^x)^y = a^{xy}$.

$(ab)^x = a^x b^x$.

$\left(\frac{a}{b}\right)^x = \frac{a^x}{b^x}$.

If $a \neq 1$, $a^x = a^y$ if and only if $x = y$.

If $x \neq 0$, $a^x = b^x$ if and only if $a = b$.

$a^0 = 1$ if $a \neq 0$.

Exercise Set 2

1. Sketch a graph of the following functions. Show the asymptote and the y-intercept. After you have sketched the curve, state the range.

a. $y = 3^x$

b. $y = 0.25^x$

c. $y = 10^x$

d. $y = 1.05^x$

2. For each function in the preceding exercise, describe how the y-values change when x-values increase by one unit. For example, in part (a) when x increases from $x = 1$ to $x = 2$, y-values increase from $y = 3$ to $y = 9$; when x-values increase from $x = 4$ to $x = 5$, y-values increase from $y = 81$ to $y = 243$. In general, y-values are multiplied by a factor of 3 when x-values increase by one unit.

3. In defining the exponential function as $y = b^x$, we stated that the base b must be greater than zero.

a. What problems are associated with negative values of b? To help answer this question, consider $m(x) = (-2)^x$ and make a table of values for $x = -3, -2, -1, -0.5, 0, 1, 1.5, 2,$

and 3. (What happens if you try to sketch a graph of $y = m(x)$?)

b. Consider the function $n(x) = 0^x$. Why does $n(0)$ present a dilemma?

4. In the definition of the exponential function $y = b^x$ we stated that the base b should not equal one.

   a. Consider the function $f(x) = (1+k)^x$ where k assumes the values $0.1, 0.05$, and 0.01. Graph each function.

   b. Now let $k = -0.1, -0.05$, and -0.01. Graph each function.

   c. Describe what happens to the graph of $y = b^x$ as the base gets close to one.

   d. Now consider the function $p(x) = 1^x$. What general characteristics of exponential functions would not apply to this function?

5. The value of a car depreciates approximately 20% each year. Complete the chart below. C represents the value of the car and a the age in years. Graph the relationship between C and a.

a	0	1	2	3	4	5
C	\$13,000					

6. If gasoline prices increase 0.5% a month for a whole year, how much would gasoline that costs \$1.15 per gallon at the beginning of the year cost at the end of the year?

7. We have seen that future value is a function of the amount of money invested, the yearly interest rate, and the number of years that the interest is allowed to compound. Assume an initial investment of \$1,000, and use computer- or calculator-generated graphs to further investigate how future value changes over time.

   a. Draw graphs for annual interest rates of 6%, 8%, 10%, and 12%, and discuss how the account balance varies with time when the interest rate is fixed.

   b. Use your graphs to determine how long it takes for the initial \$1,000 deposit to double, triple, and quadruple at each interest rate.

8. Use graphs to compare the growth of the functions $f(x) = x^2$ and $g(x) = 2^x$. How many times do these graphs intersect? Find the coordinates of all points of intersection.

9. Suppose a square piece of paper with sides of 8 inches is 0.003 inches thick. Fold the paper once, then once more (now it has four layers), and continue folding it over upon itself 50 times. (Of course we know that this is physically impossible—just use your imagination!) How thick is the paper after 50 foldings? The distance from the earth to the moon is $2.39(10^5)$ miles and from the earth to the sun is $9.3(10^7)$ miles. How does your answer compare with these distances?

10. Use the laws of exponents to solve over the real numbers.

    a. $3^{2x+4} = 27^{x-1}$

    b. $8^x = 16^{x+2}$

    c. $3^{x+1} = \frac{1}{9^x}$

    d. $\sqrt{2^{2x+1}} = 4$

11. If $f(x) = a(2^{bx})$, $f(0) = 5$, and $f(3) = 20\sqrt{2}$, find a and b.

3

Compound Interest

We began this chapter with a compound interest example in which Maria deposited \$1000 in a savings account that earned interest at a rate of 8%. This interest was compounded annually, which means that the interest was credited to Maria's account at the end of each year. Today it is more common to encounter quarterly, monthly, or daily compounding. Instead of adding all of the annual interest at the end of each year, a bank that compounds quarterly would apply one-fourth of the annual interest rate four times a year, or every three months. What effect does more frequent compounding have on the future value? Suppose Maria deposits \$1000 for ten years in a bank which pays a yearly interest rate

Table 8 Future Value of $1,000 Single Deposit for Various Compounding Frequencies and Time Periods, 8% Yearly Interest Rate

	Frequency of Compounding						
	Yearly	Quarterly	Monthly	Weekly	Daily	Hourly	Minutely
Years(N)	$k = 1$	$k = 4$	$k = 12$	$k = 52$	$k = 365$	$k = 8760$	$k = 525600$
5	$1,469	$1,486	$1,490	$1,491	$1,492	$1,492	$1,492
20	$4,661	$4,875	$4,927	$4,947	$4,952	$4,953	$4,953
30	$10,063	$10,765	$10,936	$11,003	$11,020	$11,023	$11,023
50	$46,902	$52,485	$53,878	$54,431	$54,574	$54,597	$54,598
75	$321,205	$380,235	$395,475	$401,573	$403,164	$403,418	$403,429

of 8% with quarterly compounding. Interest is credited to her account four times each year, but only one-fourth of 8%, or 2%, is used for each computation. At the end of the first quarter Maria has a balance of

$$\$1000(1.02) = \$1020.$$

At the end of the second quarter Maria has a balance of

$$\$1020(1.02) = \$1000(1.02)^2 = \$1040.40.$$

At the end of the first year, which contains four compounding periods, the balance will be

$$(\$1040.40(1.02))(1.02) = \$1000(1.02)^4 = \$1082.43.$$

Notice that this amount is slightly higher than the corresponding balance of $1080.00 at the end of one year using yearly compounding. We can continue this iterative process to determine the balance at the end of ten years. A computer spreadsheet can quickly perform forty multiplications and truncate the balance at the end of each quarter to accurately reflect the procedure used by the bank. As an alternative, we could ignore quarterly truncation and calculate the balance at the end of ten years, or forty compounding periods, as

$$\$1000((1.02)^4)^{10} = \$1000(1.02)^{40} = \$2208.04.$$

Comparing this result to the previous one of $2158.92 with yearly compounding shows that quarterly compounding makes some difference. The key is that interest begins to earn interest sooner.

Class Practice

1. Write a formula for the future value F of an initial deposit A_0 after N years with quarterly compounding. Let the yearly interest rate be represented by r.
2. Write a formula as described above, but use monthly compounding.
3. Write a general formula that can be used to find the future value when interest is compounded k times each year.
4. Study Table 8 to compare the balance in Maria's account for various compounding frequencies. Use a yearly interest rate of 8% and an initial deposit of $1,000 to verify several entries in the table. (All table entries are rounded to the nearest dollar.) Write a paragraph describing how the future value changes as the frequency of compounding increases.

Continuous Compounding

Look carefully at the values in Table 8. Reading down the columns, you should observe that the

Table 9 Values of $(1+\frac{0.08}{k})^k$

	Frequency of Compounding						
	Yearly $k=1$	Quarterly $k=4$	Monthly $k=12$	Weekly $k=52$	Daily $k=365$	Hourly $k=8760$	Minutely $k=525600$
$(1+\frac{0.08}{k})^k$	1.080000	1.082432	1.083000	1.083220	1.083278	1.083287	1.083287

balance in the account increases noticeably as the number of years increases. Now read across the rows. Are you surprised that the balance seems to level off as the frequency of compounding increases? For each *term*, or length of deposit, there appears to be an upper limit to the future value.

Each entry in Table 8 was calculated with the formula

$$\text{Future Value} = A_0\left(1+\frac{r}{k}\right)^{kN}$$

where A_0 represents the original deposit, r is the yearly interest rate, k is the number of compounding periods in a year, and N is the term of the deposit in years. In our table, $A_0 = \$1000$ and $r = 0.08$, so the values are based on the formula

$$\begin{aligned}\text{Future Value} &= \$1000(1+\tfrac{0.08}{k})^{Nk}\\ &= \$1000((1+\tfrac{0.08}{k})^k)^N.\end{aligned}$$

The value of N is constant in each row, so increases in the future values in any particular row result solely from the quantity $(1+\frac{0.08}{k})^k$. Since the balances associated with each value of N appear to have a limit, we can conclude that the quantity $(1+\frac{0.08}{k})^k$ must have some limiting value as k becomes very large. Table 9 provides values of this quantity for k-values associated with increasing frequency of compounding. Correct to six decimal places, the limiting value appears to be 1.083287.

Use a calculator to compute the value of $e^{0.08}$. By comparing this number with the values in Table 9 you should observe a relationship between the values of $e^{0.08}$ and $(1+\frac{0.08}{k})^k$ for large values of k. Does this relationship hold for other values of r? Let's look at the case where $r = 1$. We can use a calculator or computer to evaluate $(1+\frac{1}{k})^k$ for increasing values of k. Table 10 provides some of these values.

Again it seems clear that the quantity has a limiting value. Use your calculator to compare the value of e^1 with the values in the table. Mathematicians began exploring this particular relationship early in the eighteenth century. The essence of their work is that the values of $(1+\frac{1}{k})^k$ approach a limiting value as k gets larger and larger. Mathematicians defined this limiting value, which is $2.71828\ldots$, as the number e in honor of the Swiss mathematician Leonhard Euler. In mathematical terms, we can write:

$$\lim_{k\to\infty}\left(1+\frac{1}{k}\right)^k = e,$$

which is read, "the limit of $(1+\frac{1}{k})^k$ as k increases without bound is e."

We can now generalize this relationship for any value of r. We want to consider

$$\lim_{k\to\infty}\left(1+\frac{r}{k}\right)^k,$$

where r is a constant. Use your calculator to compare the values of $e^{0.06}$ and $(1+\frac{0.06}{k})^k$ for large values of k. Repeat this process for several more values of r. You should feel fairly confident that:

$$\lim_{k\to\infty}\left(1+\frac{r}{k}\right)^k = e^r.$$

The exponential function with base e, $f(x) = e^x$ (sometimes written as $f(x) = \exp(x)$), is important in many applications where a quantity grows or decays at a constant rate. Based on the relationship we have observed, we can use the exponential function with base e to arrive at fairly

Table 10 Values of $(1+\frac{1}{k})^k$

	$k=10$	$k=100$	$k=1,000$	$k=10,000$	$k=100,000$	$k=1,000,000$
$(1+\frac{1}{k})^k$	2.593742	2.704814	2.716924	2.718146	2.718268	2.718280

accurate approximations for future values when interest is compounded frequently. For example, if interest is compounded frequently so that k is large, future values given by $\$1000(1+\frac{0.08}{k})^{kN}$ can be approximated by $\$1000e^{0.08N}$.

This approximation using e is called *continuous compounding*. The reason for the name should be clear: as the frequency of the compounding increases, one can imagine compounding at every instant. The approximation

$$\text{Future Value} \approx A_0(e^r)^N$$

is reasonably accurate for quarterly compounding and even more accurate for monthly and daily compounding. Use computer or calculator graphs to compare values of $f(x) = 1000(e^{0.08})^x$, $g(x) = 1000(1+\frac{0.08}{12})^{12x}$, and $h(x) = 1000(1+\frac{0.08}{4})^{4x}$. (When using the computer, you may need to enter $e^{0.08}$ as exp(0.08).) We know that $f(x)$ provides future values based on continuous compounding and $g(x)$ and $h(x)$ yield future values based on monthly and quarterly compounding, respectively. Since values of $f(x)$ are greater than values of $g(x)$ and $h(x)$, these graphs help verify that approximations based on continuous compounding will be higher than the exact values based on frequent compounding.

Exercise Set 3

1. Verify that the entries in Table 8 can be obtained by raising the corresponding entries in Table 9 to the Nth power (where N represents the number of years) and multiplying by $1000.

2. **a.** Sketch an accurate graph of $y = e^x$. Use your calculator as needed to find exact points.

b. How does this graph compare to the graphs of $y = 2^x$ and $y = 3^x$?

3. **a.** Compute the balance that results when $2,000 is deposited for one year in an account paying 8% annual interest compounded quarterly.

b. How does this balance compare to the continuous compounding approximation?

4. If Jack invests $250 at an annual interest rate of 7.5%, what is the future value after a term of two years if the interest is compounded

a. quarterly?

b. monthly?

c. weekly?

d. daily?

e. continuously?

5. **a.** Which has the greater future value after 5 years, $1000 invested at 8% with yearly compounding or $1000 invested at 7.75% with quarterly compounding? Use graphs to determine if the number of years affects which deposit has a greater future value.

b. Which has the greater future value after 5 years, $1000 invested at 8% compounded yearly or $800 invested at 9% compounded yearly? Does the number of years affect which deposit has the greater future value?

6. It is said that the island of Manhattan was purchased for $24 in 1626. Suppose the $24 had been invested at 6% annual interest compounded quarterly. What would it be worth today? For comparison, note that many of the famous buildings in Manhattan are valued today at between 300 and 600 million dollars.

7. Banks usually offer a variety of investment accounts from which their customers can choose. Rates and frequency of compounding vary from bank to bank, and even within banks, for different types of accounts, so we need a way to compare alternative investments.

a. If you deposit \$100 in an account that pays 8% compounded quarterly, what is the balance in the account at the end of one year?

b. If a competitive bank compounds only once each year, what would their interest rate need to be to yield the same balance on your \$100 deposit?

c. The answer to part b is called the *effective annual yield*, or *effective annual interest rate.* It is the interest rate that if earned for one year and compounded only once would yield the same balance as a one-year account with frequent compounding. The effective rate provides a way to compare different interest rates that have different compounding frequencies. Suppose another bank offers 7.75% interest compounded monthly. What is the effective annual rate?

8. If we denote the effective annual rate by R, then by definition R satisfies

$$A(1+R)^1 = A\left(1+\frac{r}{k}\right)^{1k}$$

where k represents the number of compounding periods per year.

a. Verify algebraically that $R = (1+\frac{r}{k})^k - 1$.

b. Use the concept of effective annual interest rate to compare a 7.25% certificate with quarterly compounded interest to a 7% certificate with monthly compounded interest. Which option will provide a better return on your investment?

9. Suppose that when you were born your parents estimated they would need \$20,000 for college expenses. The best interest rate they could find was offered on a certificate of deposit paying 6% compounded monthly.

a. How much money should your parents invest to have a balance of \$20,000 on your eighteenth birthday?

b. What is the effective annual interest rate for this account?

10. The amount of money that would have to be invested today to yield some specified amount in the future is called the *present value* of that future amount. (In the preceding exercise you were asked to calculate the present value of \$20,000 to be paid in eighteen years.) Suppose an athlete signs a one-year contract for \$1,000,000 and agrees to be paid over a period of five years. At the beginning of each of the next five years he or she will be paid \$200,000. What is the present value of this contract assuming a 9% interest rate compounded yearly? Another way to ask this question is: how much money should the team management have deposited in an account earning 9% annual interest when the contract is signed in order to guarantee that they can pay this five-year deal? To answer this question, separately consider the present value of each of the five payments. Since the athlete is paid at the beginning of each year, you can assume that in the first year the present value of \$200,000 is \$200,000.

11. Suppose the interest rate had been 7% in the athletic contract discussed in the preceding exercise. Now what would the present value be? Does this make sense? What would you expect to be true for an interest rate of 11% ?

12. Describe precisely how present value can be used to compare two contracts similar to those for the athlete described above.

13. a. Julie will need \$3000 in five years to buy her first car. She has found that she can purchase a certificate of deposit which will pay 7.5% compounded yearly. How large should the certificate be to enable her to purchase the car? (That is, find the present value of \$3000.)

b. Julie finds that another bank offers a 7.25% certificate with quarterly compounding. How large would this certificate need to be? Which certificate is the better investment?

c. Describe precisely how the concept of present value can be used to compare investment options like those Julie faces. In such a comparison, does the higher or lower present value represent a better option?

14. A court settlement requires Ms. Jones to pay Mr. Murphy \$4000 within six months. Ms. Jones plans to wait until the end of the six months to pay Mr. Murphy, but Mr. Murphy is anxious to settle the case and will accept less money if Ms. Jones will pay immediately. How much money should Mr. Murphy accept for an immediate settlement? Base your answer on the present value of the \$4000. Assume an interest rate of 10% compounded quarterly.

15. On the same day, Chris deposits \$500 in account A that pays 10% interest compounded yearly and \$1000 in account B that earns 6% interest compounded quarterly.

a. Use a computer or calculator to draw graphs to illustrate the balance in the two accounts over time.

b. After how many years will the two accounts have the same balance?

16. Verify the following statement: The number e is the balance which results when \$1 is invested for one year at an interest rate of 100 percent compounded continuously.

17. Imagine that you have been selected as the winner of the Wonderful Mathematics Student Award and now you have to choose between two prize options. The first option is to receive a \$50,000 certificate of deposit which pays 10% interest compounded continuously. The second option is to place one cent in a fund that is guaranteed to double every six months.

a. Which option would you choose if you actually will receive the award five years from now?

b. Which option would you choose if you will receive the award in twenty years time?

c. Find the time at which the two options are equivalent.

4

Graphing Exponential Functions

In Chapter 2, we developed graphing techniques based on the ideas of transformation and composition. We will now apply these techniques to graph exponential functions. The basic shape and characteristics of the exponential curve will be the key to our graphs. You may want to review the special characteristics of exponential functions of the form $f(x) = b^x, b > 0, b \neq 1$, that are listed in Table 7 on page 109.

Example 1

Graph $y = 3e^x$, $y = e^x + 3$, $y = e^{x-3}$, and $y = e^{3x}$. Describe how the special characteristics change for each graph.

Solution The graph of $y = 3e^x$ is shown in Figure 11. It is similar to the graph of $y = e^x$, but it has been stretched vertically by a factor of 3. The y-value of the point $(0, 1)$ is multiplied by 3 so that $(0, 1)$ is stretched to $(0, 3)$. Similarly, the point $(1, e)$ is transformed to $(1, 3e)$. These changes can be verified algebraically by substituting $x = 0$ and

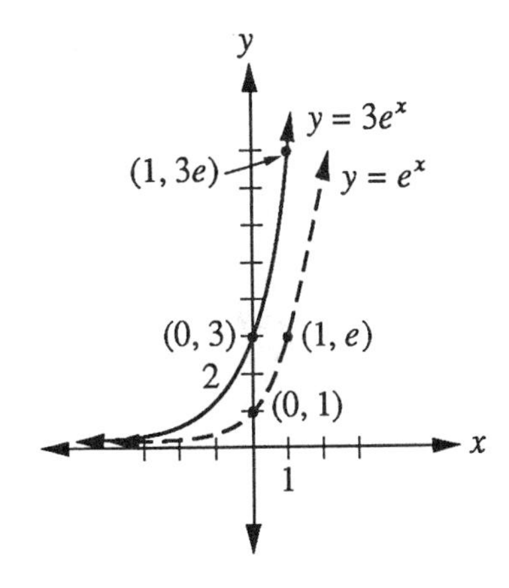

Figure 11 Graphs of $y = e^x$ and $y = 3e^x$

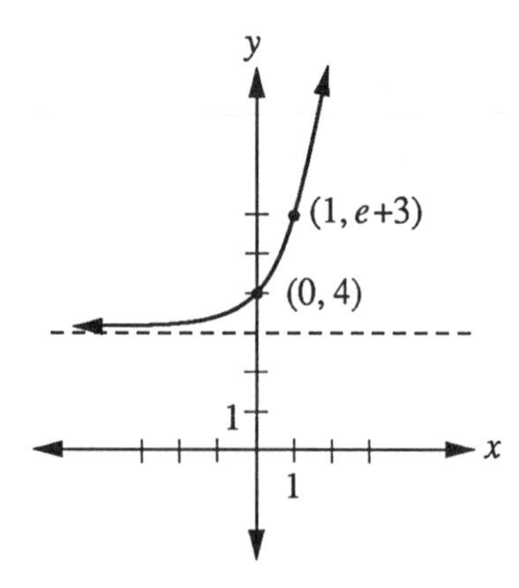

Figure 12 Graph of $y = e^x + 3$

$x = 1$ into the equation $y = 3e^x$. Use your calculator to verify that $y = 0$ is still the asymptote for this graph. If you graph $y = 3e^x$ together with $y = e^x$, you will observe that the graph of $y = 3e^x$ approaches the horizontal asymptote more slowly.

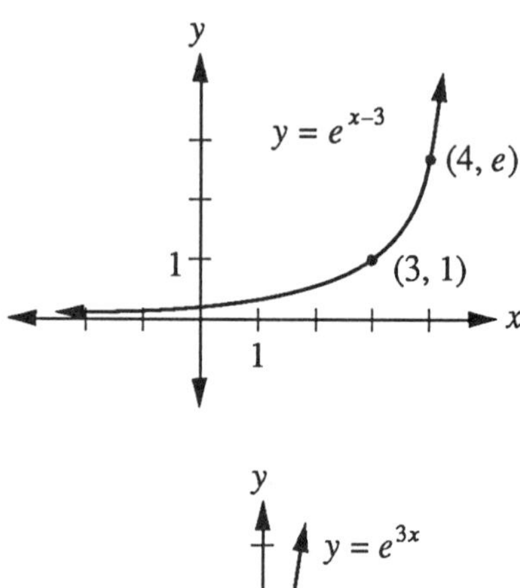

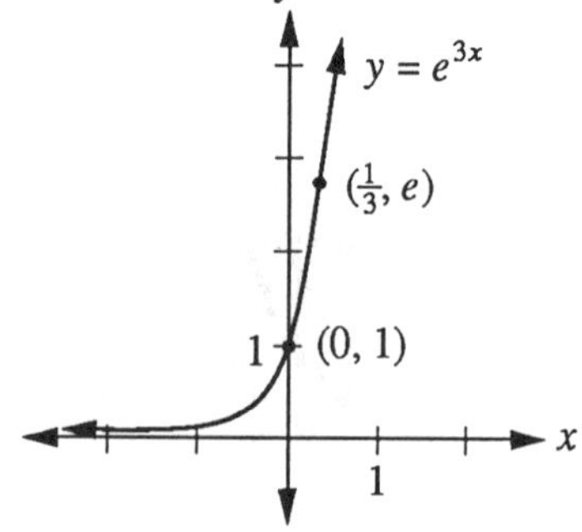

Figure 13 Graphs of $y = e^{x-3}$ and $y = e^{3x}$

The graph of $y = e^x + 3$ is provided in Figure 12. It has exactly the same shape as the graph of $y = e^x$, but it has been shifted vertically up three units. The special points change from $(0, 1)$ to $(0, 1+3)$, or $(0, 4)$, and from $(1, e)$ to $(1, e+3)$; the asymptote is also shifted up three units to $y = 3$.

The graph of $y = e^{x-3}$ is shown first in Figure 13. This graph has the same shape as $y = e^x$, but it has been shifted horizontally three units to the right. The point $(0, 1)$ is shifted to $(3, 1)$ while the point $(1, e)$ is shifted to $(4, e)$. Substitute $x = 3$ and $x = 4$ into the equation $y = e^{x-3}$ to verify that these points are on the graph.

The graph of $y = e^{3x}$, also shown in Figure 13, is obtained by horizontally compressing the graph of $y = e^x$ by a factor of $\frac{1}{3}$. The point $(1, e)$ becomes $(\frac{1}{3}, e)$; the point $(0, 1)$ does not change since $\frac{1}{3} \cdot 0 = 0$. ■

Class Practice

1. Graph each function and describe how the special characteristics of the exponential function are transformed for each.
   a. $y = (-4)2^x$
   b. $y = 3^x - 2$
   c. $y = 2^{x+5}$
   d. $y = e^{0.5x}$

Example 2

Graph $y = 4 + 2^{x+1}$.

Solution The graph of $y = 4+2^{x+1}$ has two transformations. If we let $f(x) = 2^x$, then

$$y = 4 + 2^{x+1} = f(x+1) + 4.$$

This curve will have the same shape as the graph of $y = 2^x$, but it will be shifted horizontally one unit to the left and vertically four units up. The horizontal asymptote will be $y = 4$ (rather than $y = 0$). The point $(0, 1)$ of $f(x) = 2^x$ will shift one unit to the left and four units up, so it becomes $(0 - 1, 1 + 4)$, or $(-1, 5)$. What happens to the

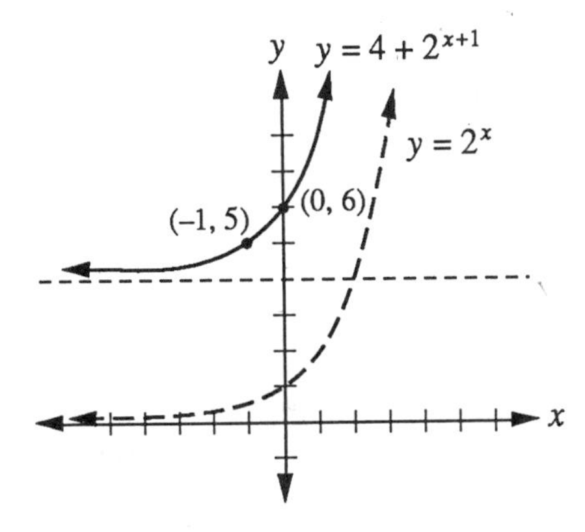

Figure 14 Graphs of $y = 2^x$ and $y = 4 + 2^{x+1}$

point $(1, 2)$ of $f(x) = 2^x$? Figure 14 shows the graph of $y = 4 + 2^{x+1}$. ■

Example 3

In a laboratory experiment, the growth of a tumor in a mouse was monitored over time. Using data analysis techniques, the function $s = 0.32e^{0.11d}$ was found to be a good model for the growth. In this equation, d represents the number of days since monitoring began, and s represents the size of the tumor in cubic centimeters. Graph the function over an appropriate domain. What is the initial size of the tumor? Determine the size after ten days and after twenty days.

Solution The graph of $y = e^x$ has been compressed vertically by a factor of 0.32 and stretched horizontally by a factor of $\frac{1}{0.11} \approx 9$ to obtain the graph shown in Figure 15. The graph contains the point $(0, 0.32)$. We can estimate that another point has coordinates of approximately $(9, 0.32e) \approx (9, 0.9)$, or we can use a calculator to obtain coordinates of another point on the graph. The appropriate domain is nonnegative values of d. Though we would expect there to be physical limitations on the size of the tumor, it is not known how large the tumor can grow, so we have not attempted to show this in the graph.

The initial size of the tumor is 0.32 cc since this is the s-value associated with $d = 0$. Substituting

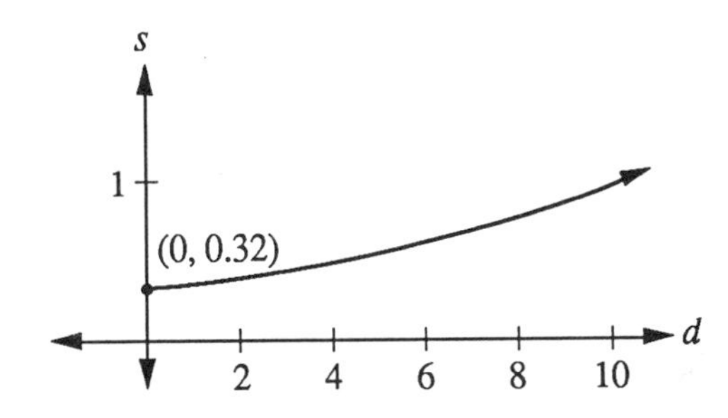

Figure 15 Graph of $s = 0.32e^{0.11d}$

$d = 10$ and $d = 20$ into the given equation reveals that the size of the tumor is 0.96 cc after ten days and 2.89 cc after twenty days. ■

Example 4

Referring to the experiment in Example 3, at what daily rate is the tumor growing?

Solution The daily growth rate is the percentage by which the size of the tumor increases each day. If we represent the size of the tumor at day n by $0.32e^{0.11n}$ and we represent the daily growth rate by g, then multiplying g times $0.32e^{0.11n}$ gives the increase in tumor size from day n to day $n+1$. Thus the tumor size at day $n+1$ equals the tumor size at day n plus this increase. We can express this relationship with the equation

$$0.32e^{0.11(n+1)} = 0.32e^{0.11n} + (0.32e^{0.11n})g.$$

Solving for g gives

$$0.32e^{0.11(n+1)} = 0.32e^{0.11n}(1+g)$$
$$e^{0.11} = (1+g)$$

so

$$(1+g) \approx 1.116$$
$$g \approx 0.116, \text{ or } 11.6\%.$$

Since the value of g does not depend on the value of n, we conclude that the growth of the tumor occurs at the constant rate of approximately 11.6% per day.

Notice that the growth rate is approximately equal to the 0.11 in the exponent of the original equation. We will see later that in an exponential equation with base e, the coefficient of the variable in the exponent can be used to approximate the growth rate whenever the growth rate is small. ■

Exercise Set 4

1. Often we can simplify graphing exponential functions by applying laws of exponents. For example, since $e^{x-3} = e^x e^{-3} \approx 0.05e^x$, we could graph $y = e^{x-3}$ either by shifting $y = e^x$ three units to the right or by compressing the graph of $y = e^x$ by a factor of 0.05. Discuss two different ways to graph the following equations as transformations of an exponential function. Which way seems easier to you? Rewrite each equation to make your choice of transformation obvious, and sketch a graph. Identify the horizontal asymptote and the points that correspond to $(0, 1)$ and $(1, b)$.

   a. $y = e^{2x}$

   b. $y = 4 \cdot 2^{x-1}$

2. **a.** Explain why these expressions are all equal. $3^{3x-2} = 3^{3(x-\frac{2}{3})} = 27^{x-\frac{2}{3}} = \frac{1}{9}(27)^x$.

   b. Graph $y = 3^{3x-2}$ using one of the expressions for 3^{3x-2}. Label the coordinates of the points that correspond to $(0, 1)$ and $(1, b)$.

   c. Which form did you choose to graph? Why?

3. Graph the following functions. Label asymptote(s) and important points.

   a. $y = 5^{x-7}$

   b. $y = e^{1.61x}$

   c. $y = 5 + 3^{2x}$

   d. $y = 1 + 2e^{-x}$

   e. $y = -8 + 0.5^x$

   f. $y = 3 - 6^x$

   g. $y = 8^{2x-2}$

   h. $y = 7(3^x)$

   i. $y = -3(2^{2x}) + 1$

   j. $y = |-5 + 3^{x+2}|$

4. A typical worker at a supermarket bakery can decorate $f(t)$ cakes per hour after t days on the job, where

$$f(t) = 10(1 - e^{-\frac{1}{4}t}).$$

   a. Sketch a graph of $f(t)$. Restrict your domain to meaningful values of t.

   b. How many cakes can a newly-employed worker decorate in an hour?

   c. After eight days, how many cakes can a worker decorate in an hour?

   d. Based on this graph, after a worker has decorated cakes for a very long time, how many cakes can he or she decorate in an hour?

5. Rheumatoid arthritis patients are treated with large doses of aspirin. Research has shown that the concentration of aspirin in the bloodstream increases for a period of time after the drug is administered and then decreases exponentially. For a typical patient, this relationship is given by $a = 14.91e^{-0.18t}$, where t represents the number of hours since peak concentration and a represents the concentration of aspirin measured in milligrams per cubic centimeter of blood.

   a. Graph the function over an appropriate domain. Label the coordinates of the points that correspond to $(0, 1)$ and $(1, e)$.

   b. Determine the peak concentration of aspirin.

   c. Determine the amount of aspirin remaining four hours after peak concentration.

   d. Use a computer- or calculator-generated graph to determine how long it takes for the concentration to decrease to 5 mg per cc of blood.

6. In a classroom experiment students made fudge the old-fashioned way. They cooked sugar, milk, and chocolate until it reached 234° Fahrenheit. Then they removed the mixture from the heat, added butter and vanilla, and let the mixture cool to 110°. As they monitored the temperature, they kept records and used data analysis techniques to determine a mathematical model for the temperature of the fudge as a function of time. Their model is

$F = 154e^{-0.00063t} + 77$, where F is the temperature of the fudge in degrees Fahrenheit and t represents the number of seconds elapsed since stirring in the butter.

a. Graph this function over an appropriate domain. Interpret the y-intercept and the horizontal asymptote.

b. Determine the temperature of the mixture after thirty minutes.

c. Use a computer- or calculator-generated graph to determine how long it takes for the temperature to reach 110°.

---

5 Introduction to Logarithms

In the preceding exercise set, a problem about aspirin dosage provided the function $a = 14.91e^{-0.18t}$ and asked you to determine the value of t when $a = 5$. What happens when we attempt to solve the equation $5 = 14.91e^{-0.18t}$ algebraically? Dividing both sides by 14.91 gives $0.335 = e^{-0.18t}$. How do we isolate the variable t? To answer this question we need to define *logarithm*. Logarithms allow us to rewrite an exponential equation so that the exponent is isolated. In this case, we can rewrite

$$0.335 = e^{-0.18t} \text{ as } \log_e 0.335 = -0.18t.$$

We read $\log_e 0.335$ as *logarithm base e of 0.335*. Both of these equations convey exactly the same information: $-0.18t$ is the exponent that we put on base e to yield 0.335. In general, $a = b^c$ and $c = \log_b(a)$ are equivalent statements. Thus, $\log_b(a)$ is defined to be the number c so that $b^c = a$.

The logarithm with base e is a special one; it is called the *natural logarithm*. We usually write $\log_e x$ as $\ln x$. We can now use the ln key on the calculator to solve for t:

$$-0.18t = \log_e(0.335) = \ln(0.335) \approx -1.094.$$

So

$$t \approx \frac{-1.094}{-0.18} = 6.08.$$

You might notice that there is only one key on the calculator marked log. It represents $\log_{10} x$. The logarithm with base 10 is called the *common logarithm* and often is written with no base indicated, a style that will be used in this text as well. We require that the base be positive and not equal one for the logarithmic function, just as we did for the base of the exponential function.

Example 1

Convert each equation to an equivalent logarithmic or exponential form.

a. $10^{-3} = 0.001$ **b.** $x^y = z$

c. $\log_2 \frac{1}{16} = -4$ **d.** $\log_a b = c$

Solution We use the definition of logarithm to rewrite each of these equations:

a. $10^{-3} = 0.001$ is equivalent to $\log_{10}(0.001) = -3$.

b. $x^y = z$ is equivalent to $\log_x z = y$.

c. $\log_2 \frac{1}{16} = -4$ is equivalent to $2^{-4} = \frac{1}{16}$.

a. $\log_a b = c$ is equivalent to $a^c = b$.

Notice that the value of the logarithm is the exponent when the equation is expressed in exponential form. ■

In the example at the beginning of this section we were given an output for an exponential function and asked to find the corresponding input. You probably recognize this process; it is the one we use to find inverse functions. To find the inverse of an exponential function $y = b^x$, we must solve for x. Since

$$y = b^x \text{ is equivalent to } x = \log_b y \text{ for } b > 0,$$

solving $y = b^x$ for x yields $x = \log_b y$. So given $f(x) = b^x$, the inverse function is $f^{-1}(x) = \log_b x$. The domain of the logarithmic function is the range of the exponential function, and the range of the logarithmic function is the domain of the exponential function. Thus, the logarithmic function has the positive real numbers as its domain and all real numbers as its range. These characteristics will become more apparent when we graph logarithmic functions in the next section.

Example 2

Find the inverse of $f(x) = 10^x + 3$. Specify the domain and range of $f(x)$ and $f^{-1}(x)$.

Solution The function $f(x)$ is one-to-one with domain all real numbers and range $y > 3$. To find $f^{-1}(x)$, we first solve for x in the equation $y = 10^x + 3$ as follows:

$$y = 10^x + 3$$
$$y - 3 = 10^x$$
$$\log_{10}(y - 3) = x.$$

Now we can write $f^{-1}(x) = \log_{10}(x - 3)$. The domain of $f^{-1}(x)$ is the range of $f(x)$, or $x > 3$; the range of $f^{-1}(x)$ is the domain of $f(x)$, or all real numbers. ■

Exercise Set 5

1. Convert each equation to exponential form.
   a. $\log_4 64 = 3$
   b. $\log_5 1 = 0$
   c. $\log 10,000 = 4$
2. Convert each equation to logarithmic form.
   a. $e^1 = e$
   b. $99^0 = 1$
   c. $\sqrt{16} = 4$
3. Evaluate the following without a calculator.
   a. $\log 0.1$
   b. $\log_3 81$
   c. $\ln \frac{1}{e^3}$
   d. $\log_2 \sqrt[5]{2^3}$
4. Estimate values of the logarithms below. Check your estimates with a calculator.
   a. $\log 12$
   b. $\log 1.2$
   c. $\log 1200$
   d. $\ln 3$
   e. $\ln 27$
   f. $\ln(\frac{1}{3})$
5. Find the inverse of the following functions and specify the domain of each.
   a. $f(x) = e^x$
   b. $f(x) = 3^{x-1}$
   c. $f(x) = \ln(2x)$
   d. $f(x) = \log_2(x - 3)$

6
Properties of Logarithmic Functions

Graphing Logarithmic Functions

The graph of $y = \log_b x$ can be obtained by first graphing its inverse $y = b^x$ $(b > 0)$ and then sketching its reflection about the line $y = x$. The graphs of these two functions are shown in Figure 16. In this figure the base b is greater than one.

The graph of the logarithmic function has

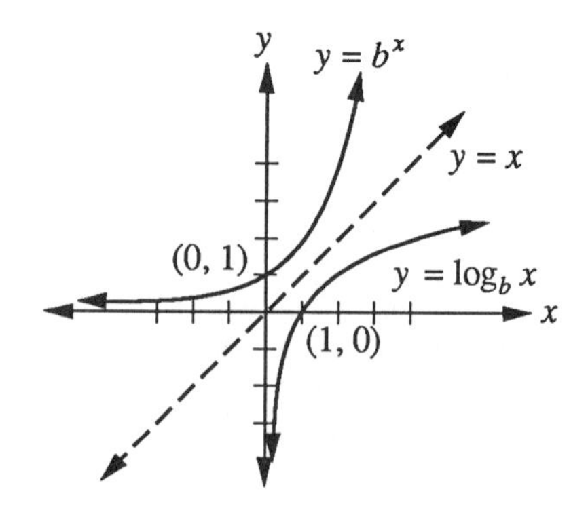

Figure 16 Exponential and Logarithmic Functions with $b > 1$

Table 17 Characteristics of Logarithmic Functions of the form $y = \log_b x,\ b > 0,\ b \neq 1$

1. The point $(1, 0)$ is on every graph.
2. The point $(b, 1)$ is on every graph.
3. The y-axis $(x = 0)$ is an asymptote to each curve.
4. The domain of each function is the set of real numbers greater than zero.
5. The range of each function is the set of real numbers.
6. Each function is one-to-one.

special characteristics that correspond to those of the exponential function that are listed in Table 7 on page 109. Use the graph of $y = \log_b x$ to verify the special characteristics of logarithmic functions that are listed in Table 17.

Several more pairs of exponential and logarithmic functions are graphed in Figure 18. Compare the graphs of $y = \log_3 x$ and $y = \log x$ to see that the larger base produces a curve that increases more slowly. Examine the graph of $y = \log_{\frac{1}{3}} x$.

Make a conjecture about the graph of $y = \log_b x$ if $0 < b < 1$. Recall that the graphs of $y = 3^x$ and $y = (\frac{1}{3})^x$ are mirror images of each other about the y-axis. How are the graphs of $y = \log_3 x$ and $y = \log_{\frac{1}{3}} x$ related?

The logarithmic function $f(x) = \log_b x$ should now be added to our function toolkit. To graph an equation that is a transformation of a logarithmic function, we will apply the same techniques that we used most recently for exponential functions. The most useful characteristics for graphing transformations of a logarithmic function are the points (1,0) and $(b, 1)$ and the vertical asymptote $x = 0$.

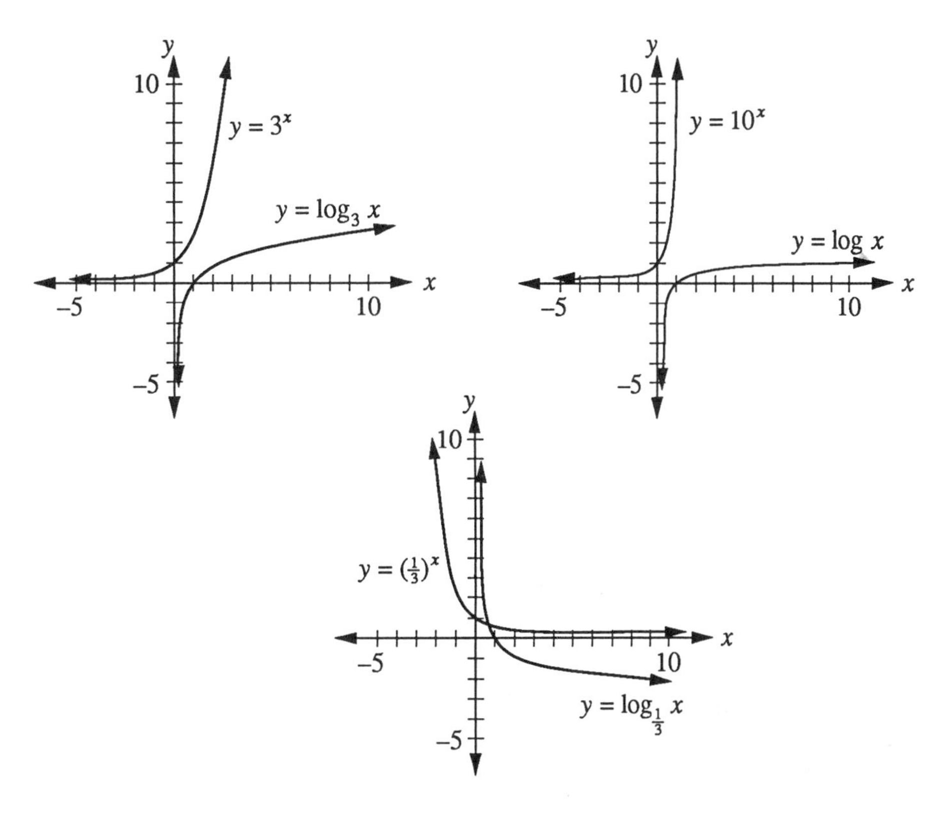

Figure 18 Exponential and Logarithmic Functions

Example 1

Graph $y = \ln(x+1)$.

Solution This graph is provided in Figure 19. It has exactly the same shape as the graph of $y = \ln x$, but the graph has been shifted to the left one unit. The vertical asymptote shifts one unit left to $x = -1$, and the special point $(1, 0)$ becomes $(0, 0)$. The point $(e, 1)$ is shifted to $(e-1, 1) \approx (1.7, 1)$. ■

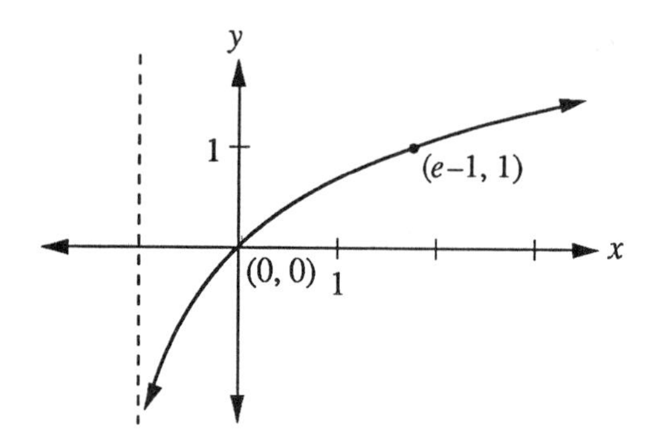

Figure 19 Graph of $y = \ln(x+1)$

Example 2

Graph $y = 4\log_2(x) + 5$.

Solution If we define $g(x) = \log_2(x)$, then $y = 4\log_2(x) + 5$ can be rewritten as $y = 4g(x) + 5$. The graph of $g(x) = \log_2(x)$ has undergone two transformations. These include a vertical stretch by a factor of four and a vertical shift five units up. The point $(1, 0)$ of $g(x)$ is transformed to $(1, 5)$ on $y = 4\log_2(x) + 5$. Note that the vertical stretch does not affect the point $(1, 0)$. The point $(2, 1)$ on $g(x)$ becomes $(2, 9)$ on $y = 4\log_2(x) + 5$. The asymptote remains $x = 0$ because there are no horizontal transformations. The graph of this function is shown in Figure 20. ■

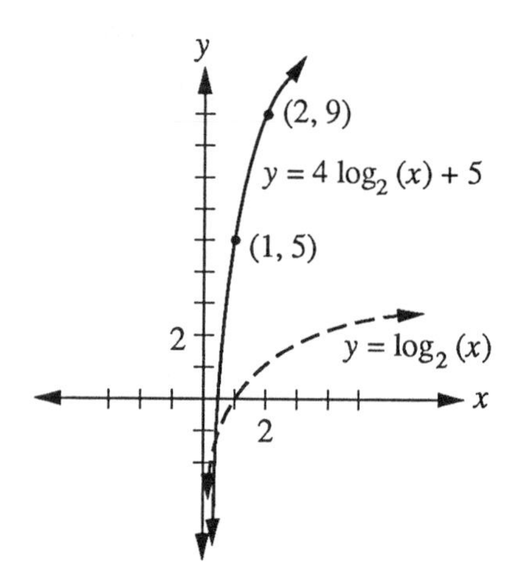

Figure 20 Graphs of $y = \log_2(x)$ and $y = 4\log_2(x) + 5$

Compositions of Exponential and Logarithmic Functions

Since the exponential and logarithmic functions are inverses and both are one-to-one functions, the compositions of these two functions should be straightforward. If $f(x) = 10^x$, then $f^{-1}(x) = \log x$, and we expect that

$$f(f^{-1}(x)) = x \text{ and } f^{-1}(f(x)) = x.$$

For what values of x are these equations true?

First consider

$$f(f^{-1}(x)) = 10^{\log x}.$$

This is a function composition with domain restrictions on the inner function $f^{-1}(x) = \log(x)$. For $f^{-1}(x) = \log x$ to be defined, we must have $x > 0$. So we can conclude that

$$f(f^{-1}(x)) = 10^{\log x} = x, \text{ if } x > 0.$$

You might want to verify this statement with some specific examples. For example, according to this statement $10^{\log 100} = 100$. Is this true? We know that $\log 100 = 2$, so $10^{\log 100} = 10^2 = 100$ as expected Use your calculator to verify that $e^{\ln 4} = 4$ and that $10^{\log(-2)}$ is undefined.

Now consider

$$f^{-1}(f(x)) = \log(10^x).$$

There are no domain restrictions on the inner function $f(x) = 10^x$ of this composition because $f(x) = 10^x$ is defined for all real numbers. Furthermore, since values of $f(x)$ are always positive, we can take the common logarithm of any value obtained from 10^x. So we can conclude that

$$f^{-1}(f(x)) = \log(10^x) = x, \text{ if } x \text{ is any real number.}$$

According to this statement, $\log(10^3) = 3$. We can verify this easily since $\log(10^3) = \log 1000 = 3$. Use your calculator to verify that $\ln e^5 = 5$. Now use your calculator to evaluate $\log(e^3)$. Are you surprised that the result is not 3? The bases of the logarithmic function and the exponential function must be the same for their composition to be the identity function. In general:

$$b^{\log_b x} = x, \text{ if } x > 0,$$

and

$$\log_b b^x = x, \text{ if } x \text{ is any real number.}$$

Laws of Logarithms

Since the logarithmic function is the inverse of the exponential function, you will not be surprised that the laws of logarithms are directly related to the laws of exponents.

For all positive real numbers a, x, and y,

$$\log_a(xy) = \log_a(x) + \log_a(y).$$
$$\log_a(\tfrac{x}{y}) = \log_a(x) - \log_a(y).$$
$$\log_a(x^y) = y\log_a(x).$$

We can verify the first of these laws if we let $x = a^u$ and $y = a^v$. We know that $\log_a(x) = u$ and $\log_a(y) = v$. So,

$$\begin{aligned}\log_a(xy) &= \log_a(a^u \cdot a^v) = \log_a(a^{u+v}) \\ &= u + v = \log_a(x) + \log_a(y).\end{aligned}$$

Note that we have expressed the logarithm of a product as a sum by using one of the laws of exponents.

Historically the laws of logarithms were very useful in simplifying complicated arithmetic computations. Notice in the second law of logarithms that the logarithm of a quotient is equal to a difference. In "pre-calculator" days this significantly simplified calculations, because a division problem could be replaced by a subtraction problem. Each of these laws was as important a part of the toolkit of a mathematician or scientist in the time before calculators as your calculator is to you today. For us, these laws are still useful in simplifying expressions. We will examine their role in graphing logarithmic functions in the following examples.

Example 3

Graph $y = \log_2 \sqrt{\frac{x}{8}}$.

Solution At first glance this example may appear to require the graphing techniques we used for compositions of functions. The laws of logarithms, however, enable us to simplify this expression so that we can graph a transformation of $y = \log_2(x)$.

The third law of logarithms allows us to rewrite this function as

$$y = \log_2 \sqrt{\frac{x}{8}} = \log_2 \left(\frac{x}{8}\right)^{\frac{1}{2}} = \frac{1}{2}\log_2\left(\frac{x}{8}\right).$$

We can now graph this function by stretching the graph of $y = \log_2(x)$ horizontally by a factor of eight and compressing it vertically by a factor of one-half. The point $(1,0)$ is transformed to $(8(1), \frac{1}{2}(0)) = (8,0)$, and the point $(2,1)$ becomes $(16, \frac{1}{2})$. The graph of this function is provided in Figure 21. ■

Example 4

Graph $y = \log(x^2)$.

Solution The function $y = \log(x^2)$ could be graphed using composition techniques by letting

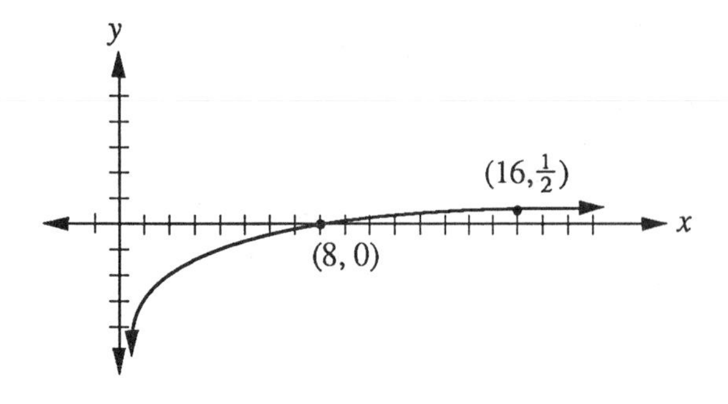

Figure 21 Graph of $y = \log_2(\frac{x}{8})^{\frac{1}{2}}$

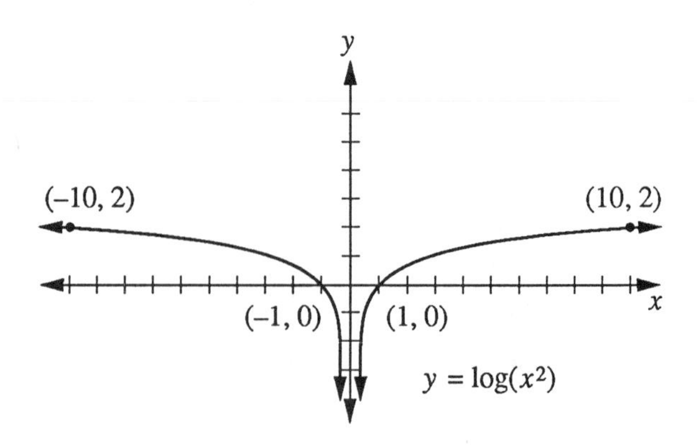

Figure 22 Graph of $y = \log(x^2)$

$f(x) = \log x$ and $g(x) = x^2$, and graphing $y = f(g(x))$. Alternatively, we could use a law of logarithms to rewrite the function as

$$y = 2\log x.$$

The function now appears to be simply a vertical stretch of $y = \log x$. There is a pitfall, however, to this second technique. Since we can take logarithms only of numbers greater than zero, the domains of the two functions are not the same. For $y = \log(x^2)$, the domain is all real numbers except zero. For $y = 2\log x$, the domain is $x > 0$. To avoid losing part of the domain of the original function, we should rewrite the function as

$$y = \log(x^2) = 2\log|x|,$$

or

$$y = \begin{cases} 2\log x, & x > 0 \\ 2\log(-x), & x < 0. \end{cases}$$

The graph will consist of two branches which are mirror images of each other across the y-axis. The y-axis will be a vertical asymptote, and the graph will contain the points $(1, 0)$, $(-1, 0)$, $(10, 2)$, and $(-10, 2)$. The graph of this function is provided in Figure 22. ■

Exercise Set 6

1. For what values of b is $f(x) = \log_b(x)$ an increasing function? a decreasing function?
2. Use a computer or calculator to graph $y = \log x$ and $y = \ln x$ on the same axes.

   a. Write a paragraph in which you compare these graphs. What special features do they share? In what ways are they different?

   b. Use graphs as needed to complete the following.

      i. For what values of x are values of $\log(x)$ negative? zero? positive?

      ii. For what values of x are values of $\ln(x)$ negative? zero? positive?

      iii. For what values of x are values of $\log(x)$ between zero and one? equal to one? greater than one?

      iv. For what values of x are values of $\ln(x)$ between zero and one? equal to one? greater than one?
3. Verify several entries in Table 23. Use the information in the table to compare the rate of growth of $f(x) = \sqrt{x}$ and $g(x) = \ln(x)$.
4. The functions $f(x) = \sqrt{x}$ and $g(x) = \ln(x)$ are the inverses of the functions $h(x) = x^2$ and $k(x) = e^x$, respectively. Graph both $h(x)$ and $k(x)$ on the same axes and compare the

Table 23 Comparison of $f(x) = x^{\frac{1}{2}}$ and $g(x) = \ln(x)$

x	$f(x) = \sqrt{x}$	$g(x) = \ln(x)$
0.1	0.316	−2.303
0.7	0.837	−0.357
1	1.000	0
1.5	1.225	0.405
5	2.236	1.609
10	3.162	2.303
25	5.000	3.219
100	10.000	4.605
1000	31.623	6.908
1000000	1000.000	13.816

rate at which they are increasing when $x > 0$. How can these graphs be used to support the conclusions you made in the previous exercise?

5. Computer scientists compare the speeds of some algorithms by finding the time required to complete the task as a function of the number of input items. Suppose there are three algorithms for solving one type of problem. The amount of time each of the algorithms A, B, and C takes is a function of the number of input items, n.

$$T_A(n) = 10000 + 2\ln(n)$$
$$T_B(n) = 100 + 3n^2$$
$$T_C(n) = (0.001)e^n$$

a. Rank the speeds of the algorithms for $n = 10$ and for $n = 300$.

b. When these algorithms are used in industry they often have thousands of input items. Which algorithm would be best for industrial use? Write a few sentences to justify your answer.

6. Graph each function. Label the points that correspond to $(1, 0)$ and $(b, 1)$ and the asymptote that corresponds to $x = 0$. Identify the domain and range of each.

a. $y = 3 + \log x$

b. $y = \log_{\frac{1}{2}} x$

c. $y = 4\log_3(x)$

d. $y = -3 + \log_{\frac{1}{2}}(x)$

e. $y = \ln(x - 5)$

f. $y = \ln(5x)$

g. $y = \log(-x)$

h. $y = |\log_4(x)|$

i. $y = \ln(4x - 3)$

j. $y = -3\log(x) + 5$

k. $y = -(3\log(x) + 5)$

7. Evaluate each of the following. State any restrictions on the domain of variables.

a. $\log_2 2^5$

b. $7^{\log_7(a+3)}$

c. $\log_{64} 8$

d. $(-3)^{\log_{-3} 5}$

8. In Example 3 of this section we graphed $y = \log_2 \sqrt{\frac{x}{8}}$ by rewriting this function as $y = \frac{1}{2}\log_2(\frac{x}{8})$. We can use another law of logarithms to write this function in the form $y = \frac{1}{2}(\log_2 x - \log_2 8) = \frac{1}{2}\log_2 x - \frac{1}{2}(3) = \frac{1}{2}\log_2 x - \frac{3}{2}$. Use this form of the equation to graph the function. Which form seems easier for graphing?

9. Use properties of logarithms to rewrite the following functions. Sketch a graph of each.

a. $y = \log(\frac{1}{x})$

b. $y = \ln \sqrt[3]{x}$

10. Most calculators give an error message when you try to evaluate expressions with large exponents, such as 75^{200}. Use laws of logarithms to evaluate 75^{200}.

7

Solving Exponential and Logarithmic Equations

Solving equations that involve exponents or logarithms, such as $2^x = 6$ or $\ln(4x) = 5$, often requires techniques that involve composition and laws of exponents and logarithms.

Example 1

Solve $10^x = 150$.

Solution This equation is similar to the one we solved when we defined the logarithm function in section 5. We need to move x out of its position as an exponent. In equations of this type the idea of composition is important. We know that $\log(10^x) = x$, so we can isolate the variable x by taking the common logarithm of both sides of the equation

$$\begin{aligned} \log(10^x) &= \log 150 \\ x &= \log 150 \end{aligned}$$

and use the calculator to evaluate

$$x \approx 2.176.$$

In the solution above we chose to take the common logarithm in order to apply the property $\log_b b^x = x$. We can obtain the same answer by taking the natural logarithm of both sides and using the third law of logarithms. That is,

$$\begin{aligned} \ln(10^x) &= \ln 150 \\ x \ln 10 &= \ln 150 \\ x &= \frac{\ln 150}{\ln 10} \approx 2.176. \end{aligned}$$

■

We can now generalize the procedure for solving for a variable in an exponent position. First take the logarithm (any base) of both sides of the equation. Then apply the law of logarithms,

$$\log_a(c^x) = x \log_a c.$$

Be sure that the expressions on both sides of the equation are positive whenever you take the logarithm of both sides.

Example 2

Solve $\ln(4x) = 5$.

Solution In this equation we need to move x from its position as the argument of a logarithmic function. We use the definition of logarithm to change from logarithmic to exponential form. Since $\ln(4x) = 5$, we can write

$$\begin{aligned} 4x &= e^5 \\ x &= \frac{e^5}{4} \approx 37.1. \end{aligned}$$

Another way to approach this equation is to apply ideas of composition and use both sides of the equation as exponents on e. (We informally refer to this process as exponentiating both sides of the equation.) Thus, $\ln(4x) = 5$ implies that

$$e^{\ln(4x)} = e^5.$$

Since we know $a^{\log_a x} = x$ for $x > 0$, we can simplify $e^{\ln(4x)}$ to get

$$\begin{aligned} 4x &= e^5 \\ x &= \frac{e^5}{4} \approx 37.1. \end{aligned}$$

■

We can now generalize the procedure for solving for a variable that is "trapped" in the argument of a logarithm: either change the equation to exponential form or exponentiate both sides of the equation.

Change in Base for Exponential and Logarithmic Functions

We can verify that $3^4 = 9^2$ and that $2^{-2} = (\frac{1}{16})^{\frac{1}{2}}$. Using a calculator, we can show that $5^2 \approx e^{3.22}$ and that $5^2 \approx 10^{1.398}$. How do we change from one exponential base to another? When is it valid to do so? These questions will be very important in subsequent sections when we write exponential functions as models of real-world phenomena. Equivalent functions may look different because of a different choice of base.

Let's investigate how we can express $f(x) = 2^x$ as an exponential function with base e. We need to express 2 as a power of e, say e^k. So we must solve the equation $2 = e^k$ for k. Both sides of this equation are positive, so we can use the technique of Example 1 to solve for k and take logarithms as follows:

$$2 = e^k$$
$$\ln 2 = \ln(e^k) = k \ln e = k$$
$$0.693 \approx k.$$

So

$$2 = e^k = e^{\ln 2} \approx e^{0.693},$$

and we can write

$$f(x) = 2^x = (e^{\ln 2})^x = e^{(\ln 2)x}$$

or

$$f(x) = 2^x \approx (e^{0.693})^x \approx e^{0.693x}.$$

Since exponential and logarithmic functions are inverses, you might suspect that we can also change the base of logarithmic functions. Let's try to change $g(x) = \log_2(x)$ to a natural logarithm. Let

$$g(x) = y = \log_2(x),$$

and then rewrite this equation as

$$2^y = x.$$

We now take the natural logarithm of both sides (which are positive):

$$\ln(2^y) = \ln x$$
$$y \ln 2 = \ln x$$
$$y = \frac{\ln x}{\ln 2} \approx 1.44 \ln(x).$$

Therefore,

$$g(x) = \log_2(x) \approx 1.44 \ln x.$$

Example 3

Evaluate $\log_3(7)$.

Solution Neither the calculator nor the computer can directly evaluate $\log_3(7)$. But we can use the techniques above to change the base and then evaluate. Let $y = \log_3 7$, and rewrite this expression in exponential form, $3^y = 7$.

Since we now have an exponential equation, we can take either the common log or the natural log of both sides of the equation as shown below.

$$\begin{aligned} 3^y &= 7 \\ \log(3^y) &= \log(7) \\ y\log(3) &= \log(7) \\ y &= \frac{\log 7}{\log 3} \\ y &\approx 1.77 \end{aligned} \qquad \begin{aligned} 3^y &= 7 \\ \ln(3^y) &= \ln(7) \\ y\ln(3) &= \ln(7) \\ y &= \frac{\ln 7}{\ln 3} \\ y &\approx 1.77 \end{aligned}$$

■

The technique illustrated in the preceding example can be used to change the base of any logarithm. To find $\log_a b$, let

$$y = \log_a b.$$

Then

$$a^y = b$$
$$\log_c a^y = \log_c b$$
$$y \log_c a = \log_c b$$
$$y = \frac{\log_c b}{\log_c a}.$$

So,

$$\log_a b = \frac{\log_c b}{\log_c a}.$$

Exercise Set 7

1. Express $g(x) = (\frac{1}{2})^x$ as an exponential function with base 10.

2. Rewrite the following numbers using base e and then using base 10.

a. 3^5

b. $12^{0.1}$

3. Dot and Don plan to invest \$5,000 in an account that pays 9% interest compounded monthly. Write three closed form expressions for the future value of the account after N years. Use $(1 + \frac{0.09}{12})$, 2, and e as bases.

4. Change $\log_b(a)$ to an expression involving logarithms with base c.

5. Estimate the value of the following. Then use your calculator to evaluate each to three significant digits.

a. $\log_4(5)$
b. $\log_7(21)$
c. $\log_{0.5}\left(\frac{1}{7}\right)$
d. $\log_{0.5}(10)$

6. Rewrite $\log x$ using the natural logarithm.

7. Show that $(\log_a b)(\log_b a) = 1$.

8. Solve the following equations for x. Give exact values whenever possible.

a. $\log_x(27) = -3$
b. $\log_{64} x = \frac{3}{2}$
c. $x = 5^{2\log_5 6}$
d. $\log_6(x^2) = -2$
e. $\log_2(x^2 + 5x + 10) = 4$
f. $5^x = 3$
g. $2^{x^2-x} = 4$
h. $5^{x^2-x} = 7$
i. $\log_2(x+4) + \log_2(x+2) = \log_2(3)$ (Be careful with domain!)
j. $3^{x+1} = 9^x 27^{x+1}$
k. $5^x + (12)5^{-x} = 7$
l. $\log(10x+5) - \log(x-4) = \log(2)$
m. $\ln(\ln(\ln(x+6))) = 0$
n. $10^{2x+1} = 4^{x-1}$
o. $\log_x(16) = 3$
p. $\log(x^4) = (\log x)^3$

9. a. Solve for t: $e^{rt} = 2$.

b. Solve for c: $y = \dfrac{A}{1 + Be^{-Acx}}$.

10. Find the inverse of the following functions and identify the domain of each.

a. $f(x) = 2^{x+1} - 3$
b. $f(x) = 3\ln(2x)$
c. $f(x) = 3^x + 3^{-x}$

11. What exponential function of the form $y = ae^{bx}$ goes through the points (3,10) and (6,50)?

12. A house purchased four years ago for \$40,000 was sold for \$60,000. If the value of the house continues to increase exponentially at the same rate, how much will it be worth next year?

13. Find the continuously compounded interest rate that is equivalent to 10% compounded quarterly.

14. Let $f(x) = 2^x - x^{10}$. Carry out several iterations of the bisection algorithm to find the zero in the interval [58,59]. Discuss which stopping criterion is most appropriate to produce an accurate estimate of the zero of $f(x)$.

8

A Model Describing Growth

We began this chapter by examining the balance in a savings account as interest compounded. This is a common example of exponential growth. Exponential growth can be described by the general equation $f(t) = ab^{kt}$, where t represents time, a is a constant, and b is the exponential base. When t increases by one, $f(t)$ is multiplied by b^k.

As you know, the function $f(t) = ab^{kt}$ is defined for all real values of t when considered strictly as a mathematical function. When this function is used to model growth, there may be restrictions on the values of t that are meaningful. For example, if we model bacterial growth or radioactive decay with the equation $f(t) = ab^{kt}$, the domain would be $t \geq 0$ (or perhaps all real numbers if we can look back in time and thus use negative values of t). Bacterial growth is continuous, and its graphical representation would be an unbroken increasing exponential curve. On the other hand, if we use $f(t)$ as a model for the balance in a savings account where interest is compounded monthly or quarterly, the domain would consist only of integer values of t. We showed a graphical representation for this exponential growth at the beginning of the chapter in Figure 2 on page 107. Since interest is compounded periodically, the growth is not continuous; instead, it occurs in discrete increments. The exponential function is a valuable mathematical model if we restrict our attention to meaningful values of t.

With the assumption that b is greater than 1, the value of k determines whether the function describes growth or decay. If k is greater than zero, the values of $f(t)$ are increasing and the function describes growth. On the other hand, if k is less than zero, the values of $f(t)$ are decreasing and the function describes decay. The coefficient a describes the initial value of the population at time $t = 0$. For ease in use of the calculator or computer, the bases of e and 10 are often used. The base of two may be useful when we think of a population doubling.

Example 1

Population Growth

The population of the earth is increasing by approximately 2% each year. Given that the population in 1990 is estimated to be 5.2485 billion, develop a mathematical statement that can be used to determine the future population of the earth.

Solution According to the information we are given, the population in 1991 can be obtained by multiplying the 1990 population by 1.02. Population estimates for subsequent years can be obtained by repeated multiplication by this factor. So we can estimate future population with the equation

$$p(x) = 5.2485(1.02)^x$$

where x represents the number of years after 1990 and p represents the population in billions.

Since the base of the exponent is not one of our "usual" bases, it might be useful to change to base e or 10. To change 1.02 to a power of e, we need to solve the equation $1.02 = e^k$. We can take the natural logarithm of both sides to obtain $\ln(1.02) = \ln(e^k) = k$, or $k \approx 0.02$. So $1.02 \approx e^{0.02}$, and the function can be rewritten as

$$p(x) \approx 5.2485e^{0.02x}. \qquad \blacksquare$$

Notice that when we use base e in Example 1, the coefficient of x in the exponent is approximately equal to the rate of growth. This occurs when using base e with small growth rates. Use your calculator to verify that $1.04 \approx e^{0.04}$ and that $1.05 \approx e^{0.05}$. We can use computer- or calculator-generated graphs to investigate this relationship. Specifically, we can graph $y = 1 + x$ and $y = e^x$ on the same axes. These functions are graphed in Figure 24. For values of x close to zero, the curves are indistinguishable. Use your own graph to estimate the largest positive value of x for which the values of $(1 + x)$ and e^x differ by less than 0.01. What does this imply about letting k equal the growth rate when using the equation $P(t) = ae^{kt}$?

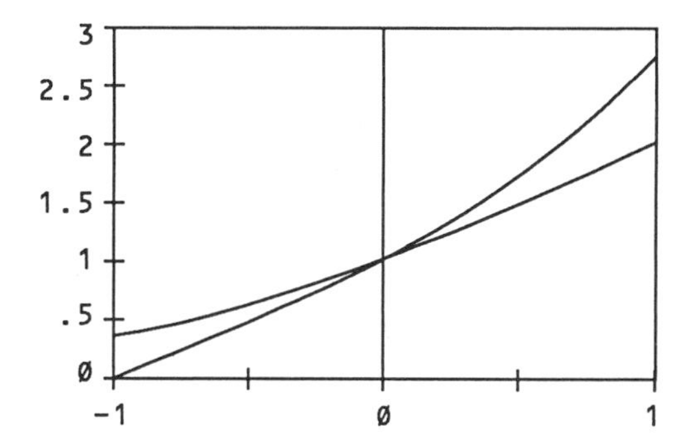

Figure 24 Graphs of $y = 1 + x$ and $y = e^x$

It should not be surprising that mathematicians view e as a very special number. Though functions to model growth can be written with any positive base, e is the unique number for which the coefficient in the exponent is approximately equal to the growth rate.

Doubling Time and Half-Life

Doubling time and half-life are special ways to measure rates of exponential growth and decay. When a quantity or population is growing exponentially, *doubling time* is the time required for the quantity or population to double. In situations where a population is decreasing exponentially, the *half-life* is the time over which the population is cut in half.

Example 2

Suppose the population of China has an annual growth rate of 1.3%. Find the length of time over which its population will double. How long will it take for the population to quadruple?

Solution Since the population is assumed to grow at a constant annual growth rate, each year the population is 1.013 times the previous year's population. If the current population of China is some number N, when the population doubles it will be $2N$. We need to solve the equation $2N = N(1.013)^t$, where t is the number of years necessary for doubling.

This yields

$$\begin{aligned} 2 &= (1.013)^t \\ \ln(2) &= \ln(1.013)^t \\ \ln(2) &= t(\ln(1.013)) \\ t &= \tfrac{\ln 2}{\ln 1.013} \\ t &\approx 53.7. \end{aligned}$$

From this solution, we see that the population in China will double in approximately 53.7 years, assuming a continuation of the current annual growth rate of 1.3%. If this growth rate does not change, the population should quadruple in two doubling times, or approximately 107 years. This can be verified algebraically by solving the equation $4N = N(1.013)^t$ for t. ∎

Class Practice

1. Rework Example 2 by first writing an equation for the population of China in the form $P(t) = ae^{kt}$ and then solving for the doubling time.
2. Verify from your work in the preceding exercise that $\frac{\ln 2}{k}$ is a general expression for doubling time when the population function is written in the form $P(t) = ae^{kt}$.
3. Let g represent the growth rate of a population. Explain why the expression $\frac{\ln 2}{g}$ provides a good approximation of doubling time whenever g is small.

The decay of radioactive nuclei over time can be modeled by a decreasing exponential function. Since we know the rate of decay of the radioactive carbon isotope carbon-14, we can date organic materials and fossils; the ages of terrestrial, lunar, and meteoritic rocks are determined using uranium, thorium, and other long-lived radioactive nuclei; and blood and fluid flow in living humans are studied using a variety of short-lived radioactive isotopes. Exponential decay can be described by the equation:

$$P(t) = P_0 e^{kt},$$

where $P(t)$ represents the number of radioactive atoms present at time t, P_0 represents the number of radioactive atoms present initially (that is, at time $t = 0$), and the constant k is negative.

A useful measure of the rate at which a radioactive substance decays is the half-life of that element. This is the time necessary for half the amount of a radioactive material to decay. For example, the half-life of carbon-14 is 5,730 years. In living matter, the relative amount of carbon-14 is constant. When the matter dies, the carbon-14 atoms decay into nitrogen-14 with an emission of radiation. The decay occurs in such a way that after approximately 5,730 years there will be only half as many carbon-14 atoms present; after approximately 11,460 years there will be only half of this half, or one fourth, of the original amount of carbon-14 remaining. Since the half-life is so long, carbon-14 is useful in dating archaeological objects. By measuring the emission of radiation, scientists can infer the population of carbon-14 atoms, compare this to the population of carbon-14 atoms in similar living matter, and estimate when the matter died. Since we know the half-life of carbon-14, we can find the constant k in $P(t) = P_0 e^{kt}$. Using the fact that after one half-life (5,730 years) one-half of the original amount of carbon-14 remains, we can substitute into the equation describing decay and solve for k as follows:

$$\begin{aligned} 0.5P_0 &= P_0 e^{5730k} \\ 0.5 &= e^{5730k} \\ \ln(0.5) &= 5730k \\ k = \tfrac{\ln(0.5)}{5730} &\approx -0.000121. \end{aligned}$$

Therefore the value of the constant k for the decay in carbon-14 is approximately -0.000121.

Example 3

Suppose we find a piece of human bone that contains 43% of the amount of radioactive carbon-14 normally found in the bone of a living person. How old is the bone?

Solution We begin with the general equation $P(t) = P_0e^{kt}$ and use the value of k found above to write $P(t) = P_0e^{-0.000121t}$. We want to determine t when $P(t) = 0.43P_0$, so we need to solve

$$\begin{aligned} 0.43P_0 &= P_0e^{-0.000121t} \\ 0.43 &= e^{-0.000121t} \\ \ln(0.43) &= -0.000121t \\ t &= \frac{\ln 0.43}{-0.000121} \\ t &\approx 6970 \text{ years.} \end{aligned}$$

We conclude, therefore, that the bone is about 6,970 years old. ■

Carbon-14 is useful in the dating of relatively old artifacts and fossils because its half-life is relatively long. If we wanted to measure the age of objects that were millions of years old, however, the carbon-14 dating process would be useless, because virtually all of the carbon-14 present in the organism at death would have decayed away. Scientists use isotopes of uranium and thorium to date very old objects. Similarly, carbon-14 dating is not useful in dating material from the last few decades, since the abundance of carbon-14 in the material is virtually indistinguishable from the amount of carbon-14 in living matter. Tritium, with a half-life of 12.1 years, is often used for dating more recent matter such as wines.

Exercise Set 8

1. Use calculator- or computer-generated graphs to investigate the relationship between the values of $\ln(1 + x)$ and x close to zero. Then use this relationship to estimate the values of $\ln(1.001)$, $\ln(1.05)$, and $\ln(0.93)$. Use your calculator to determine the accuracy of your estimates.
2. Show that $\frac{\ln(0.5)}{k}$ is a general expression for half-life when the decay function is written as $P(t) = P_0e^{kt}$. Is this expression for half-life valid if the decay function is written as $P(t) = P_02^{kt}$? Why or why not?
3. When a certain drug enters the blood stream, it is absorbed gradually by body tissue, and its concentration decreases exponentially with a half-life of 3 days. If the initial concentration of the drug in the blood stream is A, what will the concentration be 30 days later? Assume that no additional doses are taken.
4. Given a population of 10 million and an annual growth rate of 3%, how long will it take this population to double? What is the size of the population in triple the doubling time?
5. A population of bacteria is growing exponentially. A researcher determines the population of the bacteria to be 2000 at 2:00 and to be 4000 at 2:10. What is the doubling time of the population? What is the estimated population at 2:05?
6. Carbon-11 decays into boron at the rate of roughly 3.5% per minute. What is the half-life of carbon-11?
7. Since the money deposited in a savings account increases exponentially, the time required for it to double is constant.

   a. To determine a general formula for doubling time, solve the equation

   $$2A = A\left(1 + \frac{r}{k}\right)^{kN}$$

   for N in terms of r and k.

   b. Use this formula to verify several doubling times in Table 25.

   c. Describe how doubling time changes as interest rate and frequency of compounding change.
8. Bankers often approximate doubling time with the Rules of 69, 70, and 72. In the previous

Table 25 Doubling Time in Years

Yearly Interest Rate r	Frequency of Compounding Yearly	Quarterly	Daily	Continuous
.05	14.21	13.95	13.86	13.86
.08	9.01	8.75	8.67	8.66
.10	7.27	7.02	6.93	6.93
.12	6.12	5.86	5.78	5.78
.15	4.96	4.71	4.62	4.62

Table 26 Estimates for Doubling Time in Years

Yearly Interest Rate r	$\frac{69}{100r}$	$\frac{70}{100r}$	$\frac{72}{100r}$
.05	13.80	14.00	14.40
.08	8.63	8.75	9.00
.10	6.90	7.00	7.20
.12	5.75	5.83	6.00
.15	4.60	4.67	4.80

exercise you should have found that exact doubling time is given by

$$N = \frac{\ln(2)}{\ln(1+\frac{r}{k})^k}.$$

In the case of continuous compounding, where $k \to \infty$ and $(1+\frac{r}{k})^k \to e^r$,

$$N = \frac{\ln(2)}{\ln(e^r)} = \frac{\ln(2)}{r} \approx \frac{0.6931472}{r} \approx \frac{69}{100r}.$$

So bankers can approximate actual doubling time with $N \approx \frac{69}{100r}$ which is commonly referred to as the Rule of 69.

a. Use the Rule of 69 to estimate the doubling time for money invested at a rate of 8% compounded continuously.

b. Explain why replacing 69 with 70 or 72 improves the accuracy of the approximation for quarterly and yearly compounding.

c. Verify several estimates for doubling time in Table 26.

d. Compare the entries in Table 25 and Table 26 to determine which approximation is more appropriate for yearly compounding. Which is more appropriate for quarterly compounding? for daily compounding?

9. The Rule of 69 can be used to estimate the time needed for the balance in an account to double. Develop a similar rule which can be used to estimate the time needed for the balance in an account to triple.

10. A local bank advertises that if you deposit at least \$5000 in an investment account your money will double in 8 years. Assuming that interest is compounded monthly, find the interest rate on the account.

11. The surface area of the Earth is approximately 197 million square miles, including oceans. If the population of the Earth is approximately 5.2 billion people, how much room is there per person? Suppose the doubling time for the world's population is approximately 41 years. When did each person have 1 square mile?

12. The woods north of town have two types of rabbits. The light brown fuzzy ones have a current population of 300, and their population doubles every 1.4 years. The second kind of rabbit is dark brown with pink ears. Its initial population is 1560, and the population doubles every 2.5 years. How long will it be before the populations are the same size?

13. Recently in the newspapers it was stated that the number of people with AIDS would double in 13 months. What is the annual growth rate of the number of people with AIDS? At the end of February, 1988, the United States had 55,167 cases of AIDS. How many cases did the U.S. have by the end of December 1988?

14. The half-life of a radioactive substance is 8000 years. If the initial amount of the substance is m, then the quantity n remaining after t years is $n = m2^{kt}$. Find k.

15. The Mesozoic Era began approximately 225

million years ago. Can carbon-14 be used to date organic substances from that era? Explain your answer.

16. Charcoal in an ancient fire pit on Java was found to contain 1/16 the amount of carbon-14 that is present in a contemporary living sample of the same size. Estimate the age of the charcoal.

17. Cesium-137, one of the dangerous radionucleides produced in the fallout of the Chernobyl disaster, has a half-life of 30.3 years. How much time must pass before the radiation emitted by the radioactive cesium (and, therefore, the number of cesium atoms) is reduced to 10% of the amount produced in the Chernobyl explosion?

9

Logarithmic Scales

We know that $\log 0.01 = -2$, $\log 0.1 = -1$, $\log 1 = 0$, $\log 10 = 1$, $\log 100 = 2$, and so forth. Each time the argument is multiplied by ten, the value of the logarithm increases by one. That is, when x-values are in the ratio 10:1, y-values differ by one. Use your calculator and the graph of $y = \log x$ shown in Figure 27 to verify that the points $(6, \log 6)$ and $(60, \log 60)$ have x-values that are in the ratio 10:1 and y-values that differ by 1. In general, equal ratios of x-values correspond to equal increments of y-values. This property of logarithms is the basis of logarithmic scales developed by scientists. In measuring certain phenomena, scientists are often more interested in comparing the ratios of different levels of the phenomenon than they are in determining the absolute level of the phenomenon itself.

For example, astronomers use *magnitudes* to rate the apparent brightness of any given star by comparing its brightness to that of the very brightest-appearing stars. Physicists use *decibels* (dB) to rate the loudness level of a given sound by comparing its intensity to that of a just-barely-

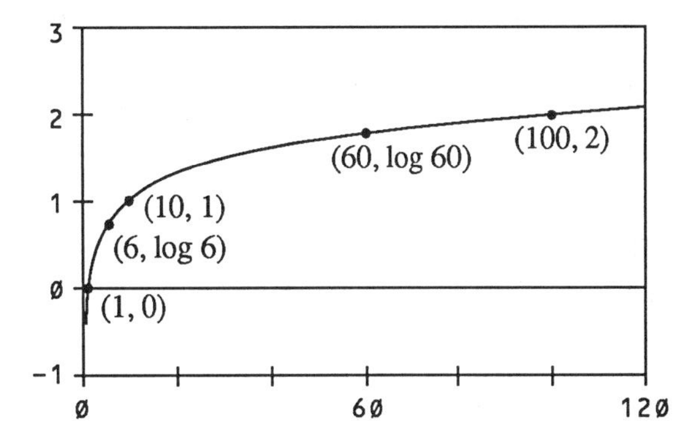

Figure 27 Graph of $y = \log x$

audible sound. In both cases, equal intervals of the measured sound or light level actually correspond to equal ratios of sound or light intensity. Similarly, geologists use the Richter scale to rate the relative energy produced in an earthquake; each number on the scale represents an earthquake ten times as strong as one of the next lower number. The scale used by each of these scientists is logarithmic. Logarithmic scales, like logarithms, convert multiplications to additions.

Sound

The intensity of a sound is dependent on the actual energy carried in the wave and is measured in units of power per unit area, or watts/(meter)^2. The greater the intensity, the louder the perceived sound. The intensities of some common sounds are given in Table 28. Physicists define a loudness level (measured in decibels) that is related to the actual intensity (I) of the sound by the following equation:

$$\text{Loudness Level (dB)} = 10 \log \left(\frac{I \text{ (W/m}^2)}{10^{-12} \text{ (W/m}^2)} \right).$$

The threshold of human hearing is a sound with intensity 10^{-12} W/m^2. It is defined to be of loudness level 0 dB and serves as a baseline by which other sounds are measured. You can use the equation above to verify that the loudness is 0 dB when the intensity is 10^{-12} W/m^2.

Table 28 Intensity and Loudness Level of Some Sounds

Sound	Intensity (W/m^2)	Loudness Level (dB)
Threshold of hearing	10^{-12}	0
Normal breathing	10^{-11}	10
Rustling leaves	10^{-10}	20
Whisper	10^{-9}	30
Quiet library	10^{-8}	40
Quiet radio	10^{-7}	50
Ordinary conversation	10^{-6}	60
Busy street traffic	10^{-5}	70
Factory	10^{-4}	80
Niagara Falls	10^{-3}	90
Siren (at 30 m)	10^{-2}	100
Loud thunder	10^{-1}	110
Rock concert (at 2 m)	1	120
Jet plane takeoff (at 30 m)	10	130
Rupture of eardrum	10^4	160

A sound of 10 decibels has ten times the intensity of a 0 decibel sound; a sound of 20 decibels has one hundred times the intensity of a 0 dB sound, or ten times the intensity of a 10 dB sound. Similarly, a sound of 30 dB has one thousand times the intensity of a 0 dB sound or ten times the intensity of a 20 dB sound. When the loudness level increases by an increment of 10 decibels, the actual sound intensity is multiplied by a factor of ten. It should be clear, therefore, that the intensity of a 60 dB sound (ordinary conversation) is *not* double the intensity of a 30 dB sound (a whisper). Rather, the sound intensity of conversation is three factors of ten, or 1000, times the intensity of a whisper, and the intensity of a whisper is 1000 times that of the least audible sound. Equal *differences* in loudness level are equivalent to equal *ratios* of sound intensity.

Example 1

Compare the decibel level of two sirens at a distance of 30 meters to the decibel level of a single siren at the same distance.

Solution The intensity of two sirens is simply twice that of one siren:

$$\begin{aligned} I_{\text{two sirens}} &= 2I_{\text{one siren}} \\ &= 2(10^{-2}) \approx 10^{0.301}(10^{-2}) = 10^{-1.699}. \end{aligned}$$

Therefore, the loudness of two sirens is:

$$\begin{aligned} \text{Loudness Level (2 sirens)} &= 10\log(\frac{10^{-1.699}}{10^{-12}}) \\ &= 10\log 10^{10.301} = 10(10.301) \approx 103 \text{ dB} \end{aligned}$$

The important lesson to be learned from this example is that the total intensity of two simultaneous sounds is the sum of the individual intensities of the two sounds, but the individual decibel levels do not add to produce the total decibel level. When the intensity level doubles, the decibel level increases by $10 \log 2 \approx 3$ decibels. ■

Light

The basic astronomical unit of brightness is the magnitude. The perceived magnitudes of two stars (m_1 and m_2) are related to their actual intensities (I_1 and I_2) by the equation

$$m_1 - m_2 = 2.5 \log\left(\frac{I_2}{I_1}\right).$$

If a bright star has intensity I_1 and a dim star has intensity I_2, then $I_2 < I_1$ and $0 < \frac{I_2}{I_1} < 1$. Since the logarithm of a number between zero and one is negative, $(m_1 - m_2)$ will be negative. Thus $m_1 < m_2$, and this implies that the magnitude of the first, brighter star is smaller. We see, then, that small magnitudes are associated with brighter, more intense light sources.

In 1856, Norman Pogson defined the zero-point of the magnitude scale such that first magnitude, $m = 1$, corresponds to the brightest-appearing stars and the sixth magnitude, $m = 6$, corresponds to the faintest stars visible to the naked eye. We can verify that this difference of five magnitudes corresponds to a factor of 100 in intensity. If $(m_2 - m_1) = 5$, we have

$$5 = 2.5 \log\left(\frac{I_1}{I_2}\right)$$
$$2 = \log\left(\frac{I_1}{I_2}\right)$$
$$10^2 = \left(\frac{I_1}{I_2}\right)$$
$$I_1 = 10^2 I_2 = 100 I_2.$$

So the intensity of a star of magnitude one is 100 times the intensity of a star with magnitude six. The magnitude system has been extended to negative magnitudes for bright things like the sun and the moon and to positive magnitudes beyond the sixth for stars visible only with telescopes. Table 29 illustrates the magnitude scale.

Table 29 Levels of Brightness

Object	**Apparent Magnitude**
Sun	−26
Full moon	−13
Venus at brightest	− 4.6
Jupiter at brightest	− 2.9
Mars at brightest	− 2.6
Sirius, the brightest star	− 1.5
Polaris, the north star	2
Faintest star visible with 7×35 binoculars	8
Faintest star visible with 8-inch telescope	14
Stars barely visible in largest telescopes	28

Consider the magnitudes of Venus and of the full moon. Using the scale of the magnitudes in Table 29, we can see that these magnitudes differ by 8.4. We have said that each difference of 5 magnitudes corresponds to a factor of 100 in brightness, so a difference in 10 magnitudes corresponds to a factor of $(100)^2$ in brightness. Since $5 < 8.4 < 10$, a difference in magnitudes of 8.4 indicates that the moon is between $(100)^1$ and $(100)^2$ times brighter than Venus. Using the equation

$$m_1 - m_2 = 2.5 \log \frac{I_2}{I_1},$$

we can compare the intensity of the full moon to the intensity of Venus:

$$-4.6 - (-13) = 2.5 \log \frac{I_2}{I_1}$$
$$8.4 = 2.5 \log \frac{I_2}{I_1}$$
$$3.36 = \log \frac{I_2}{I_1}$$
$$10^{3.36} = \frac{I_2}{I_1}$$
$$I_2 = 10^{3.36} I_1 \approx (2291) I_1.$$

We see that the full moon is approximately 2300 times brighter than Venus. As expected, 2300 is between $(100)^1$ and $(100)^2$.

Exercise Set 9

1. A light magnitude of zero ($m = 0$) corresponds to an intensity of 2.48×10^{-8} W/m^2. What is the intensity of a star of third magnitude?
2. At a party with 25 people, everyone is talking at once. What is the sound level in decibels?
3. The sound intensity level in large cities has been increasing by about one decibel annually. To what percent increase in intensity does one decibel correspond? If this annual increase continues, in how many years would the sound intensity double?
4. Find the apparent magnitude of the first quarter moon (when only the right half of the moon is sunlit). Assume that all parts of the sunlit moon contribute equally to its intensity.
5. How many times brighter (more intense) is the sun than the full moon?
6. The Richter scale measures the magnitude of an earthquake in terms of the total energy

released by the earthquake. One form of Richter's equation is

$$M = 0.67 \log E - 2.9,$$

where M is the magnitude and E is the energy in joules of the earthquake.

a. If an earthquake releases 10^{13} joules of energy, what is its magnitude on the Richter scale?

b. A very powerful earthquake occurred in Colombia on January 31, 1906, and measured 8.6 on the Richter scale. Approximately how many joules of energy were released?

7. According to seismologists, an earthquake which registers 2 on the Richter scale is hardly perceptible while an earthquake which measures 5 on this scale is capable of shattering windows and dishes and is generally classified as "minor." The San Francisco earthquake in 1989 caused great damage and registered 7.1 on the Richter scale.

a. Compare the intensity of the San Francisco earthquake to a level 5 minor earthquake. That is, calculate the ratio of their intensities.

b. If an earthquake is ten times as intense as the San Francisco earthquake, what would it measure on the Richter scale?

c. If an earthquake is twice as intense as the San Francisco earthquake, what would it measure on the Richter scale?

8. To observe stars beyond the sixth magnitude, a telescope is required. Telescopes, however, also have limitations. The *limiting magnitude* of a telescope is the magnitude of the faintest star that can be seen with the telescope. A telescope with lens diameter D meters has a limiting magnitude L, given by the formula:

$$L = 17.1 + 5.1 \log D.$$

Find the lens diameter of a telescope with limiting magnitude of 11.1.

9. An empty auditorium has a sound level of 40 dB (due to heating and air-conditioning and outside noise). On Saturday, 100 students are taking the SAT. While they are working on the test, the only sounds are labored breathing and pencils rapidly moving across the paper. The noise level then rises to 60 dB (not counting the groans). If each student contributes equally to the total noise, what would be the noise level if only 25 students were taking the test?

10. Suppose that a cluster of stars contains 100,000 stars. What is the magnitude of the entire cluster if 1000 of the stars have magnitude 15 and each of the others has magnitude 20?

11. The North Star, Polaris, is a variable star. Periodically it pulses, becoming larger and smaller, and therefore changing in brightness. If Polaris varies between 1.9 and 2.1 in magnitude, how many times more intense is it at its brightest than at its faintest?

10

Picard's Algorithm for Finding Roots

In Chapter 3, the bisection and secant algorithms were introduced. These algorithms are used to solve equations of the form $f(x) = 0$. That is, they are used to find zeros of a function. We will now study Picard's algorithm which can be used to find where $y = f(x)$ intersects the line $y = x$. These points are called *fixed points* of the function $f(x)$. Algebraically, finding fixed points is equivalent to solving equations of the form $f(x) = x$. Though Picard's method does not appear to find the roots of a function, we can use it to solve $g(x) = 0$ by first transforming this equation to $g(x) + x = x$.

First we will describe Picard's algorithm for solving the equation $f(x) = x$. Then we will illustrate this method with an example. Our first step is to guess a value for x, call it x_0. (There are no restrictions on this first guess other than being in the domain of f.) Compute $f(x_0)$, and let

$x_1 = f(x_0)$. Our second guess is x_1. Find $f(x_1)$, and let $x_2 = f(x_1)$ be the third guess. Continue this iterative process of letting $x_n = f(x_{n-1})$ until x_n and x_{n-1} are nearly equal. Then x_n is an approximate solution to the equation $f(x) = x$.

Several special restrictions must be noted. Throughout the process, x_n is equal to $f(x_{n-1})$, so the range of $f(x)$ must be a subset of the domain of $f(x)$. For the method to work, these values of x must steadily close in (converge) on a value of x for which $x = f(x)$.

Example 1

Solve the equation $x = 0.5x + 5$.

Solution We can easily solve this problem algebraically and find that $x = 10$. However, we will use Picard's algorithm to illustrate the process described above. First we need an initial guess. Suppose we let $x_0 = 8$. Then $f(8) = 9$. Next we let $x_1 = 9$ and compute $f(9)$ to find x_2. The table below shows the results of our calculations. Verify these numbers with your calculator to see how easy it is to carry out the computations. We simply use the output of one iteration as the input of the next.

x	$f(x) = 0.5x + 5$
$x_0 = 8$	$f(8) = 9$
$x_1 = 9$	$f(9) = 9.5$
$x_2 = 9.5$	$f(9.5) = 9.75$
$x_3 = 9.75$	$f(9.75) = 9.875$
$x_4 = 9.875$	$f(9.875) = 9.9375$
$x_5 = 9.9375$	$f(9.9375) = 9.96875$
$x_6 = 9.96875$	$f(9.96875) = 9.984375$
$x_7 = 9.984375$	$f(9.984375) = 9.9921875$
$x_8 = 9.9921875$	$f(9.9921875) = 9.99609375$
$x_9 = 9.99609375$	$f(9.99609375) = 9.998046875$

Accurate to two decimal places, $x_9 = f(x_9)$, and the x-value for which this occurs rounds to $x = 10$.

The geometry of this process will help us see what is happening. The graphs of $y = x$ and $f(x) = 0.5x + 5$ are drawn on the same axes in Figure 30. Their point of intersection is the solution of $x = 0.5x + 5$.

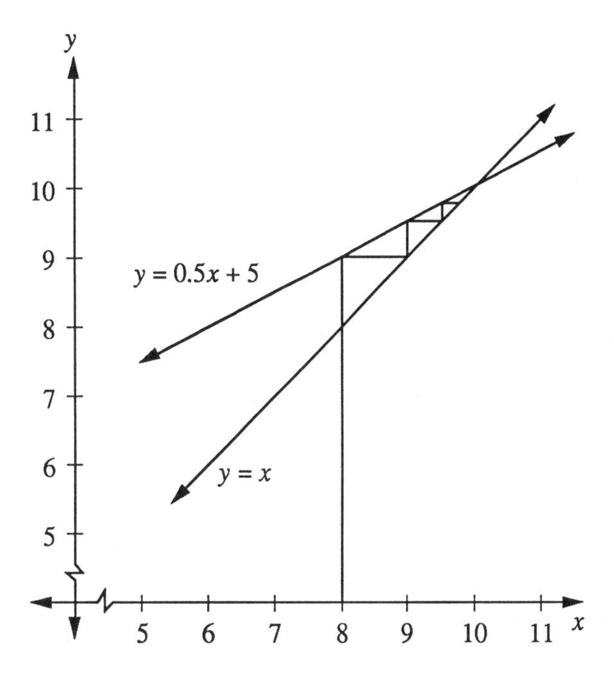

Figure 30 Graphs of $y = x$ and $y = 0.5x + 5$

Our first guess was $x = 8$. We move vertically from the point $x = 8$ on the x-axis to the line $y = 0.5x + 5$ to locate $f(8) = 9$. Now move horizontally to the line $y = x$, where we can locate $x = 9$. (Using the line $y = x$ makes the exchange of x and y easy to picture graphically.) From here we move vertically to find $f(9) = 9.5$ on the line. By repeating this process, we are "stepping in to" the point of intersection of $f(x)$ and x. ■

Class Practice

1. Use Picard's algorithm with an initial guess of $x_0 = 15$ to solve $x = 0.5x + 5$. Illustrate graphically how the process converges on $x = 10$. How many iterations are required for x_n to equal x_{n-1} with accuracy to two decimal places?

2. Repeat the preceding exercise with several different choices of x_0.

Example 2

Solve the equation $0.25e^x = x$.

Solution The graphs of $y = x$ and $y = 0.25e^x$ show that there are two solutions (see Figure 31). If we make our initial guess $x_0 = 0$ and use Picard's method, our solution is 0.357. Use your calculator to confirm this.

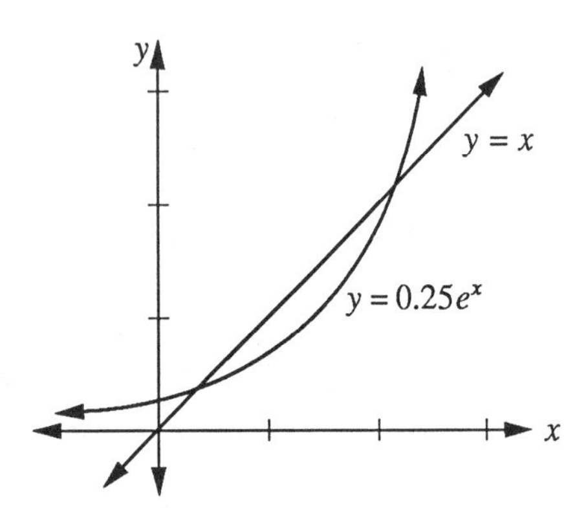

Figure 31 Graphs of $y = x$ and $y = 0.25e^x$

Now let's try to find the second solution. Choose an initial value of $x_0 = 1$. Again use the calculator to find values.

x	$f(x) = 0.25e^x$
1	.679570457
.679570457	.493257512
.493257512	.409410545

Something seems wrong. Look again at the graph. The fixed point we seek is a number between 2 and 3, but we seem to be going backwards. Use the geometry of the process to help you understand what is happening. Will starting with an initial guess larger than $x = 1$ work? Try an initial guess of $x = 2$ and then $x = 3$ on your calculator. In all cases we go either to the first solution or off to infinity.

Apparently Picard's method does not work in all situations, so we must rethink the question and attempt to find another method for finding where $y = f(x)$ and $y = x$ intersect. We know that a function and its inverse intersect on the line $y = x$. Therefore, the points where $y = f(x)$ and $y = x$ intersect are the same as the points where $y = f^{-1}(x)$ and $y = x$ intersect (see Figure 32).

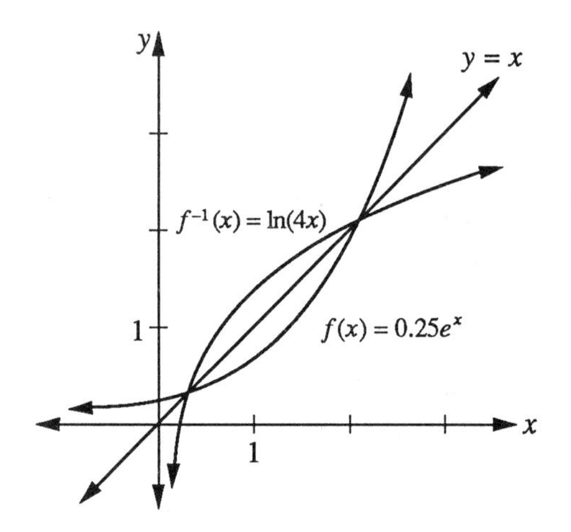

Figure 32 f and f^{-1} Intersect on $y = x$

Therefore we can look for the points of intersection of $y = x$ and $g(x) = \ln(4x)$, since $g(x)$ is the inverse of $f(x)$. The table below shows the results of the calculations.

x	$g(x) = \ln(4x)$
2	2.079441542
2.079441542	2.118393729
2.118393729	2.136952488
2.136952488	2.145675104
2.145675104	2.149748598
2.149748598	2.151645265
2.151645265	2.152527150

Therefore, the second solution for $f(x) = x$ is at $x \approx 2.15$. Can the point of intersection be found using the inverse function? ∎

Exercise Set 10

1. Write a description of Picard's algorithm; include the three components of an algorithm and the adjustment necessary if the method does not converge on a root.
2. **a.** Use Picard's method to verify that $\sqrt{x} = x$ for $x = 1$.

   b. Use Picard's method to verify that $x^3 = x$ for $x = 1$. Illustrate graphically what is happening here.
3. Solve the following equations using Picard's method. You may need to rewrite the equation in the form $f(x) = x$.

   a. $\sqrt{x+2} = x$

   b. $2^{-x} = x$

   c. $e^x = \frac{1}{x}$

   d. $x^3 + 1 = (x-1)^{(\frac{1}{3})}$

   e. $x^3 + x - 1 = 0$

   f. $e^x = x^2$

11
Graphing Using Composition

The techniques of graphing by composition that were developed in Chapter 2 can be applied when one of the composed functions is an exponential or logarithmic function.

Example 1

Sketch the graph of $y = 5^{|x|}$.

Solution This graph is shown in Figure 33. It can be graphed as a composition $y = f(g(x))$ where $f(x) = 5^x$ and $g(x) = |x|$. Notice that $f(g(x)) = 5^{|x|}$ is an even function, so the graph is symmetric about the y-axis. It contains the points $(0,1)$, $(1,5)$, and $(-1,5)$. ∎

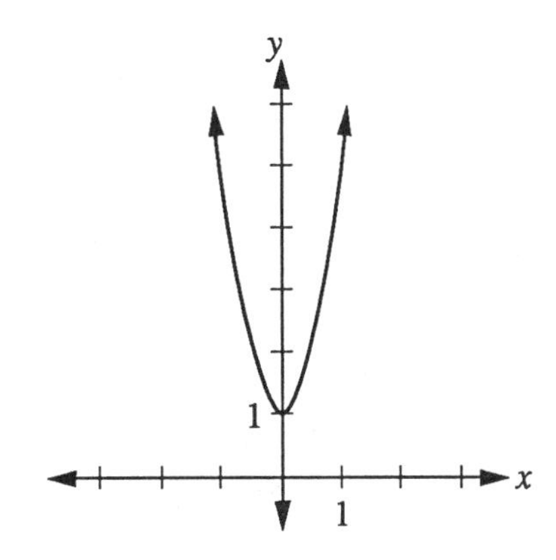

Figure 33 Graph of $y = 5^{|x|}$

Example 2

Sketch the graph of $f(x) = \ln(x^2 - 3x - 4)$.

Solution We can approach this graph as the composition $f(x) = g(h(x))$ where $h(x) = x^2 - 3x - 4$ and $g(x) = \ln(x)$. Domain plays an important role in this composition. The graphs of $h(x)$ and $g(x)$ are shown in Figure 34 and can be used to analyze the domain of $f(x)$ and the flow of the composition. Since $\ln(x)$ is defined only for $x > 0$, the domain of the composition is restricted to those values of x for which $h(x) = x^2 - 3x - 4$ is positive. The domain of $f(x)$, therefore, is $x > 4$ or $x < -1$, and the graph will have vertical asymptotes at $x = 4$ and at $x = -1$.

The graph of this composition is shown in Figure 35. Since $\ln(1) = 0$, the values of x for which $x^2 - 3x - 4 = 1$ are zeros of the composition; these are $x = \frac{3 \pm \sqrt{29}}{2}$. Notice that the y-values of the composition are negative for those x-intervals where $0 < x^2 - 3x - 4 < 1$. For the x-intervals where $x^2 - 3x - 4 > 1$, the effect of the logarithm is to decrease the values of $h(x)$. This means that the graph of $y = \ln(h(x))$ rises less steeply than the graph of $y = h(x)$. Notice that the symmetry of $h(x) = x^2 - 3x - 4$ about $x = \frac{3}{2}$ is carried into the composition. ■

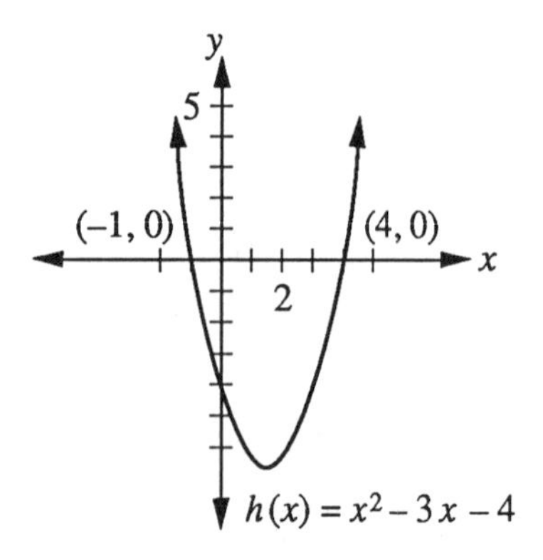

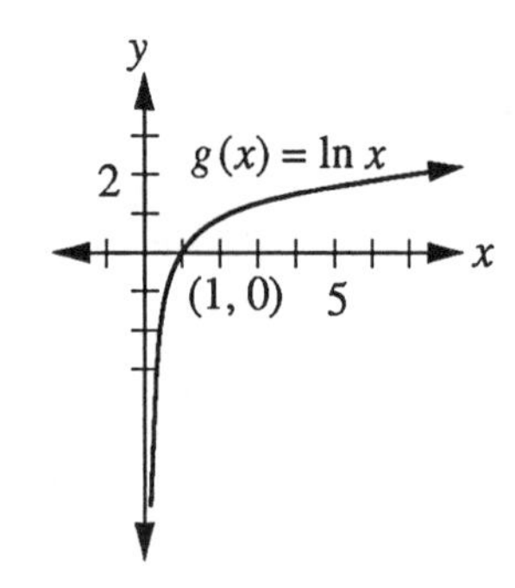

Figure 34 Graphs of $h(x) = x^2 - 3x - 4$ and $g(x) = \ln(x)$

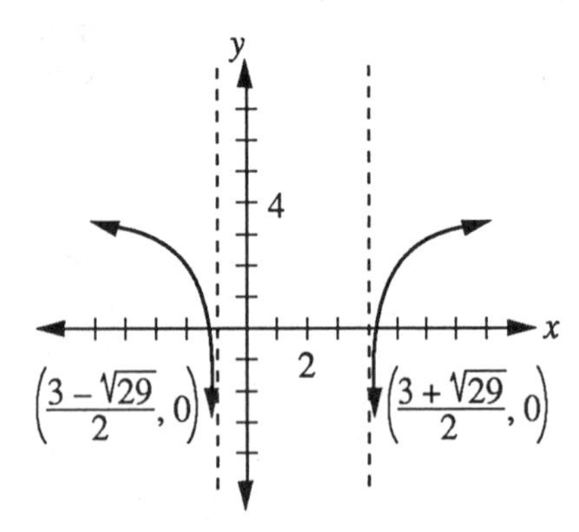

Figure 35 Graph of $f(x) = \ln(x^2 - 3x - 4)$

Example 3

Sketch the graph of $f(x) = 3^{\frac{1}{x-1}}$.

Solution The function f is the composition of $m(x) = 3^x$ and $k(x) = \frac{1}{x-1}$. Since $f(x) = m(k(x))$ and $m(x)$ has no domain restrictions, the domain of the composite function $f(x)$ is the same as the domain of $k(x)$. Therefore x cannot equal 1. Also, since $m(x) > 0$ for all values of x, we know that $f(x) = m(k(x)) > 0$ for all x. The graphs of $y = m(x)$ and $y = k(x)$ are shown in Figure 36.

Since $k(x)$ never equals zero, the input to $m(x)$ is never zero, so the y-value of $m(k(x))$ is never one. It is also important that $k(x)$ has $y = 0$ as a horizontal asymptote. As x-values increase or decrease without bound, $k(x) = \frac{1}{x-1}$ approaches zero; so $f(x) = 3^{\frac{1}{x-1}}$ approaches 3^0, or 1. This produces a horizontal asymptote in the composition at $y = 1$.

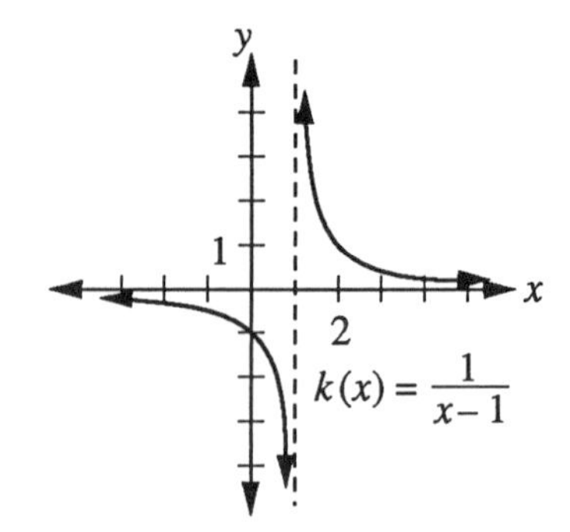

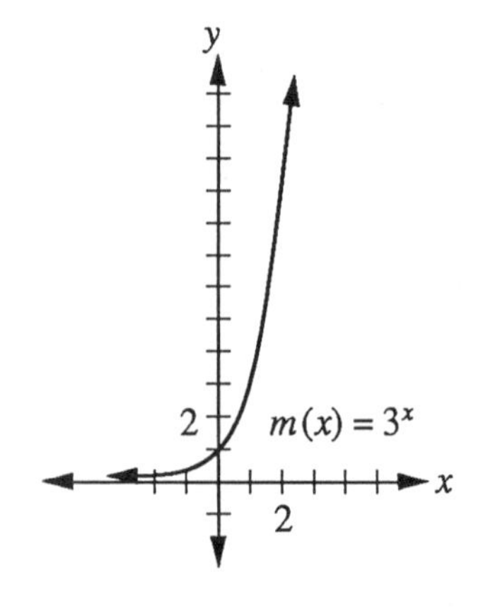

Figure 36 Graphs of $k(x) = \frac{1}{x-1}$ and $m(x) = 3^x$

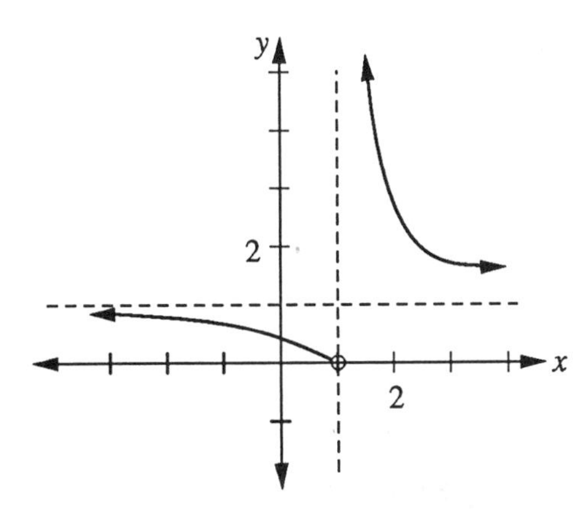

Figure 37 Graph of $f(x) = 3^{\frac{1}{x-1}}$

Now consider the behavior of $f(x)$ close to $x = 1$. As x approaches 1 from the left, $\frac{1}{x-1}$ decreases without bound, and the value of $f(x) = 3^{\frac{1}{x-1}}$ approaches 0 from above. The open circle on the graph of the composition is used to signify this. As x approaches 1 from the right, $\frac{1}{x-1}$ increases without bound, and the value of $3^{\frac{1}{x-1}}$ also increases without bound. The graph of the composition is shown in Figure 37. ■

Exercise Set 11

Use the idea of decomposition to help you sketch the graph of each function. Be careful with domain.

1. $f(x) = \dfrac{1}{e^x}$

2. $f(x) = \dfrac{1}{\ln x}$

3. $f(x) = 3^{|x|}$

4. $f(x) = \log |x|$

5. $f(x) = \ln \dfrac{1}{x}$

6. $f(x) = 2^{(\frac{1}{x})}$

7. $f(x) = 2^{(x^2)}$

8. $f(x) = e^{x^2-2x}$

9. $f(x) = 2^{\frac{x}{x-2}}$

10. $f(x) = \ln(x^2 - 5)$

11. $f(x) = \ln(x^3 - 5x)$

12. $f(x) = 3^{\sin x}$

13. $f(x) = \ln(\sin x)$

14. $f(x) = e^{-x^2}$

15. $f(x) = \dfrac{1}{1 + 2e^{-x}}$

16. $f(x) = \log(10^{-x^2+2})$

17. $f(x) = 10^{\log(-x^2+2)}$

Mathematica
Spoken Here
Length
Time

CHAPTER 5

Data Analysis Two

1

Introduction

Our work in previous chapters has given us considerable experience with the process of creating mathematical models. This chapter will build on that experience and provide some new and useful tools for creating *empirical models*, or models based on data.

The data in Table 1 were obtained as follows. A thermometer was placed in a cup of hot water, and the cup was placed in a refrigerator. The thermometer was read periodically to determine the temperature of the water.

Can you determine what the temperature of the water was after 100 minutes?

Before we try to answer this question, let's examine the scatter plot provided with Table 1 and think about why it looks the way it does. Notice that temperature is plotted on the vertical axis, and time is plotted on the horizontal axis. First, the graph is decreasing, since the water is losing heat to the cooler air in the refrigerator. Second, the graph eventually levels off. The water loses heat only until its temperature is the same as that of the air in the refrigerator, then its temperature stabilizes. Third, the rate of decrease is steep at first and then becomes more gradual. This is because the water loses heat more quickly when it is still hot; when its temperature is close to that of the surrounding air it loses heat more slowly.

Our ability to answer the question posed above depends on finding a mathematical model for the cooling process — this means that we need to look for a functional relationship between time and temperature. Based on our work in Chapter 1, we know how to fit either a line or a parabola to a data set. However, the shape of the scatter plot in Table 1 might indicate that neither of these curves will provide a good fit. We need to develop some additional tools in order to analyze this data.

2

Straightening Curves

When we try to fit a curve to a set of data, our goal is to find a function whose graph fits the data well. The type of function we choose may be indicated by theory or by past experience. Often we lack such knowledge and must choose a function based on the scatter plot of the data. If the scatter plot indicates that a straight line will not provide a satisfactory fit, we are faced with having to decide just how "curved" the graph of the data is. This is not an easy decision. It is difficult to

Table 1 Cooling Data

Time (min)	Temperature (degrees Fahrenheit)
0	124
5	118
10	114
16	109
20	106
35	97
50	89
65	82
85	74
128	64
144	62
178	59
208	55
244	51
299	50
331	49
391	47
overnight	45

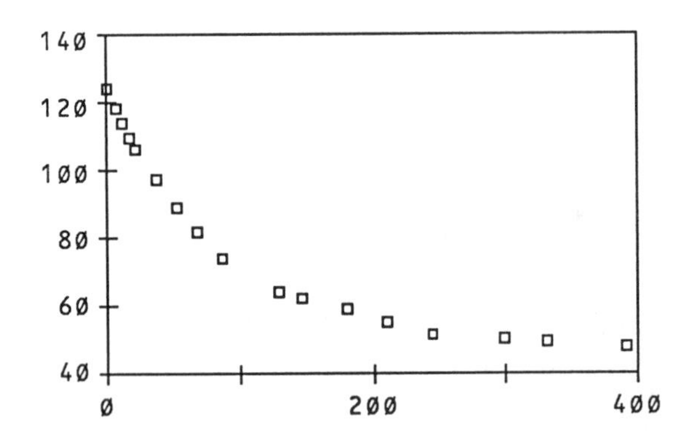

determine the curvature by looking at the scatter plot. Most of us cannot easily distinguish the difference in curvature of quadratic, cubic, and exponential graphs. In contrast, we can usually determine visually whether or not a graph is linear, since we need only to decide whether or not the points line up. Therefore, if we can re-express the original data so that the points lie along a line rather than a curve, we can then determine the function that best models a nonlinear relationship. This process of *re-expressing data in order to make it linear* will be illustrated in the next several examples.

Example 1

Free-Fall

In Chapter 1 we found a model for a data set that involved a freely falling object. Table 2 shows the same data set and scatter plot. The distance the object has fallen is graphed on the vertical axis, and the amount of time since release is

Table 2 Free-Fall Data

Time (sec)	Distance (cm)	Time (sec)	Distance (cm)
0.16	12.1	0.57	150.2
0.24	29.8	0.61	182.2
0.25	32.7	0.61	189.4
0.30	42.8	0.68	220.4
0.30	44.2	0.72	254.0
0.32	55.8	0.72	261.0
0.36	63.5	0.83	334.6
0.36	65.1	0.88	375.5
0.50	124.6	0.89	399.1
0.50	129.7		

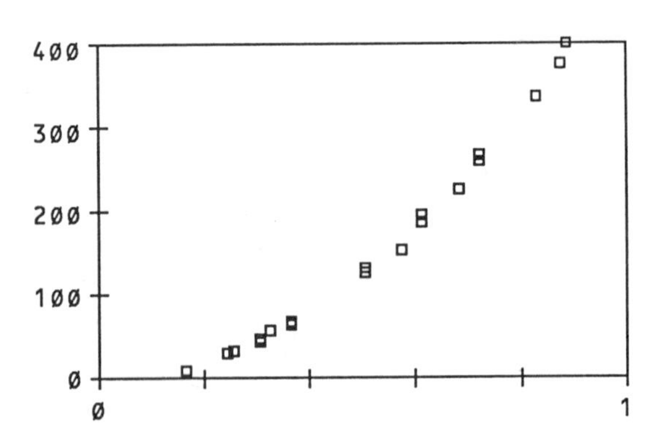

plotted on the horizontal axis. The scatter plot indicates that the distance depends on time in a nonlinear way. The points show upward curvature; the slopes of segments between consecutive points increase as x-values increase. When we worked with this data set before, we found that the y-values are a quadratic function of the x-values. Recall that we made the data set linear by taking the square root of the y-values. It is useful to think of taking the square root as "undoing" the squaring that was done by a quadratic function. Taking the square root has the effect of bringing down the y-values since they are all greater than 1 and straightening the curved scatter plot. Instead of fitting a curve to the ordered pairs (x, y), we fit a line through the ordered pairs $(x, \sqrt{y})$ (see Figure 3). The median-median line we obtain is

$$\sqrt{y} = 22.19x + 0.12.$$

When we square both sides of the equation to solve for y, we obtain the quadratic model

$$y = 492.40x^2 + 5.33x + 0.01.$$

Figure 4 shows how this model fits the original ordered pairs (x, y).

The residual plots in Figures 3 and 4 provide different, but important, information. The residuals plotted in Figure 3 are associated with the square root of the y-values. This residual plot helps us decide whether our choice of re-expression is a good one. In this example, the residuals are randomly distributed and indicate that taking the square root of y-values is a good choice. If the re-expression technique does not succeed in making the data linear, there will usually be a noticeable pattern in this residual plot. Once we are satisfied with the re-expression, we should examine the model superimposed on the original data as we have done in Figure 4. Notice that the pattern of the residuals is the same in both residual plots. However, the size of the residuals in Figure 4 provides additional information. Since these residuals are associated with the actual y-values, they indicate the magnitude of the error to be expected when we use the model to make predictions. ■

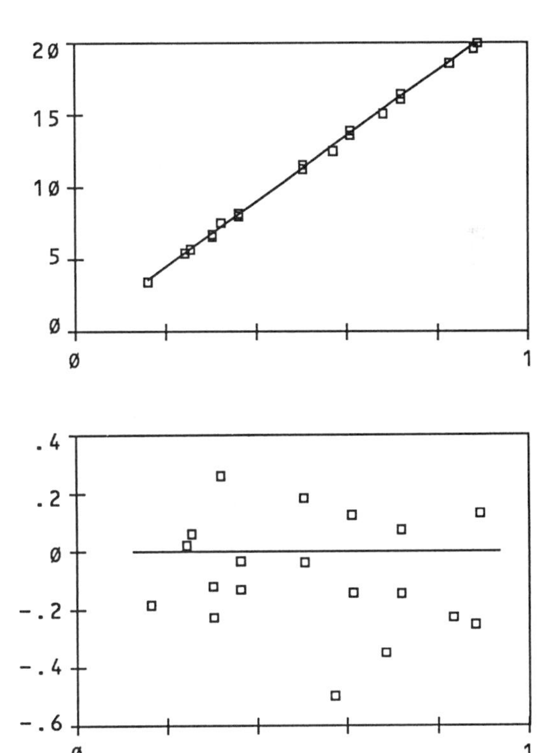

Figure 3 Median-Median Line and Residuals for Re-expressed Free-Fall Data

What we have done in Example 1 is re-express the data to make it linear. This is an important technique in data analysis. Do you understand why changing from ordered pairs (x, y) to ordered pairs $(x, \sqrt{y})$ makes the data linear? One way to think about this choice of re-expression is to recall that the y-values in the original data set do not increase steadily; instead, the y-values increase more quickly for large x-values than for small x-values. These y-values need to be lowered. There are many ways to re-express the y-values to pull the points down — square roots, cube roots, or fourth roots would all have this effect. Since the relationship between these variables is quadratic, taking the square root will lower the points just enough so that they will lie along a line. Note that taking the square root is the *inverse* of squaring.

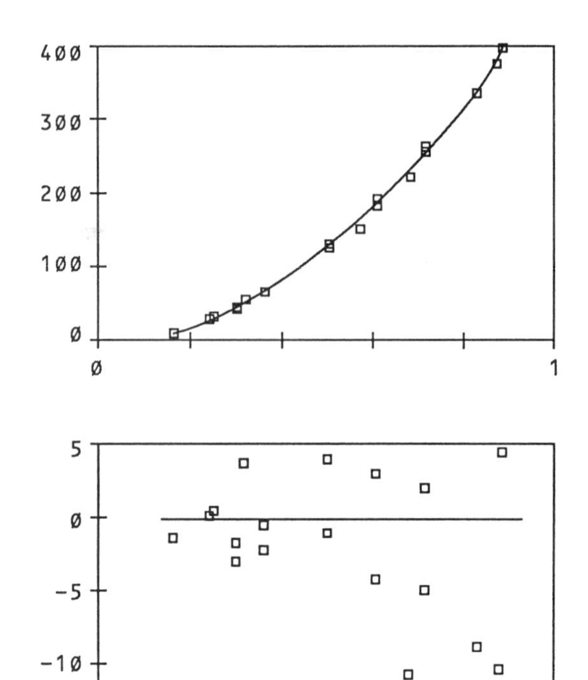

Figure 4 Quadratic Model and Residuals for Free-Fall Data

Class Practice

1. There are other re-expression techniques that could be used to linearize a data set when the relationship between the variables is quadratic. Consider the effect of squaring the x-values to obtain ordered pairs of the form (x^2, y). Consider also the effect of dividing y-values by x-values to obtain ordered pairs of the form $(x, \frac{y}{x})$. In general, how will the choice of transformation affect the quadratic model? Which transformation will force the parabola to pass through the point $(0, 0)$? Which one ensures that the vertex is located on the vertical axis?

Example 2

Population Growth

The data in Table 5 show how the population per square mile in the United States has changed over a period of years. On the scatter plot the horizontal axis denotes the number of years since 1700. What was the population density in 1887?

Table 5 Population Data

Year	People Per Square Mile	Year	People Per Square Mile
1790	4.5	1890	17.8
1800	6.1	1900	21.5
1810	4.3	1910	26.0
1820	5.5	1920	29.9
1830	7.4	1930	34.7
1840	9.8	1940	37.2
1850	7.9	1950	42.6
1860	10.6	1960	50.6
1870	10.9	1970	57.5
1880	14.2	1980	64.0

Source: *The World Almanac, 1989*

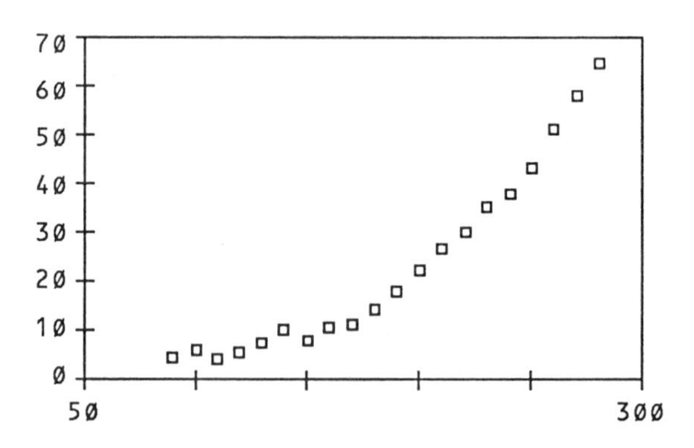

Solution It is difficult to judge the curvature shown in the scatter plot; just how curved is the graph? We can benefit from our previous knowledge that population growth is often exponential; this should give us some idea about how to re-express the y-values so that the data will become linear. By reasoning similar to that used for the free-fall data, the transformation that we anticipate will work is the inverse of the exponential function. Thus, we should try taking the logarithm of the y-values, since logarithms "undo" exponentials. If the ordered pairs (x, y) are indeed on an exponential curve, then the ordered pairs $(x, \ln y)$ will be on a straight line. The converse is also true—if the ordered pairs $(x, \ln y)$ are linear, then the ordered pairs (x, y) are exponential. Figure 6 shows the median-median line for the re-expressed data points $(x, \ln y)$.

The equation of the median-median line is

$$\ln y = 0.015x + 0.010.$$

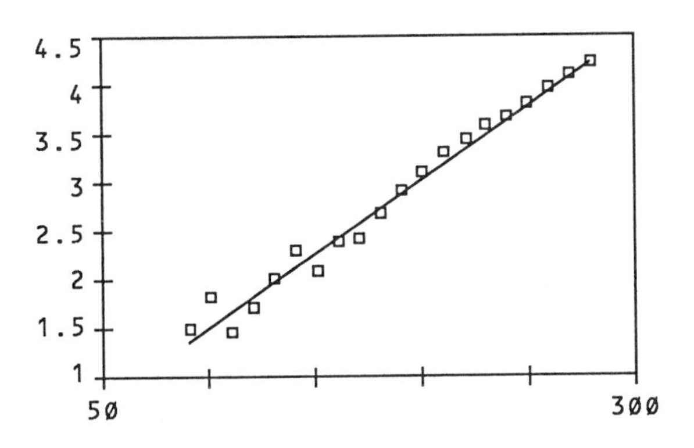

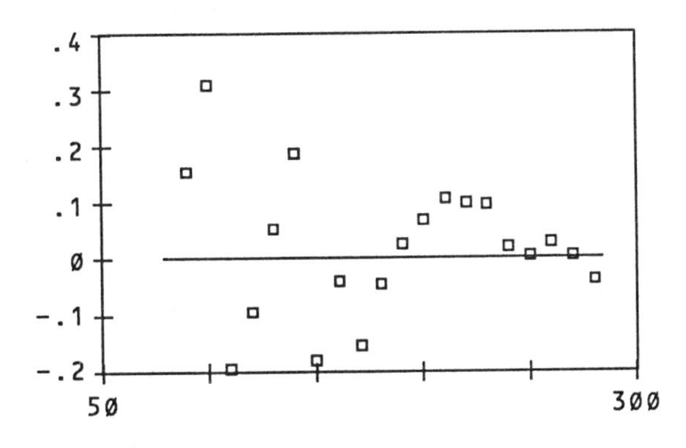

Figure 6 Median-Median Line and Residuals for Re-expressed Population Data

To find a model for the original data (x, y) we exponentiate and simplify as follows:

$$y = e^{(0.015x+0.010)}$$
$$y = e^{0.015x}e^{0.010}$$
$$y = 1.010e^{0.015x}$$

where x represents the number of years since 1700. This exponential function models the relationship between time and population. Figure 7 shows this model graphed with the original data points.

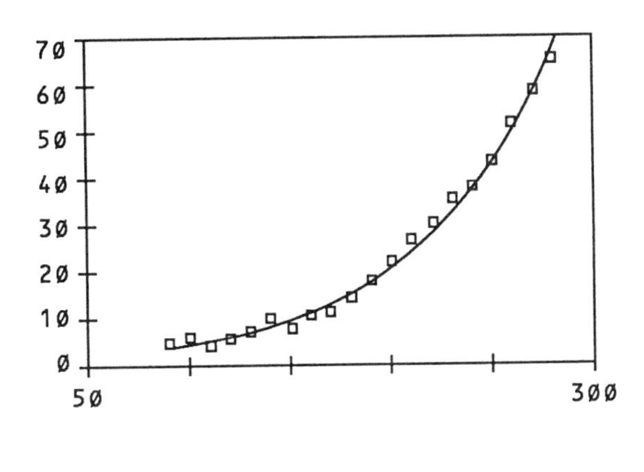

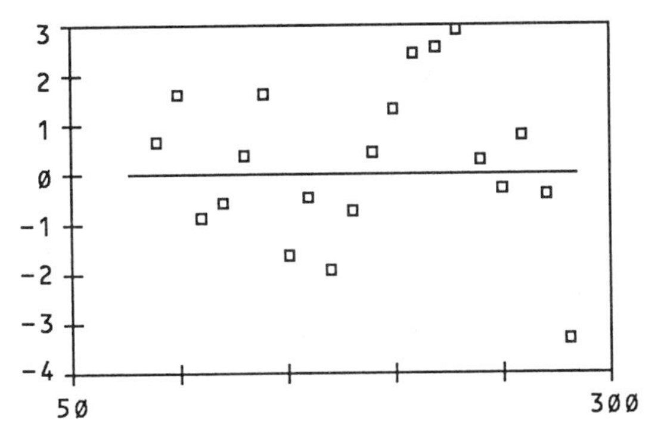

Figure 7 Exponential Model and Residuals for Population Data

We can now use this model to determine the population density in 1887. Since 1887 − 1700 = 187, we evaluate $y = 1.010e^{0.015(187)}$ to estimate that in 1887 the population density was approximately 16.7 people per square mile. ∎

The technique of linearizing data by taking the logarithm of the y-values is frequently used; there are many data sets that can be linearized in this way. This method of re-expressing data is called *semi-log re-expression*, and a graph of ordered pairs $(x, \ln y)$ is called a *semi-log plot*. Semi-log re-expression will linearize any data set that can be modeled by an exponential function of the form $y = Ae^{Bx}$.

You should study the residuals shown in Figures 6 and 7. The random pattern of the residuals from the median-median line indicates that the line is a good fit for the re-expressed data. However, the size of these residuals is deceptive. Since we took the logarithm of the y-values, these residuals represent errors in the logarithms of the y-values rather than in the actual y-values. They appear small because logarithms tend to be small numbers. In contrast, the residuals from the exponential model are larger; they represent the actual error we would encounter when using the model to predict population density.

Example 3

Tree Volume

The data in Table 8 represent the age and volume of several hardwood trees of the same species. Can you predict the volume of wood in a 55-year-old tree?

Table 8 Tree Volume Data

Age (years)	Volume (100's of board feet)
20	1
40	6
80	33
100	56
120	88
160	182
200	320

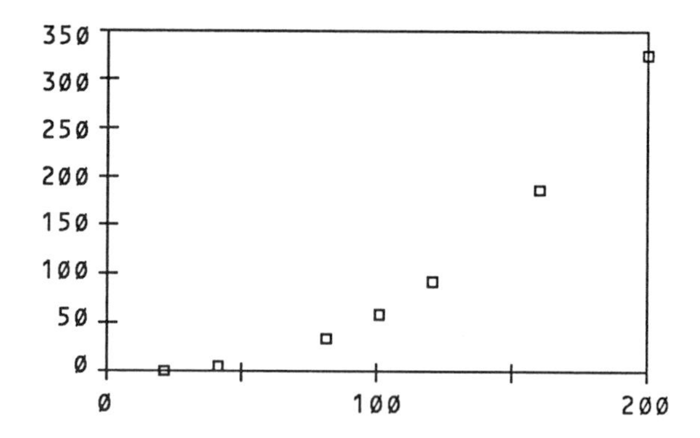

Solution In order to make a reasonable prediction we need to find a mathematical model for the relationship between age and volume. The scatter plot of the data shown in Table 8 has the tree age plotted on the horizontal axis and the volume of wood on the vertical axis. The volume of wood measured in board feet assumes all boards are 1 foot wide and 1 inch thick. Inspection of the scatter plot indicates that the relationship between these variables is not linear, but it is difficult to judge what kind of curvature is displayed. Since there is no underlying theory that would lead us to expect a particular relationship between age and volume, we have no real choice but to guess. We have seen in Example 1 that taking the square root of the y-values can straighten a curve if the functional relationship is quadratic, so we can try re-expressing the data and looking at a scatter plot of ordered pairs $(x, \sqrt{y})$. Figure 9 shows a median-median line fit to these re-expressed ordered pairs, as well as the corresponding residuals.

The scatter plot looks reasonably linear, but there is a clear pattern in the residual plot. This pattern of positive residuals at each end and negative residuals in the middle indicates that the ordered pairs $(x, \sqrt{y})$ are not linear, but actually curve upward. Since the re-expressed ordered pairs are not linear, the original ordered pairs (x, y) are not quadratic. Our attempt to straighten the original scatter plot by taking the square root of the y-values has not brought down the y-values far enough; the implication is that the curvature of the original pairs (x, y) is more than the curvature of a parabola.

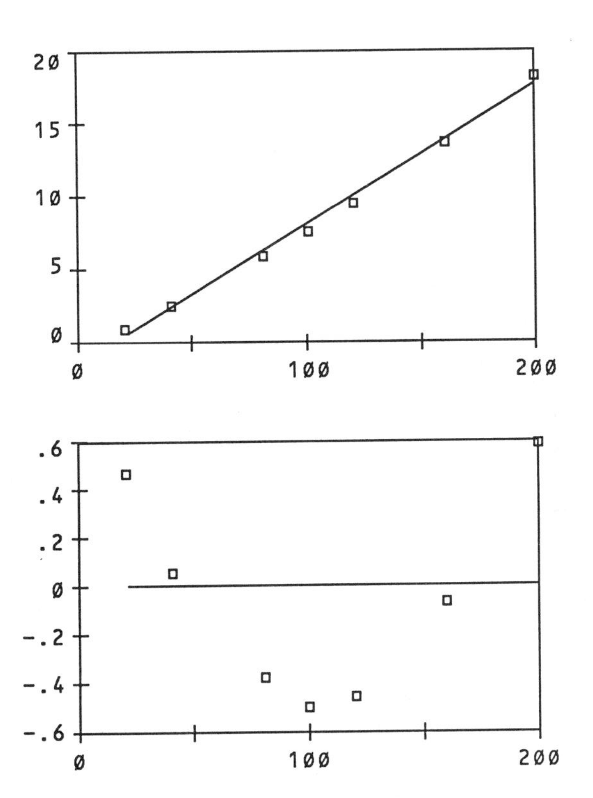

Figure 9 Median-Median Line and Residuals for Re-expressed Tree Volume Data, $(x, y^{\frac{1}{2}})$

You should now use semi-log re-expression to verify that the relationship between tree age and volume is not exponential.

At this point, we know that the relationship between tree age and volume is neither exponential, linear, nor quadratic. Perhaps the relationship can be modeled by a *power function* with an equation of the form $y = Ax^B$. Proceeding under this assumption, we need to find a way to re-express the data so that we can fit a median-median line. Properties of logarithms provide the appropriate strategy.

Suppose you have a data set that can be modeled by the power function $y = Ax^B$. Then the following will be true:

$$
\begin{aligned}
y &= Ax^B \\
\ln y &= \ln(Ax^B) \\
\ln y &= \ln A + \ln x^B \\
\ln y &= \ln A + B \ln x.
\end{aligned}
$$

You should notice that $\ln y$ is a linear function of $\ln x$; the slope is B and the y-intercept is $\ln A$. This means that you will be able to fit a median-median line to the ordered pairs $(\ln x, \ln y)$.

Let's try this technique on the tree volume data; we will transform the ordered pairs (x, y) to ordered pairs $(\ln x, \ln y)$ and see if the re-expressed data displays a linear relationship. Figure 10 shows a median-median line fit to the re-expressed data; the equation of the line is $\ln y = 2.49 \ln x - 7.43$. The residuals indicate that the linear fit for $(\ln x, \ln y)$ is good, so we

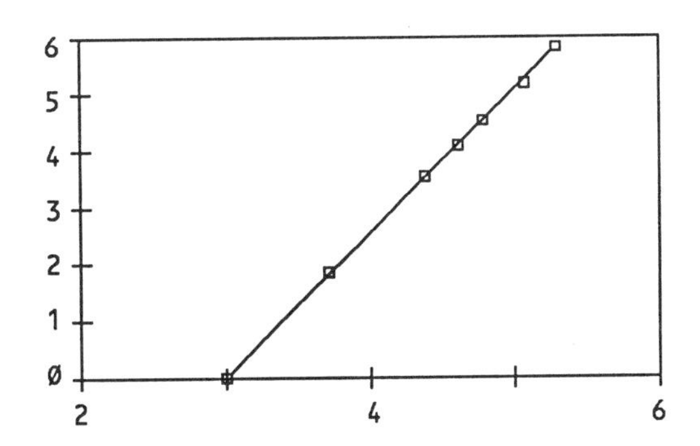

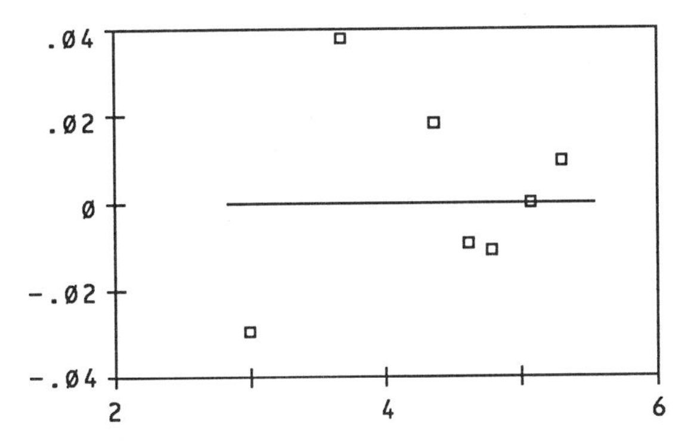

Figure 10 Median-Median Line and Residuals for Re-expressed Tree Volume Data, $(\ln x, \ln y)$

conclude that the original ordered pairs (x, y) can be modeled by a power function.

We can find the equation of the power function as follows:

$$\begin{aligned} \ln y &= 2.49 \ln x - 7.43 \\ y &= e^{(2.49 \ln x - 7.43)} \\ y &= e^{2.49 \ln x} e^{-7.43} \\ y &= (0.000593) x^{2.49}. \end{aligned}$$

Figure 11 shows a graph of this power function superimposed on the original ordered pairs. The residuals indicate that this function does a good job of modeling the relationship between tree age and volume.

To answer our question about the volume of a 55-year-old tree, let $x = 55$ and find the corresponding value of y:

$$\begin{aligned} y &= (0.000593) 55^{2.49} \\ y &\approx 13. \end{aligned}$$

On the basis of our model, we would predict that a 55-year-old tree would have a volume of about 1300 board feet. ■

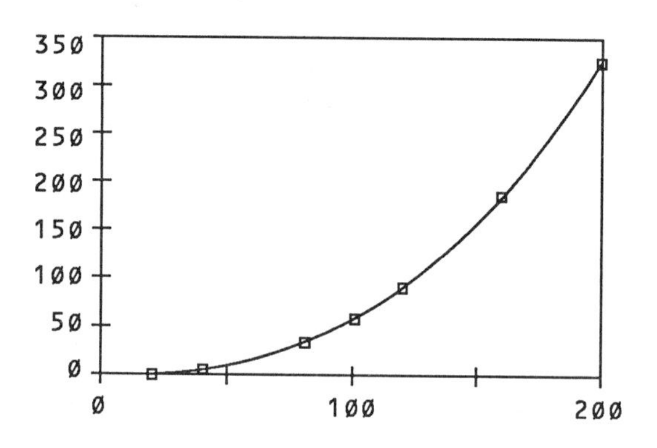

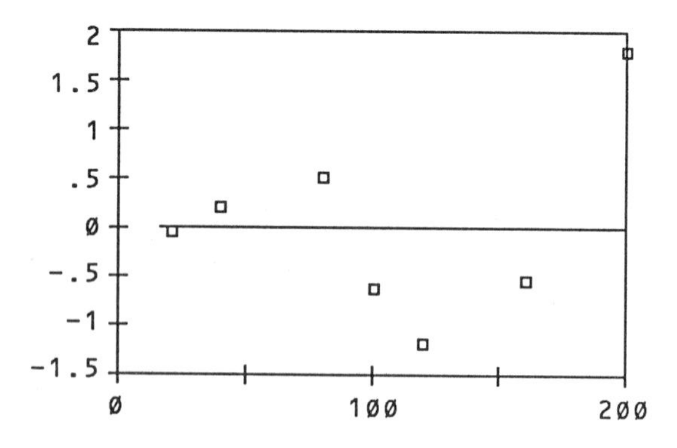

Figure 11 Power Function Model and Residuals for Tree Volume Data

The technique of linearizing data by taking the logarithm of both the x-values and the y-values is frequently used; there are many data sets that can be linearized in this way. This method of re-expressing data is called *log-log re-expression*, and a graph of ordered pairs $(\ln x, \ln y)$ is called a *log-log plot*. Log-log re-expression will linearize any data set that can be modeled by a power function of the form $y = Ax^B$.

An important observation regarding semi-log re-expression and log-log re-expression should be emphasized here. The linear equation obtained for the median-median line will involve re-expressed variables. Thus, the residuals from a semi-log plot or a log-log plot indicate only how well the re-expression has linearized the data. They do not accurately represent the errors you may encounter when you use the model to make predictions. To determine the size of the errors to be expected when using the model, you need to convert the equation to one in which the dependent variable is isolated and graph this function with the original ordered pairs. Then you will be able to analyze the residuals associated with the original ordered pairs.

The preceding examples illustrate only a few of the techniques available for straightening curves — inverse functions, semi-log re-expression, and log-log re-expression. These techniques are not exhaustive, and they all have limitations. Re-expression to produce linearity is often an experimental process involving a great deal of trial and error. Your knowledge of functions, inverses, and special properties of functions should guide you as you try different transformations and examine scatter plots. Using a computer or calculator to perform the re-expressions and to plot the resulting ordered pairs makes it easy to experiment until you are satisfied.

Exercise Set 2

1. Think back to the scatter plot of the free-fall data. If you had not known that the relationship between distance and time was quadratic, you might have thought the scatter plot indicated an exponential relationship. If you had performed semi-log re-expression of this data, what would you expect to have observed in the semi-log plot? Explain your answer. Verify your response by doing the re-expression.

2. The scatter plot in Figure 12 curves downward; the y-values increase more slowly for large x-values than they do for small x-values. Suggest ways to re-express this data in order to linearize it.

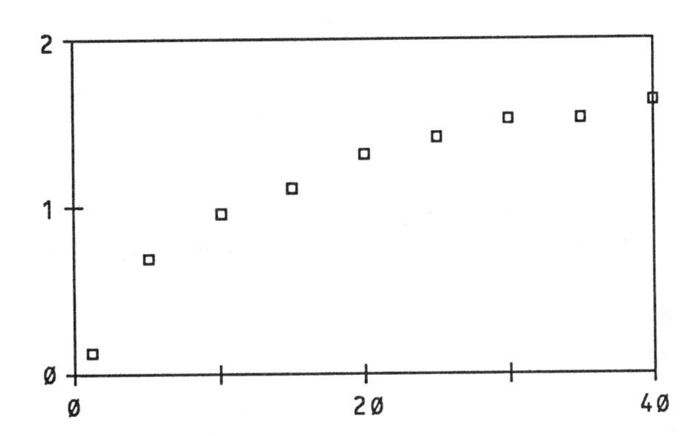

Figure 12 Scatter Plot for Exercise 2

3. Suppose you want to find the equation of the curve that best fits these points: $(0,0), (1,3.85), (3,5.7), (7,17)$. What problem will you encounter if you try to do log-log re-expression? How can you avoid or correct this problem?

4. Suppose two students have previous knowledge that indicates that a data set might be modeled by an exponential function. Both students decide to use semi-log re-expression to help them analyze the data; student A uses natural logarithms and student B uses common logarithms. How will the models they find for the data compare to each other?

5. In the exercises at the end of Chapter 1, you analyzed data gathered in a physics laboratory as an object slides down an inclined plane. This data set is provided again in Table 13.

Table 13 Inclined Plane Data

Time (sec)	Position (cm)	Time (sec)	Position (cm)
0	0	1.0	12.5
0.1	0.2	1.1	15.0
0.2	0.6	1.2	17.7
0.3	1.3	1.3	20.6
0.4	2.2	1.4	23.8
0.5	3.3	1.5	27.2
0.6	4.7	1.6	30.8
0.7	6.3	1.7	34.8
0.8	8.1	1.8	38.9
0.9	10.2	1.9	43.1

a. Find a quadratic model for this data by fitting a median-median line to re-expressed ordered pairs $(x, \sqrt{y})$.

b. Find a quadratic model for this data by fitting a median-median line to re-expressed ordered pairs (x^2, y).

c. Find a quadratic model for this data by fitting a median-median line to re-expressed ordered pairs $(x, \frac{y}{x})$.

d. Use log-log re-expression to find a power function model for this data.

e. Write a paragraph comparing these models and explaining which you think is best. Include in your discussion any advantages or disadvantages of each re-expression based on characteristics of the phenomenon you are trying to model.

6. **a.** Make a scatter plot of the data set in Table 14.

b. Predict what will happen if you fit a line to the data set in Table 14. Predict what will happen if you fit a line to the re-expressed data set $(x, \sqrt{y})$. Check your predictions by

Table 14 Data Set for Exercise 6

x	2	6	10	14	20
y	40	1,080	5,000	13,720	40,000
x	24	40	50	55	
y	69,120	320,000	625,000	831,875	

fitting a median-median line to each data set and examining the residuals.

c. The ordered pairs in Table 14 were generated by a third-degree power function. Describe three different techniques that will linearize the data.

d. Re-express the data in one of the ways you described above and fit a median-median line to the re-expressed data. If necessary, solve your equation for y.

7. Rheumatoid arthritis patients are treated with large quantities of aspirin. The concentration of aspirin in the bloodstream increases for a period of time after the drug is administered and then decreases in such a way that the amount of aspirin remaining is a function of the amount of time that has elapsed since peak concentration. Table 15 gives data for a particular arthritis patient after taking a large dose of aspirin. Use this data to predict how much aspirin remains in the patient's bloodstream ten hours after the 15 mg reading.

Table 15 Drug Concentration Data

Hours Elapsed Since Peak Concentration	0	1	2	3	4
Mg of Aspirin Per 100 cc of Blood	15	12.5	10.5	8.7	7.3
Hours Elapsed Since Peak Concentration	5	6	7	8	
Mg of Aspirin Per 100 cc of Blood	6.1	5.1	4.3	3.6	

8. Use the graph of the function $y = \sqrt{x}$ provided in Figure 16 to estimate the area of the region that is under the graph, above the x-axis, and between the vertical lines $x = 0$ and $x = 2$. This area appears to be a little less than two square units, since it contains one complete block and two partial blocks.

a. Repeat this process for the regions between $x = 0$ and $x = t$ for integer values of t from 1 to 10. The data set you create should consist of ordered pairs of the form (t, A), where A represents the area under the graph between $x = 0$ and $x = t$ and above the x-axis.

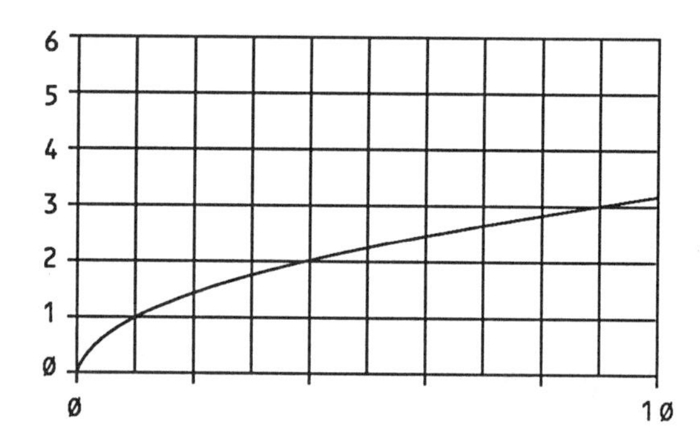

Figure 16 Graph of $y = x^{1/2}$

b. Make a scatter plot of your ordered pairs. Use the scatter plot to make a guess about the relationship between t and A. Check your guess by re-expressing the data to make it linear. If you are not satisfied with your choice of re-expression, experiment with others until your data appears linear.

c. Find an equation of the median-median line through the linearized data. Solve your equation for the dependent variable A, and examine the model superimposed on the original data.

d. Techniques from calculus allow us to determine that the area under $y = \sqrt{x}$ between $x = 0$ and $x = t$ is given exactly by the equation $A = (2/3)t^{3/2}$. How close to this equation is your empirical model?

e. How well does the theoretical model stated in part **d** fit the data you gathered?

9. Refer to the cooling data and scatter plot in Table 1. Recall that we wanted to find a model for this data in order to predict the temperature of the water after 100 minutes.

a. On the basis of an examination of the scatter plot, what functions might be good models? Why?

b. What characteristic of this data set makes it impossible to linearize the data with log-log or semi-log re-expression?

3

Linear Least Squares

What Is the Best-Fitting Line?

Up to this point our approach to curve fitting has been to re-express data in a way that will create a linear relationship. Then we can find the equation of a median-median line through the linear data set. As the examples and exercises have shown, the median-median line rarely passes exactly through any of the data points. Every data point (x, y) has associated with it a residual; the residuals serve as an indication of how far the line is from being an exact fit. The fact that the median-median line does not achieve an exact fit does not weaken its effectiveness as a model, provided the residuals are small in size and randomly distributed.

You may have wondered, however, if some other line might do a better job than the median-median line of fitting the data. If so, what line would be better? What line would be best? If this question were posed for class discussion, you would probably get many different opinions, some of which are listed below:

The best-fitting line passes exactly through the greatest number of data points.

The best-fitting line results in equal numbers of data points above the line and below the line; that is, the number of positive and negative residuals must be equal.

The best-fitting line is influenced most by data points near the middle of the data set and does not give much weight to extreme values.

The best-fitting line minimizes the sum of all the perpendicular distances between the data points and the line.

The best-fitting line minimizes the sum of all the vertical distances between the data points and the line.

A discussion about a line of best fit cannot be resolved without a definition of the word "best". Since mathematical models are often used for interpolating and extrapolating, mathematicians have defined the best line to be the line that minimizes errors in predictions. This means that the best line is the one that minimizes residuals. If we consider several lines through a data set, we would probably agree that small residuals are preferable to large ones.

The data shown in Table 17 were collected by chemistry students. The students varied the temperature of a gas in a closed container and recorded the pressure exerted by the gas. A scatter plot of the data, with pressure graphed on the vertical axis, suggests that when volume is held

Table 17 Temperature/Pressure Data

Temperature (degrees Kelvin)	**Pressure (mm of mercury)**
263	752.2
268	755.3
278	776.5
293	811.3
298	834.4
303	839.7
308	853.6
318	891.5
323	906.1

constant, pressure of a gas is related to temperature in a linear way. Two different lines through the temperature/pressure data are shown in Figures 18 and 19. Figure 18 shows the median-median line and a residual plot; the equation of the median-median line is $y = 2.72x + 24.39$. Figure 19 shows a slightly different line, $y = 2.61x + 56.75$, and a residual plot.

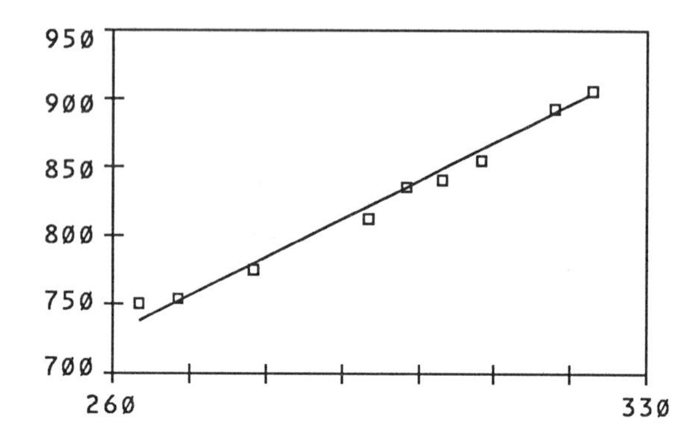

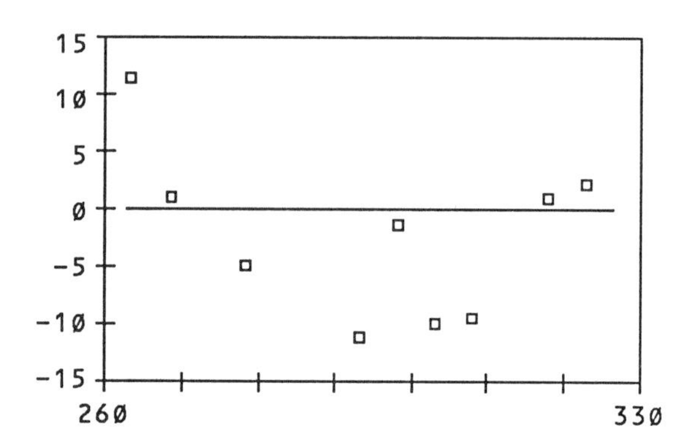

Figure 18 Median-Median Line and Residuals for Temperature/Pressure Data

Class Practice

1. For the line shown in Figure 18 ($y = 2.72x + 24.39$), calculate the residual associated with each point. Make a table that displays x-values, y-values from the data set, y-values predicted by the median-median line, and residuals. Calculate the sum of the residuals.

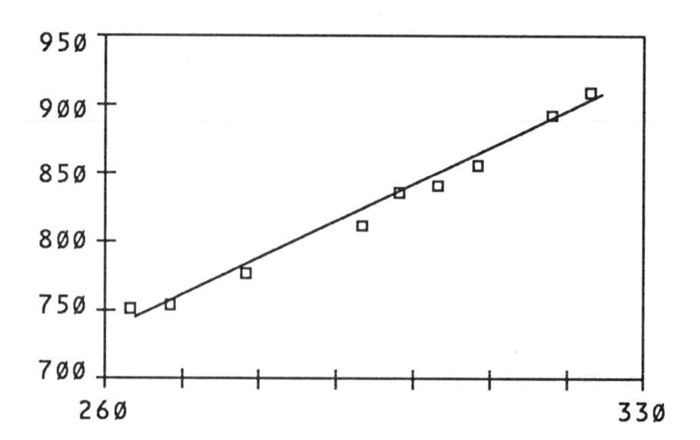

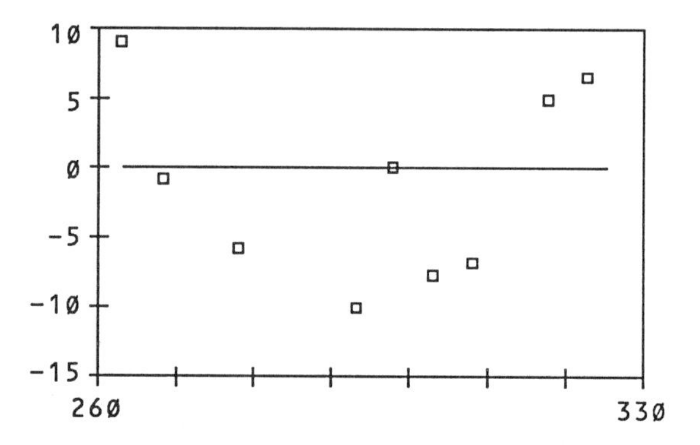

Figure 19 Alternate Linear Model and Residuals for Temperature/Pressure Data

(A computer spreadsheet is a useful tool for these calculations.)

2. Repeat the previous exercise using the line shown in Figure 19 ($y = 2.61x + 56.75$).
   a. Interpret the slope and y-intercept of the median-median line for the temperature/pressure data.
   b. What change in pressure is brought about by a one-degree change in temperature? What change in temperature would cause a one-millimeter change in pressure?

The Least Squares Principle

Finding the best-fitting line requires that we keep the residuals as small as possible, since they represent the errors in predictions. How can we accomplish this? Your first thought may be to minimize the sum of the residuals. However, the preceding class practice exercises should make it clear that residuals can be either positive or negative, and therefore their sum could be close to zero even for a poorly fitting line. Your next thought probably would be to minimize the sum of the absolute values of the residuals. Since absolute value is inconvenient to work with, we can avoid messy arithmetic and still accomplish our goal by minimizing the sum of *squared* residuals. This criterion, called the *principle of least squares*, dates back to the nineteenth-century work of Adrien Legendre. According to the least squares principle, the best-fitting line is defined to be the one that minimizes the sum of the squares of the residuals. Such a line must have residuals close to zero so that their squares will be small, and therefore the sum of their squares will be as small as possible. This line is referred to as the *least squares line* or the *regression line*. The second line through the temperature/pressure data, $y = 2.61x + 56.75$, is the least squares line. You should notice that it is close to, but not identical to, the median-median line.

Class Practice

1. According to the principle of least squares, can the sum of squared residuals be zero? If so, what would that mean about the fit?
2. Expand the tables you created in the previous class practice exercises to include columns for squared residuals. Verify that the sum of squared residuals is smaller for the least squares line than for the median-median line.

Techniques from calculus and algebra can be used to find the slope and intercept of the least squares line. The calculations required to find the least squares line without calculus are usually quite tedious, and they become more and more time-consuming as the number of data points increases. Most mathematicians and statisticians use computer programs to perform this analysis for them. Many calculators will also compute the slope and intercept of the least squares line. We will assume that you have technology available to find the equation of the least squares line for a data set. Our attention will focus on analyzing rather than on calculating this line.

As we proceed with the process of fitting lines and curves to data sets, we need to remember to examine a scatter plot of each data set. Examination of the scatter plot of the data allows us to check for linearity. This check is important, because it is possible, though not wise, to find the least squares line for data sets in which the relationship between the variables is not linear. Examination of a scatter plot will also allow us to observe outliers, if they exist. As a general rule we should try to explain outliers. They may indicate errors in the data that can be corrected, or draw our attention to an important characteristic of the phenomenon we are trying to model. After fitting a least squares line, we should examine the residual plot to make sure that the residuals are small and random.

Comparison of Median-Median and Least Squares Lines

Now that you have two methods of modeling a data set with a line, you should be wondering about the advantages offered by each method. The least squares line is the more popular one; it is the line generally included in statistics textbooks and computer software packages. The obvious advantage of the least squares criterion is that it offers a reasonable standard by which the line is developed. By minimizing residuals we minimize the total error in the line. However, the least squares line does not always capture the major thrust of the data. The computational formulas used to determine the slope and y-intercept of the least squares line involve every point. Therefore, if the data set has outliers, the least squares line may be "thrown off" a great deal. The median-median line, in contrast, is highly resistant to the effects of outliers; it is often called the *resistant line*.

Keep in mind that linear models involve two parameters, the slope and the y-intercept of the line. These two values are not of equal importance in the model. Of the two, the slope gives the most information about the relationship between the two variables. A slope of -3 indicates that an increase in the independent variable results in a threefold decrease in the dependent variable. The intercept, on the other hand, simply gives an initial condition or starting point. Knowing that the intercept is 5 tells us almost nothing about the relation between the variables. As a result, errors in the slope of a linear model are much more costly in terms of understanding the relationship than are errors in the intercept.

Every point in a data set contributes directly to the values of the slope and intercept in the least squares line. In contrast, each data point affects the median-median line only indirectly, via the three summary points. It is easy to illustrate this phenomenon by using data points that lie on the line $y = x$ and altering one or two points.

Table 20 Data Set and Least Squares Line for Example 1

x	1	2	3	4	5	6	7	8	9
y	1	2	3	4	5	6	7	8	5

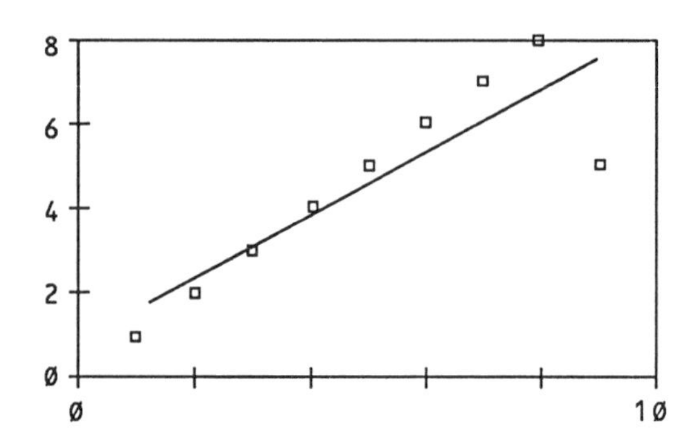

Example 1

A least squares line is fit to the data set provided in Table 20. Note that all of the data points except $(9, 5)$ lie on the line $y = x$. You should see that the least squares line is pulled down from the mass of data by the last data point (9, 5); its equation is $y = 0.73x + 0.89$.

The summary points for the median-median line are (2,2), (5,5), and (8,7); the equation of the median-median line is $y = 0.83x + 0.50$. Make your own scatter plot of this data set and sketch the median-median line on the scatter plot. Does it do a better job than the least squares line of "staying with the data"? Compare the way the point $(9, 5)$ influences the median-median line with the way it influences the least squares line. Suppose the last point were $(9, 2)$ instead of $(9, 5)$. How would this point affect the least squares and median-median lines? ■

Example 2

A least squares line is fit to the data set in Table 21. Note that all points except $(5, 9)$ lie on the line $y = x$.

The equation of the least squares line is $y = x + 0.44$. Notice that the slope of the least squares line is 1; changing from the point $(5, 5)$ to the point $(5, 9)$ has not influenced the slope. This is because the variation occurs in the middle of the data. Variations near the average x-value have little effect on the slope of a least squares line. They do, however, alter the y-intercept.

The summary points for the median-median line are (2,2), (5,6), and (8,8). What is the median-median line for this data set? Compare the way the point $(5, 9)$ influences the median-median line with the way it influences the least squares line. ■

Table 21 Data Set and Least Squares Line for Example 2

x	1	2	3	4	5	6	7	8	9
y	1	2	3	4	9	6	7	8	9

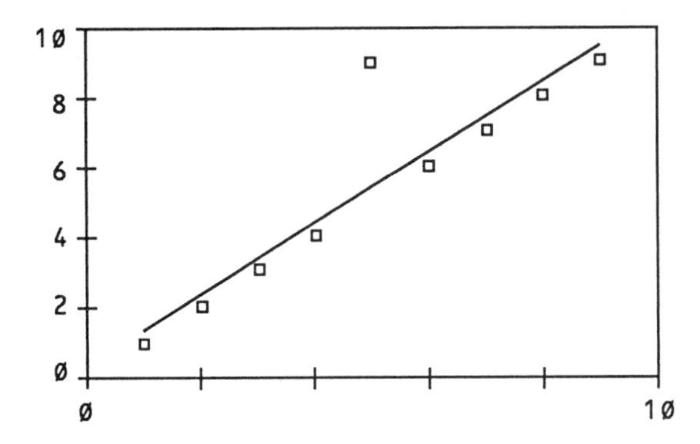

Table 22 Data Set and Scatter Plot for Class Practice Exercise

x	1	2	3	4	5	6	7	8	9
y	1	2	3	4	5	6	7	9	8

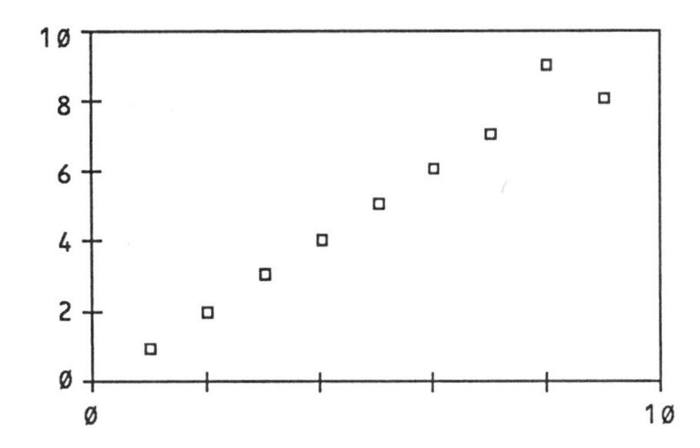

Class Practice

1. Most of the data points in Table 22 lie on the line $y = x$. The point $(8, 9)$ is above the line, and the point $(9, 8)$ is below the line.
   a. You should expect one point to pull the least squares line up and another point to pull the line down. Predict whether or not these two effects will offset each other.

   b. Use a computer or calculator to find the equation of the least squares line.

   c. Have the effects on the least squares line of the two variant points offset each other? If not, which point wins the "tug of war"? Can you explain why?

   d. What are the summary points for the median-median line?

   e. What effect, if any, do the variant points have on the median-median line?

As we have seen from the preceding examples, the least squares line is very sensitive to outliers, particularly at the extremes of a data set. You should notice that outliers near the middle of a data set tend to affect the intercept of the least squares line, and outliers near the extremes of a data set tend to affect the slope. The farther an outlier is from the middle of a data set, the more influence it has on the slope. In contrast, the median-median line is relatively resistant to the effects of outliers. If you want extreme points to influence your linear model, the least squares line is preferred over the median-median line. If, on the other hand, you prefer to fit a line that is not influenced by extreme points, the median-median line is a better choice.

There are other important differences between the two lines. The median-median line has nothing like the least squares criterion as its basis. The least squares criterion is a particularly good rationale for obtaining a linear model if the data contains no wildly erratic points. In practice both lines can be drawn quite quickly with the computer, and the extent to which outliers affect each line can be readily determined. Unlike the median-median line the least squares line carries with it a lot of statistical theory that is very useful for more in-depth statistical investigations. Under certain conditions important statistical inferences may be drawn from the least squares line. A complete description of these is beyond the scope of this course. It should be understood, however, that techniques to support such inferences have not yet been developed for the median-median line. In summary, the least squares line gives valuable information that is essential in more advanced statistical applications and is thus useful far beyond the linear model we use here: but watch out for outliers!

Exercise Set 3

1. **a.** Study the data in Table 23. Note that all of the points except $(6, 2)$ lie on the line $y = x$. Predict how the point $(6, 2)$ will influence the median-median line and how it will influence the least squares line.

Table 23 Data Set for Exercise 1

x	1	2	3	4	5	6	7	8	9
y	1	2	3	4	5	2	7	8	9

b. Use a computer or calculator to check your predictions. You should notice that the effect of the point $(6, 2)$ is greater than the effect of the point $(5, 9)$ in Example 2. Can you explain why?

2. **a.** Use a computer or calculator to fit a least squares line to the data set in Table 24. Does the slope of this line make an accurate statement about the general nature of the data?

b. Find the equation of the median-median line. How accurately does the slope of this line convey information about the general nature of the data?

3. We have seen that the median-median line and least squares line for the temperature/pressure data provided in Table 17, page 153, are quite close to each other. One important difference between the two lines has to do with how the lines would change if one or more of the data points were to change.

Table 24 Data Set for Exercise 2

x	1	2	3	4	5	6	7	8	9
y	1	2	3	4	5	6	7	8	50

a. Suppose the data point (323, 906.1) were replaced with the point (323, 946.1). What effect does this have on the median-median line? (You should be able to answer this question without doing any calculations.)

b. Use a computer or calculator to find out how changing (323, 906.1) to (323, 946.1) affects the least squares line. Explain what you find.

c. Predict what would happen to the least squares line if the point (298, 834.4) were changed to (298, 814.4).

4. In Chapter 1 we found a median-median line to model the relationship between the rate of cancer deaths and an index of exposure to radioactive contamination near Hanford, Washington. The same data set is provided in Table 25.

Table 25 Cancer Data

Exposure Index	2.5	2.6	3.4	1.3	1.6
Death Rate	147	130	130	114	138
Exposure Index	3.8	11.6	6.4	8.3	
Death Rate	162	208	178	210	

a. Use a computer or calculator to find the equation of the least squares line through this data set.

b. Is this a good model? Analyze the residual plot and write several sentences to support your answer.

c. The equation of the median-median line is $y = 10.4x + 112.4$. Compare it to the equation of the least squares line. Why do you think they are different?

d. The symbol $\overline{x}$ represents the average, or *mean*, of the x-values, and $\overline{y}$ represents the mean of the y-values. Verify that the least squares line passes through the point $(\overline{x}, \overline{y})$.

e. Interchange the x- and y-values in the original ordered pairs. What do you think will be the equation of the new least squares line? Check your prediction.

f. In part e, did you predict that the equation would be the inverse of the one obtained in part a? In view of the least squares criterion, why is this not the case?

5. A biology student noticed that crickets seemed to chirp faster in the summer than in the spring or fall. Her grandmother had always told her that she could determine the temperature by listening to the crickets. Over the next season she counted the chirps per minute of a cricket and recorded the temperature. Her data is provided in Table 26.

Table 26 Cricket Data

Chirps (Per Minute)	55	67	75	83	91	99
Temperature (Fahrenheit)	50	54	55	58	58	60
Chirps (Per Minute)	119	134	140	149	164	178
Temperature (Fahrenheit)	67	69	70	74	77	79

a. Find a linear least squares model that the student can use to estimate the temperature by listening to the crickets.

b. Interpret the slope and y-intercept in terms of the phenomenon.

c. Explain how this model could be used to estimate the temperature very quickly by counting chirps for only 15 seconds.

d. If you wanted to describe mathematically the relationship between temperature and cricket chirps, which variable is more appropriate to consider as the dependent variable? Is this the same variable that you treated as the dependent variable in part a? If not, find a new model. Interpret the slope and y-intercept.

6. According to the *MINITAB Handbook* (Second Edition by Barbara F. Ryan, Brian L. Joiner, and Thomas A. Ryan, Jr., PWS-KENT Publishing Company, copyright 1985), a statistician named Frank Anscombe constructed the data sets listed in Table 27.

a. Find the equation of the least squares line and median-median line for each of the four data sets composed of ordered pairs $(A, B), (A, C), (A, D)$, and (E, F).

i. Compare the median-median and least squares lines you obtain for each data set.

ii. Compare the least squares lines for the four data sets. If you knew nothing but the equations of the least squares lines, how much information would you really have about the data sets?

b. Closely examine the scatter plots for the four data sets.

i. For which data sets is a linear model appropriate?

ii. Discuss problems that might occur in fitting a line to each data set.

iii. Anticipate what the residual plots will look like when the linear models are fit to each data set. Use a computer or calculator to check your predictions.

Table 27 Anscombe's Data Sets

A	10.00	8.00	13.00	9.00	11.00	14.00	6.00	4.00	12.00	7.00	5.00
B	8.04	6.95	7.58	8.81	8.33	9.96	7.24	4.26	10.84	4.82	5.68
C	9.14	8.14	8.74	8.77	9.26	8.10	6.13	3.10	9.13	7.26	4.74
D	7.46	6.77	12.74	7.11	7.81	8.84	6.08	5.39	8.15	6.42	5.73
E	8.00	8.00	8.00	8.00	8.00	8.00	8.00	19.00	8.00	8.00	8.00
F	6.58	5.76	7.71	8.84	8.47	7.04	5.25	12.50	5.56	7.91	6.89

4

Re-expression and Least Squares Lines

We frequently want to find a mathematical model when the relationship between variables is not linear. We have discussed ways to re-express data in order to straighten the scatter plot. If we succeed, we can fit a line to the linearized data and then transform our linear equation so that it expresses the relationship between the original variables.

Our previous work involved fitting a median-median line to data after linearizing. In an analogous way, we can re-express data and then fit a least squares line. The following examples illustrate this process. Keep in mind as you study the examples that the least squares principle will be applied to the re-expressed data and that the resulting line will be influenced by every point in the data set.

Example 1

Cooling

Refer to the data and scatter plot shown in Table 1, page 144. Recall that we are looking for a function to model the relationship between time and temperature so that we can predict the temperature of the water after 100 minutes.

In Exercise 9 in Section 2 you tried to find a model for this data set. What functions did you consider as possible models? Does the scatter plot seem to indicate a quadratic, an exponential, or a reciprocal-power function? More than likely you recognized the horizontal asymptote and rejected the quadratic model. You may have tried semi-log or log-log re-expression. If so, your residual plot should have indicated a poor linear fit for the re-expressed ordered pairs. The reason for this poor fit is that semi-log re-expression will linearize a data set only if it can be modeled by an equation of the form $y = Ae^{Bx}$. Such a data set will exhibit a horizontal asymptote at the line $y = 0$. Similarly, log-log re-expression will linearize a data set only if it can be modeled by an equation of the form $y = Ax^B$. When we use this re-expression for a data set that we suspect can be modeled by a reciprocal-power function, the horizontal asymptote will likewise be $y = 0$. However, the scatter plot of the cooling data exhibits a horizontal asymptote at $y = 45$. If the cooling phenomenon can be modeled by an exponential or reciprocal-power function, its equation will need to account for a vertical shift of 45 units.

The techniques of semi-log and log-log re-expression are still valuable to us in our attempt to find a model for the cooling data. Our knowledge of functions and transformations should indicate that the ordered pairs $(x, y - 45)$ lie on a curve asymptotic to $y = 0$. Thus, if the relationship is exponential, then semi-log re-expression performed on ordered pairs $(x, y - 45)$ will linearize the cooling data. Figure 28 shows the least squares line through the re-expressed data $(x, \ln(y - 45))$ and the corresponding residual plot. The least squares line appears to fit the re-expressed data well, and, as expected, the residuals are small. There appears to be a cyclical

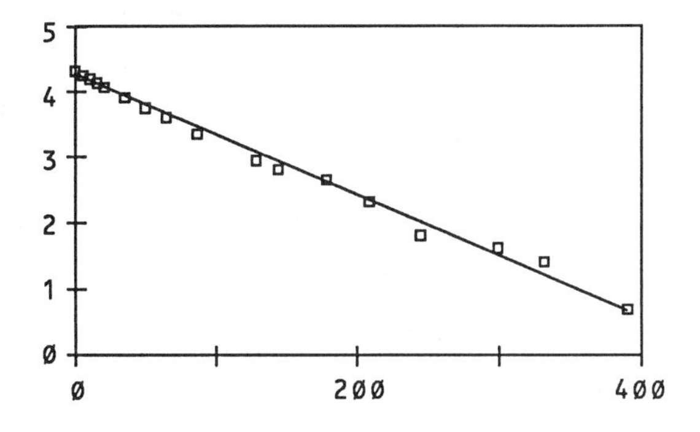

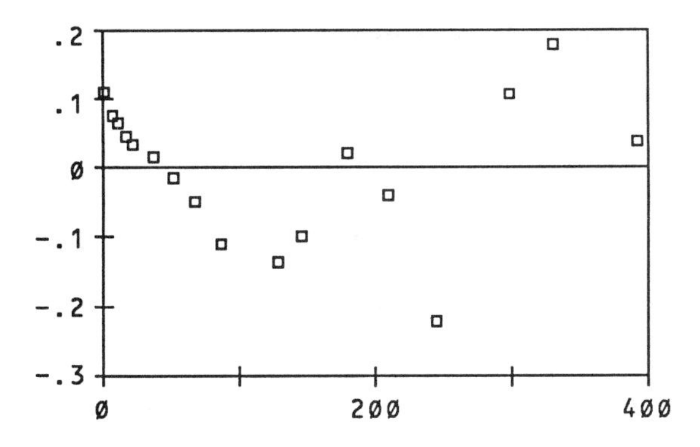

Figure 28 Least Squares Line and Residuals for Re-expressed Cooling Data

pattern in the residual plot, however. Patterns in the residuals at this stage of the analysis often indicate that the selected re-expression technique has not satisfactorily linearized the data. This conclusion should not be made too hastily, however. The pattern we observe could be associated with the data rather than with the model. In this example we might surmise that the fluctuations indicate interruptions in the cooling process, perhaps from compressor cycles of the refrigerator or from changes in the refrigerator temperature as a result of opening the door.

The equation of the least squares line through the ordered pairs $(x, \ln(y-45))$ is

$$\ln(y-45) = -0.00921x + 4.26.$$

Solving for y gives:

$$\begin{aligned} \ln(y-45) &= -0.00921x + 4.26 \\ y - 45 &= e^{-0.00921x+4.26} \\ y - 45 &= e^{4.26}e^{-0.00921x} \\ y &= 70.81e^{-0.00921x} + 45. \end{aligned}$$

Figure 29 shows a graph of this exponential function superimposed on the original scatter plot. The residuals again display a cyclical pattern but are consistently small relative to the observed temperatures.

In our discussion of mathematical models, we have explained that a good model is one that shares the essential features of the phenomenon it describes without all the clutter. We expect there to be some deviation between the model and the observed data points, and our decision concerning the adequacy of a model is a subjective one. In this example, we feel that the function $y = 70.81e^{-0.00921x} + 45$ seems to fit the general trend of the data and that the decreasing exponential function is a satisfactory model for the cooling phenomenon.

The equation $y = 70.81e^{-0.00921x} + 45$ expresses the relationship between temperature and time as the liquid cools. We can now answer the question posed at the beginning of this chapter: "What was the temperature of the liquid after 100 minutes?" Substituting $x = 100$, we find that $y = 73$. This means that after 100 minutes the temperature was about 73 degrees. ∎

Class Practice

1. Study the residual plot in Figure 29. Notice that the largest residuals are the ones associated with early observations. How can you account for this pattern?

2. Verify that a reciprocal-power function (shifted 45 units) would not be a good model for the cooling phenomenon by re-expressing the data and examining the scatter plot that should be linear if this model were appropriate.

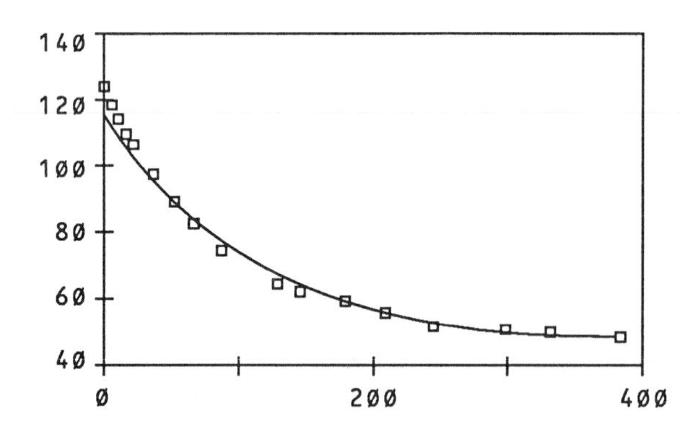

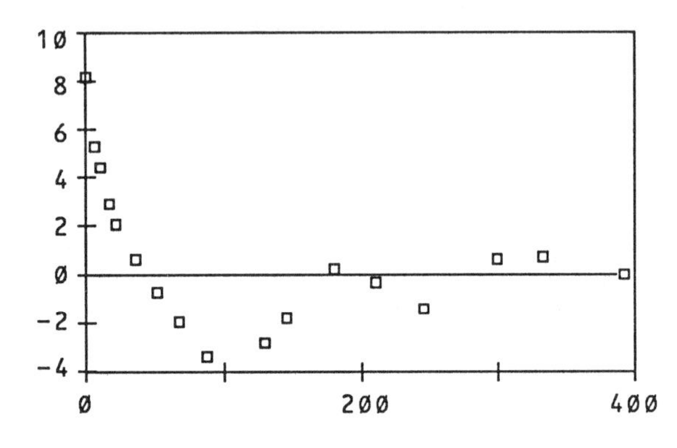

Figure 29 Exponential Model and Residuals for Cooling Data

Table 30 Planet Data

Planet	Distance	Years
Mercury	36.0	0.241
Venus	67.0	0.615
Earth	93.0	1.000
Mars	141.5	1.880
Jupiter	483.0	11.900
Saturn	886.0	29.500
Uranus	1782.0	84.000
Neptune	2793.0	165.000
Pluto	3670.0	248.000

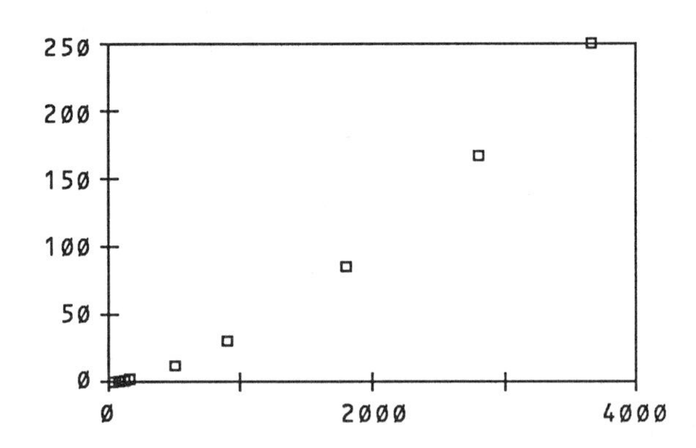

Example 2

Planet Orbits

Suppose a tenth planet were discovered at a distance of 1.4 billion miles from the sun. How long do you expect it would take this planet to complete its orbit around the sun?

Solution In order to answer this question, we need to know something about the relationship between a planet's distance from the sun and the time it takes to complete an orbit. Table 30 provides data on the nine planets. The x-values represent the planet's mean distance from the sun, measured in millions of miles. The y-values represent the number of years required for the planet to complete its revolution around the sun; this is called the *sidereal year*. If we can find an equation to model the relationship between x and y, we will be able to use this model to answer the question posed above. The scatter plot in Table 30 has distinct curvature, so it is clear that the relationship is not linear. It is difficult to judge from the plot, however, just what the curvature is.

Since we have no previous knowledge about the relationship between mean distance and sidereal year, we have no real choice but to do experimental re-expression. We can try both semi-log and log-log re-expression, but we should bear in mind that these techniques will linearize only exponential or power function relationships that have not been shifted vertically. Figure 31 shows the results of trying both semi-log and log-log re-expression. Note that common logarithms

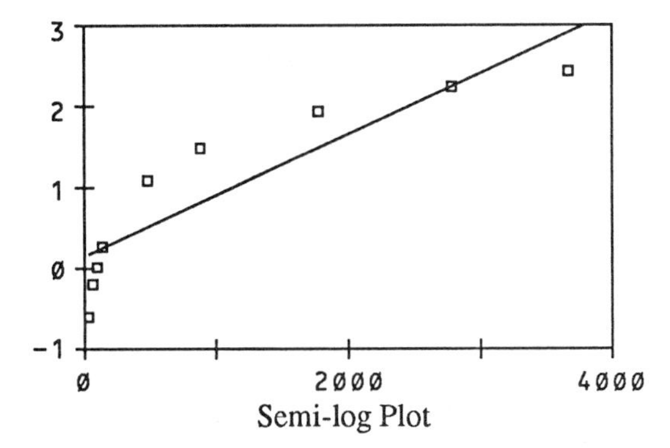

Semi-log Plot

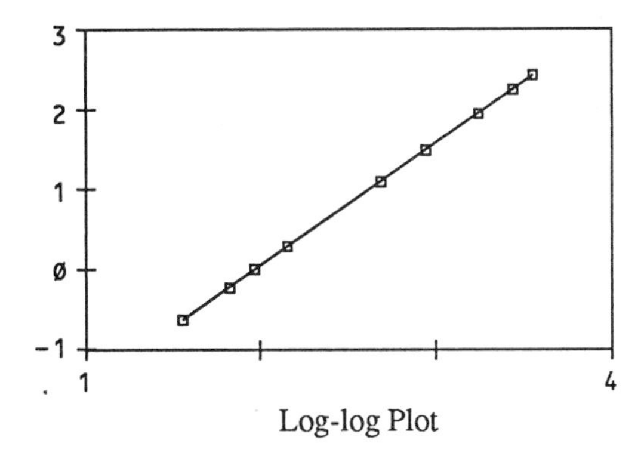

Log-log Plot

Figure 31 Semi-Log and Log-log Least Squares on Planet Data

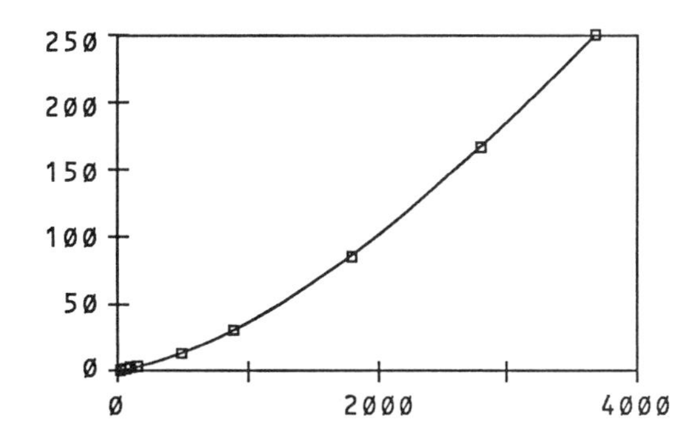

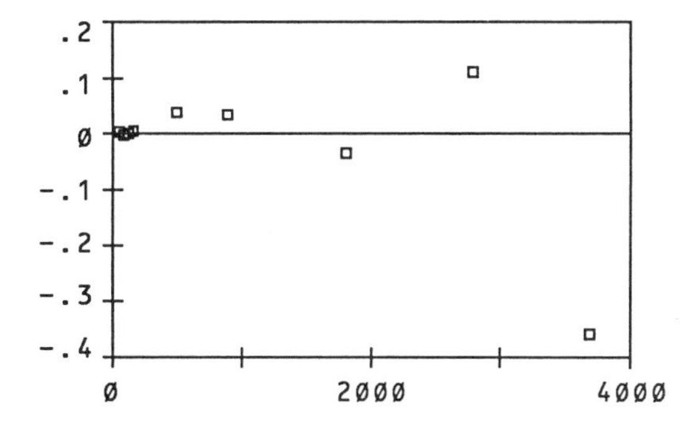

Figure 32 Power Function Model and Residuals for Planet Data

have been used. The semi-log plot shown first curves downward, but the log-log plot next appears to be linear. To check this model, we will fit the power function model $y = 0.001118x^{1.4999}$ to the original ordered pairs and examine the residuals (see Figure 32).

The equation of the least squares line through the ordered pairs $(\log_{10} x, \log_{10} y)$ is

$$\log_{10} y = 1.4999 \log_{10} x - 2.9515.$$

This line appears to be a good model for the re-expressed data, so we can conclude that a power function is a good model for the ordered pairs (x, y). Solving for y results in the following:

$$\begin{aligned} \log_{10} y &= 1.4999 \log_{10} x - 2.9515 \\ y &= 10^{1.4999 \log_{10} x - 2.9515} \\ y &= x^{1.4999} \cdot 10^{-2.9515} \\ y &= 0.001118x^{1.4999}. \end{aligned}$$

Our analysis of the data associated with nine planets reveals that $y = 0.001118x^{1.4999}$, or $y \approx 0.001x^{1.5}$. This equation can also be written as

$$y^2 \approx 10^{-6}x^3,$$

which implies that the square of the sidereal year is proportional to the cube of the mean distance from the sun. This relationship is called *Kepler's Third Law*; it was discovered empirically by Johann Kepler in 1618. He experimented, without the aid of data analysis or computers, until he found the mathematical equation to describe this relationship.

We can now use the model $y \approx 0.001x^{1.5}$ to predict the time that would be needed for a planet 1.4 billion miles from the sun to complete its orbit. Substituting $x = 1400.0$, we find that $y \approx 52.383$. Approximately 52.4 years would be needed for such a planet to orbit the sun. ∎

After linearizing data, it is very important to solve the linear equation for the dependent variable and then graph this model with the original data set. This allows you to realistically assess the size of errors to be expected when you use the model to make predictions. When you re-express data and then fit a least squares line to the linearized data, the least squares principle is applied to re-expressed y-values, not to the original y-values. For instance, suppose you have taken the logarithm of the y-values and then fit a least squares line to ordered pairs $(x, \log y)$. You will have minimized the error in the logarithms of the y-values but not of the actual y-values. This can lead to deceptive information about the size of residuals, since logarithms tend to be small numbers.

Exercise Set 4

1. In a laboratory experiment, a tumor was induced in a plant, and the growth of the tumor was recorded over time. The data set is given in Table 33.

a. Use re-expression and least squares analysis to find a model for this data.

Table 33 Growth of Tumor

Number of Days After Induction	14	19	23	26	28
Size of Tumor (cc)	1.85	4.25	7.5	11	14.5

Number of Days After Induction	30	33	35	37	41
Size of Tumor (cc)	18.95	28.65	38	49.75	84.5

b. At what rate is the tumor growing?

2. Researchers in anthropology are interested in the way urban growth affects lifestyle and stress levels. They measured the average walking speed of persons living in various cities; their data set is shown in Table 34.

Table 34 Walking Speed versus Population

Population (thousands)	341.9	5.5	0.4	78.2	867.0	14.0
Walking Speed (ft/sec)	4.8	3.3	2.8	3.9	5.2	3.7

Population (thousands)	23.7	70.7	304.5	138.0	2602.0
Walking Speed (ft/sec)	3.3	4.3	4.4	4.4	5.1

(Source: *UMAP Unit 551,* COMAP, Inc., 1983)

a. Find an equation to model the relationship between walking speed and population. (Re-express the data and use a calculator or computer to find the equation of a least squares line.)

b. According to your model, in what sized city would the walking speed be zero?

c. What does your model imply about walking speed as population increases? Be specific. Is there an upper bound to walking speed?

d. What does your model imply about the relationship between urban growth and stress?

3. The biology student who found a model for predicting temperature by listening to crickets stirred much interest among her friends. One decided to repeat the experiment, but he gathered his data in a different way. He used a stopwatch to determine the time required to hear 50 chirps. The data in Table 35 show the number of seconds required to count 50 cricket chirps at various temperatures.

Table 35 Cricket Chirps and Temperature

Number of Seconds to Count 50 Chirps	94	59	42	36	32	32
Temperature (°F)	45	52	55	58	60	62
Number of Seconds to Count 50 Chirps	26	24	21	20	17	16
Temperature (°F)	65	68	70	75	82	83

a. Make a scatter plot of the data and re-express the data to produce linearity.

b. Fit a least squares line to the linearized data. Write an equation that will allow you to predict the temperature based on a count of cricket chirps.

c. Suppose you find that it takes 25 seconds to count 50 chirps. Use your model to predict the temperature.

4. A ball is thrown upward from the top of an 80-foot building. Its height above ground at various times is given in Table 36.

Table 36 Height of Thrown Object

Time (sec)	0	0.2	0.6	1.0	1.2
Height (feet)	80	92	110	130	134
Time (sec)	1.5	2.0	2.5	2.8	3.0
Height (feet)	142	144	140	132	129
Time (sec)	3.4	3.8	4.0	4.5	
Height (feet)	112	90	82	44	

a. Find a model for this data set. (Re-express the data and use a calculator or computer to find the equation of a least squares line.)

b. According to your model, when will the ball hit the ground?

CHAPTER 6

Modeling

1 *Introduction*

There are many problems in the world around us that are solved with concepts and skills acquired through mathematical training. The solutions to some problems are direct and exact, but more often we are required to think more deeply and develop mathematical models in our search for solutions. Mathematical modeling is a process that requires creative thinking, experimentation, setting priorities, and making choices. We present several examples to illustrate methods for developing mathematical models to solve a variety of problems.

2 *The Tape Erasure Problem*

A tape recording is made of a meeting between two managers. Their conversation starts at the 21st minute on the tape, and it lasts for 8 minutes. The tape records for 60 minutes. While playing back the tape one of the managers accidentally erases 15 consecutive minutes of the tape, but doesn't know which 15 minutes were erased.

1. What is the probability that the entire conversation was erased?
2. What is the probability that some part of the conversation was erased?
3. Suppose the exact position of the conversation on the tape is not known, except that it began sometime after the 21st minute. What is the probability that the entire conversation was erased?

You may have solved probability problems in previous mathematics courses by computing the ratio of the number of successful outcomes to the number of all possible outcomes. This approach is limited to those situations where there are a finite number of outcomes. The questions above require a different approach since there are infinitely many locations for the erasure. We will use geometric models to find these probabilities.

To answer question 1 we need to analyze the starting time of the 15-minute erasure. We can get acquainted with the problem by looking at a few specific starting times. Suppose the erasure begins 10 minutes from the beginning of the tape. Then the portion of the tape between the 10th and 25th minutes is erased; this section includes a portion of the conversation but not the entire conversation. What if the erasure begins at the 15th minute or the 25th minute? If the erasure begins at the 15th minute it will end at the 30th minute and thus erase the entire conversation. If the erasure begins at the 25th minute, it will erase only a portion of the conversation. If we let x be the number of minutes from the beginning of the tape to the start of the erasure, then x can vary from 0 to 45. We are using x to represent every possible way that the tape erasure can occur; we can then analyze this problem by looking at x-values. Using a portion of a number line to represent these values, we can see that the set of all possible outcomes is an interval 45 units in length. This interval is called the *sample space.* An illustration is shown in Figure 1. An interval on the number line is a simple model of this problem; every possible starting time is represented by a point in this interval.

The conversation lasts from the 21st minute to the 29th minute, so the erasure can start as early as the 14th minute and as late as the 21st minute in order to erase the entire conversation. Thus, the value of x must be between 14 and 21 in order to be in the region we call the *event space.* The event space is the part of the sample space that consists of all successful outcomes for the problem. In this case a successful outcome is the erasure of the entire conversation. The *probability of a success* is the ratio of the length of the event space to the length of the sample space. The event space is 7 units long, the sample space is 45 units long, so the probability that the entire conversation was erased is $\frac{7}{45}$, or about 0.16.

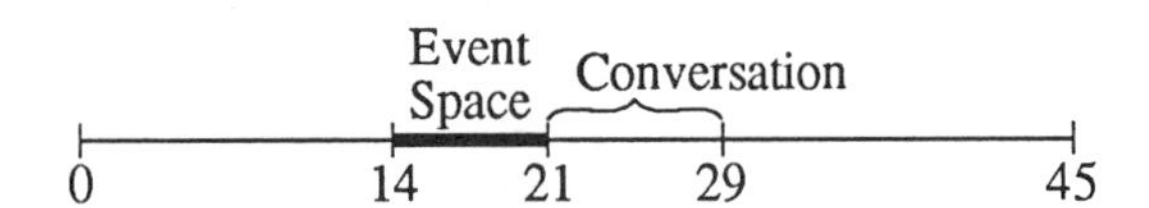

Figure 1 Model for Question 1 of the Tape Erasure Problem

To answer question 2, we can still let x represent the starting time of the erasure. The sample space is the same as it was in part 1, but the event space is larger. The event space now includes values of x that result in any portion of the conversation being erased. In order to identify the event space, the situation can be analyzed algebraically, or you can use trial and error. You should find that a portion of the conversation will be erased if x is between 6 minutes and 29 minutes. So the event space is 23 units long, and the probability that some part of the conversation was erased is $\frac{23}{45}$, or about 0.51.

For question 3, we need to consider both the starting time of the erasure and the starting time of the conversation. Let x represent the starting time of the erasure, and let y be the starting time of the conversation. Since the erasure is 15 minutes long, x can vary from 0 to 45. Since the conversation starts after the 21st minute and is 8 minutes long, y can vary from 21 to 52. The sample space consists of all ordered pairs (x, y) in the rectangle shown in Figure 2. Our sample space is now two-dimensional with an area of 1395 square units.

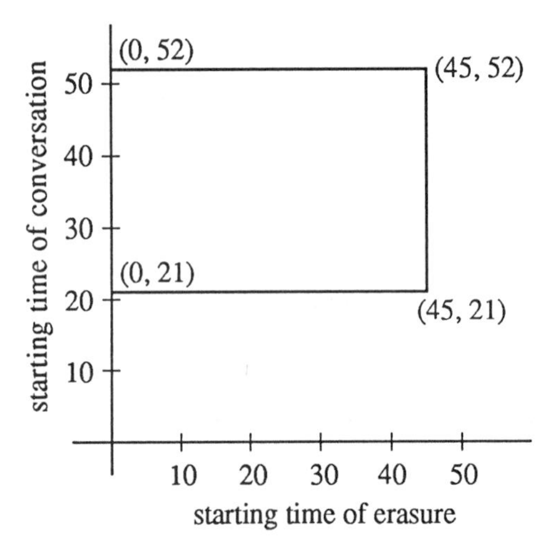

Figure 2 Sample Space for Question 3 of the Tape Erasure Problem

Which points in the sample space are in the event space? One way to investigate the event space is by testing individual points. For example, the point $(15, 30)$ corresponds to the erasure starting at the 15th minute and the conversation starting at the 30th minute. The erasure would be over before the conversation began, so this point is not in the event space. By similar reasoning, you should be able to decide that $(17, 23)$ and $(20, 27)$ are in the event space, whereas $(30, 25)$ is not. In general, it is clear that the conversation must start after the erasure; thus, $y \geq x$. Also, the conversation must start by the time 7 minutes of the erasure have gone by, so that there will still be 8 minutes of erasure left to erase the 8-minute conversation; thus, $y \leq x + 7$. This means that a point is in the event space if its coordinates satisfy the inequalities $x \leq y \leq x + 7$, which is the region shaded in Figure 3. Its area is 192.5 square units. The probability we want is the ratio of the area of the event space to the area of the sample space, or

$$\frac{192.5}{1395} \approx 0.14.$$

This means that the probability that the entire conversation was erased is 0.14 when we know only that the conversation started after the 21st minute.

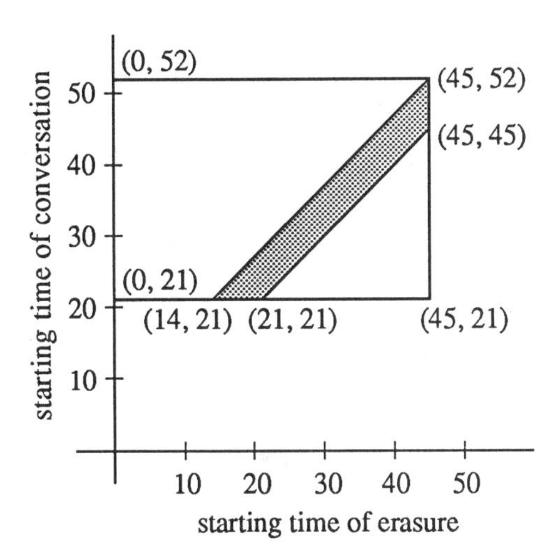

Figure 3 Event Space for Question 3 of the Tape Erasure Problem

A significant aspect of the tape erasure problem is the geometric model we use to represent it. This modeling process can be applied to many probability problems where counting is not possible. Once the model is created and the sample and event spaces are identified, probabilities are calculated as the ratio of two geometric lengths or areas.

The following exercise set consists of more problems that can be solved by geometric probability techniques. The geometric aspects of some problems are more obvious than others. The two keys to each problem are the identification of what events are occurring randomly and the establishment of a model that represents these events by points.

Exercise Set 2

1. Suppose 100 darts are thrown at random at a dart board, all of them hitting the board, and 26 of the darts land in a region that is painted red. Make a sketch to show what the dart board might look like and how much of it is painted red.
2. An electric power line is suspended between two poles that are 1200 feet apart. High winds result in a break at a random point on the line. What is the probability that the break is less than 30 feet from a pole?
3. The Inner Circle Sandwich Shop has a square dart board measuring 18 inches by 18 inches that hangs on a wall. Customers can win free sandwiches by throwing darts at the board. The dart board contains three concentric circles with their centers at the center of the square. The radii of the circles are 3 inches, 4 inches, and 6 inches, respectively. The prize for throwing a dart into the inner circle (the bull's eye) is a free large sandwich. Customers win a medium sandwich if their dart lands in the circle of radius 4 (excluding the bull's eye). The prize for throwing a dart inside the circle of radius 6 (excluding the two inner circles) is a free small sandwich. A customer wins nothing if his or her dart lands outside the largest circle.

   a. If you want to find the probability of winning a sandwich, you need to make two assumptions. What are those assumptions?

b. Find the probability that a customer will win a medium sandwich.

c. The management has decided that too many people are winning sandwiches, so they want to change the chances of winning. What would the radius of the largest of the three circles have to be so that the probability of *not* winning a sandwich would be 0.95?

4. A surveillance company has detection devices that illuminate a region shaped like a quarter circle with a radius of 100 yards. These devices will be distributed along a border pass that is 1000 yards wide (see Figure 4).

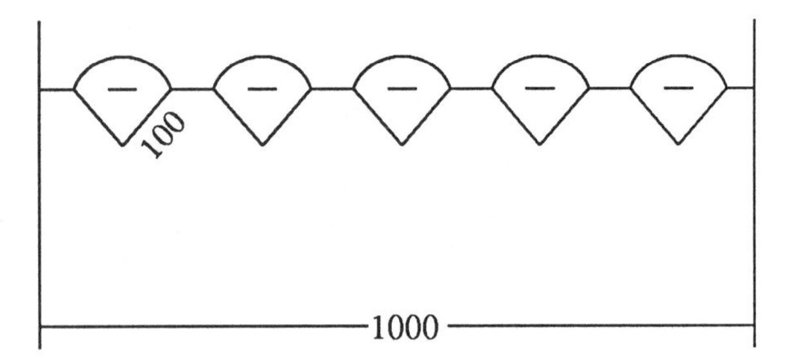

Figure 4 Surveillance Problem

a. Suppose 5 of the devices are used to survey the pass. If an intruder enters the pass at a random point, what is the probability of detection?

b. How many devices are needed so that the probability of detecting an intruder is greater than 0.9?

5. Max and Yvonne want to meet at the ice-cream shop to enjoy their favorite after-dinner treat. Both agree to arrive sometime between 8:00 and 8:30; they also agree that the first person to arrive will buy two cones and then wait for the other person. If the second person has not arrived within 12 minutes, the first person will start to eat that person's cone and will then continue to wait for the friend. If they arrive at the same time, each will buy his or her cone. Find the probability that each person eats only one ice-cream cone.

6. The manager of a busy computer center is expecting two telephone calls; one will be from an IBM salesperson and the other from an Apple representative. The IBM person will call sometime between 2:00 and 4:00 p.m. The Apple person will call sometime between 2:30 and 3:15. Assume that both calls can occur with equal chance at any time within these periods.

a. Suppose that the manager needs to get some information from the IBM salesperson before talking to the Apple representative. What is the probability that he or she will be able to do this (*i.e.*, that the IBM salesperson will call first)? Assume that the manager has two phone lines.

b. Suppose instead that immediately after each call, the manager needs to discuss the results of the call with his or her supervisor. This discussion will take 10 minutes. What is the probability that the manager will miss the second call during this discussion? Assume that each phone call lasts 5 minutes.

7. Suppose two numbers between 0 and 4 are chosen at random and then added. For what value of k will 50% of the sums be less than or equal to k? For what values of k will 100% of the sums be less than or equal to k?

3
Free Throw Percentages

Imagine that you are sitting in front of the television one Saturday afternoon watching the Chicago Bulls play the Boston Celtics. The Bulls' star player drives to the basket and is fouled. As he stands at the free throw line, the announcer states that he is hitting 78 percent of his free throws this year. He misses the first shot but makes the second. Later in the game, the player is fouled for the second time. As he moves to the free throw line, the announcer states that he has made 76 percent of his free throws so far this year. Can you determine how many free throws this player has attempted and how many he has made this year?

This appears to be a straightforward question. If we let x represent the number of shots made and y the number of attempts, then we can represent the problem analytically by the system of equations

$$\frac{x}{y} = \frac{78}{100}$$

and

$$\frac{x+1}{y+2} = \frac{76}{100}.$$

Now we can solve for x and y. Since $100x = 78y$ and $100x+100 = 76y+152$, we find that $x = 20.28$ and $y = 26$.

We recognize that since x and y represent numbers of shots, only integer values are meaningful. Do we round down to 20 shots out of 26 attempts? Unfortunately, hitting 20 of 26 shots gives the player a percentage of 77, and not the reported 78. Hitting 21 out of 26 gives a percentage of 81. Did the announcer make a mistake? No. Before the shots, the player had indeed hit 78 percent of his free throws and afterward he had hit 76 percent. What's going on here?

Our analytic model using the system of linear equations needs to be modified. The key is the expression 78%. If the announcer followed general rounding practice, the value 78% can actually represent any percentage that is at least 77.5% and less than 78.5%. Similarly, a reported percentage of 76% means the actual percentage is at least 75.5% and less than 76.5%. We really have a system of inequalities to solve, rather than a system of equations.

Our analytic description of the problem should be

$$\frac{775}{1000} \leq \frac{x}{y} < \frac{785}{1000}$$

and

$$\frac{755}{1000} \leq \frac{x+1}{y+2} < \frac{765}{1000}.$$

We are searching for a solution (or solutions) where x and y are both positive integers. We can restate the compound inequalities as follows:

$$\frac{775}{1000} \leq \frac{x}{y} \text{ and } \frac{x}{y} < \frac{785}{1000}$$

and

$$\frac{755}{1000} \leq \frac{x+1}{y+2} \text{ and } \frac{x+1}{y+2} < \frac{765}{1000}.$$

Since x and y are positive we can multiply across these inequalities and isolate y in each inequality to obtain:

$$y \leq \frac{1000x}{775} \text{ and } y > \frac{1000x}{785}$$

and

$$y \leq \frac{1000x}{755} - \frac{510}{755} \text{ and } y > \frac{1000x}{765} - \frac{530}{765}.$$

We wish to find the *lattice points* (points whose coordinates are integers) in the region bounded by the four lines described by the inequalities above.

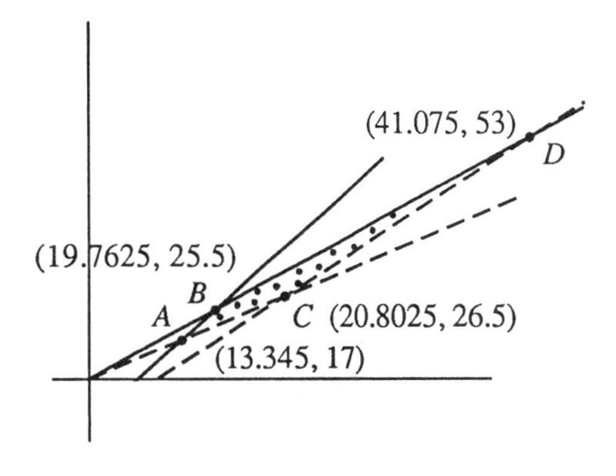

Figure 5 Lattice Points in the Region

This region is sketched in Figure 5. It is not drawn to scale in order to show the constraints more clearly. The intersection points of the lines are easily determined algebraically. You should verify that these points are $A = (13.345, 17)$, $B = (19.7625, 25.5)$, $C = (20.8025, 26.5)$, and

$D = (41.075, 53)$. We need to find the lattice points in the quadrangular region determined by these points.

To find the lattice points, you might examine a computer- or calculator-generated graph of the region bounded by the four lines. If you try this approach, you will see why it is not used. Identifying integer coordinates in the region is very difficult from the graph.

There are several efficient ways to perform this search analytically. If you examine the graph in Figure 5, it should be clear that between points A and B, integer x-values are in the interval $14 \leq x \leq 19$ and y-values are bounded above by $y = \frac{1000}{755}x - \frac{510}{755}$ and below by $y = \frac{1000x}{785}$. Substituting $x = 14$ into both of these expressions yields $y = 17.9$ and $y = 17.8$. Since there are no integers between these y-values, there are no lattice points associated with $x = 14$. Similar results occur for $x = 15$, $x = 16$, and $x = 17$. Substituting $x = 18$ into the expressions yields $y = 23.2$ and $y = 22.9$, so we have a lattice point in the region at $(18, 23)$. Between points B and C and points C and D the boundary lines change. We continue this process with appropriate x-values and boundary lines to identify the remaining lattice points. There are six lattice points in the region defined by our inequalities:

$$(18, 23), (21, 27), (25, 32), (28, 36), (31, 40), (38, 49).$$

Another method to solve this problem is to write a simple program and have a computer search for solutions in the rectangular region where $14 \leq x \leq 41$ and $17 \leq y \leq 53$ (see Figure 6). You may want to write your own program, or you can use the BASIC program listed below:

```
10 PRINT
"X","Y","X/Y","(X+1)/(Y+2)","X+1","Y+2"
20 FOR Y = 17 TO 53
30 FOR X = 14 TO 41
40 LET A = X/Y : LET B = (X+1)/(Y+2)
50 IF A < .785 AND (A > .775 OR A =
.775) AND B < .765 AND (B > .755 OR B =
.755) THEN PRINT X, Y, A, B, X+1, Y+2
60 NEXT X
70 NEXT Y
80 END
```

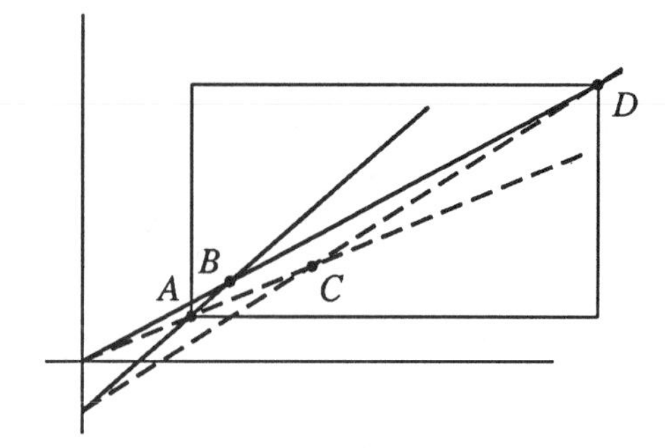

Figure 6 Checking the Rectangular Region

The output of the program is the following:

X	Y	X/Y	(X+1)/(Y+2)	X+1	Y+2
18	23	.78261	.76000	19	25
21	27	.77778	.75862	22	29
25	32	.78125	.76471	26	34
28	36	.77778	.76316	29	38
31	40	.77500	.76190	32	42
38	49	.77551	.76471	39	51

Once again we see that there are six different ways that the Bulls' star player could have satisfied the stated conditions. Can we determine which answer is correct?

Exercise Set 3

1. A player is making 50% of her shots before a game, and after going 8 for 14 is now hitting 55% of her shots. What is the fewest number of shots she could have taken so far this season?
2. Suppose that a player is batting .299. On his next time at bat, he gets a hit. Is it possible for his batting average now to be .306? Explain.
3. Explain the difference between saying that a basketball player hit 20.28 shots out of 26 and saying that the average American family has 2.4 children.

4

Determining the "Best" Speed Limit[1]

A stretch of Interstate 40 is being widened just outside the Research Triangle Park in North Carolina to accommodate increasing traffic from Raleigh. Unfortunately, to widen the highway, only one lane is left open to traffic for several hours each day. The Department of Transportation would like to have the traffic move as quickly as possible along the one available lane, but does not want to have a lot of rear-end collisions. What speed limit should they set in order to maximize the flow of traffic and still ensure safe travel? Does such a speed limit even exist?

Before we can answer these questions, we must make sure we understand them. What is meant by the flow of traffic? Suppose you were to set up an observation post along the highway. The number of cars that travel past this post in a unit of time is one measure of traffic flow. If three cars pass the observation post in one minute when traffic is dense, you would probably feel that the traffic is moving very slowly. We can express this traffic flow by the quantity 3 cars per minute or equivalently by the quantity 0.05 cars per second. If we assume that cars travel at the posted speed limit, then we suspect that there is a relationship between the traffic flow and the speed limit. We would like to find a mathematical expression for this relationship. Then we can analyze the relationship to determine what speed limit maximizes traffic flow.

Consider the situation in which a car in dense traffic has just passed the observation post. How much time is required before the next car passes the post? Since distance equals the product of rate and time, we know that time T is the ratio of the distance the car must travel to the speed at which it is moving. The distance the car must travel is composed of two parts: the following distance, or the distance between this car and the preceding one, and the car length (since we are trying to find the time required for the next car to pass the post). At this point we will make some assumptions to simplify our task. First, assume that all cars are approximately the same length. We will let C represent this length; the units for the car length C are feet. We will also assume that all cars follow each other at about the same distance D, measured in feet. Making these assumptions provides uniform traffic flow with all cars traveling at the same speed V, measured in miles per hour. Figure 7 illustrates the uniform traffic flow that results from these assumptions.

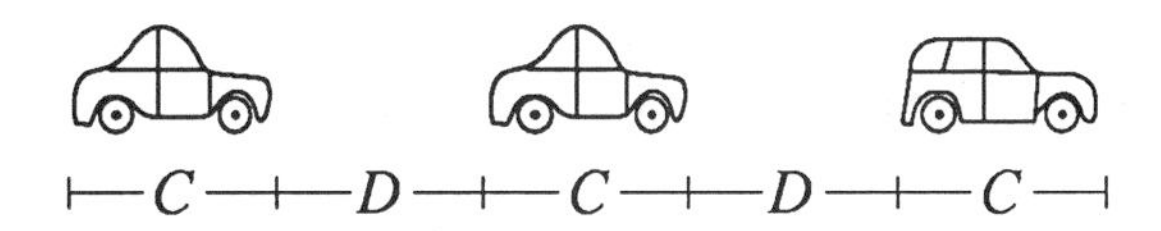

Figure 7 Uniform Traffic Flow

With these assumptions we can now write an expression for T, the time between cars:

$$T = \frac{C+D}{V},$$

where C and D are measured in feet and V is measured in miles per hour. We have a mixture of linear units in this expression. To eliminate this problem we can convert miles per hour to feet per second as follows:

$$\begin{aligned} V\frac{\text{miles}}{\text{hour}} &= V\frac{\text{miles}}{\text{hour}} \cdot \frac{1 \text{ hour}}{3600 \text{ sec}} \cdot \frac{5280 \text{ ft}}{1 \text{ mile}} \\ &= \frac{22}{15}V\frac{\text{ft}}{\text{sec}} \\ &\approx 1.467V \text{ ft/sec.} \end{aligned}$$

So the time between cars (in seconds) is given by

$$T = \frac{C+D}{1.467V},$$

---

[1]Based in part, with permission, upon ideas in Stone, Alan, and Ian Huntley, "Easing the Traffic Jam," *Solving Real Problems with Mathematics*, The Spode Group, Cranfield Press, Cranfield, United Kingdom, 1982.

where 1.467 converts speeds from miles per hour to feet per second.

We now have an expression for the number of seconds that elapse between the arrival of cars at the observation post. We would like an expression for traffic flow F which is measured in cars per second. This is simply the reciprocal of T:

$$F = \frac{1}{T} = \frac{1.467V}{C + D}.$$

Evidently traffic flow depends upon both velocity and the distance between cars; however, the distance between cars is also related to the velocity. As a safe driving practice, we know that the faster cars travel, the greater the separation should be. This separation (D) should be the stopping distance required for a car traveling V miles per hour. As you know, there are two components to stopping distance. One is the distance a car travels as the driver reacts to a situation and places his or her foot on the brake. We will denote this distance by R. The second component of stopping distance is braking distance, denoted by B, which is the distance a car travels once the brake has been applied.

We would like to determine the relationship between D, which equals $R + B$, and the velocity V. Such a relationship would allow us to express traffic flow, F, in terms of the constant car length C and one variable, the velocity V.

Before proceeding, let's assess what we have done thus far. Since the problem asks that we determine the speed limit that will maximize traffic flow, our first task was to define what we mean by traffic flow. The definition of traffic flow as the number of cars per unit of time led to our consideration of time per car and thus to the mathematical expression for traffic flow as a function of car speed, car length, and distance between cars. Several assumptions were made along the way. Specifically, we assumed that all cars are the same length and follow each other at equal safe distances. In reality, we know that neither of these assumptions is true. Cars vary in size, and all drivers do not maintain safe separation distance. The assumption of equal car length allows us to represent a variable quantity with a constant in the mathematical expression for traffic flow. Similarly, the assumption of equal separation distances will allow us to develop a submodel that will reduce to one the number of variables in the expression. Making assumptions to simplify a problem is an important, and often necessary, step in the process of mathematical modeling. Later we will re-examine these assumptions and perhaps modify them to ensure that our solution is reasonable.

What we will now attempt to do is develop a submodel for D to use as a component of our larger model for traffic flow; that is, we will try to develop an expression for D in terms of the variable V. We can develop this submodel empirically using the information in Table 8 from the Ministry of Transportation Code used in England.

This table was written as an instructional listing to show the relative relationship between velocity and stopping distance. First we consider the relationship between velocity (V) and reaction distance (R). It is clear from Table 8 that this relationship is simply $R = V$, with R measured in feet and V in miles per hour. If the car is traveling at K mph, it will travel K feet while the driver reacts and initiates the braking.

Next we turn our attention to the braking distance B. In Figure 9, first we have graphed braking distance on the vertical axis and velocity on the horizontal axis. We see that the graph is concave up; clearly B is not a linear function of V.

Table 8 Reaction and Braking Distances

Velocity (mph)	Reaction Distance (ft)	Braking Distance (ft)
10	10	5
20	20	20
30	30	45
40	40	80
50	50	125
60	60	180
70	70	245
80	80	320
90	90	405

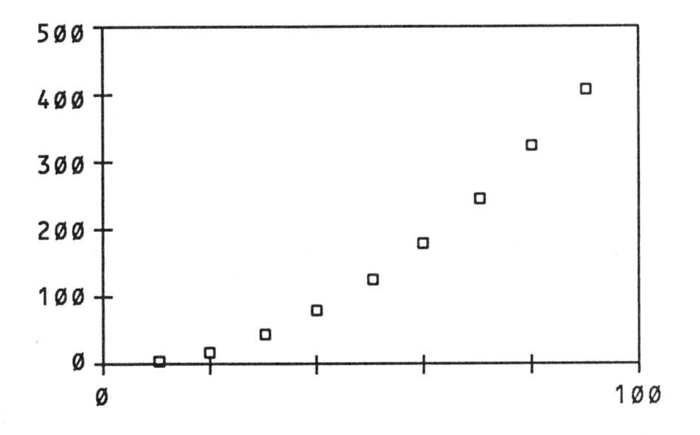

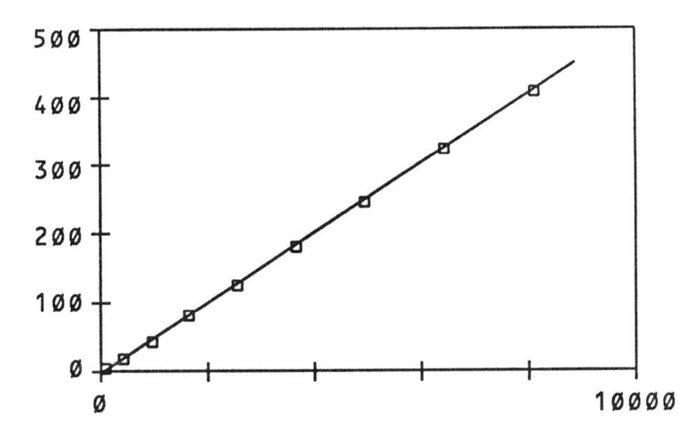

Figure 9 Braking Distance as a Function of Velocity

Using the techniques of data analysis we can see that the relationship is quadratic and re-express the data by squaring values of the velocity. Looking at the graph of ordered pairs (V^2, B) in Figure 9, we see that the re-expressed data is linear. By fitting a least squares line through the data, we determine that the relationship between braking distance and velocity is $B = 0.05V^2$. This relationship is based on braking distance in feet and velocity in miles per hour.

The important aspect of the data is that the reaction distance is a linear function of velocity while the braking distance is a quadratic function of velocity. Doubling the speed from 30 mph to 60 mph doubles the reaction distance from 30 feet to 60 feet. However, the braking distance quadruples from 45 feet to 180 feet.

Our empirical submodel allows us now to write the distance D as the sum of the reaction distance and the braking distance, both in terms of velocity:

$$D = R + B = V + 0.05V^2.$$

Based on the information provided in the Ministry of Transportation chart, we can express traffic flow as

$$F = \frac{1.467V}{C + D} = \frac{1.467V}{C + V + 0.05V^2}.$$

We now have a mathematical expression relating traffic flow and the speed of the traffic. This expression is our model, and we can use it to answer our original question. Our goal is to find the value of V that maximizes the value of F. We stated earlier that we would assume that all cars are of the same approximate length; let's suppose now that the average car is 12 feet long so that $C = 12$. We can now use a calculator or computer to generate a graph and use it to determine the speed that yields the greatest traffic flow. Graphing over the domain $10 \leq V \leq 50$, we find that the maximum traffic flow is 0.575 cars per second when the cars travel at a speed of approximately 15 miles per hour. This graph is shown in Figure 10.

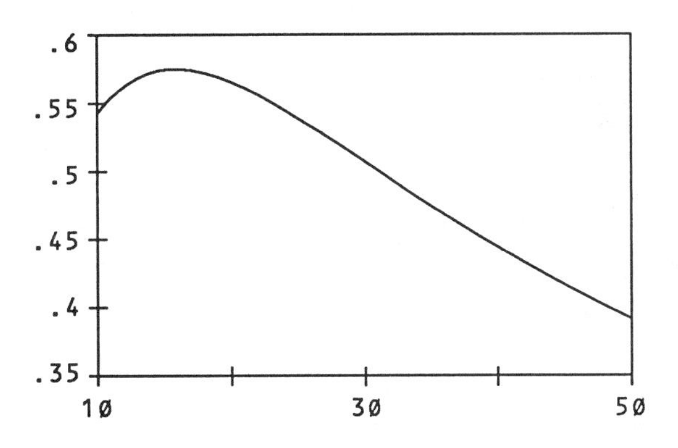

Figure 10 Traffic Flow as a Function of Velocity

Class Practice

1. In developing this model we assumed that all cars follow each other at the safe driving

distance specified by the Ministry of Transportation in England. When cars travel at 15 miles per hour, what is this safe driving distance?

2. When cars travel at 15 miles per hour and follow each other as specified by the Ministry of Transportation, how many cars will pass the observation post in one hour?

3. Do you think that cars will travel through a construction area at a speed of 15 mph? How is traffic flow affected if the cars travel at 20 mph? at 25 mph?

4. Suppose the cars are compact cars and have an average length of 10 feet. Are the specified speed and traffic flow altered significantly?

5. Suppose a lane on a section of Interstate 40 is reserved for large trucks which average 50 feet in length. According to this model, what speed is required for maximum flow in this lane?

6. Let C vary from 6 to 51 in increments of 3. Find the maximum speed for each car (or truck) length. How does the maximum speed relate to car length?

7. Do you feel that the model we developed is appropriate for all car lengths? Did we make assumptions that might not be valid for long trucks?

Improving the Model

How realistic is the model just developed? Did we make assumptions that compromise the usefulness of the model or the validity of the answer? This model assumes that every driver allows sufficient distance for his or her car to stop without hitting the one in front of it. In your experience on highways, does it seem that drivers really drive at a distance which will prevent accidents? Perhaps a more realistic model would have the distance represented by $D = (R + sB)$ where s is a parameter whose value lies between 0 and 1. If $s = 1$, we have the model we analyzed above; if $s = 0$ we have the extreme situation where drivers leave no space for braking distance. More typically, the value of s will fall somewhere in between. Making this adjustment to the model assumes that drivers will consistently maintain a separation required for reaction distance but often "fudge" on the braking distance.

The function for traffic flow under this variation in the model becomes

$$F = \frac{1.467V}{12 + V + s(0.05V^2)}.$$

Class Practice

1. Determine the speed and distance required to maximize traffic flow with $s = 0.1$, 0.25, 0.5, 0.75, and 0.9.

2. What value of s do you think is the most realistic? What speed and distance would you recommend for traffic on Interstate 40?

Another Variation

Do drivers really separate their following distance into two pieces, reaction distance and braking distance? Another model for the following distance is that the drivers simply leave some portion of the required distance between them, not just a portion of the braking distance. This translates to the model

$$F = \frac{1.467V}{12 + s(V + 0.05V^2)}$$

where the parameter s varies from 0 to 1 as before. This model can now be analyzed in the same way as the two above.

Class Practice

1. How does this formulation alter the speed and distance requirements with $s = 0.1$, 0.25, 0.5, 0.75 and 0.9?

2. Since speed limits are always given in five miles per hour increments, a speed of 31.8 mph and a speed of 28.3 mph most likely

would both result in a posted speed limit of 30 mph. In this case, the difference in the two formulations would be insignificant to the problem. Is the difference in the two formulations with the parameter s significant? Do high or low values of s seem to have more effect on the result?

3. The general rule many drivers follow is to always leave one car length for each 10 mile/hour in speed. Does this linear model appear to be a safe model to follow?

This example illustrates another important aspect of mathematical modeling. Once a model is developed, it is not necessarily complete. The model should be refined and modified to obtain a more accurate reflection of reality. From the initial model, which had an empirical submodel, two refinements were proposed as more realistic models. Which of these models do you think is the best? Can you recommend a modification that you feel is even more realistic?

5
The Tape Counter Problem

Tape recorders, tape players, and VCR's are common in today's society. Frequently people will use one of these machines to record several different shows or songs on the same tape. One of the difficulties in using tape players is finding the position of a particular song or show on the tape. Most tape players have a counter to help determine positions on the tape. The counter reading increases as the tape is being played, but the numbers on the counter can be misleading. If the counter goes from 0 to 500, 250 is *not* halfway through the tape; and when you have 50 counts left, you do *not* have one tenth of the tape remaining. Just how does the counter work? What is the relationship between the counter reading and the time that the tape has played? Does the counter speed up or slow down as the tape is played? To answer these questions we need a mathematical model which expresses the relationship between the length of time the tape has played and the counter reading.

To gain more understanding of the operation of such a counter, watch a reel-to-reel player. Notice that as the tape plays, the supply reel decreases in size while the take-up reel increases in size. Each time the take-up reel rotates, its radius increases, taking up more tape as it goes (see Figure 11).

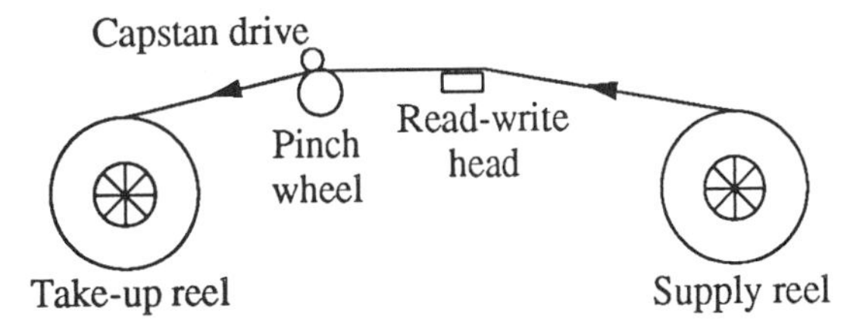

Figure 11 Movement of a Tape Player

One way to answer the question about the counter and time is to collect data and develop an empirical model. A tape player was played for 30 minutes and the counter readings were recorded at various times. The data are provided in Table 12. Since we would like to predict the time from the counter reading, we choose the counter reading c as the independent variable and attempt to express time t as a function of the counter reading.

Table 12 Counter Reading and Elapsed Time

Counter	0	22	43	63	100	180
Time (sec)	0	60	120	180	300	600

Counter	250	311	368	409	420
Time (sec)	900	1200	1500	1740	1800

Examination of a scatter plot of the data (see Figure 13) reveals that the relationship between time and the counter reading is not linear. Is it quadratic? Since the problem involves the increasing radius of the take-up reel, a quadratic relationship

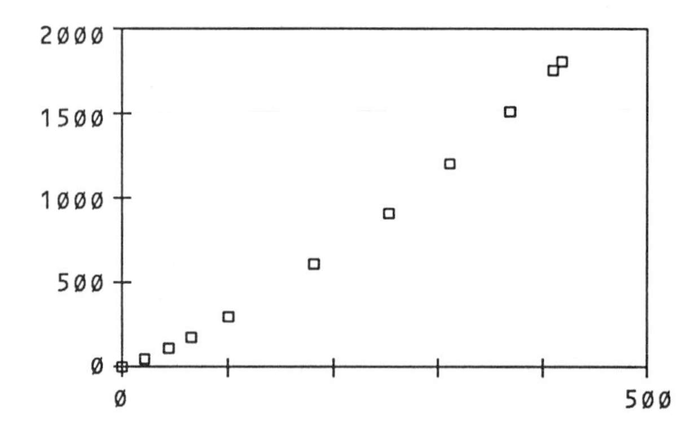

Figure 13 Scatter Plot of Time versus Counter Reading

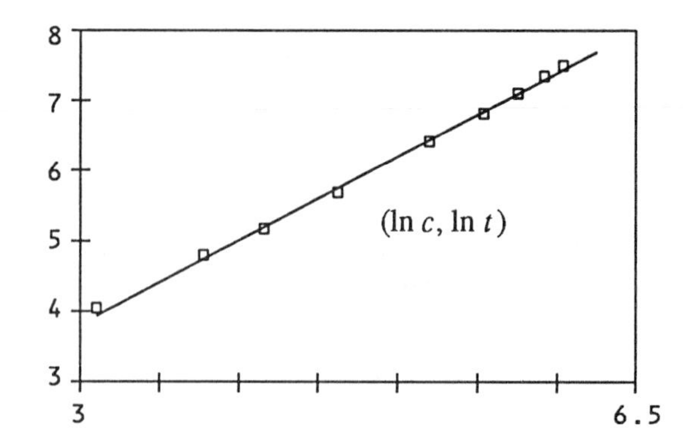

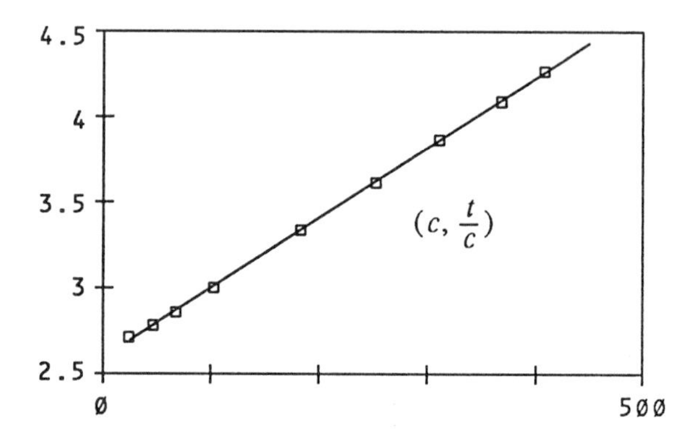

Figure 14 Median-Median Line Fits for $(\ln c, \ln t)$ and $(c, \frac{t}{c})$

seems likely. After all, as the tape rolls onto the reel, the area defined by the tape is a quadratic function of the radius, namely πr^2.

If we start the counter at zero, then the time played is initially zero. This gives us the point $(0, 0)$ automatically. The consequence of this point is significant in determining a model. If the relationship is indeed quadratic, the point (0,0) requires that there be no constant term in the model, so the form of the quadratic is either $t = Kc^2$ or $t = K_1c^2 + K_2c$. If the relationship can be modeled by $t = Kc^2$, re-expressing the data from (c, t) to $(\ln c, \ln t)$ should linearize it; if $t = K_1c^2 + K_2c$ is the correct model, then the re-expression $(c, \frac{t}{c})$ should linearize the data. (We need to delete the point $(0, 0)$ before performing these re-expressions.) Figure 14 shows the median-median line fits to both re-expressions of the data.

Analysis of the residual plots corresponding to these fits indicates that the second one is better. The equation of the median-median line is $\frac{t}{c} = 0.004c + 2.61$, so we conclude that the counter reading and time played are related by the quadratic equation $t = 0.004c^2 + 2.61c$. We have written the equation so that time is considered a function of the counter reading because the reading is the variable whose value we know. If we are playing a 30 minute tape, and the counter reading is 220, then we estimate that the tape has been playing for approximately $0.004(220)^2 + 2.61(220) = 767.8$ seconds. This means there are about 1032 seconds, or 17 minutes and 12 seconds, left of the tape.

The empirical model was developed using data from a 30 minute tape. Is the model the same for a 45 minute or 60 minute tape? Are those tapes thinner? Does it matter? Will this model be appropriate for all tape players? We could construct an empirical model for each size cassette and for several different tape players; however, there is a more fundamental question unresolved. Just what makes the coefficients in our model 0.004 and 2.61? Where do these numbers come from? The empirical model argues for a quadratic relationship, but it does not explain what the components are. It is not an *explicative* model, one which explains what the coefficients represent.

What are the important features of the tape recorder? Look again at Figure 11. The tape is pulled across the read-write head at a constant speed s by the capstan drive. (If you have doubts about the constant speed, think about how we can splice tapes together.) The counter is turned by either the take-up reel or the supply reel. For our argument, we will assume that the take-up reel controls the counter. We would like to relate the counter reading c with the time t the tape has been playing.

As usual, we need to make several assumptions to simplify the problem. First, let's assume that the counter reading is proportional to the number of turns T of the take-up reel. Mathematically, we say that $c = kT$. In this expression, k expresses the number of counts on the counter per turn of the reel. For example, if $k = \frac{1}{2}$ then the counter increases one-half unit for each turn of the take-up reel. Also, let's assume that the tape wraps uniformly on the reel. This means that the tape does not stretch significantly as it winds. The thickness of the tape h stays the same from the first wrap to the last.

Now let l represent the length of tape that has wrapped around the take-up reel. Since l represents the length of circular wrappings, we know that l is related to circumference. Circumference is given by $2\pi r$; unfortunately, r is not constant. The radius increases with each rotation. By how much does it increase? Since the tape wraps around the reel each time the take-up reel rotates, the increase in r is one thickness of tape for each rotation. We can describe this mathematically with the expression $r = r_0 + r_T$. The radius of the hub of the take-up reel is denoted r_0. It does not change. However, the outer portion of the radius, denoted r_T, increases by one thickness of tape h with each turn. After T turns of the take-up reel, $r_T = hT$.

Let's summarize what we know at this point. First, we assume that $c = kT$. Also,

$$r = r_0 + r_T, \text{ and } r_T = hT.$$

We can write an expression for the length of the tape wrapped around the take-up reel, l, in terms of the number of rotations of the take-up reel, T. After T rotations, there are T circumference measures contributing to the total length of tape l, so:

$$l = T(2\pi r)$$

$$l = T(2\pi(r_0 + hT)).$$

Notice that we have neither c nor t in the equation.

Remember that we also assumed that the tape moves across the read-write head at a constant speed s. So another expression for l is $l = st$. Solving for t, we find that

$$t = \frac{l}{s} = \frac{T(2\pi(r_0 + hT))}{s}.$$

Based on the assumption that $c = kT$, we have $T = \frac{c}{k}$. This gives the preceding equation the form

$$t = \frac{2\pi r_0 c}{ks} + \frac{2\pi h c^2}{k^2 s}.$$

This is exactly the quadratic form of our empirical model. And in this model, we know what the coefficients mean. In this sense the model is explicative. The linear coefficient represents the product of the hub radius, the constant 2π, and the reciprocal of the tape speed s and the constant of proportionality representing the number of counts per rotation of the take-up reel. The quadratic coefficient is the product of the tape thickness and the constant 2π divided by the product of the tape speed s and the square of the constant of proportionality.

Is this model correct? An important aspect of developing models is the verification of the model. We can verify this model by measuring the variables s, h, and r_0, finding k, and substituting the values into the equation to see if they agree with our empirical model. On the machine which generated the initial data, the value of s is 4.76 cm/sec, the tape thickness h is 0.0018 cm, and the hub radius of the tape r_0 is 1.1 cm. The value of k can be found by graphing the counter reading c against the number of turns T; k is the slope of that line. For this tape recorder, k

Table 15 Counter Reading and Elapsed Time (in Seconds)

Counter Reading	Observed Time	Predicted Time Empirical Model	Predicted Time Explicative Model
0	0	0	0
22	60	59	62
43	120	120	128
63	180	180	197
100	300	301	343
180	600	599	730
250	900	903	1151
311	1200	1199	1580
368	1500	1502	2035
409	1740	1737	2393
420	1800	1802	2494

is 0.55. Substituting into the equation for t, we find that

$$t = \frac{(2\pi)(1.1)c}{(0.55)(4.76)} + \frac{(2\pi)(0.0018)c^2}{(0.55)^2(4.76)}.$$

Simplifying yields $t = 2.64c + 0.00785c^2$.

Recall that our empirical model was $t = 2.61c + 0.004c^2$. The linear coefficients are very close in both models, but the coefficient of the quadratic term is nearly twice as large in the new model. Which model is correct? Table 15 gives the values of time calculated using the empirical and explicative models for the counter readings observed in the example. Notice that both models work reasonably well initially. This is because the linear term in the model dominates for small values of c. Since the linear terms in both models are essentially the same, the results are very close. However, as c increases the quadratic term becomes increasingly important, and the estimates from the two models begin to differ. Notice that for the largest values of c, the explicative model is very poor.

It seems obvious now that our explicative model is not correct. When a model cannot be verified, the model builder must re-examine his model. Where is the error? The first place to look is the simplifying assumptions used in the beginning. Initially we assumed that the tape wraps uniformly and does not stretch. Perhaps it does, in fact, stretch and the stretching creates the error. Or perhaps we made a more fundamental error in our analysis.

Look again at the expression for r_T. We have said that $r_T = hT$ and that $r = r_0 + r_T$. This gives us $r = r_0 + hT$, which in turn gives us an expression for l which is $l = T(2\pi(r_0 + hT))$. Let's look closely at this expression in light of how the tape accumulates on the reel. After one complete rotation, the length of tape that has accumulated on the take-up reel is simply $2\pi r_0$, since on the first turn there was no tape already on the reel. At this point $T = 1$, so according to our expression for l, we have $l = 1(2\pi(r_0 + 1h)) = 2\pi(r_0 + h)$. It appears that we have introduced h in our expression one rotation too soon.

If $T = 2$, then $l = 2(2\pi(r_0 + 2h))$ according to our expression for l. But this is the value we would get if the tape traveled *twice* around the circumference of a circle with radius $(r_0 + 2h)$. In reality, the tape travels only *once* around a circle with radius $(r_0 + 2h)$ and *once* around a circle of radius $(r_0 + h)$. After three rotations, when $T = 3$, the length of the tape is given by $l = 3(2\pi(r_0 + 3h))$, indicating that the circumference $2\pi(r_0 + 3h)$ is traversed *three* times. In reality, the tape winds only *once* around a circumference $2\pi(r_0 + 3h)$, *once* around $2\pi(r_0 + 2h)$, and so forth. By the time we have fifty rotations, we are using $l = 50(2\pi(r_0 + 50h))$, which suggests that we have fifty wrappings with circumference $2\pi(r_0 + 50h)$! Do you see why the error in our model kept getting larger and larger?

Let's rework the length l more carefully. When the tape starts initially, it winds approximately the circumference of a circle whose radius is equal to the hub of the take-up reel. That is,

$$\text{if } T = 1 \text{ then } l_1 = 2\pi r_0.$$

The length of tape added on the second rotation, l_2, is approximately the circumference of a circle whose radius is the hub plus one thickness of tape h. So,

$$\text{if } T = 2 \text{ then } l_2 = 2\pi(r_0 + h).$$

Continuing in this fashion, we find

$$T = 3 \longrightarrow l_3 = 2\pi(r_0 + 2h)$$

$$T = 4 \longrightarrow l_4 = 2\pi(r_0 + 3h)$$

$$T = 5 \longrightarrow l_5 = 2\pi(r_0 + 4h)$$

and in general

$$T = n \longrightarrow l_n = 2\pi(r_0 + (n-1)h).$$

The total length of tape accumulated by this process, l, is the sum of all the l_i's. That is,

$$l = 2\pi r_0 T + 2\pi h\left[1 + 2 + 3 + \cdots + (T-1)\right].$$

This can be simplified to

$$l = 2\pi\left[r_0 T + h\left[1 + 2 + 3 + \cdots + (T-1)\right]\right]$$

and using

$$1 + 2 + 3 + \cdots + (T-1) = \frac{T(T-1)}{2},$$

$$\begin{aligned} l &= 2\pi\left[r_0 T + h\frac{T(T-1)}{2}\right] \\ &= 2\pi\left(r_0 - \frac{h}{2}\right)T + (\pi h)T^2. \end{aligned}$$

Substituting $T = \frac{c}{k}$ into this new expression gives

$$l = (2\pi)\left(r_0 - \frac{h}{2}\right)\frac{c}{k} + (\pi h)\frac{c^2}{k^2}.$$

Using this new expression and substituting from the equation $l = st$ gives the equation

$$t = \left(\frac{2\pi r_0}{ks} - \frac{\pi h}{ks}\right)c + \frac{\pi h}{k^2 s}c^2.$$

Notice that the coefficient of the quadratic term is indeed one-half the quadratic coefficient in the first explicative model, and the linear coefficient has been reduced slightly. How does this revised model fit the data? Using $h = 0.0018$ cm, $s = 4.76$ cm/sec, $k = 0.55$, and $r_0 = 1.1$ cm as before, we find that the new expression becomes

$$\begin{aligned} t &= \frac{(2\pi)(1.1) - 0.0018\pi}{(0.55)(4.76)}c + \frac{0.0018\pi}{(0.55)^2(4.76)}c^2 \\ &= 2.64c + 0.004c^2. \end{aligned}$$

If we compare this new result to the empirical model, we see that we now have an explicative model which can be verified. The coefficients are products of the initial conditions of the problem, namely the hub radius, the speed of the tape across the read-write head, and the constant of proportionality between the number of turns of the take-up reel and the counter. Had we not verified our initial model, we would never have known that we were in error. More importantly, we now have a model that should work for all tape lengths (and thicknesses), and for all tape players. The explicative model is much more general and therefore more useful than the empirical model.

This example illustrates one of the most important aspects of developing mathematical models. Once a model has been developed, it must be verified and, if necessary, improved. In this example we developed an explicative model that was incorrect. By comparing the empirical model and our first explicative model, we were able to focus on the aspect of the model that produced the error. We then proceeded to make the necessary changes and to develop and to verify a second explicative model.

Exercise Set 5

1. Use the least squares technique to fit a curve to the data given in this section. How does it compare to the quadratic used in the discussion?
2. Choose several tape players and take measurements for h, s, and r_0. Plot the counter reading against the number of turns of the take-up reel and the number of turns of the supply reel. One scatter plot should be linear and the other curved. The slope of the linear graph is k. Substitute these values into the general explicative model. How accurate is the model?

6 *Developing a Mathematical Model*

The process of developing mathematical models is inventive and creative. The concept is quite simple, though in practice the process itself is often difficult. Essentially, the mathematician wants to create a mathematical representation of some phenomenon that he or she wishes to understand more completely. Whatever the phenomenon, the model is developed by stripping away the extraneous aspects of the problem to allow a more focused study of its essential aspects. In this way, a model is a caricature of the problem. Certain characteristics are accentuated while others are de-emphasized. What is essential in one model may be irrelevant in another. In developing a model for projectile motion, friction might be ignored because its role in the phenomenon is minor. However, to the engineers and mathematicians designing the brakes for new aircraft, the model must include the effects of friction as an important feature.

The process of developing a mathematical model can be modeled as well. The model builder begins with a question or curiosity about some perceived phenomenon in the world. As the modeler focuses on the phenomenon, he or she isolates some particular aspect of the phenomenon to model. By isolating a particular aspect to consider, the mathematician can develop several submodels which can be combined to produce a larger, more complex model. We did this in the tape counter problem. We looked at the winding of the tape around the spool as a separate problem and developed a mathematical representation for it. Later, we included this submodel as one piece of the larger model.

Once the phenomenon has been selected, the modeler must idealize the problem or simplify it so that it is manageable and less complex. All simplifying assumptions must be carefully noted so that everyone recognizes the deviations from reality inherent in the model. In the tape counter problem, we assumed uniform thickness of the tape with no thinning as the tape wound around the spool, while in the problem of determining the best speed for single lane travel, we assumed that all cars are the same length and that all cars follow at the same distance. Neither of these assumptions is totally accurate; the tape does in fact stretch as it winds, and cars are of different lengths and follow each other at varying distances. However, the errors introduced by these assumptions do not invalidate the conclusions reached. If, on the other hand, the assumptions are too general, so that the error introduced is significant, the model may not represent reality, and the assumptions will need to be modified so that a more accurate, more useful model can be developed. There is an inherent trade-off between the simplicity and the accuracy of a model.

After the problem has been idealized, a mathematical representation is developed by determining the relationships between the variables remaining in the problem. For example, in the speed limit problem, we described the relationship between the stopping distance and the two components, braking distance and reaction distance. Also, we developed an empirical submodel describing how following distance is related to speed. An empirical relationship was also determined between the counter reading and the number of turns of the spool on the tape recorder. If the relationships cannot be determined, stronger assumptions must be made to simplify the problem even more.

Once the model is constructed, it must be analyzed or interpreted. After developing a mathematical function to model traffic flow, we needed to analyze its graph to find the best speed and following distance. The model itself produces no answers or solutions to the original problem. Its power lies in the insight and illumination it offers to help answer questions about the phenomenon being modeled.

Interpretations and conclusions drawn from the model must be checked for accuracy and utility. That is, the model must be verified. Do the conclusions make sense? Does the prediction of the model fit the observations of the phenomenon? In the tape counter problem, we had developed an inaccurate model. We needed to investigate the model to determine the source of the inaccuracy and correct it. If the model cannot be verified, then the model must be inspected step-by-step. Are there mathematical errors? Are the assumptions reasonable? Are they useful? Relax the assumptions a bit. Is the new model of greater utility? Continuing to vary the problem, relaxing assumptions, and adding variables until the model accurately reflects the aspects of the phenomenon under study is a large part of the modeling process.

Never stop too soon when developing a model. What questions does your solution pose? Is it possible to generalize your results? Questioning your results is essential if the model is to be improved.

Another important aspect of model building is analyzing the sensitivity of the model to small changes in the variables. In the problem of best speed limit, how much does the speed change if the car lengths are 10 feet rather than 12 feet? In the tape counter problem, how much will the counter reading be off if our measurements are off by 0.01 cm? Which measurements are most crucial? Can the differences in the predictions of the model be attributed to small errors in measurement, or are the differences too great to be explained that way?

The process of developing a mathematical model is dynamic. As new insight is gained, the model is improved and the process begins again. An existing model can be modified much more easily than a new model can be developed from scratch. If the model developed is too simple, it can be improved by introducing the missing components into the existing model. At times the model will collapse under the weight of these additions and must be constructed with new initial assumptions. If so, we've learned something. That is the essential ingredient — learning something about the phenomenon under study. If we do, then the process of mathematical modeling is successful.

7

Some Problems to Model

In previous sections of this chapter, situations were described and models were developed, verified, and modified. In this section, the problems will be described and a few hints given so that you can develop models on your own.

Tennis Serves

Tall players seem to have an advantage when serving in tennis. In matches shown on TV, it seems that short players must use spin serves and are less able to hit hard, flat serves. Why?

Here are questions to consider:

1. At what heights are serves usually hit?

2. How high is the net? How long is the service box? How far is the ball from the net when it is hit?

3. Is there a minimum height from which a flat, driving serve can be hit?

4. Is there an advantage in serving from far behind the service line? Is there an advantage in serving from the corner of the service area?

5. Should the server jump when serving?

Elevators

In some buildings, all elevators go to all floors. In others, certain elevators go only to the lower floors while the rest go only to the top floors. Why? What is the advantage of having elevators which travel only between certain floors?

Suppose a building has 5 floors (1–5) that are occupied. The ground floor (0) is not used for business purposes. Each floor has 80 people working on it, and there are 4 elevators available. Each elevator can hold 10 people. The elevators take 3 seconds to travel between floors and average 22 seconds on each floor when someone enters or exits. If all of the people arrive at work at about the same time and enter the elevator on the ground floor, how should the elevators be used to get the people to their offices as quickly as possible?

Here are questions to consider:

1. How many trips must each elevator make?

2. On each trip, how much time is spent traveling from floor to floor and how much time is spent waiting for passengers to exit on each floor?

3. How much time does it take if all elevators go to all floors?

4. How much time does it take if two elevators serve only the top floors and the other two serve only the bottom floors?

5. How much time does it take if one elevator serves only the top floors, one only the bottom floor, and the other two serve all floors?

6. If a fifth elevator were built, to which floors should it go?

7. Suppose the bottom three floors have 100 people working on each and the top two floors have only 50 people each. How should the four elevators be used?

8. If the travel time were increased to 5 seconds, would your solutions change? What if the travel time were decreased to 2 seconds?

9. If the time needed to exit at each floor were increased to 40 seconds, would the solutions change?

Ten-Speed Bicycles

The faster you pedal a bicycle, the faster it goes. Analyze the gearing on a standard ten-speed bike and construct a model which will give the speed traveled as a function of the number of revolutions per minute of the pedals for various sizes of the front and rear sprockets.

Here are some questions to consider:

1. What is the circumference of the tires on a standard ten-speed? How far does the bike go with each revolution of the wheels?

2. If you wish to go 20 mph, how fast must the wheels rotate?

3. The teeth on the sprockets are the same distance apart. Can the number of teeth on a sprocket be an indicator of its size?

4. If the front and rear sprockets are of equal size, how fast does the rear wheel turn if the pedals turn 30 times each minute? How far does the bicycle go in one minute?

5. If the front sprocket is twice the size of the rear sprocket and the pedals turn 60 times each minute, how many revolutions will the wheels make? How fast is the bike going?

6. What happens to the number of revolutions of the wheels if the rear sprocket size is constant but the front sprocket gets larger?

7. What happens to the number of revolutions of the wheels if the front sprocket size is constant but the rear sprocket gets larger?

Hot Dog Stand Location

A map of a portion of a college campus is given in Figure 16. The map shows the walking paths and dormitories in this section of campus. Your roommate has convinced you to open a hot dog stand at one of the intersections along the walkways. You would like the stand to be as convenient as possible for the students. Where on campus should you set up your stand?

Here are some questions to consider:

1. What would make the hot dog stand convenient? How far do students have to walk to reach the stand from their dorms?
2. What do you want to optimize when placing the stand? Do you want the shortest average distance for students or the smallest maximum distance to the stand? Are these measures different?
3. What is the average walk from the dorms if the stand is placed at A? at F? What is the maximum walk to each?
4. Suppose the dormitories are located at positions A, C, D, E, F, G, and K and the numbers of students in each dorm are 400 in A, 300 in C, D, E, and F, and 800 in both G and K. Does the position of the stand change with either minimum average distance or minimum maximum distance as the placement criterion?
5. Rather than place the stand at an intersection, would it be better to place it along one of the paths, halfway between B and G for example?
6. What paths could be added that would alter the placement of the stand?
7. Suppose two stands were to be used. Where should they be placed?

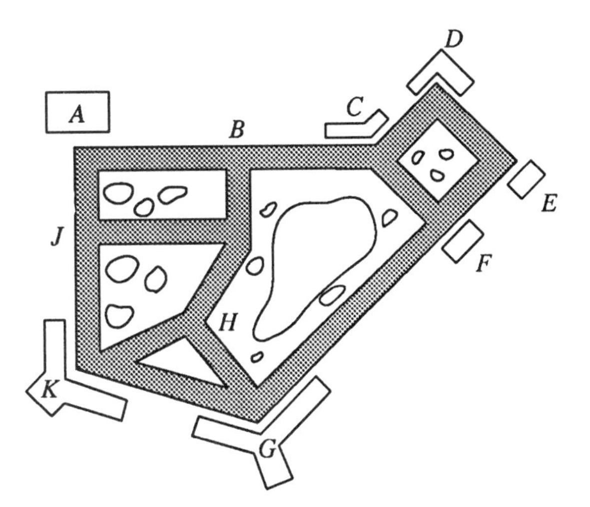

Figure 16 Map of Campus

Highway Patrol Location

The North Carolina highway patrol is responsible for patrolling the stretch of Interstate 40 between Cary and the Research Triangle Park. This stretch is approximately 15 miles long. With the new construction along this part of the highway, it has become particularly dangerous. Accidents are likely to happen anywhere in the region. The highway patrol would like to have as rapid a response to an accident as possible. Is it better to have a car driving the highway from end to end or to have a patrol car sitting in the middle waiting for an accident to happen?

Here are some questions to consider:

1. What do you want to optimize?
2. What determines the length of time it will take the patrol car to respond to a distress call?
3. Is a moving patrol car more likely to be closer to an accident than a stationary one?
4. If you assume that the accidents are uniformly distributed along the length of the highway, will geometric probability help?
5. Suppose improvements are made first in the middle 8 miles of road. If no accidents happen on this stretch, which car placement is better?

6. Suppose there is no place in the middle of this stretch of highway for the patrol car to wait on the shoulder. If the closest to the middle that the patrol car can wait is 3 miles from Cary or 4 miles from the Triangle, what should the patrol car do?

Airplane Engines

If an airplane engine has a probability p of working correctly, which is safer, a single engine or a multi-engined plane?

Here are some questions to consider:

1. How many engines are needed for each plane to fly? Can a three engine plane fly with one engine, or must at least two work for it to fly?

2. Suppose that multi-engined planes can fly if at least two engines work correctly. Which planes are safer? Does your answer depend on the value of p? Explain.

Barbers

Estimate the number of barbers needed in your town. Here are some questions to consider:

1. What is the population of your town? What portion of the population has its hair cut by a barber?

2. How often does one generally get a haircut? How many times a year does one go to the barber?

3. How long does it take to get a hair cut? How many hours a day does a barber work? How many heads can one barber cut in a year?

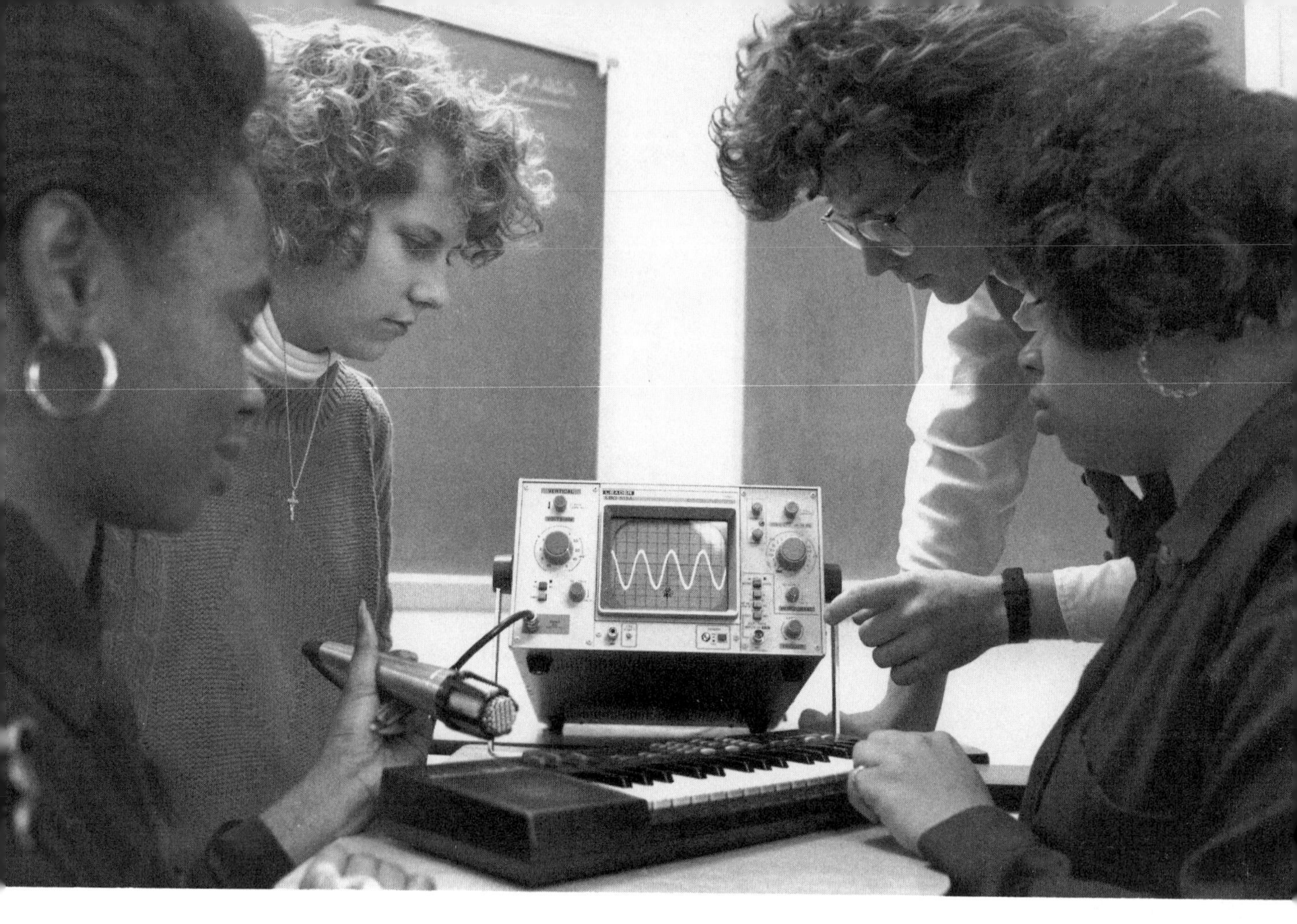

CHAPTER 7

Trigonometry

1

The Curves of Trigonometry

The hands of the clock on the wall move in a predictable way. As time passes, the distance between the ceiling and the tip of the hour hand changes. Suppose that the hour hand is 15 cm long and that at 12 o'clock the distance between the tip of the hour hand and the ceiling is 23 cm. At 3 o'clock the distance will have increased to 38 cm, and at 6 o'clock the distance will have increased to 53 cm. This is the maximum possible distance; after 6 o'clock the distance will decrease. At 9 o'clock the distance will be 38 cm, and at 12 o'clock the distance will again be 23 cm. Note that 23 cm is the minimum possible distance. The same sequence of values will be repeated every 12 hours. Figure 1 shows several (time, distance) ordered pairs; the t-coordinate represents the number of hours elapsed since we began recording distances at 12 o'clock.

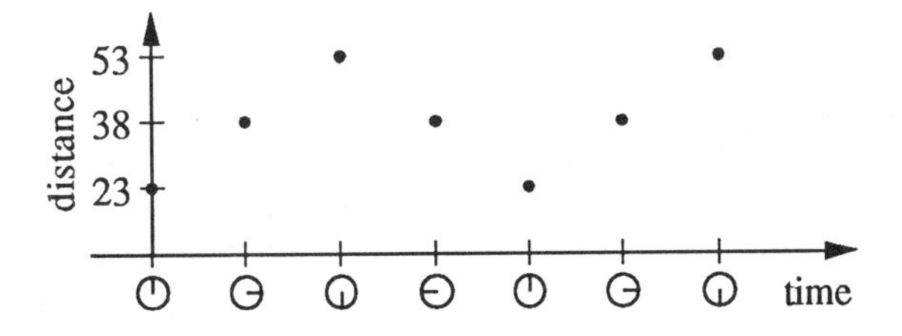

Figure 1 Distance from Ceiling versus Time

How do you think the maximum and minimum points in Figure 1 should be connected? One possibility is to connect the points with line segments. Linearity implies that the distance from the tip of the hour hand to the ceiling changes at a constant rate. However, the distance changes more quickly around 9 o'clock and 3 o'clock and more slowly around 6 o'clock and 12 o'clock. To display these facts on a graph, the points in Figure 1 should be connected with a curve; the curve is somewhat flatter near its turning points and somewhat steeper away from its turning points. A graph with points connected is shown in Figure 2.

The graph in Figure 2 should remind you of the sine function from the toolkit; it is a continuous curve that oscillates periodically between maximum and minimum values. Such a curve is called a *sinusoid.* How could you extend the graph to show a 48 hour time period?

Now suppose we began collecting (time, distance) ordered pairs at 3 o'clock. Our collection of ordered pairs would include (0, 38), (3, 53), (6, 38), (9, 23), and (12, 38). The graph beginning at 3 o'clock is shown first in Figure 3; in this graph t represents the number of hours after 3 o'clock. You should recognize a transformation of the sine curve.

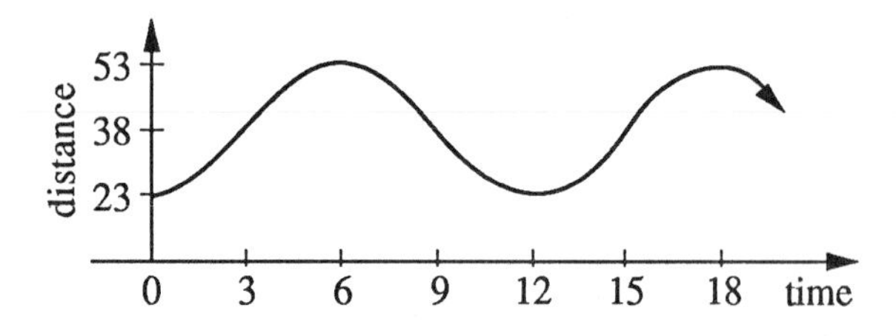

Figure 2 Distance from Ceiling versus Time Beginning at 12 O'Clock

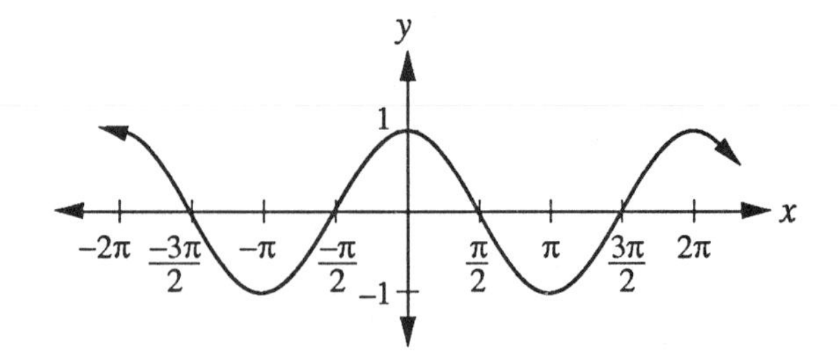

Figure 4 Graph of $y = \cos x$

Suppose now that the distance to the ceiling is measured beginning at 6 o'clock. If t represents the number of hours after 6 o'clock, then the graph will begin at a maximum. This graph is also shown in Figure 3. The graph has a shape that is characteristic of another trigonometric function called *cosine.*

The cosine function, written $y = \cos x$, has a graph that is very similar to the graph of the sine function. The graph of $y = \cos x$ is shown in Figure 4; the graph is periodic with period 2π. The graph also oscillates between a maximum of 1 and a minimum of -1. The *amplitude* of the graph

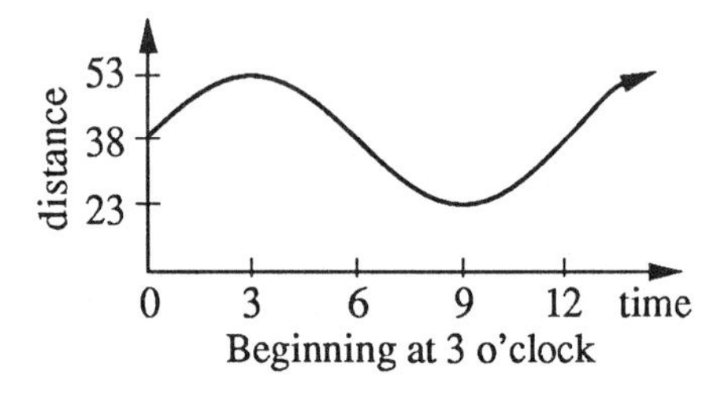

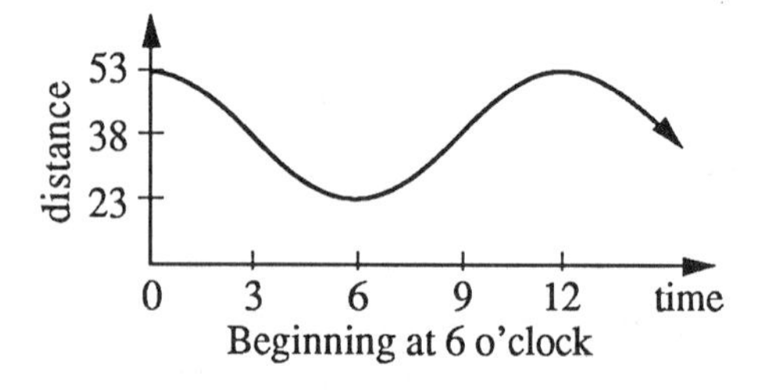

Figure 3 Distance from Ceiling versus Time Beginning at 3 O'Clock and at 6 O'Clock

is defined as

$$\text{amplitude} = \frac{\text{maximum value} - \text{minimum value}}{2}.$$

Thus the amplitude of $y = \cos x$ is $\frac{1-(-1)}{2} = 1$. What is the amplitude of the graph in Figure 2?

Note that the graph of $y = \cos x$ is symmetric about the y-axis; it is an even function and $\cos x = \cos(-x)$. The x-intercepts of $y = \cos x$ occur at $-\frac{3\pi}{2}$, $-\frac{\pi}{2}$, $\frac{\pi}{2}$, and $\frac{3\pi}{2}$, or generally at odd multiples of $\frac{\pi}{2}$; this can be written as $\frac{\pi}{2} + k\pi$ or $\frac{(2k+1)\pi}{2}$, where k is any integer. The maximum values of the function occur at even multiples of π, and the minimum values occur at the odd multiples of π. You should add the cosine function to your toolkit and should become very familiar with the basic shape, x-intercepts, and maxima and minima of its graph.

If you study Figure 4 you should see that the graph of the cosine function is a horizontal shift of the sine curve $\frac{\pi}{2}$ units to the left. This implies that the equation $\cos x = \sin(x + \frac{\pi}{2})$ is true for all values of x.

Example 1

Use the periodicity of the sine and cosine functions to evaluate each of the following: $\sin(-\frac{7\pi}{2})$, $\sin(19\pi)$, $\cos(7\pi)$, and $\cos(\frac{5\pi}{2})$.

Solution Since the period of the sine function is 2π, we can add any multiple of 2π to the argument

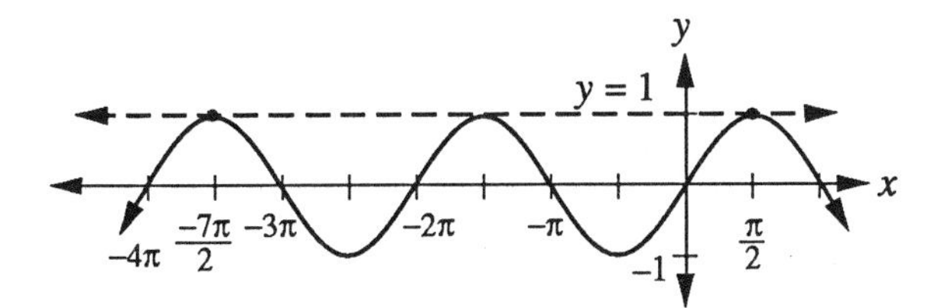

Figure 5 $\sin(-\frac{7\pi}{2}) = \sin(\frac{\pi}{2})$

of the function without changing the sine (see Figure 5). Therefore,

$$\sin\left(-\frac{7\pi}{2}\right) = \sin\left(-\frac{7\pi}{2} + 4\pi\right) = \sin\left(\frac{\pi}{2}\right) = 1.$$

Similarly,

$$\sin(19\pi) = \sin(19\pi - 18\pi) = \sin(\pi) = 0.$$

The period of the cosine function is also 2π, so

$$\cos(7\pi) = \cos(7\pi - 6\pi) = \cos(\pi) = -1.$$

Finally,

$$\cos\left(\frac{5\pi}{2}\right) = \cos\left(\frac{5\pi}{2} - 2\pi\right) = \cos\left(\frac{\pi}{2}\right) = 0. \quad \blacksquare$$

Four Other Trigonometric Functions

Four combinations of the sine and cosine functions occur frequently and are given special names. These trigonometric functions are defined as follows:

$$\text{tangent}(x) = \tan x = \frac{\sin x}{\cos x}$$

$$\text{cotangent}(x) = \cot x = \frac{\cos x}{\sin x} = \frac{1}{\tan x}$$

$$\text{secant}(x) = \sec x = \frac{1}{\cos x}$$

$$\text{cosecant}(x) = \csc x = \frac{1}{\sin x}.$$

Knowledge of the sine and cosine functions enables us to evaluate these four new functions, and also to investigate their graphs. The domains of these functions will be examined later.

Example 2

Evaluate $\tan\pi$, $\cot\pi$, $\sec\pi$, and $\csc\pi$.

Solution

$$\tan\pi = \frac{\sin\pi}{\cos\pi} = \frac{0}{-1} = 0$$

$$\cot\pi = \frac{\cos\pi}{\sin\pi} = \frac{-1}{0} \text{ undefined}$$

$$\sec\pi = \frac{1}{\cos\pi} = \frac{1}{-1} = -1$$

$$\csc\pi = \frac{1}{\sin\pi} = \frac{1}{0} \text{ undefined} \quad \blacksquare$$

Example 3

Graph $y = \sec x$.

Solution Since $\sec x = \frac{1}{\cos x}$, graphing the secant function is the same as graphing the reciprocal of the cosine function. Thus, we need to graph $y = \frac{1}{f(x)}$ where $f(x) = \cos x$. Wherever $y = \cos x$ is equal to 1 or -1, $y = \sec x$ will have this same y-value. Wherever $y = \cos x$ has an x-intercept, $y = \sec x$ will have a vertical asymptote. Also, $y = \sec x$ will be positive when $y = \cos x$ is positive and will be negative when $y = \cos x$ is negative. A graph of $y = \sec x$ is shown in Figure 6. Notice that the periodicity and symmetry of the cosine function carry over to the graph of the secant function. ■

You will be asked to graph the cosecant function in the next exercise set; when doing so you should think of $y = \csc x$ as $y = \frac{1}{\sin x}$.

Graphing $y = \tan x$ requires a different method. Since $\tan x = \frac{\sin x}{\cos x}$, we will analyze the ratio of y-values associated with $y = \sin x$ and $y = \cos x$ to determine values of $y = \tan x$. Graphs of the sine and cosine functions are provided in Figure 7. We know that $\tan x = 0$ wherever $\sin x = 0$; this

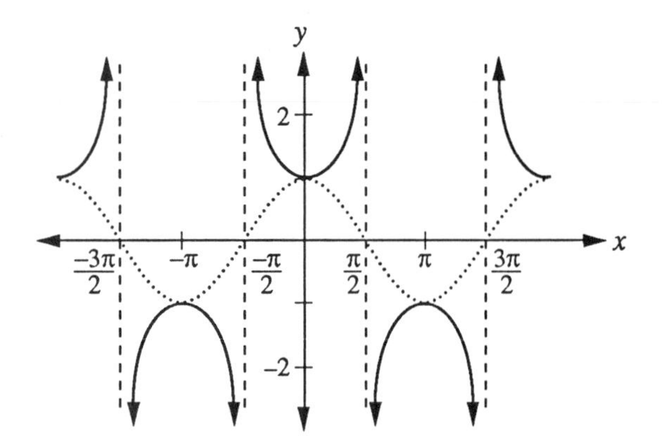

Figure 6 Graph of $y = \sec x$

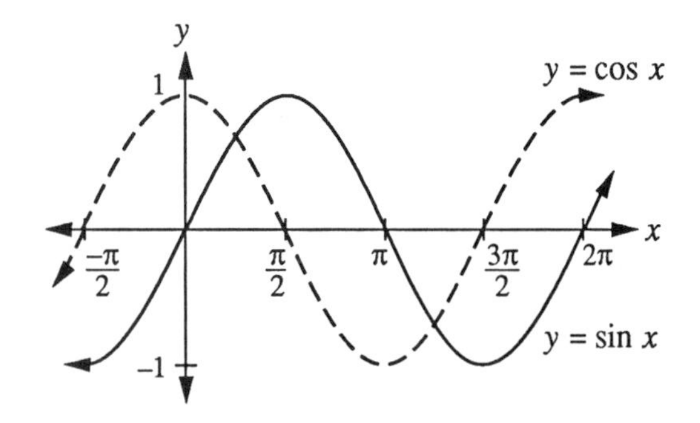

Figure 7 Graphs of $y = \sin x$ and $y = \cos x$

occurs when $x = k\pi$, k any integer. We also know that $\tan x$ will be undefined wherever $\cos x = 0$. This occurs when x is equal to an odd multiple of $\frac{\pi}{2}$, that is, $\pm\frac{\pi}{2}, \pm\frac{3\pi}{2}, \pm\frac{5\pi}{2}$, and so forth. Thus, the domain of the tangent function is all real numbers except $\frac{(2k+1)\pi}{2}$, k any integer.

To continue our investigation of the tangent function we will use the sine and cosine curves shown in Figure 7 and focus first on the interval $0 \le x < \frac{\pi}{2}$. When $x = 0$, $\sin x = 0$, and so $\tan x = 0$. The graphs in Figure 7 show that when $0 < x < \frac{\pi}{2}$ both $\sin x$ and $\cos x$ are positive. Therefore $\tan x$ is positive in this interval. When x is close to zero, $\sin x$ is close to zero and $\cos x$ is close to 1. Therefore, $\tan x$ is close to 0. As x-values increase in this interval, $\sin x$ gets larger and $\cos x$ gets smaller. Therefore, $\tan x$ increases. Figure 7 shows that halfway from 0 to $\frac{\pi}{2}$, $\sin x$ is equal to $\cos x$; thus $\tan\frac{\pi}{4} = 1$. Near $x = \frac{\pi}{2}$, $\sin x$ is approaching 1 and $\cos x$ is approaching 0. Therefore, $\tan x$ increases without bound as x nears $\frac{\pi}{2}$. Recall that $\tan\frac{\pi}{2}$ is not defined since $\cos\frac{\pi}{2} = 0$.

Now let's investigate the interval $\frac{\pi}{2} < x \le \pi$. In this interval $\sin x$ is positive and $\cos x$ is negative, so $\tan x$ is negative. Near $x = \frac{\pi}{2}$, $\sin x$ is approaching 1 and $\cos x$ is approaching 0. Since $\tan x$ is negative in this interval, $\tan x$ decreases without bound as x nears $\frac{\pi}{2}$. At the x-value midway between $\frac{\pi}{2}$ and π, $\sin x$ is equal to the opposite of $\cos x$; thus $\tan\frac{3\pi}{4} = -1$. We also know that $\tan\pi = 0$ since $\sin\pi = 0$.

In the interval $\pi < x < \frac{3\pi}{2}$ the sine and cosine functions are both negative so the tangent is positive. The graphs of sine and cosine in Figure 7 show that $\sin x = \cos x$ midway between π and $\frac{3\pi}{2}$, so $\tan\frac{5\pi}{4} = 1$. As x-values approach $\frac{3\pi}{2}$, $\sin x$ approaches -1 and $\cos x$ approaches 0 through negative values, so $\tan x$ increases without bound.

How do you think the graph of $y = \tan x$ looks in the interval $\frac{3\pi}{2} < x \le 2\pi$? Verify that $\tan x$ is negative in this interval. As x approaches $\frac{3\pi}{2}$ from the right, $\sin x$ approaches -1 and $\cos x$ approaches 0 through positive values so $\tan x$ decreases without bound. Figure 7 shows that $\sin x$ and $\cos x$ are opposites halfway between $\frac{3\pi}{2}$ and 2π; thus $\tan\frac{7\pi}{4} = -1$. We also know that $\tan 2\pi = 0$ since $\sin 2\pi = 0$.

A sketch of the graph of $y = \tan x$ on the interval $0 \le x \le 2\pi$ is shown in Figure 8.

We can use knowledge of symmetries to complete the graph of $y = \tan x$. We first determine whether $\tan x$ is odd or even by investigating $\tan(-x)$.

$$\begin{aligned}\tan(-x) &= \frac{\sin(-x)}{\cos(-x)} \\ &= \frac{-\sin x}{\cos x} \\ &= -\tan x\end{aligned}$$

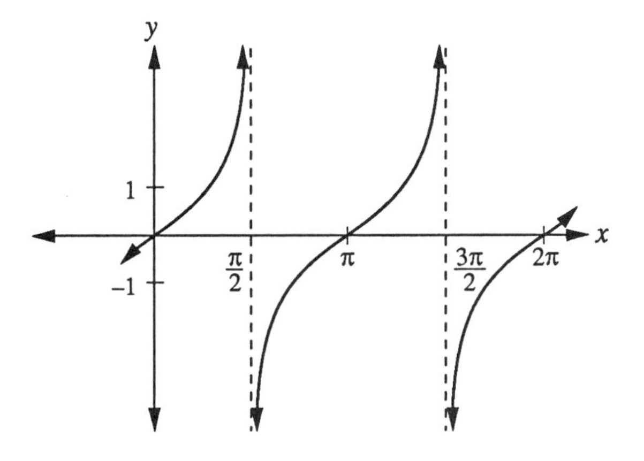

Figure 8 Graph of $y = \tan x$ for $0 \leq x \leq 2\pi$

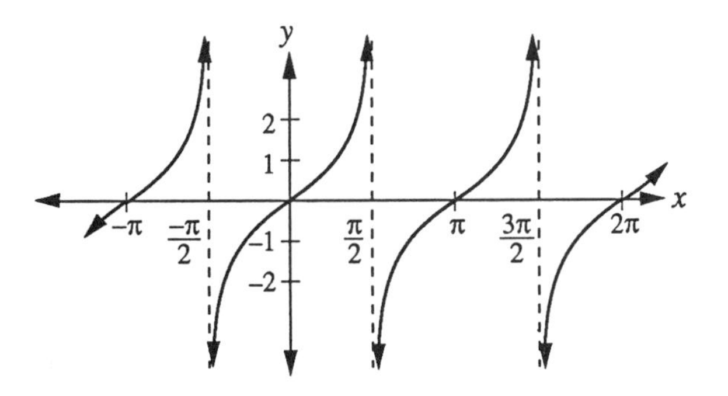

Figure 9 Graph of $y = \tan x$

This means that $\tan x$ is an odd function and its graph is symmetric about the origin. A more complete graph of the tangent function is shown in Figure 9. The graph shows that the period of the tangent function is π. The concept of amplitude does not apply to the tangent function since values of $\tan x$ get infinitely large. The domain of the tangent function excludes the odd multiples of $\frac{\pi}{2}$, and the range is all real numbers. You should add the tangent function to your toolkit and should become very familiar with the basic shape, intercepts, and asymptotes of its graph.

Exercise Set 1

1. Use the fact that both the sine and cosine functions have period 2π to evaluate the following.

a. $\cos(-7\pi)$

b. $\sin(\frac{-9\pi}{2})$

c. $\cos(\frac{-\pi}{2})$

d. $\sin(29\pi)$

e. $\sin(-4\pi)$

f. $\cos(\frac{101\pi}{2})$

g. $\sin(\frac{9\pi}{2})$

h. $\cos(10\pi)$

2. Sketch the graph of $y = \cot x$ for $-7 < x < 7$. Identify the domain, the range, the intercepts, the asymptotes, and the period of the function. Use a calculator or computer to check your graph.

3. Sketch the graph of $y = \csc x$ for $-7 < x < 7$. Identify the domain, the range, the intercepts, the asymptotes, and the period of the function. Use a calculator or computer to check your graph.

4. Identify the trigonometric functions that are even and those that are odd.

5. Sketch what you think would be a reasonable graph of the following situations.

a. Imagine a diving board just after someone has dived off. Assume there is no friction or air resistance so that the diving board does not slow down over time. Sketch a graph of the motion of the diving board if it vibrates 10 inches in each direction per second.

b. Estimate and then graph the number of hours of daylight per day in your hometown over a two year period.

c. A pendulum that consists of a small weight suspended by a length of string hangs vertically when it is at rest. If the pendulum's bob is displaced by a small amount from its equilibrium position and then released, the bob will swing back and forth in an arc which subtends an angle twice as large

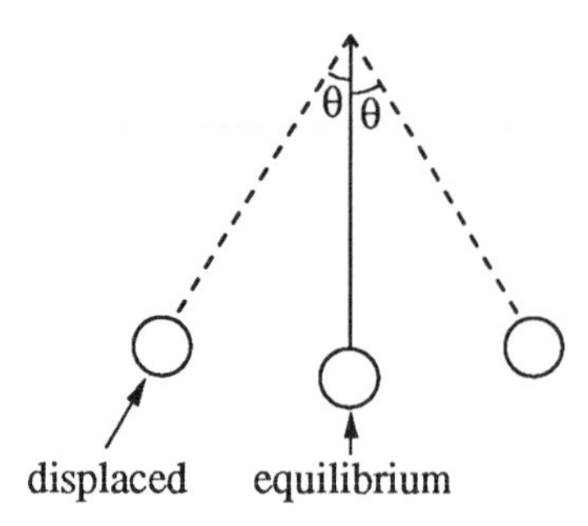

Figure 10 Pendulum

as the initial angle of displacement. That is, the pendulum sweeps out equal angles on each side of equilibrium as seen in Figure 10. Suppose that the total angle through which the pendulum moves is 10 degrees and that it takes 1 second to complete one swing from left to right. Make a graph showing the pendulum's angular displacement from equilibrium over time. Assume there is no friction or air resistance.

6. Sketch a graph of $y = \sin x$ and $y = x$ for $-\frac{\pi}{2} < x < \frac{\pi}{2}$. Use your calculator (in radian mode) to investigate the relationship between values of $\sin x$ and x for x-values close to zero. Write a few sentences about this relationship.

7. How can you transform the graph of $y = \sec x$ to obtain the graph of $y = \csc x$? Write an equation to describe the relationship between the graphs.

8. How can you transform the graph of $y = \tan x$ to obtain the graph of $y = \cot x$? Write an equation to describe the relationship between the graphs.

2

Graphing Transformations of Trigonometric Functions

Many phenomena that we observe are periodic and can be modeled by either the sine or the cosine function. However, most periodic phenomena do not have period 2π and do not oscillate between 1 and -1. Therefore, when we use the toolkit functions as models for these phenomena, we will need to incorporate horizontal and vertical stretches, compressions, and shifts into our functions.

Example 1

A weight is suspended from the ceiling on a spring. When the weight is pulled down a small distance from its equilibrium position and then released, it oscillates about that previous equilibrium position. Assume that there is no friction or air resistance, so that the weight continues to oscillate indefinitely. The weight is pulled down to 1.3 cm below its equilibrium position and then released at time $t = 0$. After one second the weight will be 1.3 cm above its equilibrium position, and when $t = 2$ it will have completed one oscillation and will again be 1.3 cm below equilibrium (see Figure 11). Make a graph to illustrate the motion of the weight over time.

Solution If we measure the *displacement* at any time as the difference between the current position and the equilibrium position, then the displacement is positive when the spring is above its equilibrium position, is negative when the spring is below its equilibrium position, and is zero when

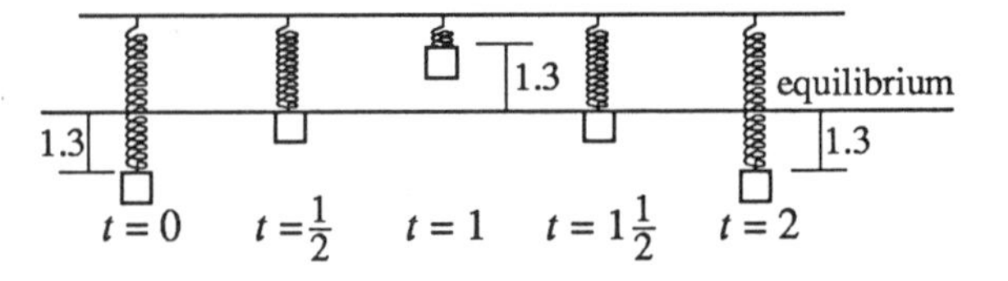

Figure 11 Motion of Weight on Spring

Table 12 Ordered Pairs for Weight on Spring

Time	0	0.5	1	1.5	2
Displacement	−1.3	0	1.3	0	−1.3

the weight is at its equilibrium position. Table 12 lists some ordered pairs that are graphed in Figure 13. Note that this graph has amplitude 1.3 and period 2. We will return to this example later in this section and write an expression for the function graphed in Figure 13. ■

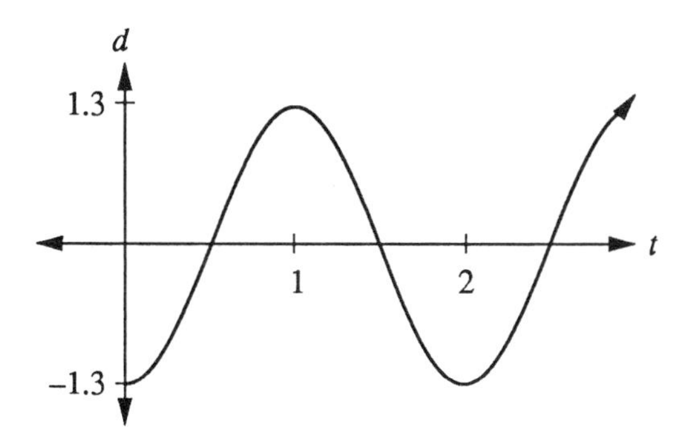

Figure 13 Spring Displacement versus Time

The motion of the weight in Example 1 is periodic. There are two useful ways to quantify periodic motion mathematically. You are already familiar with the period; this tells how long it takes to complete one cycle. In Example 1, the period is 2 since 2 seconds are required to complete each cycle. We can look at this same information in a different way and say that 1/2 cycle is completed each second. That is, the number of cycles per second is 1/2; this number is called the *frequency* of the oscillation. The frequency tells how many cycles are completed per unit of time. In Example 1 the units for the frequency are cycles per second, whereas those for the period are seconds per cycle. The frequency and period are reciprocals of each other.

Example 2

Suppose the weight in Example 1 moved more quickly so that it required 0.25 seconds to complete one oscillation. Identify the period and the frequency of this motion.

Solution Since one cycle is completed every 0.25 seconds, the period is 0.25 seconds per cycle. Four cycles will be completed each second, so the frequency is 4 cycles per second. ■

Example 3

Sketch a graph of $y = \csc(\frac{x}{3})$.

Solution If we define $f(x) = \csc x$, then $y = \csc(\frac{x}{3}) = f(\frac{1}{3}x)$, so the graph of the cosecant function is stretched horizontally by a factor of 3. Since the period of $f(x) = \csc x$ is 2π, the period of $y = \csc(\frac{x}{3})$ will be 3 times as long, or 6π. The cosecant graph has vertical asymptotes that are π units apart; on the graph of $y = \csc(\frac{x}{3})$ the vertical asymptotes will be 3π units apart. Similarly, the turning points on the graph of the cosecant function are π units apart, and on $y = \csc(\frac{x}{3})$ they will be 3π units apart. The graph of $y = \csc(\frac{x}{3})$ is shown in Figure 14. ■

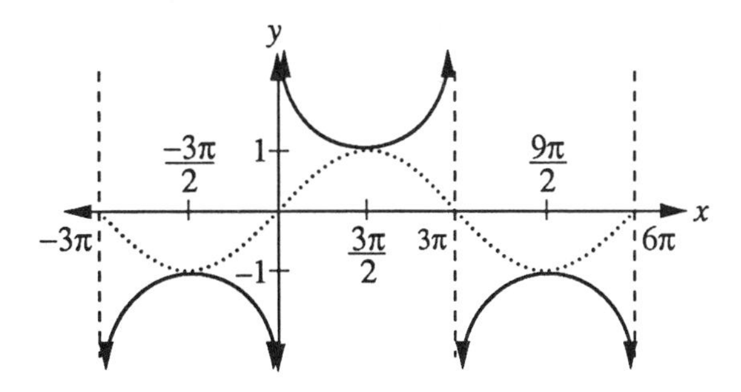

Figure 14 Graph of $y = \csc(\frac{x}{3})$

Example 4

Sketch a graph of the function $y = 4\cos(x - \frac{\pi}{2}) - 1$.

Solution If we define $f(x) = \cos x$, then our function can be expressed as $y = 4f(x - \frac{\pi}{2}) - 1$. You should recognize that the cosine function has undergone three transformations. The coefficient of 4 causes a vertical stretch by a factor of 4, so the amplitude of this transformed function will be 4. The $\frac{\pi}{2}$ causes a horizontal shift of $\frac{\pi}{2}$ units to the right. A horizontal shift of a trigonometric curve is often called a *phase shift.* The -1 causes a vertical shift one unit down. In order to sketch the graph it may be helpful to draw an intermediate graph of $y = 4\cos x$ to show the change of shape and then draw a final graph showing the change in position. The final graph is shown in Figure 15. ■

Study the graph in Figure 15 to convince yourself that it is also the graph of the function $y = 4\sin x - 1$. The equation

$$4\cos(x - \tfrac{\pi}{2}) - 1 = 4\sin x - 1$$

is true for all values of x. Equations of this type are called *identities*; they are true for all values in the domain of the variable.

Historically the study of trigonometry included lengthy proofs of identities. However, we can also use graphs to verify or to suggest that identities might be true. In order to test whether or not an equation is an identity, define a function $f(x)$ that is equal to the expression on the left side of the equal sign and define $g(x)$ equal to the expression on the right side of the equal sign. Sketch graphs of f and g in the same coordinate system. If the graphs of f and g are identical, we can conclude that an x-value substituted in either $f(x)$ or $g(x)$ will produce the same y-value. This verifies that the expressions used to define f and g are identical for the values of x on our graphs, and confirms that the original equation is an identity over the interval graphed.

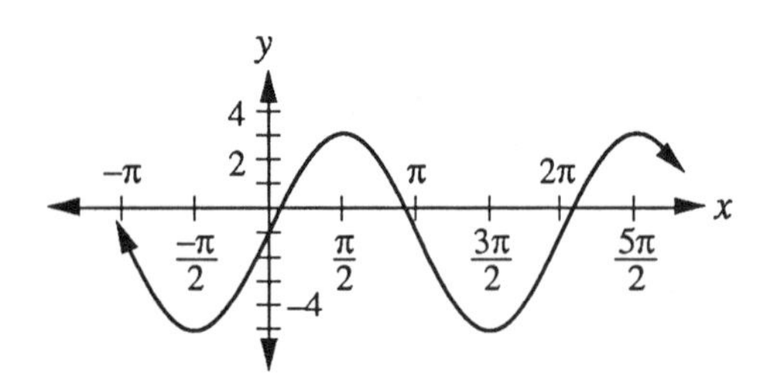

Figure 15 Graph of $y = 4\cos(x - \frac{\pi}{2}) - 1$

Example 5

Graph $f(x) = \cot(2x - \frac{\pi}{4})$.

Solution The graph will exhibit two transformations of the graph of $y = \cot x$, a horizontal compression and a horizontal shift. Rewrite the equation for f as

$$f(x) = \cot(2\left(x - \tfrac{\pi}{8}\right))$$

to see that the graph of f can by obtained from the graph of $y = \cot x$ through a horizontal compression by a factor of 2 and a shift of $\frac{\pi}{8}$ units to the right.

Figure 16 shows a graph of $y = \cot(2x)$. Note that the period is $\frac{\pi}{2}$, which is one-half the period of

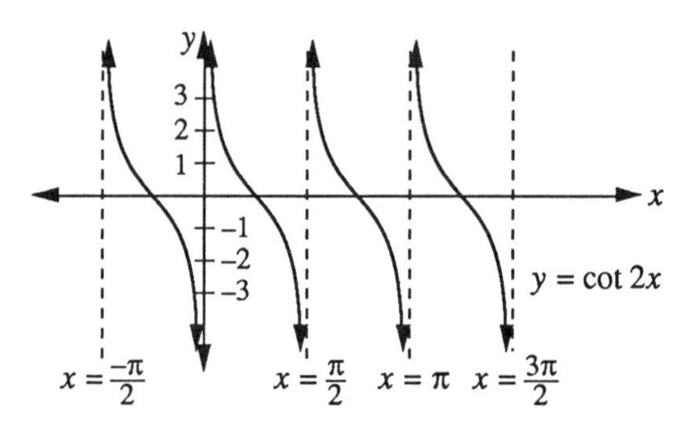

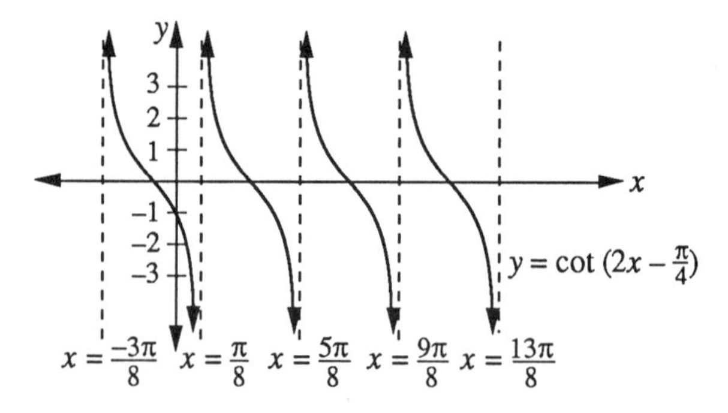

Figure 16 Graphs of $y = \cot(2x)$ and $f(x) = \cot(2x - \frac{\pi}{4})$

the cotangent function. The vertical asymptotes occur at $x = 0 + \frac{k\pi}{2}$ and the zeros occur at $x = \frac{\pi}{4} + \frac{k\pi}{2}$, where k is any integer.

The graph of $y = \cot(2x)$ is shifted $\frac{\pi}{8}$ units to the right to obtain the graph of $f(x) = \cot(2\left(x - \frac{\pi}{8}\right))$. This graph is also shown in Figure 16. The vertical asymptotes are at $x = \frac{\pi}{8} + \frac{k\pi}{2}$ and the zeros occur at $x = \frac{3\pi}{8} + \frac{k\pi}{2}$. ■

Example 6

Figure 13 shows a graph of displacement versus time for a weight suspended from a spring. Write an equation for this graph.

Solution The graph has amplitude 1.3 and period 2 and has a minimum value when $t = 0$. The function $d = 1.3\cos(t)$ has a graph with the correct amplitude. The period can be changed from 2π to 2 by doing a horizontal compression by a factor of π. Thus, $d = 1.3\cos(\pi t)$ has the correct amplitude and correct period. However, its graph has a maximum value when $t = 0$. We can achieve a minimum value when $t = 0$ by reflecting the graph over the t-axis. Thus,

$$d = -1.3\cos(\pi t)$$

is an equation for the graph in Figure 13. This equation is not unique; you will be asked to write other equations for this graph in the following exercise set. ■

Exercise Set 2

1. Write an equation that involves a horizontal translation of the cosine function for the graph in Figure 13.
2. Write an equation that involves a horizontal translation of the sine function for the graph in Figure 13.
3. Identify the period and frequency of each function.
   - **a.** $y = \sin(\frac{1}{2}x)$
   - **b.** $y = \sin(\frac{3}{4}x)$
   - **c.** $y = \sin(-4x)$
   - **d.** $y = \sin(\pi x)$
   - **e.** $y = \sin(2\pi x)$
   - **f.** $y = \sin(2\pi f x)$
4. Based on your answers to the preceding exercise, what is the period of $y = \sin(Bx)$, where B is a real number? What is the frequency?
5. Sketch at least two complete cycles of each function. Label the scale on both axes.
   - **a.** $y = 2\cos x$
   - **b.** $y = \sin(-2x)$
   - **c.** $y = \sin(x - \frac{\pi}{4})$
   - **d.** $y = \csc(x - \pi)$
   - **e.** $y = \sin(x - \frac{\pi}{2}) + 1$
   - **f.** $y = 1 - \cos x$
   - **g.** $y = \sec(\frac{x}{3})$
   - **h.** $y = \sin(3\pi x) - 2$
   - **i.** $y = \tan(\pi x)$
   - **j.** $y = \cos(2x - \pi)$
   - **k.** $y = 1 + \csc(-2x)$
   - **l.** $y = \sec(2x) + 1$
   - **m.** $y = \tan(x + \frac{\pi}{4})$
   - **n.** $y = 3\sin(2x - \frac{\pi}{2})$
   - **o.** $y = 2 + \cot x$
6. Use graphs to verify that the following identities are true.
   - **a.** $\sin(-x) = -\sin x$
   - **b.** $\sec(-x) = \sec x$
   - **c.** $\sin(x + \frac{\pi}{2}) = \cos x$
   - **d.** $\sin(\frac{\pi}{2} - x) = \cos x$ (**Hint**: $(\frac{\pi}{2} - x) = -(x - \frac{\pi}{2})$.)
7. Graph each function. After you have graphed all the functions, study your graphs to find the ones that match and use them as a basis of conjecture for three trigonometric identities.
   - **a.** $y = \cos(\frac{\pi}{2} - x)$
   - **b.** $y = \sin x$
   - **c.** $y = \sin(x - \pi)$
   - **d.** $y = -\sin x$

e. $y = \cot(\frac{\pi}{2} - x)$

f. $y = \tan x$

8. Write an equation of a sinusoid with the following characteristics.

   a. period 2π, amplitude 4, phase shift 0

   b. period $\frac{\pi}{3}$, amplitude 0.1, phase shift $-\frac{\pi}{4}$

   c. period π, amplitude $\frac{1}{3}$, phase shift 1

   d. period 2, amplitude 1, phase shift $-\pi$

9. There is a ferris wheel every year at the fair. This year the wheel has a radius of 33.2 feet and makes a complete revolution every 15 seconds. For clearance, the bottom of the ferris wheel is 4 feet above ground. Write a function that shows how one passenger's height above ground varies over time as he or she rides the ferris wheel.

10. In Los Angeles on the first day of summer (June 21) there are 14 hours 26 minutes of daylight, and on the first day of winter (December 21) there are 9 hours 54 minutes of daylight. On the average there are 12 hours 10 minutes of daylight; this average amount occurs on March 20 and September 22.

    a. Make a graph that displays this information.

    b. Write an expression for d, the number of minutes of daylight per day, as a function of t, the number of months after June 21.

11. A weight is suspended from a spring which hangs vertically from the ceiling of a room. At time $t = 0$ the weight is pulled down so that its distance from the ceiling is 24.4 cm. When released it oscillates about its previous equilibrium position. At time $t = 1/2$ it reaches its minimum distance from the ceiling, which is 21.8 cm. Assume there is no friction or air resistance. Express the weight's distance from the ceiling as a function of time.

12. At high tide the ocean generally reaches the five foot mark on a retaining wall. At low tide the water reaches the one foot mark. Assume that high tide occurs at 12 noon and at midnight and that low tide occurs at 6 p.m. and 6 a.m. Write an equation to show how the height of the water mark varies over time. What does your independent variable represent?

13. Listed below are the mean monthly temperatures, in degrees Fahrenheit, for New Orleans based on records from 1951 to 1980.

Jan. 52	Apr. 69	July 82	Oct. 69
Feb. 55	May 75	Aug. 82	Nov. 60
Mar. 61	June 80	Sept. 79	Dec. 55

Make a graph in which the monthly temperature is the dependent variable and the month is the independent variable. Write an equation to model this data. Use a computer or calculator to check how well your function fits the data.

3

Sine and Cosine as Circular Functions

Perhaps you have wondered how to generate values for the sine and cosine functions. Why is $\sin 0$ equal to 0? Why is $\cos \pi$ equal to -1? To answer these questions, we will begin with a circle of radius 1 with its center at (0, 0) in the coordinate plane. Such a circle is often called a *unit circle*. The equation of this circle is $x^2 + y^2 = 1$. The circumference of a unit circle is $2\pi r = 2\pi(1) = 2\pi$.

Imagine an ant that walks along the circumference of a unit circle. Suppose it begins at $(1, 0)$ and walks 2π units; it will have walked all the way around the circle and will end its trip back at $(1, 0)$. If it begins at $(1, 0)$ and walks π units, it will end its trip at the point $(-1, 0)$. If it begins at $(1, 0)$ and walks $\frac{\pi}{2}$ units in a counterclockwise direction, it will end at $(0, 1)$. If it begins at $(1, 0)$ and walks $\frac{\pi}{2}$ units in a clockwise direction, it will end at $(0, -1)$ (see Figure 17). Note that the lengths of all these ant trips are actually arc lengths of a unit circle.

We will define counterclockwise to be the positive direction and clockwise to be the negative

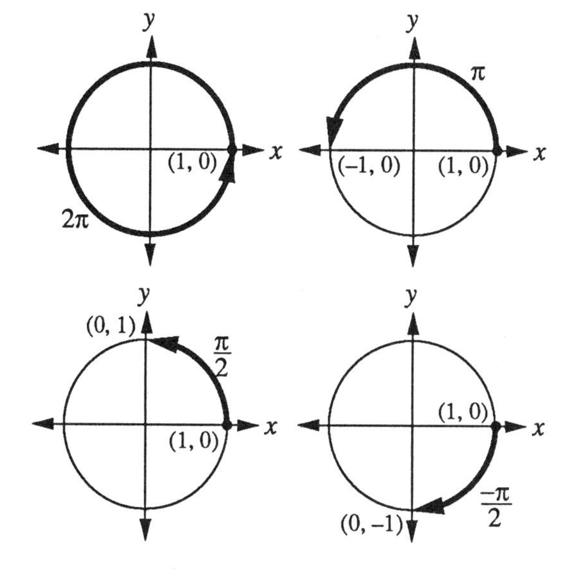

Figure 17 Ant Walking Along Circumference of Unit Circle

direction. You should assume that the ant begins from (1, 0) unless specified otherwise. Table 18 lists various lengths for ant trips, together with the coordinates of the point where the ant will stop.

As you study the entries in this table, look at ordered pairs of the form (length of trip, x-coordinate at end of trip). For instance, a length of 0 is paired with an x-coordinate of 1, a length of π is paired with an x-coordinate of -1, and a length of $\frac{3\pi}{2}$ is paired with an x-coordinate of 0. These ordered pairs are graphed in Figure 19. You

Table 18 Lengths and Endpoints of Ant Trips

Trip Length	Coordinates at End of Trip
0	(1,0)
$\frac{\pi}{2}$	(0, 1)
π	(−1, 0)
$\frac{3\pi}{2}$	(0, −1)
2π	(1,0)
$-\frac{\pi}{2}$	(0, −1)
4π	(1,0)

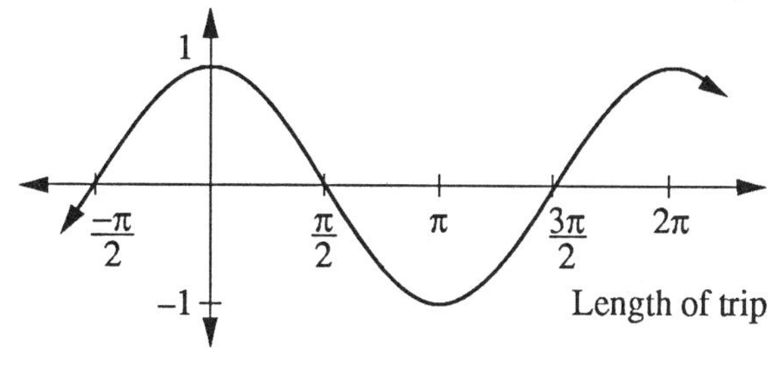

Figure 19 Ordered Pairs (Length of Trip, x-coordinate of Endpoint)

should recognize the shape of the cosine curve. The x-coordinate of the ant's location varies from 1 to -1 as it walks along the circumference of the circle; this is why the range of the cosine function contains all real numbers from 1 to -1. Whether the ant walks t units or $t + 2\pi$ units along the circumference, it ends up at the same point. This is why the range of the cosine function repeats itself every 2π units.

Now let's look at ordered pairs of the form (length of trip, y-coordinate at end of trip). These ordered pairs include (0, 0), $(\frac{\pi}{2}, 1)$, $(-\frac{\pi}{2}, -1)$, and so forth. These ordered pairs are graphed in Figure 20; you should recognize the shape of the sine curve.

If we let t represent the length of the arc that the ant walks along, then

$$\cos t = \text{the } x\text{-coordinate of the arc's endpoint}$$

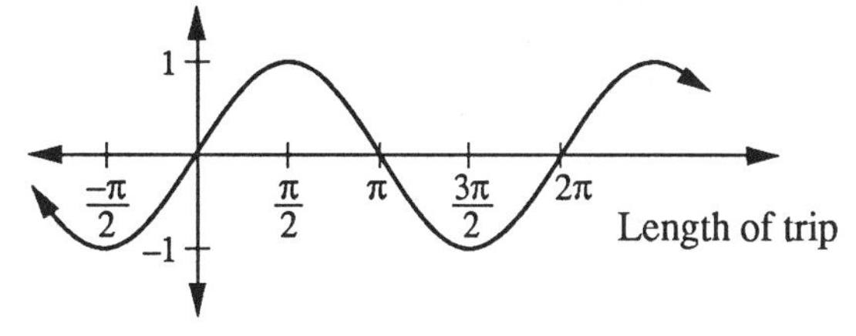

Figure 20 Ordered Pairs (Length of Trip, y-coordinate of Endpoint)

and

$$\sin t = \text{the } y\text{-coordinate of the arc's endpoint.}$$

These definitions of $\sin t$ and $\cos t$ are a precise way of describing the sine and cosine functions we have already studied. You should now be able to answer the questions posed at the beginning of this section. Why is $\sin 0 = 0$? Why is $\cos \pi = -1$?

The sine and cosine functions are often called *circular functions* since their definitions are based on a unit circle. Figure 21 shows an important geometric interpretation of sine and cosine; if point P is the endpoint of an arc of length t that begins at $(1, 0)$ on the unit circle, then the coordinates of P are $(\cos t, \sin t)$.

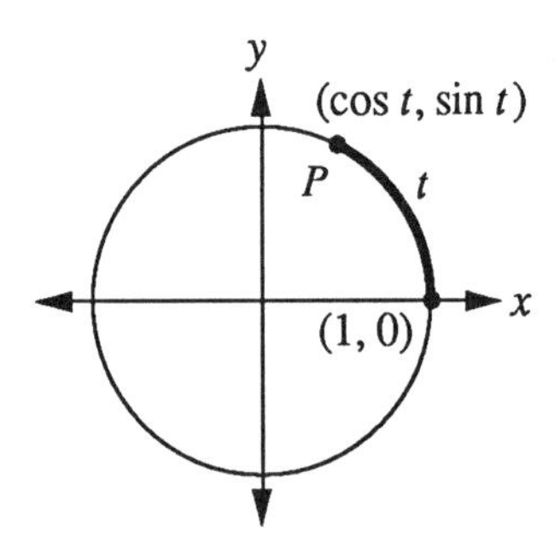

Figure 21 Geometric Interpretation of Sine and Cosine

Since the coordinates of all points on the unit circle satisfy the equation $x^2 + y^2 = 1$, we know that

$$(\cos t)^2 + (\sin t)^2 = 1.$$

Mathematicians usually write $(\cos t)^2$ as $\cos^2 t$, so the preceding equation can be written as

$$\cos^2 t + \sin^2 t = 1.$$

This equation is an identity since it is true for all values of the variable t.

Example 1

An ant starts at $(1, 0)$ and walks in a counterclockwise direction 4 units along the circumference of a unit circle. In what quadrant does the ant stop? What are the coordinates of the point where the ant stops?

Solution To determine the quadrant in which the ant stops, we need to compare the distance it walks (4 units) to other arc lengths. The ant walked more than half-way around the circle since $4 > \pi$, but it did not walk around the entire circle since $4 < 2\pi$. It would have walked three-quarters of the way around the circle if the length of its trip had been $\frac{3\pi}{2}$; since $4 < \frac{3\pi}{2}$, the ant walked less than three-quarters of the way around the circle. Therefore, it ended its trip in the third quadrant (see Figure 22).

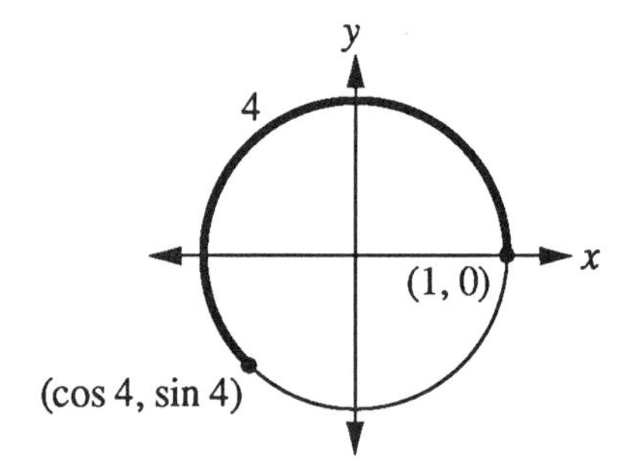

Figure 22 Arc Length of 4 Units

The coordinates of the point where the ant stops are given by $(\cos 4, \sin 4)$. Using your calculator in radian mode, you can evaluate $\cos 4$ and $\sin 4$; you should find that the ant stops at the point whose coordinates are approximately $(-0.65, -0.76)$. Verify that

$$(-0.65)^2 + (-0.76)^2 \approx 1.$$ ■

Exercise Set 3

1. Use the unit circle to evaluate each of the following.

   a. $\sin(\frac{-3\pi}{2})$

   b. $\cos(\frac{-3\pi}{2})$

   c. $\sin(\frac{133\pi}{2})$

d. $\cos(\frac{133\pi}{2})$
e. $\tan(-49\pi)$
f. $\sin(-49\pi)$
g. $\sin(7\pi)$
h. $\cos(7\pi)$
i. $\sec(\frac{-5\pi}{2})$
j. $\cot(\frac{-5\pi}{2})$
k. $\csc(\frac{7\pi}{2})$
l. $\tan(\frac{7\pi}{2})$

2. An ant starts at $(1, 0)$ and walks in a counterclockwise direction 5 units along the circumference of a unit circle. What are the coordinates of the point where the ant stops? In what quadrant does the ant stop?

3. Suppose the ant starts at $(1, 0)$ and walks in a clockwise direction 2 units along the circumference of a unit circle. What are the coordinates of the point where the ant stops? In what quadrant does the ant stop?

4. We can define two new functions that are variations of the sine and cosine functions. Replace the unit circle with a unit square whose center is at the origin (see Figure 23). The ant starts at $(\frac{1}{2}, 0)$ and walks t units along the perimeter of the square. The two new functions we define will be called square-sine and square-cosine. Let square-sin(t) be equal to the y-coordinate of the point where the ant stops on the unit square, and let square-cos(t) be equal to the x-coordinate of the point where the ant stops on the unit square. For example, square-sin(1) $= 0.5$.

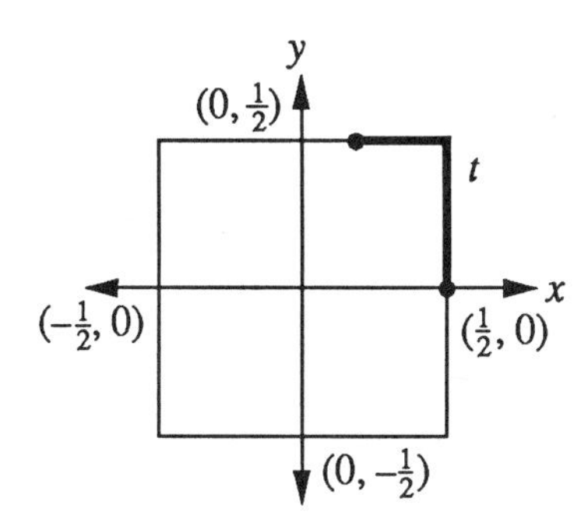

Figure 23 Unit Square

a. Find the value of square-sin($\frac{1}{2}$), square-sin($\frac{-7}{2}$), square-sin(26), square-cos(-1), square-cos($\frac{-151}{2}$), square-cos(62).

b. Make a sketch of the square-sine function.

c. Is the square-sine function periodic? If so, what is the period?

d. Sketch the square-cosine function.

e. Can the square-sine function be shifted to obtain the square-cosine function? If so, write an equation to express this relationship.

4

The Symmetry of the Circle

In the preceding section the sine and cosine functions were defined using the unit circle centered at the origin. The unit circle has many symmetries which help us evaluate values of the circular functions. In this section we will explore those symmetries and relate them to the sine and cosine functions.

The circle $x^2 + y^2 = 1$ is symmetric about the x-axis, the y-axis, and the origin. Therefore, if the coordinates of one point on the circle are known, the coordinates of three other points can easily be determined. For example, the point P shown in Figure 24 has coordinates (a, b). Because of symmetry about the y-axis, the coordinates of point Q in the second quadrant are $(-a, b)$. Because of symmetry about the origin, the coordinates of point R in the third quadrant are $(-a, -b)$. Because of symmetry about the x-axis, the coordinates of point S in the fourth quadrant are $(a, -b)$.

How does this symmetry relate to the circular functions? To explore this question, suppose an ant trip of length t begins at $(1, 0)$ and ends at point P. What length ant trip would terminate at Q? Since the ant can get to Q by walking π units in the positive direction followed by t units

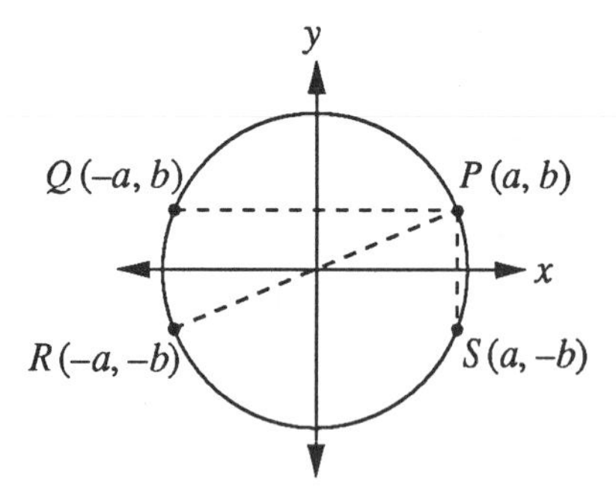

Figure 24 Symmetry of the Unit Circle

in the negative direction, a trip of length $\pi - t$ will end at Q. Similarly, a trip of length $\pi + t$ will end at R. Verify that a trip of length $2\pi - t$ and a trip of length $-t$ will both end at S (see Figure 25).

The information in Figures 24 and 25 can be combined with the definitions of sine and cosine to yield some interesting relationships. Recall that $\sin t$ is equal to the y-coordinate of P and $\cos t$ is equal to the x-coordinate of P. Since P and Q have the same y-coordinate, $\sin t$ and $\sin(\pi - t)$ are equal. Similarly, since P and S have the same x-coordinate, $\cos t$ and $\cos(2\pi - t)$ and $\cos(-t)$ are equal. Since P and R have opposite x-coordinates and opposite y-coordinates, $\sin(\pi + t) = -\sin t$ and $\cos(\pi + t) = -\cos t$. Each of these equations is an identity; each is true for all values of

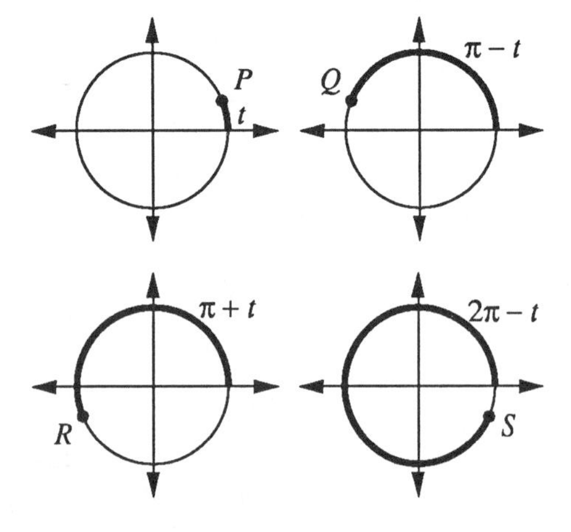

Figure 25 Arc Lengths on the Unit Circle

x. Use the symmetry of the unit circle to verify that $\cos(\pi + t) = -\cos(-t)$ is also an identity. It is more useful to understand symmetry than to memorize trigonometric identities.

This collection of identities can help you find values of the sine and cosine functions. For instance, if you know that $\sin t = 0.8$ then all of the following must also be true.

$$\sin(\pi - t) = 0.8$$

$$\sin(\pi + t) = -0.8$$

$$\sin(2\pi - t) = \sin(-t) = -0.8$$

Since $\sin t$ and $\cos t$ are the y- and x-coordinates of a point on the unit circle, $\sin^2 t + \cos^2 t = 1$. Since $\sin t = 0.8$,

$$\cos^2 t = 1 - \sin^2 t$$

$$\cos^2 t = 1 - (0.8)^2$$

and

$$\cos t = \pm\sqrt{1 - 0.64} = \pm\sqrt{0.36} = \pm 0.6.$$

You would need to know additional information in order to determine whether the cosine of t is positive or negative.

If we know that t is in the first quadrant, we can conclude that $\cos t = 0.6$. Then we also know that

$$\cos(\pi - t) = -0.6$$

$$\cos(\pi + t) = -0.6$$

$$\cos(2\pi - t) = \cos(-t) = 0.6$$

Example 1

Suppose an arc of length θ terminates in the second quadrant and $\sin\theta = 0.62$ (see Figure 26). Find $\sin(\pi - \theta)$, $\cos\theta$, and $\tan(-\theta)$.

Solution Again, we use the symmetry of the circle to recognize that

$$\sin(\pi - \theta) = \sin\theta = 0.62.$$

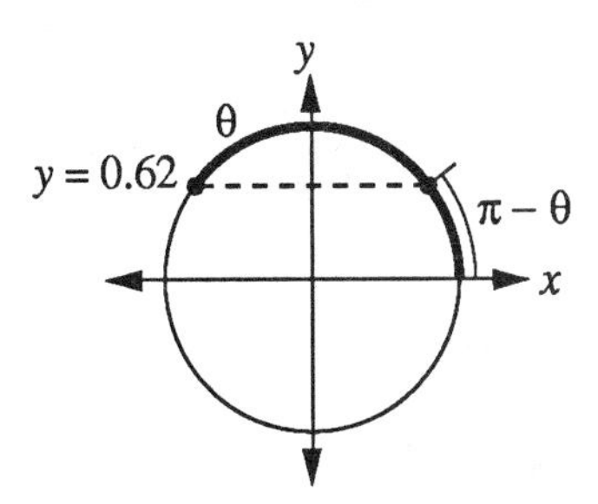

Figure 26 Arc Length for Example 1

We know that

$$\cos\theta = \pm\sqrt{1-\sin^2\theta}.$$

Since θ is in the second quadrant, its cosine must be negative, so

$$\cos\theta = -\sqrt{1-(0.62)^2} \approx -0.78.$$

Since the tangent function is odd, $\tan(-\theta) = -\tan\theta$, so

$$\tan(-\theta) = -\tan\theta = -\frac{\sin(\theta)}{\cos(\theta)} \approx -\frac{0.62}{-0.78} \approx 0.79.$$

■

Example 2

On a unit circle, an ant trip ends at a point whose x-coordinate is 0.45. Find the length of the ant trip.

Solution A bit of thought should convince you that there are many different trips that will terminate at a point whose x-coordinate is 0.45. If we use t to represent the length of the trip, then we need to find the values of t that satisfy $\cos t = 0.45$. We will use the notation $t = \cos^{-1} 0.45$ to indicate that t is a number whose cosine is 0.45; t is often called the *inverse cosine* or *arccosine* of 0.45. Use the appropriate key sequence on your calculator (in radian mode) to find a number whose cosine is 0.45. The display should read 1.104031; this is a number whose cosine is 0.45. Thus, an ant trip about 1.1 units long will terminate at a point whose x-coordinate is 0.45. Remember that if the ant walks t units or $t + 2\pi$ units or $t + 4\pi$ units it will end up at the same point. In fact, all numbers of the form $1.1 + 2k\pi$ where k is an integer are solutions of $\cos t = 0.45$. These solutions can be written as $\{t \mid t = 1.1 + 2k\pi, k \text{ any integer}\}$, read "the set of all t such that $t = 1.1 + 2k\pi$, k any integer." This notation represents the infinite set containing the numbers $1.1, 1.1+2\pi, 1.1-2\pi, 1.1+4\pi, 1.1-4\pi$, and so on.

You can find other values of t that satisfy $\cos t = 0.45$ by using the symmetry of the circle. If $\cos 1.1 = 0.45$, then it must be true that $\cos(2\pi - 1.1) = \cos 5.18 = 0.45$. Because the circumference of the circle is 2π, the numbers $5.18+2\pi$, $5.18+4\pi$, and so on, all have cosine equal to 0.45. The complete set of possible lengths for the ant trip is $\{t \mid t = 1.1+2k\pi \text{ or } t = 5.18+2k\pi, k \text{ any integer}\}$. ■

Special Values on the Unit Circle

There are many numbers whose sine and cosine we can evaluate without a calculator. For instance, we know $\sin\frac{\pi}{2} = 1$ and $\cos\pi = -1$ from the definitions of sine and cosine as circular functions and from the graphs of these functions. There are several other special numbers whose sine and cosine we can evaluate without a calculator.

Example 3

Evaluate $\sin\frac{\pi}{4}$ and $\cos\frac{\pi}{4}$.

Solution Let P with coordinates (c, s) be the endpoint of the arc that begins at (1,0) and has length $\frac{\pi}{4}$. Note that $c = \cos\frac{\pi}{4}$ and $s = \sin\frac{\pi}{4}$. Figure 27 shows that P lies on the line $y = x$, which implies that $c = s$.

Since P is on the unit circle, its coordinates must satisfy the equation of the circle, which is $x^2 + y^2 = 1$. Therefore,

$$c^2 + s^2 = 1.$$

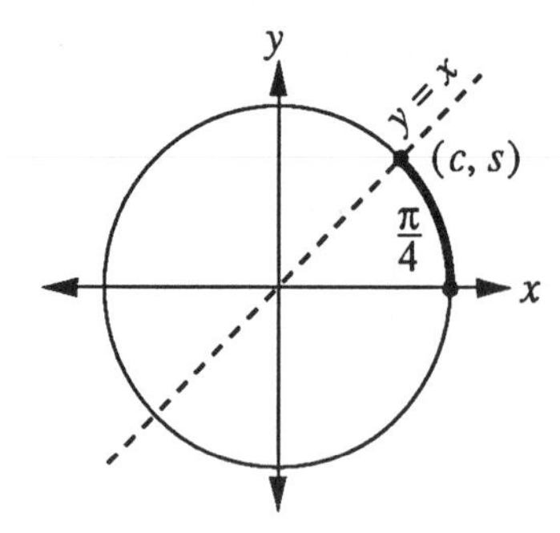

Figure 27 Arc Length $\frac{\pi}{4}$ on Unit Circle

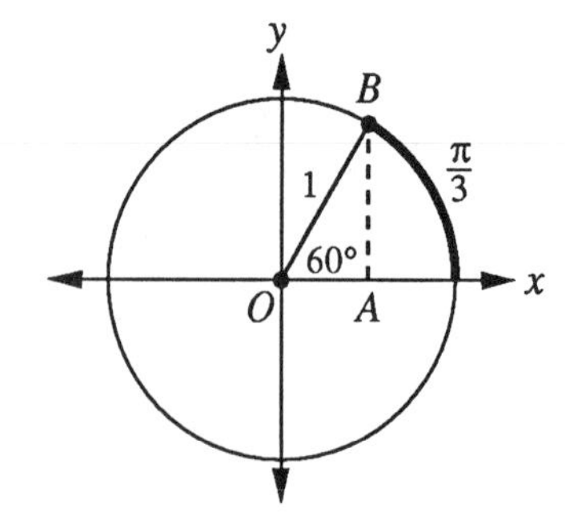

Figure 28 Arc Length $\frac{\pi}{3}$ on Unit Circle

Substituting $c = s$ gives

$$\begin{aligned} c^2 + c^2 &= 1 \\ 2c^2 &= 1 \\ c &= \sqrt{\tfrac{1}{2}} = \tfrac{\sqrt{2}}{2}. \end{aligned}$$

We chose the positive square root for the value of c because P is in the first quadrant. Since $c = s$, we know that both c and s have the value $\frac{\sqrt{2}}{2}$. This means that $\sin\frac{\pi}{4} = \frac{\sqrt{2}}{2}$ and $\cos\frac{\pi}{4} = \frac{\sqrt{2}}{2}$. Note that you can evaluate sine and cosine of $\frac{3\pi}{4}$, $\frac{5\pi}{4}$, and $\frac{7\pi}{4}$ using symmetries of the circle. ■

Example 4

Evaluate $\sin\frac{\pi}{3}$ and $\cos\frac{\pi}{3}$.

Solution Let B be the endpoint of the arc that begins at (1,0) and has length of $\frac{\pi}{3}$ (see Figure 28). We will apply some geometric techniques to find the coordinates of B. The length of BO is 1 since BO is a radius of the circle. From B draw a line segment that is perpendicular to the x-axis at A. The length of this line segment is equal to the y-coordinate of B, which is equal to $\sin\frac{\pi}{3}$. The length AO represents the x-coordinate at B, which is equal to $\cos\frac{\pi}{3}$.

If we can find the lengths of AB and AO then we will know $\sin\frac{\pi}{3}$ and $\cos\frac{\pi}{3}$. Since $\frac{\pi}{3}$ is one-sixth of 2π, $\angle BOA$ is one-sixth of 360°, or 60°. This means that triangle AOB is a 30-60-90 triangle. The side opposite the 30° angle is one-half as long as the hypotenuse, and the side opposite the 60° angle is $\frac{\sqrt{3}}{2}$ times as long as the hypotenuse. Therefore, $AO = \frac{1}{2}$ and $AB = \frac{\sqrt{3}}{2}$. We can conclude that $\sin\frac{\pi}{3} = \frac{\sqrt{3}}{2}$ and $\cos\frac{\pi}{3} = \frac{1}{2}$.

The symmetry of the unit circle can be used to evaluate the sine and cosine of $\frac{2\pi}{3}$, $\frac{4\pi}{3}$, and $\frac{5\pi}{3}$. ■

Exercise Set 4

1. Evaluate $\sin\frac{\pi}{6}$ and $\cos\frac{\pi}{6}$ using geometric techniques like those in Example 4. Do not use your calculator. You should get exact values, not decimal approximations.
2. Label a unit circle with the endpoints of the arcs which begin at (1,0) and have length $\frac{\pi}{6}$, $\frac{\pi}{4}$, $\frac{\pi}{3}$, $\frac{\pi}{2}$, $\frac{2\pi}{3}$, $\frac{3\pi}{4}$, $\frac{5\pi}{6}$, π, $\frac{7\pi}{6}$, $\frac{5\pi}{4}$, $\frac{4\pi}{3}$, $\frac{3\pi}{2}$, $\frac{5\pi}{3}$, $\frac{7\pi}{4}$, $\frac{11\pi}{6}$. Use values you know and symmetries of the unit circle to label the coordinates of each endpoint.
3. Use your knowledge of the special values and the symmetry of the circle to evaluate the following without a calculator.

   a. $\sin(-\frac{5\pi}{4})$

   b. $\sin(\frac{71\pi}{6})$

   c. $\sin(\frac{5\pi}{3})$

   d. $\cot(\frac{23\pi}{3})$

   e. $\cos(\frac{33\pi}{4})$

   f. $\csc(\frac{57\pi}{3})$

g. $\tan(\frac{-17\pi}{6})$
h. $\tan(\frac{-33\pi}{4})$
i. $\sec(\frac{-11\pi}{4})$

4. If $\cos x = -0.25$ and x is in the second quadrant, find $\sin x$ and $\tan x$ without finding x.

5. If $\sin\theta = -0.36$ and θ is in the fourth quadrant, find each of the following without finding θ.
   a. $\sin(-\theta)$
   b. $\cos\theta$
   c. $\tan(\pi - \theta)$
   d. $\tan(\pi + \theta)$

6. Use your calculator in radian mode to evaluate each of the following. Explain the relationships between your answers.
   a. $\sin(1)$
   b. $\cos(1)$
   c. $\cos(4.14)$
   d. $\sin(2.14)$
   e. $\tan(7.28)$

7. Suppose $\sin x = 0.1411$ and one value of x is known to be 3. Use symmetry to find all other possible values for x.

8. Suppose $\cos x = 0.9397$ and one value of x is known to be $\frac{\pi}{9}$. Use symmetry to find all other possible values for x.

9. An arc of length 5 begins at (1,0) on the unit circle and ends at $(0.28, -0.96)$. This fact gives you information about the endpoints of three other arcs that begin at $(1, 0)$. Give the lengths of the three arcs and the coordinates of their endpoints.

10. Suppose that an arc of length a ends at a point in the first quadrant whose x-coordinate is 0.44. Find $\sin(a)$, $\cos(a+\pi)$, and $\sin(2\pi-a)$.

11. The unit circle is symmetric about the line $y = x$. This has some interesting implications for values of circular functions. Suppose point P with coordinates (a, b) is the terminal point of an arc of length θ. We know $\sin\theta = b$ and $\cos\theta = a$. Let Q be the mirror image of P in the line $y = x$ (see Figure 29).
    a. Explain why the coordinates of Q are (b, a).

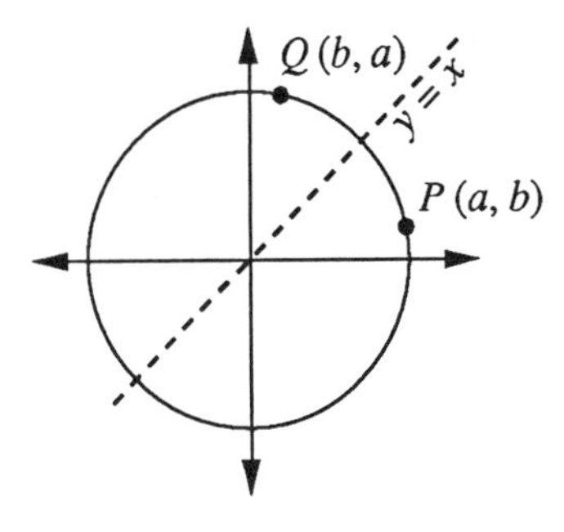

Figure 29 Symmetry about $y = x$

    b. Explain why an arc of length $\frac{\pi}{2} - \theta$ will terminate at Q.
    c. We know that $a = \sin(\frac{\pi}{2} - \theta)$ and that $a = \cos\theta$, so we can conclude that $\sin(\frac{\pi}{2} - \theta) = \cos\theta$. Write another similar identity.

12. Sine and cosine are called *cofunctions* and the identities mentioned in the preceding exercise are called cofunction identities. There are similar identities for secant and cosecant and for tangent and cotangent. Write these identities, graph and verify that they are indeed true for values in the domain of the variable.

13. Suppose we know that $\cos\theta = w$ and that θ is in the third quadrant. Express each of the following in terms of w.
    a. $\sin\theta$
    b. $\tan\theta$
    c. $\cos(\frac{\pi}{2} - \theta)$
    d. $\csc(\frac{\pi}{2} - \theta)$
    e. $\sec(\pi - \theta)$
    f. $\cot(-\theta)$
    g. $\sin(\frac{\pi}{2} + \theta)$

5

The Trigonometry of Angles

There are many applications of trigonometry that depend on angles. What does it mean to take the sine of an angle?

Mathematicians define the sine, cosine and tangent of an angle θ in the following way. First, the angle is placed in *standard position*, see Figure 30. This means that the vertex of the angle is at the origin of a coordinate system and the initial side is on the positive x-axis. Then a point P with coordinates (a, b) is chosen on the terminal side of θ one unit from the origin. The sine of θ is defined to be b, the cosine of θ is a, and the tangent of θ is $\frac{b}{a}$. The other three trigonometric functions are reciprocals of these three. The definitions of all six trigonometric functions of angles are summarized below.

If point P with coordinates (a, b) is one unit from the origin, then

$$\sin\theta = b \quad \cos\theta = a \quad \tan\theta = \tfrac{b}{a}$$

$$\csc\theta = \tfrac{1}{b} \quad \sec\theta = \tfrac{1}{a} \quad \cot\theta = \tfrac{a}{b}$$

The six trigonometric functions just defined have as their domain a set of angles. Note that the values of these trigonometric functions can be either positive or negative, depending on what quadrant the terminal side lies in.

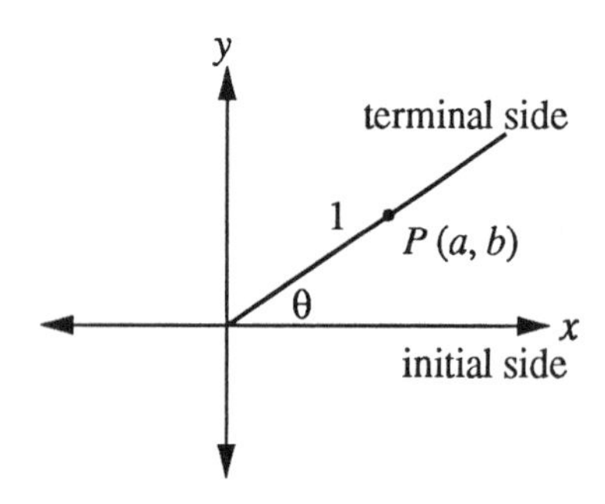

Figure 30 Angle in Standard Position

Class Practice

1. Identify the angles that are excluded from the domain of each trigonometric function.
2. Since P is one unit from the origin, a and b are coordinates of points on a unit circle. What does this imply about the range of each function?

Example 1

The point $F(5, -12)$ lies on the terminal side of θ when the angle is in standard position. Find the values of the six trigonometric functions of θ.

Solution To use the angle definitions of the trigonometric functions we need to know the coordinates of the point P that is one unit from the origin on the terminal side of θ. From each of the points P and F draw a perpendicular to the x-axis (see Figure 31). Triangles QOP and GOF are similar; therefore

$$\frac{PQ}{FG} = \frac{PO}{FO}.$$

This proportion can be rewritten as

$$\frac{PQ}{PO} = \frac{FG}{FO}.$$

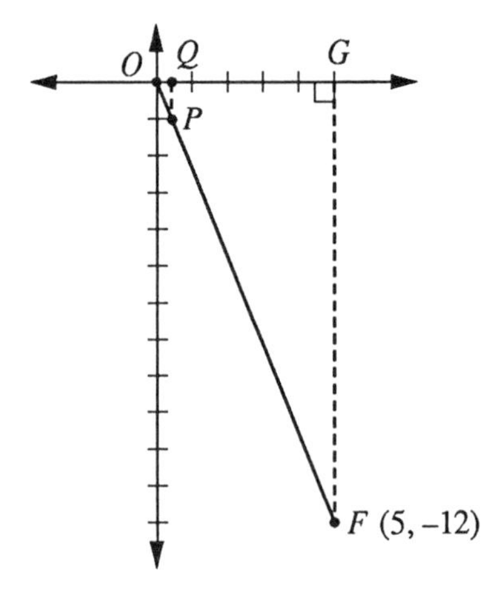

Figure 31 Similar Triangles for Example 1

Use the Pythagorean Theorem to find that $FO = 13$. Substituting known values in the equation gives

$$\frac{PQ}{1} = \frac{12}{13},$$

so $PQ = \frac{12}{13}$ and the y-coordinate of P is $-\frac{12}{13}$. Using another proportion based on similar triangles we can find that $QO = \frac{5}{13}$, so the x-coordinate of P is $\frac{5}{13}$. Since P has coordinates $(\frac{5}{13}, -\frac{12}{13})$, we can evaluate the six trigonometric functions of θ.

$$\sin\theta = \tfrac{-12}{13} \quad \cos\theta = \tfrac{5}{13} \quad \tan\theta = \tfrac{-12}{5}$$

$$\csc\theta = \tfrac{13}{-12} \quad \sec\theta = \tfrac{13}{5} \quad \cot\theta = \tfrac{5}{-12}$$

■

The results of Example 1 can be generalized as follows. If point P with coordinates (a, b) is r units from the origin on the terminal side of θ, and point Q with coordinates (x, y) is 1 unit from the origin on the terminal side of θ, then by similar triangles we know that $\frac{y}{1} = \frac{b}{r}$, $\frac{x}{1} = \frac{a}{r}$, and $\frac{y}{x} = \frac{b}{a}$ (see Figure 32). Therefore:

$$\sin\theta = \tfrac{b}{r} \quad \cos\theta = \tfrac{a}{r} \quad \tan\theta = \tfrac{b}{a}$$

$$\csc\theta = \tfrac{r}{b} \quad \sec\theta = \tfrac{r}{a} \quad \cot\theta = \tfrac{a}{b}$$

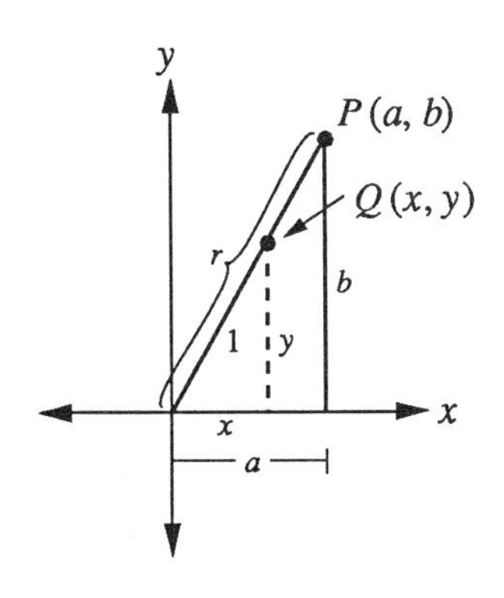

Figure 32 Similar Triangles: $\frac{y}{1} = \frac{b}{r}$, $\frac{x}{1} = \frac{a}{r}$, and $\frac{y}{x} = \frac{b}{a}$

Example 2

Evaluate cos 45°.

Solution Draw a 45° angle in standard position (see Figure 33). Choose point P on the terminal side of the angle 1 unit from the origin. To find the cosine of the angle we need to determine the x-coordinate of point P. From P drop a perpendicular to the x-axis and create a 45-45-90 right triangle. We know that in such a triangle the hypotenuse is $\sqrt{2}$ times as long as each leg. In this case the hypotenuse is 1 unit long so the two legs of the triangle are $\frac{1}{\sqrt{2}}$ units long. Thus, the x-coordinate of P is $\frac{\sqrt{2}}{2}$, and the cosine of 45° is $\frac{\sqrt{2}}{2}$. ■

Figures 32 and 33 suggest that the ratios in the definitions of the six trigonometric functions can be expressed in terms of the lengths of the sides of a right triangle.

$$\sin\theta = \frac{b}{r} = \frac{\text{side opposite } \theta}{\text{hypotenuse}} = \frac{\text{opp}}{\text{hyp}}$$

$$\cos\theta = \frac{a}{r} = \frac{\text{side adjacent to } \theta}{\text{hypotenuse}} = \frac{\text{adj}}{\text{hyp}}$$

$$\tan\theta = \frac{b}{a} = \frac{\text{side opposite } \theta}{\text{side adjacent to } \theta} = \frac{\text{opp}}{\text{adj}}$$

Note that these relationships are true only if the triangle is a right triangle. If you use the short form $\sin\theta = \frac{\text{opp}}{\text{hyp}}$, you must remember that opposite means the side opposite the angle whose sine you want.

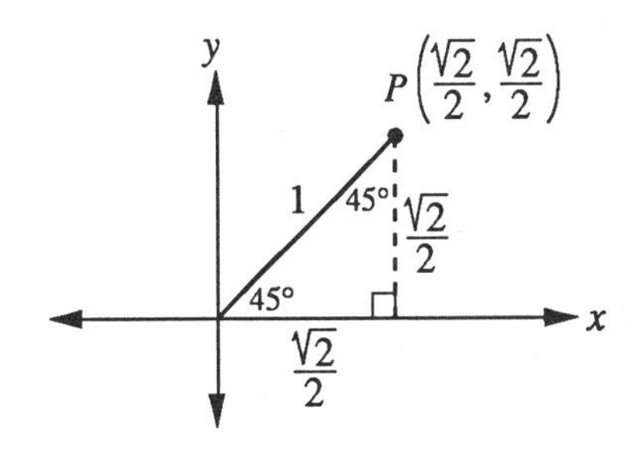

Figure 33 45-45-90 Triangle

Example 3

In right triangle ABC $a = 3$, $b = 4$, and c is the hypotenuse (see Figure 34). Find the measure of $\angle A$.

Solution This problem can be solved in several different ways. The sides whose lengths are known are opposite and adjacent to $\angle A$; this suggests using the tangent function. We know that $\tan A = \frac{\text{opp}}{\text{adj}} = \frac{3}{4} = 0.75$. Therefore we can write $A = \tan^{-1} 0.75$ and use a calculator in degree mode to find that $A = 36.869898$. This means that the measure of $\angle A$ is about 37°.

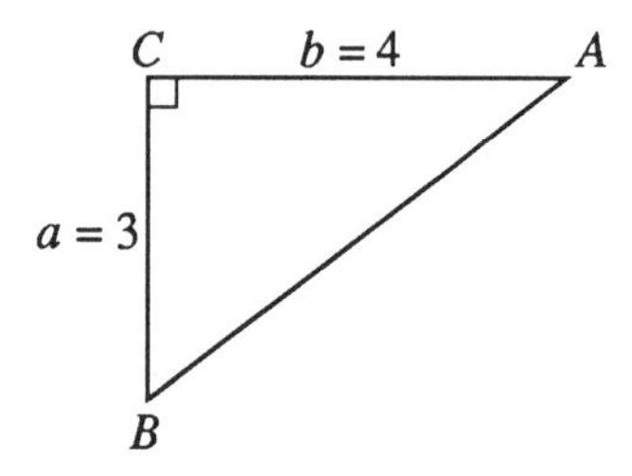

Figure 34 Right Triangle ABC for Example 3

Another way to solve this problem is to first use the Pythagorean Theorem to find that the hypotenuse of the right triangle is 5 units long. We know that $\sin A = \frac{\text{opp}}{\text{hyp}} = \frac{3}{5} = 0.60$. Thus $A = \sin^{-1} 0.60$ and a calculator in degree mode gives the display 36.869898. How could you solve this problem using the cosine function? ■

Exercise Set 5

1. Find $\sin\theta$ if $(2, 7)$ is a point on the terminal side of θ.
2. Find $\cos\theta$ if $(13, -8)$ is a point on the terminal side of θ.
3. Find $\sin\theta$ if $(5, 6)$ is a point on the terminal side of θ.
4. Find $\tan\theta$ if $(-10, 34)$ is a point on the terminal side of θ.
5. Find $\cos\theta$ if $(-44, -67)$ is a point on the terminal side of θ.
6. Find $\tan\theta$ if $(236, 101)$ is a point on the terminal side of θ.
7. A 27° angle is in standard position and its terminal side intersects the circle $x^2 + y^2 = 20$ at the point (a, b). Find a and b.
8. Suppose the terminal side of the angle T intersects the circle $x^2 + y^2 = 12$ at the point $(2, 2\sqrt{2})$. Find the measure of angle T.
9. An arc of length 6.5 centimeters begins at the point $(3, 0)$ on the circle $x^2 + y^2 = 9$. What are the possible end points of the arc?
10. A unit circle and a circle of radius $\sqrt{5}$ are drawn on the same axes. The angle T intersects the larger circle at $(1.2, 1.9)$ as shown in Figure 35.

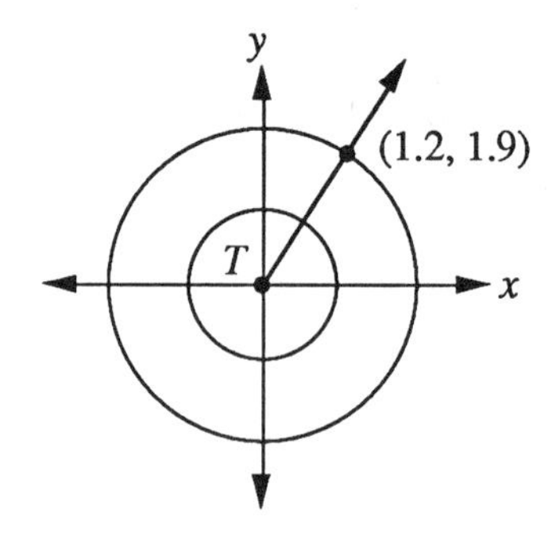

Figure 35 Circles $x^2 + y^2 = 1$ and $x^2 + y^2 = 5$

a. At what point does the terminal side of $\angle T$ intersect the unit circle? How can you interpret these coordinates?

b. Find the measure of $\angle T$.

11. Figure 36 shows a unit circle with center O. Segment OA intersects the circle at B and AP is tangent to the circle at P.

a. Express the coordinates of B in terms of θ.

b. What are the coordinates of A? How long is segment AP? (Notice that AP is a tangent to the circle.)

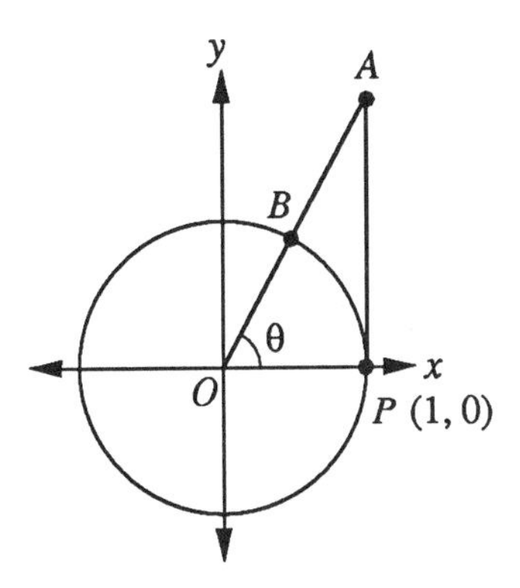

Figure 36 Unit Circle

c. How long is OA? (Notice that OA is a secant of the circle.)

6
Radian Measure

You now know two definitions of the sine function. According to the definition based on the unit circle, the domain of the sine function is the set of real numbers and $\sin 1$ means the sine of an arc of length 1 on a unit circle. According to the definition based on angles, the domain of the sine function is the set of angles and $\sin 1^\circ$ means the sine of a 1° angle.

How does the new definition of the sine of an angle compare to our original definition of the sine of a real number? How can we relate the two definitions? Understanding the relationship between the two definitions depends on an alternate way to measure angles that we will investigate in this section.

Imagine a 90° angle in standard position in the coordinate plane. This angle subtends an arc equal to one-fourth the circumference of any circle centered at the origin. In a unit circle, a 90° angle subtends an arc $\frac{\pi}{2}$ units long. In a circle of radius 5, the circumference is 10π so a 90° angle subtends an arc length of $\frac{10\pi}{4} = \frac{5\pi}{2}$. What arc length will the angle subtend in a circle of radius $\frac{1}{2}$?

Look at the ratio of the arc length subtended by a 90° angle to the radius of the circle for the various size circles mentioned.

$$\frac{\text{arc length}}{\text{radius}} = \frac{\frac{\pi}{2}}{1} = \frac{\frac{5\pi}{2}}{5} = \frac{\frac{\pi}{4}}{\frac{1}{2}}$$

These ratios are all equal to $\frac{\pi}{2}$. The fact that this ratio is constant is the basis for a unit of angle measure called the *radian*. The radian measure of an angle is defined to be the ratio of the subtended arc length to the radius of the circle.

$$\text{radian measure} = \frac{\text{arc length}}{\text{radius}}$$

We have already seen that a 90° angle has a radian measure of $\frac{\pi}{2}$.

The definition of radian means that the radian measure of an angle tells how many times the circle's radius is contained in the length of the subtended arc. The radian measure of an angle will be 1 if the length of the subtended arc is equal to the radius. Thus, in a unit circle a 1 radian angle subtends an arc of length 1. In a circle of radius 3, a one radian angle subtends an arc of length 3 and a two radian angle subtends an arc of length 6 (see Figure 37).

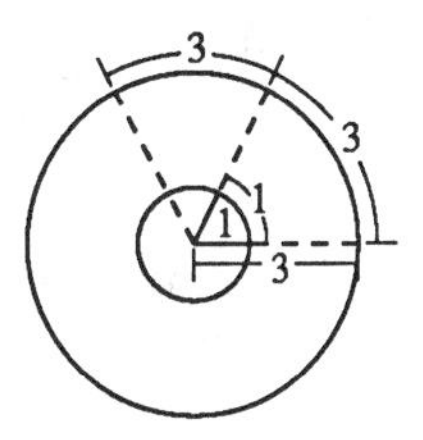

Figure 37 Radian Measure

There are certain radian measures that occur frequently; you should become familiar with these. In a unit circle, an angle of 2π radians subtends an arc length of 2π. This is equal to the circumference of the circle, so the angle must be equal to 360°. An angle of π radians in a unit circle subtends half the circumference, so the angle is equal to a 180° angle. Verify the following.

$$\tfrac{\pi}{6} \text{ radians} = 30 \text{ degrees}$$
$$\tfrac{\pi}{4} \text{ radians} = 45 \text{ degrees}$$
$$1 \text{ radian} \approx 57.3 \text{ degrees}$$
$$\tfrac{\pi}{3} \text{ radians} = 60 \text{ degrees}$$
$$\tfrac{\pi}{2} \text{ radians} = 90 \text{ degrees}$$

You should become familiar with these measures and with multiples of each, such as $\frac{2\pi}{3}$, $\frac{5\pi}{4}$, and $\frac{11\pi}{6}$.

Class Practice

1. Mark a protractor with the following radian measures: $\frac{\pi}{6}, \frac{\pi}{4}, 1, \frac{\pi}{3}, \frac{\pi}{2}, 2, \frac{2\pi}{3}, \frac{3\pi}{4}, \frac{5\pi}{6}, \pi$. Note the positions of 1 and 2 relative to the other measures.

Example 1

In Figure 38, the arc length is $\frac{5\pi}{4}$ inches and the radius of the circle is 2 inches. Find the radian measure of the central angle.

Solution Since the radian measure is equal to the ratio of the arc length to the radius, the radian measure of the central angle is $\dfrac{\frac{5\pi}{4}}{2}$, or $\frac{5\pi}{8}$ radians. ∎

Notice that the arc length and the radius were both measured in inches; the units cancel in their ratio, leaving a pure number with no units. This implies that the radian must be regarded as a quantity without units.

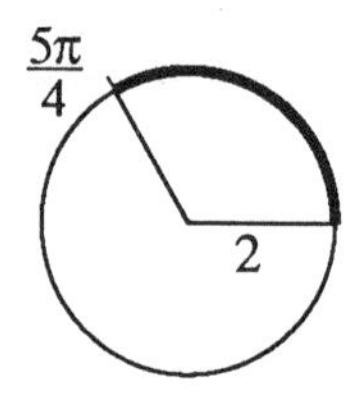

Figure 38 Arc Length in Circle of Radius 2

Example 2

In Figure 39, the arc length is 4 cm and the angle measures $\frac{\pi}{6}$ radians. Find the radius of the circle.

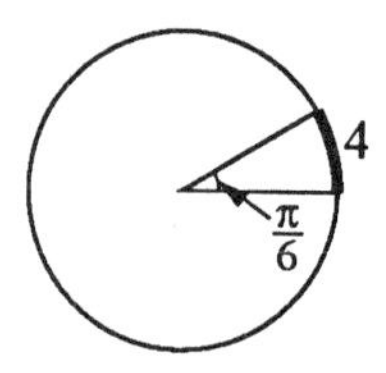

Figure 39 Angle of $\frac{\pi}{6}$ Radians

Solution Since

$$\text{radian measure} = \frac{\text{arc length}}{\text{radius}},$$

it is also true that

$$\text{radius} = \frac{\text{arc length}}{\text{radian measure}}.$$

We have

$$\text{radius} = \frac{4 \text{ cm}}{\frac{\pi}{6} \text{ radians}} = \frac{24 \text{ cm}}{\pi \text{ radians}} \approx 7.64 \text{ cm}.$$

This example again illustrates that radians must be viewed as pure, or dimensionless units, since the division of centimeters by radians yields centimeters. ∎

We can now relate the two definitions of the sine function. Originally we defined the sine of the real number t to be the y-coordinate of the end-point of an arc t units long on a unit circle. Later we defined the sine of an angle with measure t to be the y-coordinate of a point 1 unit from the origin on the terminal side of the angle in standard position. The relationship between these two definitions hinges on radian measure. In a unit circle an angle of t radians subtends an arc of length t. Therefore, the endpoint of the arc and the point on the terminal side of the angle are the same point! This is illustrated in Figure 40. When we see $\sin t$ we can think of t either as the length of an arc or as the radian measure of an angle—the two are equivalent.

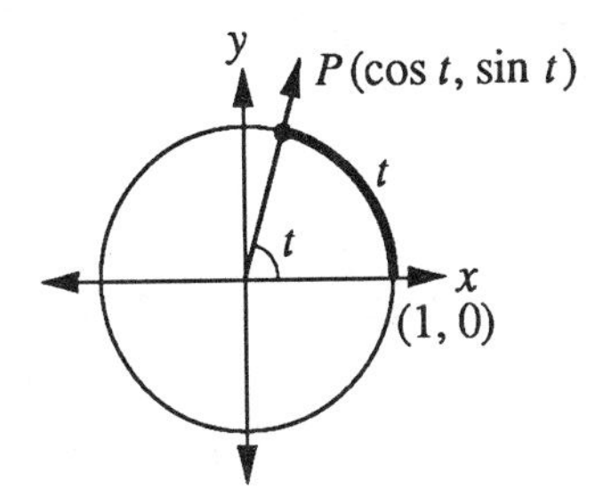

Figure 40 Radian Measure and Arc Length

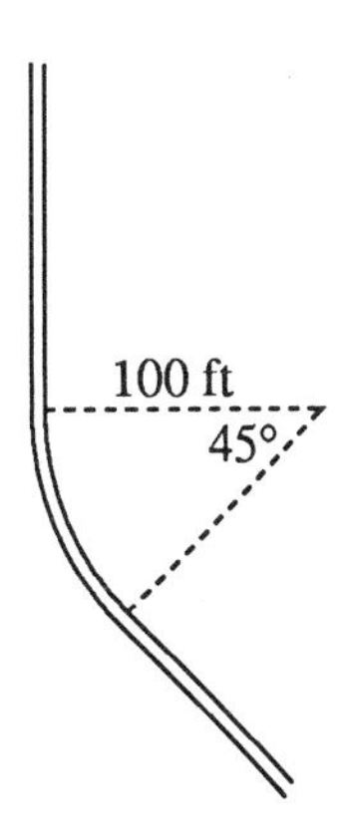

Figure 41 Turn in Southbound Road

Exercise Set 6

1. Express each angle measure in degrees.

a. $\frac{5\pi}{3}$ radians
b. $\frac{15\pi}{2}$ radians
c. 5 radians
d. $\frac{19\pi}{6}$ radians
e. $\frac{25\pi}{9}$ radians
f. r radians

2. Express each angle measure in radians.

a. 1000°
b. 89.8°
c. 720°
d. 176.3°
e. 1°
f. d degrees

3. Use symmetry to find all the angles whose cosine is equal to $\cos(\frac{2\pi}{7})$.

4. Use symmetry to find all the angles whose sine is equal to $\sin(\frac{11\pi}{12})$.

5. In a circle of radius 3.2 cm, what arc length is subtended by a central angle that measures 1.4 radians?

6. Find the measure of the central angle that subtends an arc of length 8 cm in a circle of radius 5 cm.

7. A southbound road turns southeast as shown in Figure 41. If the radius of curvature is 100 feet, what is the length of the curved portion of the road?

8. A winch with a radius of 6 inches is to lift a bucket from a well that is 40 feet deep. How many turns of the winch are required to lift the bucket (see Figure 42)? Ignore the thickness of the rope.

9. A circle of radius 2 is centered at the origin of the coordinate plane. Suppose an ant starts at the point (2, 0) and walks counterclockwise along the circumference of the circle. (Note that the circumference is 4π.) Define the function $f(t)$ = the x-coordinate of the ant's location after walking t units.

a. Evaluate $f(0), f(\pi)$, and $f(2\pi)$.
b. Identify the domain of f.
c. Identify the range of f.
d. Explain why f is a periodic function. Identify the period.

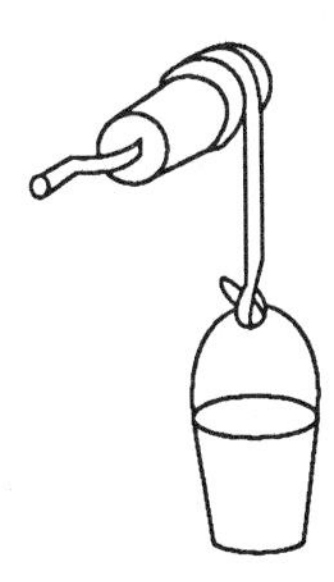

Figure 42 Winch Lifting Bucket

e. Graph f.

f. Write an expression for f.

10. Using the information in the preceding exercise, define $g(t)$ = the y-coordinate of the ant's location after walking t units along the circumference. Complete parts **a** through **f** about function g.

11. Generalize the results of the two preceding exercises for a circle of radius r. Write a sentence or two to explain why the value of r appears where it does in the equations for f and g.

7
Solving Trigonometric Equations

Applications of trigonometry frequently involve modeling a situation with a trigonometric function and then solving an equation based on that function to answer questions about the situation. The examples in this section will illustrate several techniques for solving trigonometric equations.

Example 1

Solve $\cos x = 0.3$.

Solution The solutions of this equation are numbers whose cosine is 0.3. You can think of this problem in terms of the unit circle; solutions are arc lengths that terminate at points whose x-coordinate is 0.3. You can also think of this problem in terms of the cosine curve; solutions are the x-coordinates of points of intersection of $y = \cos x$ and $y = 0.3$ (see Figure 43). Either way you visualize this problem, it should be clear that there are infinitely many numbers whose cosine is 0.3. If x is one solution, then by the symmetry of the circle $2\pi - x$ is another solution.

One solution of $\cos x = 0.3$ can be found using your calculator in radian mode: evaluating

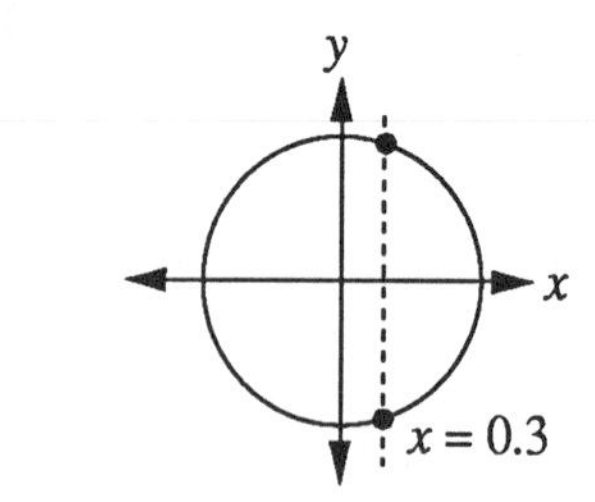

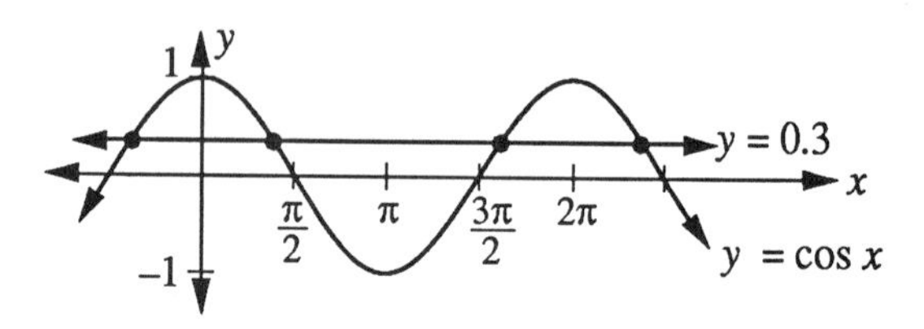

Figure 43 Solutions of $\cos x = 0.3$

$\cos^{-1} 0.3$ gives a display of 1.2661037. This is a number whose cosine is 0.3. By symmetry, $2\pi - 1.2661037 = 5.0170816$ is another number whose cosine is 0.3. Because of the periodicity of the cosine function, adding multiples of 2π to either 1.266 or 5.017 will yield more solutions. The complete solution set is $\{x \mid x = 1.266 + 2k\pi \text{ or } x = 5.017 + 2k\pi, k \text{ any integer}\}$. ■

The previous example required the use of a calculator to solve $\cos x = 0.3$. Although we know there are infinitely many real numbers whose cosine is 0.3, a calculator is capable of giving us only one such number. Think carefully about the process of solving $\cos x = 0.3$. We write $x = \cos^{-1} 0.3$ to indicate that x is a number whose cosine is 0.3. When you evaluate $\cos^{-1} 0.3$, your calculator reports a number whose cosine is 0.3; in effect this "undoes" the cosine function and allows us to isolate x in the equation $\cos x = 0.3$. This process uses the inverse of the cosine function. We will discuss this inverse in more detail later and explain how the calculator "decides" which of many possible answers it will give. For now you should understand that your calculator reports only one number whose cosine is 0.3. You must use the concepts of symmetry and periodicity to find all numbers with cosine equal to 0.3.

Example 2

Find all solutions of $\cos 2x = -0.5$ between 0 and 2π.

Solution We want to find where $y = \cos 2x$ intersects $y = -0.5$. The graph in Figure 44 shows that there are two intersections in each cycle of the cosine curve. The period of $y = \cos 2x$ is π, so two cycles are completed every 2π units. This means there are four solutions between 0 and 2π.

With your calculator in radian mode, you should find that 2.094 is a number whose cosine is -0.5. By the symmetry of the unit circle, $2\pi - 2.094 = 4.189$ is another number whose cosine is -0.5. Therefore,

$$2x = 2.094 \text{ or } 4.189$$

$$x = 1.047 \text{ or } 2.094.$$

The equation has 2 more solutions between 0 and 2π; they can be obtained by adding π (the period of $y = \cos 2x$) to the two solutions we already know. The x-values $1.047, 2.094, 4.189$ and 5.236 are all solutions of $\cos 2x = -0.5$. Verify that all four of these numbers do indeed satisfy the original equation.

You may recognize that -0.5 is a special value of the cosine function that can be determined without a calculator: $\cos \frac{2\pi}{3} = -\frac{1}{2}$. This means that $\frac{2\pi}{3}$ is one value for $2x$. By the symmetry of the unit circle another value for $2x$ is $\frac{4\pi}{3}$. Thus, $2x = \frac{2\pi}{3}$ or $\frac{4\pi}{3}$, so $x = \frac{\pi}{3}$ or $\frac{2\pi}{3}$. Other solutions can be gotten by adding π to these two solutions, resulting in $\frac{\pi}{3}, \frac{2\pi}{3}, \frac{4\pi}{3}$, and $\frac{5\pi}{3}$. ∎

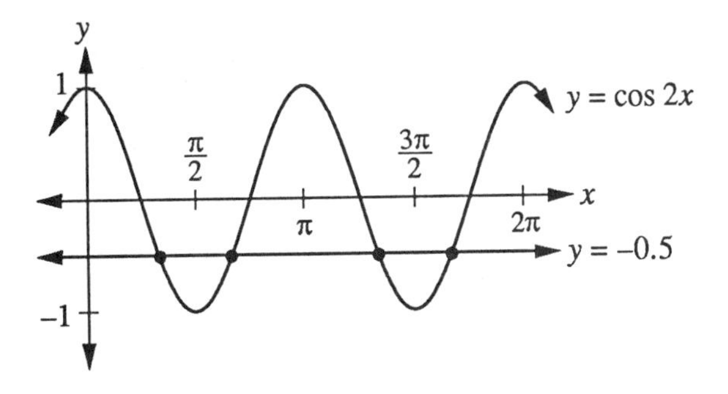

Figure 44 Solutions of $\cos 2x = -0.5$

Example 3

A weight suspended from a spring oscillates so that the displacement d is related to the elapsed time t by the equation

$$d = -1.3\cos(\pi t).$$

At what time(s) is the displacement equal to 1?

Solution To answer this question we need to find nonnegative t-values that satisfy

$$1 = -1.3\cos(\pi t).$$

Dividing both sides of the equation by -1.3 gives

$$-0.769 = \cos(\pi t).$$

You can use your calculator in radian mode to evaluate $\cos^{-1}(-0.769)$; the display should be about 2.448. By the symmetry of the unit circle, $2\pi - 2.448 = 3.835$ is another number whose cosine is -0.769. Thus,

$$\pi t = 2.448 \text{ or } \pi t = 3.835$$

$$t = 0.779 \text{ or } t = 1.221$$

We can find more solutions by adding multiples of 2, the period of $\cos(\pi t)$, to these two solutions. Therefore, the displacement of the weight is equal to 1 when $t = 0.779 + 2k$ or $t = 1.221 + 2k$, k any positive integer. ∎

Example 4

Solve $\sin x > 0.8$.

Solution Solutions of this inequality are numbers whose sine is greater than 0.8. Solutions correspond to points on the sine curve whose y-coordinate is greater than 0.8 (see Figure 45). Because the sine curve oscillates, it is useful to find where $\sin x$ is equal to 0.8 and then study the graph to see where $\sin x$ is greater than 0.8. Using your calculator in radian mode to solve $\sin x = 0.8$ gives $x = 0.927$. By symmetry another solution is $x = \pi - 0.927 = 2.214$. Therefore, the solution set of $\sin x = 0.8$ is

$\{x \mid x = 0.927 + 2k\pi$ or $x = 2.214 + 2k\pi$, k any integer$\}$. These points are marked on the graphs in Figure 45. Note that $\sin x > 0.8$ for x-values between 0.927 and 2.214. Because of periodicity, intervals containing solutions of the inequality are repeated every 2π units. Therefore, the complete solution set of $\sin x > 0.8$ is

$$\{x \mid 0.927+2k\pi < x < 2.214+2k\pi, k \text{ any integer}\}.$$

■

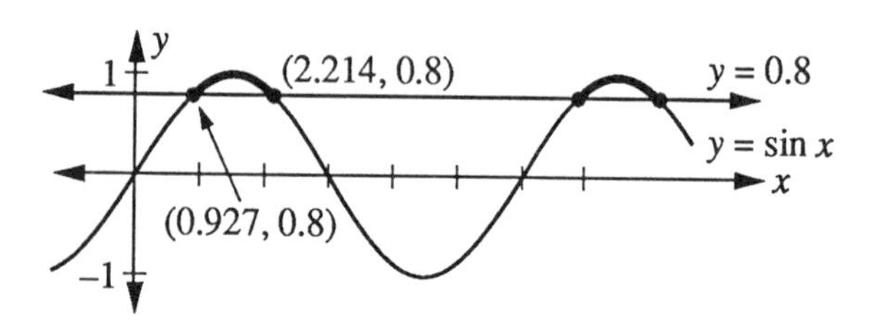

Figure 45 Solutions of $\sin x > 0.8$

Example 5

The graph in Figure 1, page 189, shows how the distance between the ceiling and the tip of the hour hand on a clock changes over time. At what times is this distance less than 30 cm?

Solution We first need to write an equation for the graph in Figure 1. The period is 12, the amplitude is 15, and the graph starts at a minimum and has been shifted up 38 units. An equation for the graph is

$$d = -15\cos\left(\frac{\pi t}{6}\right) + 38.$$

To determine when the distance is less than 30 cm we need to solve the inequality

$$-15\cos\left(\frac{\pi t}{6}\right) + 38 < 30.$$

The oscillation of the graph makes it useful to first find where $-15\cos(\frac{\pi t}{6}) + 38$ is equal to 30. Subtracting 38 from both sides and then dividing by -15 yield

$$\cos\left(\frac{\pi t}{6}\right) = \frac{8}{15} = 0.533.$$

Therefore,

$$\frac{\pi t}{6} = \cos^{-1} 0.533 = 1.009$$

or

$$\frac{\pi t}{6} = 2\pi - 1.009 = 5.274,$$

so

$$t = 1.924 \text{ or } t = 10.073.$$

These are t-values at which the distance is equal to 30 cm. The graph in Figure 46 shows that the distance is less than 30 cm when t is between 10.073 and $1.924 + 12 = 13.924$. Since $t = 10.073$ corresponds to about 10:04 and $t = 13.924$ corresponds to about 1:56, the distance is less than 30 cm between the hours of about 10:04 and 1:56. ■

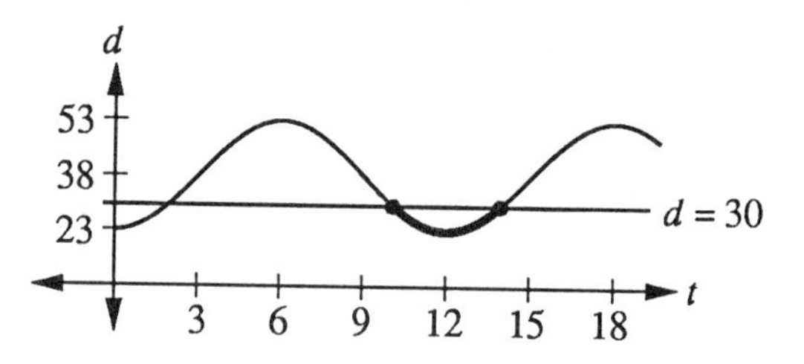

Figure 46 Solutions of $-15\cos(\frac{\pi t}{6}) + 38 < 30$

Example 6

Find all solutions of $\tan^2 x + \tan x - 1 = 0$.

Solution This is a quadratic equation in $\tan x$. If we let $a = \tan x$, the equation can be written as

$$a^2 + a - 1 = 0.$$

Use the quadratic formula to find that $a = \frac{-1\pm\sqrt{5}}{2} \approx 0.618$ or -1.618. This means that $\tan x = 0.618$ or $\tan x = -1.618$. Within one cycle of the tangent function there is exactly one x-value whose tangent is 0.618 and exactly one x-value whose tangent is -1.618. This is because within each cycle the tangent function is one-to-one; think about the graph of $y = \tan x$ to convince yourself of this.

You can solve $\tan x = 0.618$ and $\tan x = -1.618$ using your calculator in radian mode; you should find that $x = 0.554$ or $x = -1.017$. Since the period of the tangent function is π, the complete solution set is $\{x \mid x = 0.554 + k\pi \text{ or } x = -1.017 + k\pi, k \text{ any integer}\}$. ■

Example 7

Find a real number that is equal to its cosine.

Solution A real number x that is equal to its own cosine will satisfy the equation $x = \cos x$. This will be the x-coordinate of a point where the graphs of $y = x$ and $y = \cos x$ intersect (see Figure 47). There is no direct algebraic technique

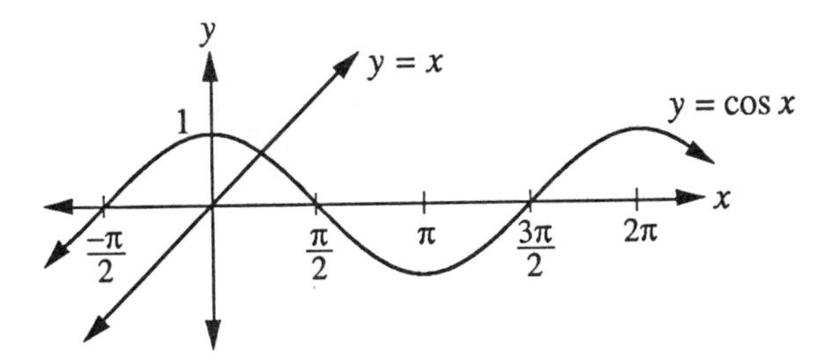

Figure 47 Graphs of $y = \cos x$ and $y = x$

that will allow us to solve this equation for x. However, it is significant to note that we are seeking a *fixed point* of the function $f(x) = \cos x$; a fixed point is defined as one for which $f(x) = x$. Recall from Chapter 4, Section 10 that Picard's method can be used to find a fixed point of a function. This method requires that we make an initial guess x_0 for the fixed point and then evaluate

$$f(x_0) = x_1, \quad f(x_1) = x_2, \quad f(x_2) = x_3,$$

and so forth until we find a value x_n such that $f(x_n) = x_{n+1} \approx x_n$. When this happens we will have found a fixed point of $f(x)$.

A list showing successive values of x_n, beginning with $x_0 = 1$, is provided in Table 48. These values were generated with a calculator in radian mode and show that $x = 0.74$ is a solution (accurate to hundredths) of $x = \cos x$. ■

Table 48 Table of Values for Solving $x = \cos x$

x	$\cos x$	x	$\cos x$
1	0.5403	0.7221	0.7504
0.5403	0.8576	0.7504	0.7314
0.8576	0.6543	0.7314	0.7442
0.6543	0.7935	0.7442	0.7356
0.7935	0.7014	0.7356	0.7414
0.7014	0.7640	0.7414	0.7375
0.7640	0.7221		

Exercise Set 7

1. Find all solutions for the following equations. Use your calculator and your knowledge of symmetry and periodicity.
   a. $\cos x = 0.75$
   b. $3\sin(2x) = 0.6$
   c. $\sin(x - \frac{\pi}{4}) = 0.4$
   d. $\cos(2x) + 1 = 3.4$
   e. $\sin(x - \frac{\pi}{4}) < 0.4$
   f. $\tan x > 0.6$
   g. $\cos(2x - \frac{\pi}{3}) = 0.8$
2. The high water mark on a beach wall is a sinusoidal function of time: $d = 6 + 4\cos((\frac{\pi}{6})t - \frac{\pi}{3})$, where t is the number of hours after midnight and d is the depth of the water in feet. What is the earliest time of day at which the water is at its highest? When is the water 6 feet up the wall?
3. Find all solutions for the following equations.
   a. $4\sin^2 x = 1$
   b. $2\cos^2 x + \cos x = 1$
   c. $2\tan^2 x + 7\tan x + 4 = 0$
   d. $4\cos^2 x - \cos x = 0$
   e. $\cos x = x^2 - 1$
   f. $\sin(x + \frac{\pi}{4}) = \sin x$
   g. $2^{\sin x} = 1$
   h. $e^{\sin x} = 7$

i. $\frac{\sin 2x}{x} = 1$

j. $\sin x = x^2$

k. $\sin x + x = 0$

l. $\cos x + x = 0$

m. $(\sqrt{\sin^2 x}) = \sin x$

4. When a mass is suspended from a spring, the formula $x = (0.1)\cos(4\pi t)$ describes the displacement (in meters) of the mass from its equilibrium position at time t. The positive direction is downward.

a. What is the maximum vertical distance through which the mass moves?

b. What is the minimum time that it takes the mass to move through the distance described in part a?

c. When is the second time (after it is released at $t = 0$) that the mass is at its equilibrium position?

d. When does the mass reach its equilibrium position for the third time (after release)?

e. Draw a graph of the mass's displacement from its equilibrium position as a function of time. Does your graph accurately reflect each of the answers to the previous questions?

5. In a tidal river the time between high tide and low tide is approximately 6.2 hours. The average depth of the water in a port on the river is 4 meters; at high tide the depth is 5 meters.

a. Sketch a graph of the depth of the water in the port over time if the relationship between time and depth is sinusoidal and there is a high tide at 12:00 noon. Write an equation for your curve. Let t represent the number of hours after 12:00 noon.

b. If a boat requires a depth of 4 meters of water in order to sail, how many minutes before noon can it enter the port and by what time must it leave to avoid being stranded?

c. If a boat requires 3.5 meters of water in order to sail, at what time, before noon, can it enter the port and by what time must it leave to avoid being stranded?

d. A boat that requires a depth of 4 meters of water is at a dock in the port. As some of the cargo is unloaded, the depth of water required to sail decreases. Suppose at noon the crew begins to unload cargo. The unloading of the cargo decreases the draft of the boat at a rate of 0.1 meters/hour. At what time must the ship stop unloading and leave the port in order to ensure that it sails before 6 pm?

8

Inverse Trigonometric Functions

Table 49 shows the results of using a calculator to evaluate $\sin^{-1}(\sin A)$ for several values of A.

For these values of A, $\sin^{-1}$ undoes the effect of sin and returns the display to A. The relationship between sin and $\sin^{-1}$ should suggest inverse functions. If we define $f(x) = \sin x$ and $g(x) = \sin^{-1} x$, then $g \circ f(x) = x$ for the values listed in Table 49.

Table 49 Calculator Results for $\sin^{-1}(\sin A)$

A	$\sin A$	$\sin^{-1}(\sin A)$
−0.26	−0.2570806	−0.26
0.94	0.8075581	0.94
1.3	0.9635582	1.3

We have already used $\sin^{-1}$ to solve equations like $\sin x = 0.8$. When we evaluate $\sin^{-1}(0.8)$ in radian mode, the display is 0.9272952. The calculator reports only one number whose sine is 0.8 even though we know there are infinitely many such numbers. How does a calculator decide which number to pick? The key to understanding your calculator's choice is the fact that $f(x) = \sin x$ is not a one-to-one function (see Figure 50). Many different x-values are paired with

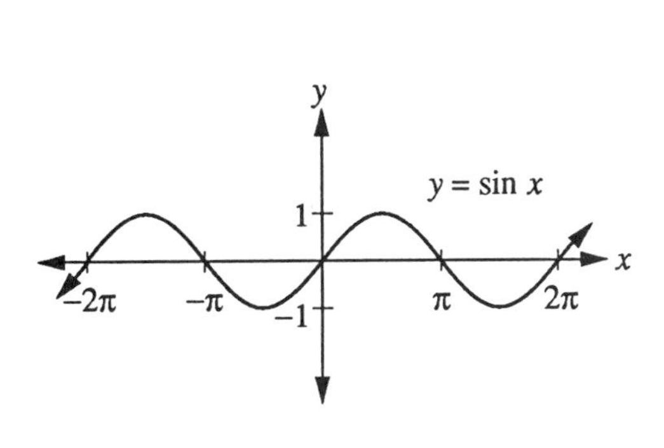

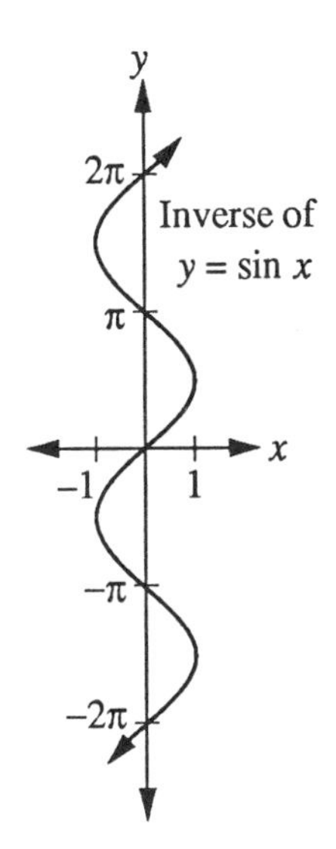

Figure 50 Graphs of $y = \sin x$ and Its Inverse

the same y-value. This is why there are infinitely many solutions to $\sin x = 0.8$.

Because $f(x) = \sin x$ is not one-to-one, its inverse will not be a function. If you reflect the graph of $f(x) = \sin x$ about the line $y = x$ you will see a graph that clearly is not a function (see Figure 50). However, we can restrict the domain of f to an interval of x-values in which f is one-to-one. There are many different ways to choose this restriction. The interval $\frac{\pi}{2} \leq x \leq \frac{3\pi}{2}$ would be satisfactory; so would the union of the intervals $-\pi \leq x \leq -\frac{\pi}{2}$ and $\frac{\pi}{2} \leq x < \pi$. Both of these choices include the entire range of f without repeating range values. However, mathematicians have agreed on restricting the domain of $f(x) = \sin x$ to the interval $-\frac{\pi}{2} \leq x \leq \frac{\pi}{2}$. This interval has the advantages that it is continuous and it contains 0. The first graph in Figure 51 shows $f(x) = \sin x$ over this restricted domain. The domain is $-\frac{\pi}{2} \leq x \leq \frac{\pi}{2}$ and the range is $-1 \leq y \leq 1$.

This choice for the domain restriction on $f(x) = \sin x$ determines the domain and range of the inverse function. The symbol $\sin^{-1} x$ is often used to represent the inverse of the sine function; another notation used is $\arcsin x$. The inverse function $f^{-1}(x) = \sin^{-1} x$ has domain $-1 \leq x \leq 1$ and range $-\frac{\pi}{2} \leq y \leq \frac{\pi}{2}$. A graph

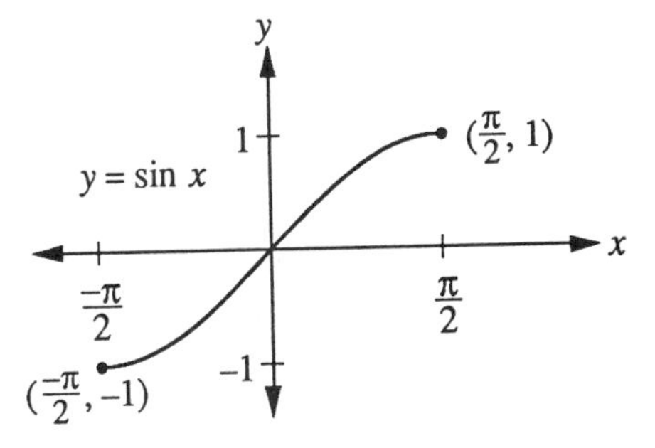

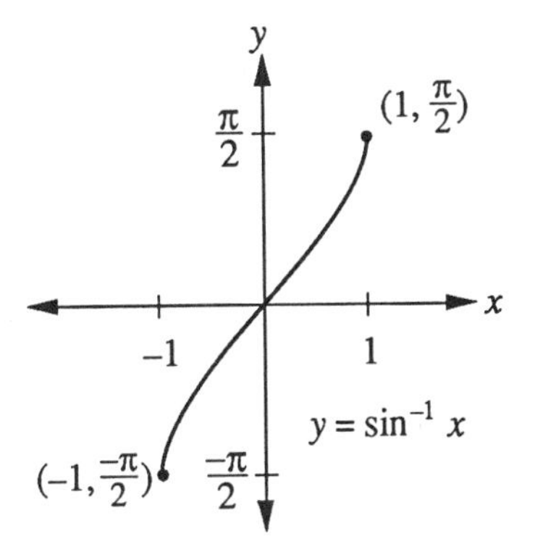

Figure 51 Graphs of $y = \sin x$ with Restricted Domain and $y = \sin^{-1} x$

of $f^{-1}(x) = \sin^{-1} x$ is also shown in Figure 51; it is obtained by reflecting the graph of $f(x) = \sin x$ with a restricted domain about the line $y = x$.

The fact that the range of f^{-1} is the interval $-\frac{\pi}{2} \le x \le \frac{\pi}{2}$ means that $\sin^{-1} x$ always lies between $-\frac{\pi}{2}$ and $\frac{\pi}{2}$. Use your calculator to verify that the inverse sine of any number is always between $-\frac{\pi}{2}$ and $\frac{\pi}{2}$ if you are working in radian mode.

The notation associated with the inverse function is important. The equation $\sin x = 0.8$ has infinitely many solutions and $\sin^{-1} 0.8$ represents only one of the solutions. The notation $\sin^{-1} 0.8$ is used to denote the solution between $-\frac{\pi}{2}$ and $\frac{\pi}{2}$.

Example 1

Find two positive numbers whose sine is -0.96.

Solution When your calculator is in radian mode, $\sin^{-1}(-0.96)$ gives the display -1.2870022, so $\sin^{-1}(-0.96)$ is approximately -1.287. This is the number between $-\frac{\pi}{2}$ and $\frac{\pi}{2}$ whose sine is -0.96. But we need to find two positive numbers with sine equal to -0.96. Figure 52 shows that an arc length of $2\pi - 1.287 = 4.996$ has the same endpoint as -1.287. By the symmetry of the circle, $\pi - (-1.287) = 4.429$ has the same sine as -1.287. Therefore, 4.996 and 4.429 are two positive numbers whose sine is -0.96. ■

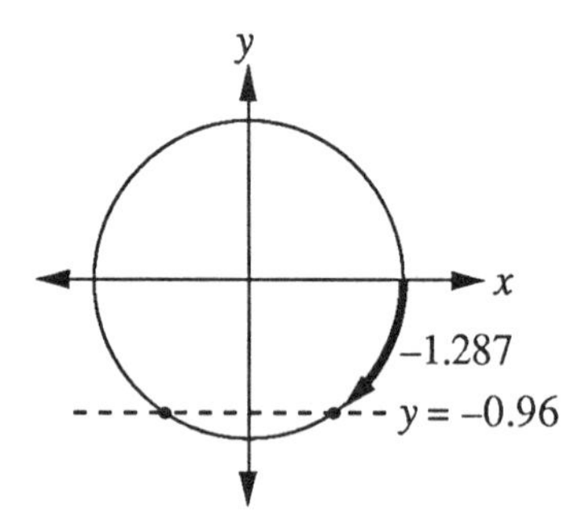

Figure 52 Diagram for Example 1

Like the sine function, the cosine function is not one-to-one. It is necessary to restrict the domain so that the inverse of the cosine function will be a function. As with the sine function, there are many choices for this domain restriction. Mathematicians have agreed to restrict x to the interval $0 \le x \le \pi$. Figure 53 shows the graphs of $g(x) = \cos x$ with restricted domain and $g^{-1}(x) = \cos^{-1} x$. Note that $g^{-1}(x) = \cos^{-1} x$ has domain $-1 \le x \le 1$; this is also the range of $g(x)$. The range of $g^{-1}(x) = \cos^{-1} x$ is $0 \le y \le \pi$; this is the same as the domain of $g(x)$.

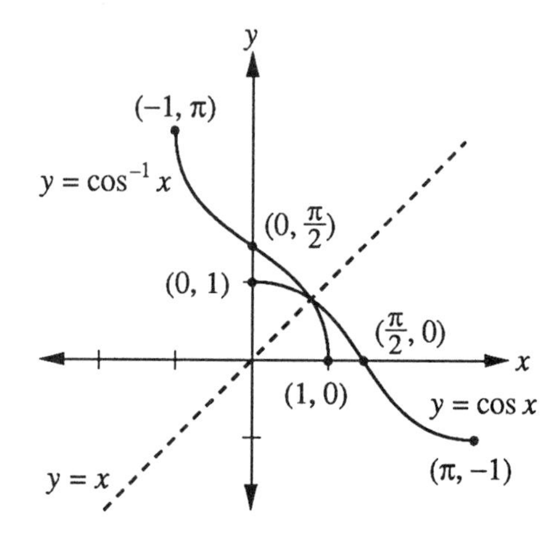

Figure 53 Graphs of $y = \cos x$ with Restricted Domain and $y = \cos^{-1} x$

There are many ways to restrict the domain of the tangent function in order to make it one-to-one. You can use your calculator to find out what the commonly accepted restriction is. Put your calculator in radian mode and find the inverse tangent of $-3, -2, -1, 0, 1, 2,$ and 3. You should notice that all the results are between $-\frac{\pi}{2}$ and $\frac{\pi}{2}$ (between about -1.6 and 1.6). This indicates that the range of the inverse tangent function is $-\frac{\pi}{2} < x < \frac{\pi}{2}$, so the accepted domain restriction to make tangent a one-to-one function must be $-\frac{\pi}{2} < x < \frac{\pi}{2}$.

Figure 54 shows the graphs of $h(x) = \tan x$ and $h^{-1}(x) = \tan^{-1} x$. The restricted domain of tangent and the range of inverse tangent are from $-\frac{\pi}{2}$ to $\frac{\pi}{2}$. The range of tangent and the domain of inverse tangent are both the set of all real numbers. Note that the graph of $h^{-1}(x) = \tan^{-1} x$ has horizontal asymptotes at $y = -\frac{\pi}{2}$ and at $y = \frac{\pi}{2}$.

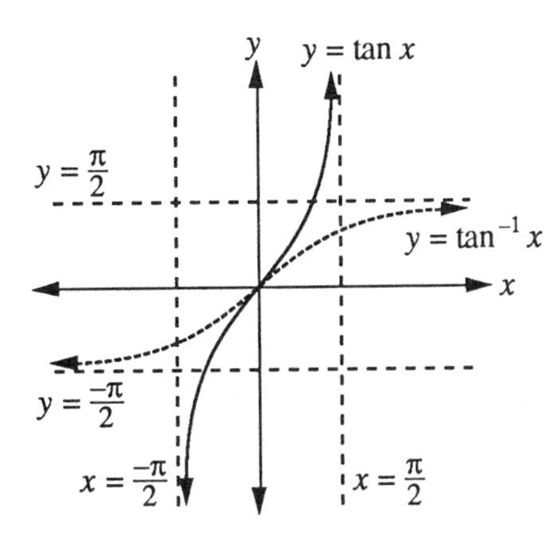

Figure 54 Graphs of $y = \tan x$ with Restricted Domain and $y = \tan^{-1} x$

Example 2

Evaluate $\tan^{-1}(-1)$ and $\cos^{-1}(-0.5)$.

Solution Because the range of the inverse tangent function includes numbers from $-\frac{\pi}{2}$ to $\frac{\pi}{2}$, $\tan^{-1}(-1)$ represents the number in this interval whose tangent is -1. When your calculator is in radian mode,

$$\tan^{-1}(-1) \approx -0.785.$$

Since the range of the inverse cosine function includes numbers from 0 to π, $\cos^{-1}(-0.5)$ represents a number between 0 and π whose cosine is -0.5. In radian mode,

$$\cos^{-1}(-0.5) \approx 2.094.$$

You may recognize that $\tan^{-1}(-1)$ and $\cos^{-1}(-0.5)$ involve special values of the trigonometric functions and can be evaluated without a calculator. For $\tan^{-1}(-1)$ we seek a number between $-\frac{\pi}{2}$ and $\frac{\pi}{2}$ whose tangent is -1. It is useful to visualize this in a unit circle; the number that meets these criteria is $-\frac{\pi}{4}$. While it is true that $\tan\frac{3\pi}{4} = -1$ and $\tan\frac{7\pi}{4} = -1$, neither $\frac{3\pi}{4}$ nor $\frac{7\pi}{4}$ is in the range of the inverse tangent function. For $\cos^{-1}(-0.5)$ we seek a number between 0 and π whose cosine is -0.5. The number that meets these criteria is $\frac{2\pi}{3}$. Again, even though it is true that $\cos\frac{4\pi}{3} = -0.5$, $\frac{4\pi}{3}$ is not in the range of the inverse cosine function. ■

Few calculators have keys for the remaining three trigonometric functions: secant, cosecant, and cotangent. Therefore, we cannot directly determine values of the inverse secant, inverse cosecant, and inverse cotangent functions. Since cotangent is the reciprocal of tangent, you may suspect that $\cot^{-1}$ is the reciprocal of $\tan^{-1}$. This is generally not the case:

$$\cot^{-1} x \neq \frac{1}{\tan^{-1} x}.$$

The symbol $\cot^{-1} x$ represents a number whose cotangent is x, and $\tan^{-1} x$ represents a number whose tangent is x. These numbers are not reciprocals of each other.

Example 3

Evaluate $\csc^{-1}(-2.29)$ and $\sec^{-1}(5.03)$.

Solution Your calculator probably doesn't have a cosecant key. However, if $x = \csc^{-1}(-2.29)$, then

$$\begin{aligned} \csc x &= -2.29 \\ \frac{1}{\csc x} &= \frac{1}{-2.29} \\ \sin x &= \frac{1}{-2.29} \\ x &= \sin^{-1}\left(\frac{1}{-2.29}\right). \end{aligned}$$

So $\csc^{-1}(-2.29) = \sin^{-1}(\frac{1}{-2.29})$. We can evaluate $\csc^{-1}(-2.29)$ by finding the inverse sine of $\frac{1}{-2.29}$. Your calculator should give a display of -0.4519063 for $\sin^{-1}(\frac{1}{-2.29})$; verify that the cosecant of this display is indeed -2.29. Since the inverse sine function on your calculator always produces values between $-\frac{\pi}{2}$ and $\frac{\pi}{2}$, and $\csc^{-1}(b)$ is equal to $\sin^{-1}(\frac{1}{b})$, your calculator will report that the inverse cosecant of any number is also between $-\frac{\pi}{2}$ and $\frac{\pi}{2}$.

We can evaluate $\sec^{-1}(5.03)$ as follows. Define $x = \sec^{-1}(5.03)$. Then

$$\sec x = 5.03$$

$$\frac{1}{\sec x} = \frac{1}{5.03}$$
$$\cos x = \frac{1}{5.03}$$
$$x = \cos^{-1}\left(\frac{1}{5.03}\right).$$

Using a calculator in radian mode to evaluate $\cos^{-1}(\frac{1}{5.03})$ displays 1.3706557; this is a number whose secant is 5.03. Note that since the inverse cosine function is used to evaluate the inverse secant function, the inverse secant of any number will be between 0 and π. ■

The results of the preceding example can be generalized in the following identities:

$$\csc^{-1}(b) = \sin^{-1}\left(\frac{1}{b}\right)$$

and

$$\sec^{-1}(b) = \cos^{-1}\left(\frac{1}{b}\right).$$

Example 4

Solve $\cot x = 3$.

Solution One solution can be represented by $x = \cot^{-1} 3$. How can you use your calculator to evaluate $\cot^{-1} 3$? Go back to the original equation and rewrite it as follows:

$$\cot x = 3$$
$$\frac{1}{\cot x} = \frac{1}{3}$$
$$\tan x = \frac{1}{3}$$
$$x = \tan^{-1}\left(\frac{1}{3}\right).$$

Using a calculator in radian mode to evaluate $\tan^{-1}(\frac{1}{3})$ gives the display 0.3217506. Verify that this is a number whose cotangent is 3. Since the cotangent function has period π, the set of all solutions of the equation $\cot x = 3$ is given by $\{x \mid x = 0.3218 + k\pi,\ k \text{ any integer}\}$. ■

Exercise Set 8

1. Evaluate the following.
   - a. $\sin^{-1} 0.95$
   - b. $\sin^{-1}(-0.95)$
   - c. $\cos^{-1} 0.02$
   - d. $\sin^{-1} 0$
   - e. $\cos^{-1}(-0.67)$
   - f. $\tan^{-1}(-3)$
   - g. $\tan^{-1} 100$
   - h. $\sin^{-1} 1.95$
   - i. $\tan^{-1} 1$
   - j. $\cot^{-1} 10$
   - k. $\sec^{-1} 1.5$
   - l. $\sec^{-1} 10$
   - m. $\sec^{-1}(-2)$
   - n. $\csc^{-1} 1.01$
   - o. $\csc^{-1}(-4.9)$
2. Use your calculator to evaluate the following. Write a sentence or two to explain your answers.
   - a. $\sin(\sin^{-1} 0.9)$
   - b. $\sin^{-1}(\sin 2)$
   - c. $\cos(\cos^{-1}(-0.39))$
   - d. $\cos^{-1}(\cos 1)$
3. Sketch a graph of each function.
   - a. $y = \sin^{-1} 2x$
   - b. $f(x) = 2\tan^{-1} x$
   - c. $g(x) = 1 + \cos^{-1} x$
   - d. $y = \sin^{-1}(x + 1)$
4. Find the inverse of each function.
   - a. $f(x) = \sin^{-1} 2x$
   - b. $g(x) = \cos^{-1}(x - \pi)$
   - c. $h(x) = 2\tan^{-1} x$
5. State the domain and range of each inverse function in the preceding exercise.
6. Sketch a graph of the function $y = \csc^{-1} x$.
7. Sketch a graph of the function $y = \sec^{-1} x$.
8. Sketch a graph of $y = \cot x$. The period and

the position of the vertical asymptotes suggest restricting the domain to $0 < x < \pi$ in order to make the function one-to-one. This is the domain restriction that is commonly used by mathematicians. With this domain restriction, the range of the inverse cotangent function will be $0 < y < \pi$. Sketch a graph of $y = \cot^{-1} x$ under this domain restriction on $y = \cot x$.

9. When the inverse tangent function is used to evaluate the inverse cotangent function on a calculator, the values of inverse cotangent will be in the same range as those of the inverse tangent, that is, from $-\frac{\pi}{2}$ to $\frac{\pi}{2}$. Therefore, the graph of $y = \cot^{-1} x$ would be as shown in Figure 55 if it were based on values obtained from a calculator.

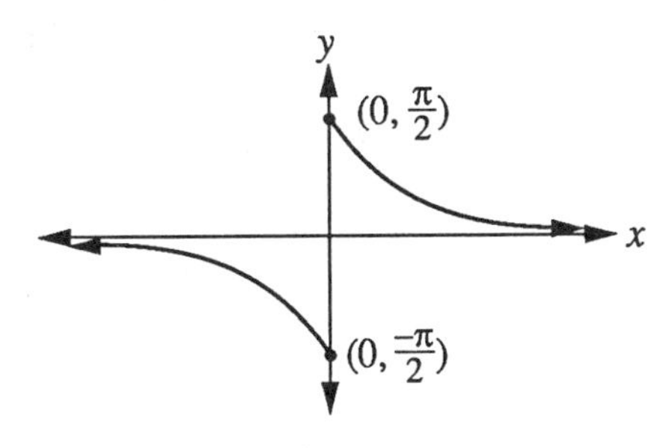

Figure 55 Graph of $y = \cot^{-1} x$ Based on Calculator Values

a. Compare the graph of $y = \cot^{-1} x$ in Figure 55 with the one you drew for the previous exercise. What disadvantages does the graph in Figure 55 have?

b. Note that whenever x is negative your calculator will give a negative value for $\tan^{-1}(\frac{1}{x})$, but we would like the range of $\cot^{-1}$ to include only numbers from 0 to π. How can we modify the calculator procedure for evaluating $\cot^{-1} x$ when x is negative to achieve this range?

9

Composition with Inverse Trigonometric Functions

In the preceding section and exercise set you investigated some examples of composition with inverse trigonometric functions. For instance, we found that

$$\sin(\sin^{-1} 0.9) = 0.9$$

and that

$$\sin^{-1}(\sin 2) = 1.1415927$$

Let $f(x) = \sin x$ and $g(x) = \sin^{-1} x$. It appears that for some values of x it is true that $f(g(x)) = x$ and $g(f(x)) = x$. However, there are x-values for which f and g do not compose to the identity function.

Example 1

Let $f(x) = \sin x$ and let $g(x) = \sin^{-1} x$. For what values of a is $f(g(a)) = a$? For what values of a is $g(f(a)) = a$?

Solution To determine when $f(g(a)) = a$ we must first consider the domain of the composition. The domain of $f \circ g$ is the subset of the domain of g for which $g(a)$ is in the domain of f. The domain of g is $-1 \leq a \leq 1$; $g(a)$ is always between $-\frac{\pi}{2}$ and $\frac{\pi}{2}$, so all values of $g(a)$ are in the domain of f. Therefore, the domain of $f \circ g$ is $-1 \leq a \leq 1$. Because g is defined as a function that undoes the sine function, $f(g(a)) = a$ for all values of a in its domain, that is, for all a between -1 and 1.

To determine when $g(f(a)) = a$ we must consider the domain of $g \circ f$. The domain of the inner function f is all real numbers; $f(a)$ is always between -1 and 1, so all values of $f(a)$ are in the domain of g. Thus, the domain of $g \circ f$ is all real numbers. To determine when $g \circ f$ is equal to the identity function, notice that the value of $g(f(a))$ is in the range of g, so $g(f(a))$ is between $-\frac{\pi}{2}$ and $\frac{\pi}{2}$. In order for $g(f(a))$ to be equal to a, the value of a

must also be in the interval from $-\frac{\pi}{2}$ to $\frac{\pi}{2}$. Therefore, $g(f(a)) = a$ for all a between $-\frac{\pi}{2}$ and $\frac{\pi}{2}$. ∎

Class Practice

1. Determine the values of a for which

$$\cos(\cos^{-1} a) = a.$$

2. Determine the values of a for which

$$\cos^{-1}(\cos a) = a.$$

3. Determine the values of a for which

$$\tan(\tan^{-1} a) = a.$$

4. Determine the values of a for which

$$\tan^{-1}(\tan a) = a.$$

Example 2

Find an exact, nondecimal value for $\sin(\cos^{-1}\frac{3}{8})$ without using your calculator.

Solution We seek the sine of an angle θ whose cosine is $\frac{3}{8}$; we can find the sine of θ without ever finding θ. Since $\theta = \cos^{-1}\frac{3}{8}$, θ must be between 0 and π. Figure 56 shows an angle θ in standard position with its terminal side intersecting the unit circle at point P. Since $\cos\theta = \frac{3}{8}$, the x-coordinate of point P is $\frac{3}{8}$ and θ is therefore a first quadrant angle. The y-coordinate of P will be $\sin\theta$, which is what we seek. The coordinates (x, y) of P satisfy $x^2 + y^2 = 1$; therefore $y = \pm\sqrt{1 - (\frac{3}{8})^2} = \pm\sqrt{\frac{64}{64} - \frac{9}{64}} = \pm\frac{\sqrt{55}}{8}$. Since θ is in the first quadrant the sine of θ is positive. Thus, $y = \frac{\sqrt{55}}{8}$ and $\sin(\cos^{-1}\frac{3}{8}) = \frac{\sqrt{55}}{8}$. You can use your calculator to confirm this result. ∎

Example 3

Express $\tan(\cos^{-1}(x+1))$ in terms of x without trigonometric functions.

Solution We seek the tangent of an angle θ whose cosine is $x + 1$. The technique of Example 2 can be used here with slight adaptation. Let $\theta = \cos^{-1}(x + 1)$. We do not know whether $x + 1$ is positive or negative, so we must consider two possible positions for angle θ; it can be in either the first or second quadrant. Draw θ in standard position and drop a perpendicular to the x-axis (see Figure 57). Since $\cos\theta = x + 1$, the side adjacent to θ has length $x + 1$ and the hypotenuse has length 1. The length of the side opposite θ can be determined using either the Pythagorean Theorem or the identity $\sin^2\theta + \cos^2\theta = 1$; the length of this side is $\sqrt{1 - (x+1)^2} = \sqrt{-x^2 - 2x}$. The tangent of θ is the ratio of the opposite side to the adjacent side, so

$$\tan(\cos^{-1}(x+1)) = \frac{\sqrt{-x^2 - 2x}}{x + 1}.$$

Note that the tangent will be positive if $x + 1$ is positive and negative if $x + 1$ is negative. ∎

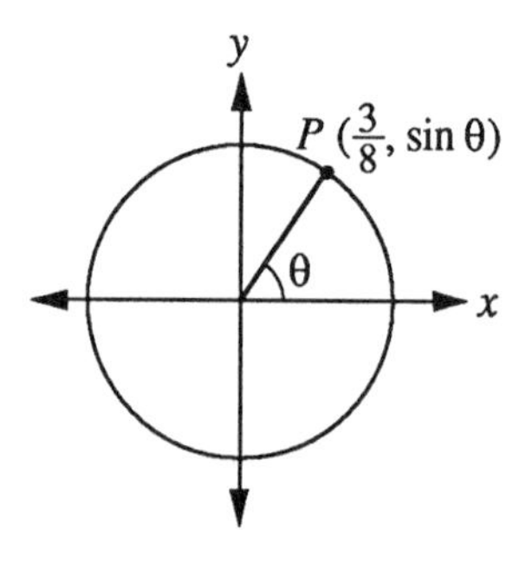

Figure 56 Diagram for Example 2

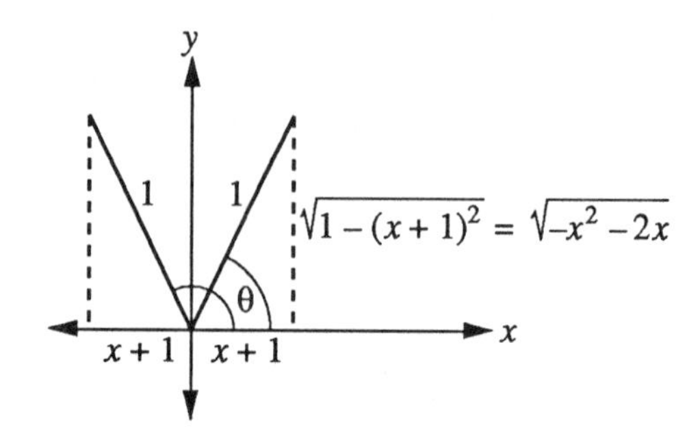

Figure 57 Diagram for Example 3

Class Practice

These questions refer to Example 3.

1. For what values of x is $\cos^{-1}(x+1)$ defined?
2. For what values of x is $\sqrt{-x^2-2x}$ a real number?
3. Compare your answers for problems 1 and 2.

Example 4

Solve $\sin^{-1} 2x = \cos^{-1} x$.

Solution We need to "undo" an inverse trigonometric function in order to isolate one of the variables in this equation. This can be accomplished by composing a function with its inverse. We choose to take the sine of both sides of the equation. This "undoes" the inverse sine function as follows:

$$\sin(\sin^{-1} 2x) = \sin(\cos^{-1} x)$$

$$2x = \sin(\cos^{-1} x).$$

We can use the technique of Example 2 to find the sine of an angle θ whose cosine is x. Since $\theta = \cos^{-1} x$, θ must be a first or second quadrant angle. In either case the sine of θ is positive. Using the triangles shown in Figure 58, we see that

$$\sin\theta = \sin(\cos^{-1} x) = \sqrt{1-x^2}.$$

We now need to solve $2x = \sqrt{1-x^2}$ for x:

$$2x = \sqrt{1-x^2}$$

$$4x^2 = 1 - x^2$$

$$5x^2 = 1$$

$$x = \pm\sqrt{\frac{1}{5}}.$$

Two possible values for x are $\sqrt{\frac{1}{5}}$ and $-\sqrt{\frac{1}{5}}$. Checking these x-values in the original equation shows that only $x = \sqrt{\frac{1}{5}}$ is a solution. The value

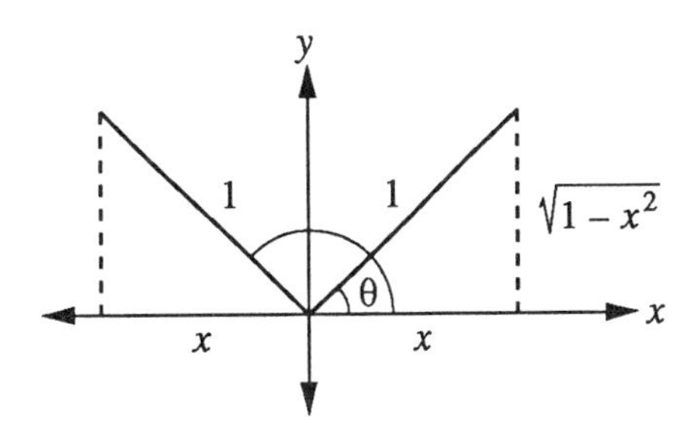

Figure 58 Diagram for Example 4

$x = -\sqrt{\frac{1}{5}}$ is an extraneous solution that was introduced when we squared both sides of our equation. The negative value for x does not work because the inverse sine of a negative number is between $-\frac{\pi}{2}$ and 0, whereas the inverse cosine of a negative number is between $\frac{\pi}{2}$ and π. ■

Exercise Set 9

1. Evaluate the following expressions without using your calculator. Give exact values. State any domain restrictions necessary for expressions that involve variables.

a. $\sin(\cos^{-1} 0.25)$

b. $\tan(\sin^{-1} 0.7)$

c. $\cos(\tan^{-1} 2x)$

d. $\sec(\sin^{-1}(-\frac{x}{x+1}))$

e. $\tan(\cos^{-1}(-\frac{1}{x-1}))$

2. Sketch a graph of each function, giving special attention to the domain and range.

a. $y = \sin(\sin^{-1} x)$

b. $f(x) = \sin^{-1}(\sin x)$

c. $v = \cos(\cos^{-1} u)$

d. $h(u) = \cos^{-1}(\cos u)$

e. $y = \tan(\tan^{-1} x)$

f. $y = \tan^{-1}(\tan x)$

3. Solve each equation.

a. $\cos^{-1} 2x = \sin^{-1} x$

b. $\cos^{-1}(3x-1) = \sin^{-1} x$

c. $\sin^{-1} 2x = \pi + \cos^{-1} x$

4. A picture 3 feet high is placed on a wall with the bottom of the picture 4 feet above the eye level of an observer. Write a function that expresses the angle θ subtended at the observer's eye in terms of the distance x between the observer and the wall (see Figure 59.) Use a computer or graphing calculator to find the distance x that maximizes the subtended angle. This distance will provide the best viewing angle for the observer.

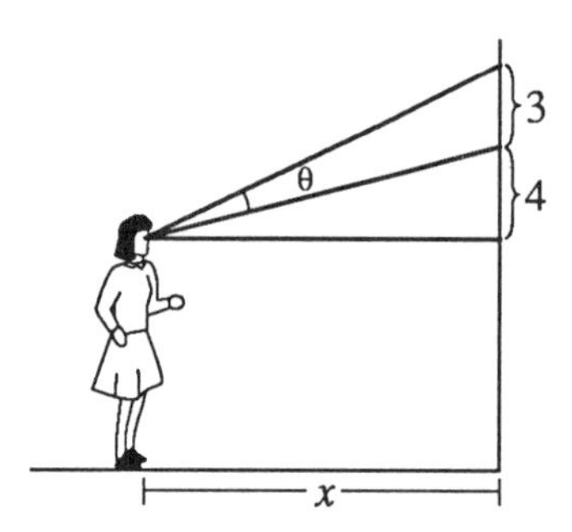

Figure 59 Angle Subtended by a Picture

5. A solar receptor is placed at ground level between two buildings whose heights are 115 meters and 70 meters, respectively (see Figure 60). The buildings are 100 meters apart. The angle θ is the angle swept out by the sun at the receptor between the time it appears over the top of the taller building and the time it disappears behind the smaller building. For the greatest efficiency, the receptor should be located where it receives the most sunlight, in other words, where θ is a maximum. Write a function that expresses the angle θ in terms of the distance x between the receptor and the taller building. Use a computer or graphing calculator to determine the most efficient location for the receptor.

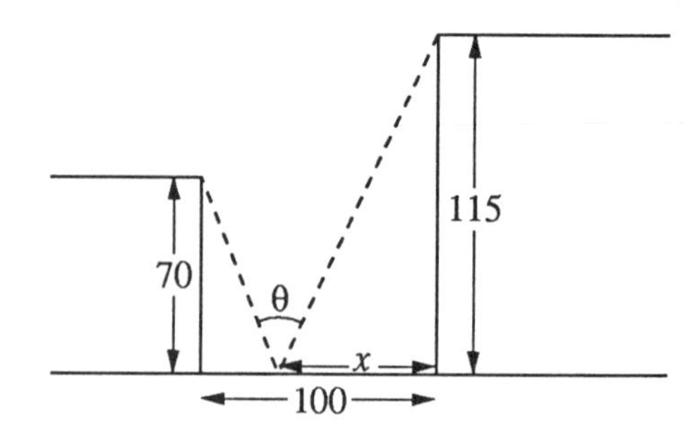

Figure 60 Location of a Solar Receptor

10
Advanced Graphing

The examples in this section will illustrate several techniques for graphing functions that are variations of the trigonometric functions. These techniques include addition of ordinates, multiplication of ordinates, and composition of functions.

Example 1

Graph $y = 1 + \sin x$ and $h(x) = x + \sin x$.

Solution The function $y = 1 + \sin x$ is a transformation of the toolkit sine curve; the curve is simply moved up one unit vertically. The graph is shown in Figure 61. The toolkit sine curve oscillates around the x-axis. Adding 1 to the y-coordinates results in a sine curve that oscillates around the horizontal line $y = 1$.

To graph $h(x) = x + \sin x$, think of h as the sum of the functions $f(x) = x$ and $g(x) = \sin x$. For any particular x-value, the y-value on the graph of h is the sum of the y-values from f and g. For example, when $x = \frac{\pi}{2}$, $f(\frac{\pi}{2}) = \frac{\pi}{2}$ and $g(\frac{\pi}{2}) = 1$; therefore, $h(\frac{\pi}{2}) = \frac{\pi}{2} + 1$. At $x = \pi$, $f(\pi) = \pi$ and $g(\pi) = 0$ so $h(\pi) = \pi$. Try some other values for x. You should find that whenever $\sin(x) = 0$ then $h(x) = x$. Notice also that $h(x)$ varies between $x - 1$ and $x + 1$ since $\sin x$ varies from -1 to 1. A graph of $h(x) = x + \sin x$ is also

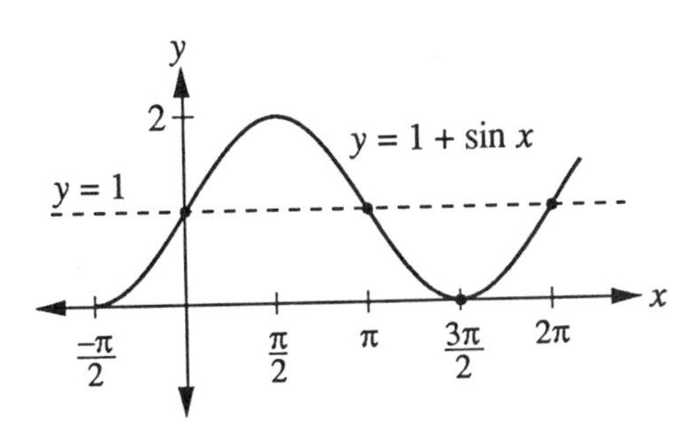

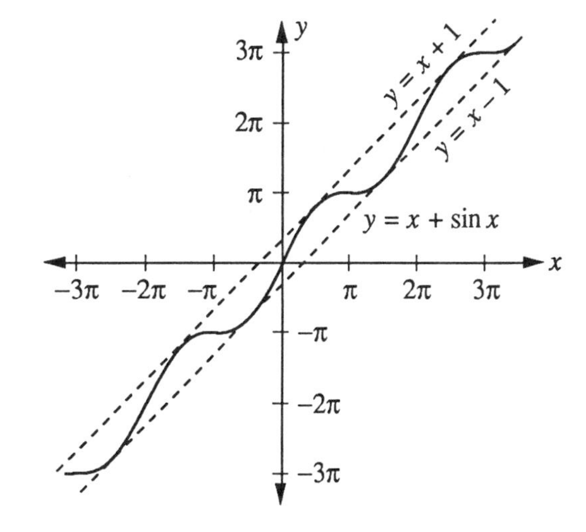

Figure 61 Graphs of $y = 1 + \sin x$ and $h(x) = x + \sin x$

shown in Figure 61. Notice that the graph of $h(x)$ is odd since $h(-x) = h(x)$. The graph can be described as a skewed sine curve oscillating around the line $y = x$. The y-values from $g(x) = \sin x$ are added to the y-values from $f(x) = x$ to obtain the y-values for $h(x) = f(x) + g(x)$. You may recall this graphing technique from Chapter 3; it is called addition of ordinates. ■

Example 2

Graph $y = 2\sin x$ and $m(x) = x \sin x$.

Solution The function $y = 2\sin x$ is a transformation of the toolkit sine curve. The coefficient of 2 causes a vertical stretch, so the graph has amplitude 2. The graph is shown in Figure 62; notice that the sine curve oscillates between the horizontal lines $y = 2$ and $y = -2$. The sine curve touches these horizontal lines whenever $\sin x$ has a maximum point or a minimum point, that is, when x is an odd multiple of $\frac{\pi}{2}$.

The function $m(x) = x \sin x$ is even:

$$m(-x) = (-x)\sin(-x) = (-x)(-\sin x)$$
$$= x \sin x = m(x).$$

Since the graph of m is symmetric about the y-axis, we can analyze its behavior for $x > 0$ and use symmetry to complete the graph for $x < 0$.

You can think of $m(x) = x \sin x$ as the product of $f(x) = x$ and $g(x) = \sin x$. The factor of x

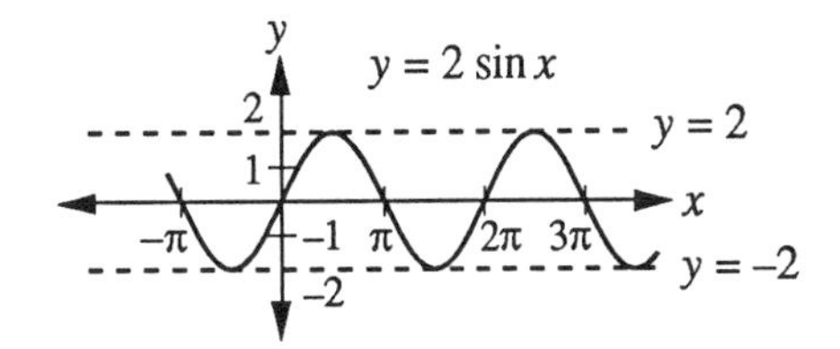

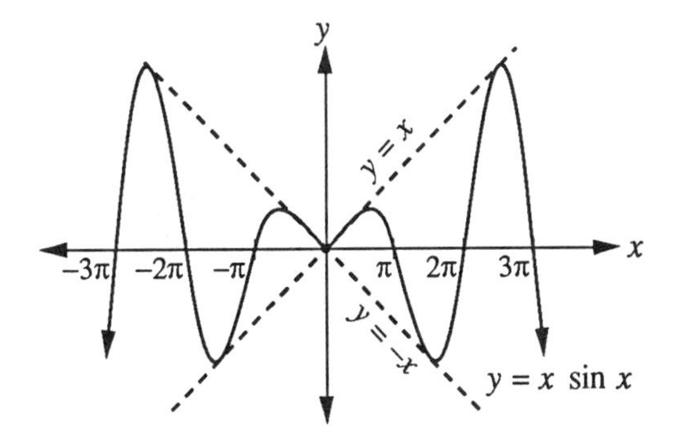

Figure 62 Graphs of $y = 2\sin x$ and $m(x) = x \sin x$

causes a vertical stretch of $\sin x$, but the stretch factor changes as x changes. The x-intercepts of m occur wherever either $f(x) = 0$ or $g(x) = 0$. This means that m has x-intercepts at $x = k\pi$, where k is any integer.

What happens to the graph of m when $g(x) = \sin x$ has a maximum point or a minimum point? When $x = \frac{\pi}{2}$, $f(\frac{\pi}{2}) = \frac{\pi}{2}$ and $g(\frac{\pi}{2}) = 1$ so $m(\frac{\pi}{2}) = \frac{\pi}{2}$. When $x = \frac{3\pi}{2}$, $f(\frac{3\pi}{2}) = \frac{3\pi}{2}$ and $g(\frac{3\pi}{2}) = -1$ so $m(\frac{3\pi}{2}) = -\frac{3\pi}{2}$. In general, $m(x)$ is equal to x whenever $g(x)$ reaches its maximum value of 1. Similarly, $m(x)$ is equal to $-x$ whenever $g(x)$ reaches its minimum value of -1. For all x-values

$$-1 \leq \sin x \leq 1,$$

so

$$-x \leq x \sin x \leq x$$

for all x-values. This means that the graph of m oscillates between the lines $y = x$ and $y = -x$; the graph is bounded by, or enclosed by, the lines $y = x$ and $y = -x$. These lines are said to be *envelopes* of the graph of m. A graph of $m(x) = x \sin x$ is shown in Figure 62. ■

Example 3

Study the graph of $f(x) = (\sin 6x)(\cos x)$ shown in Figure 63. Explain why the graph looks the way it does.

Solution The x-intercepts occur when either $\sin 6x = 0$ or $\cos x = 0$; this happens when $x = \frac{k\pi}{6}$ or when $x = \frac{(2k+1)\pi}{2}$. You can think of $y = \sin 6x$ oscillating between the envelopes $y = \cos x$ and $y = -\cos x$. Whenever $\sin 6x = \pm 1$, or $x = \frac{(2k+1)\pi}{12}$, the graph of f touches its envelope. The graph is symmetric about the origin since $f(-x) = -f(x)$. It is not useful to think of the graph of f as $y = \cos x$ oscillating between $y = -\sin 6x$ and $y = \sin 6x$ since the frequency of $y = \sin 6x$ is greater than the frequency of $y = \cos x$. ■

Example 4

Study the graph of $w(x) = \sin(x^2)$ shown in Figure 64. Explain why the graph looks the way it does.

Solution It should be clear that $w(-x) = w(x)$; this is why the graph is symmetric about the y-axis. The function w can be decomposed as $p \circ q$ where $p(x) = \sin x$ and $q(x) = x^2$. Think about the sine function operating on x^2 instead of on x to gain some insight into the graph's behavior. You can understand the rate at which w oscillates by finding x-intercepts on the graph. We know that $\sin a = 0$ when $a = 0, \pi, 2\pi, 3\pi, 4\pi$ and so forth. Therefore, $\sin(x^2) = 0$ when $x^2 = 0, \pi, 2\pi, 3\pi, 4\pi$ and so forth. This means that x-intercepts occur at $x = 0, \pm\sqrt{\pi}, \pm\sqrt{2\pi}, \pm\sqrt{3\pi}, \pm\sqrt{4\pi}$, and so forth. Decimal approximations of these x-values are

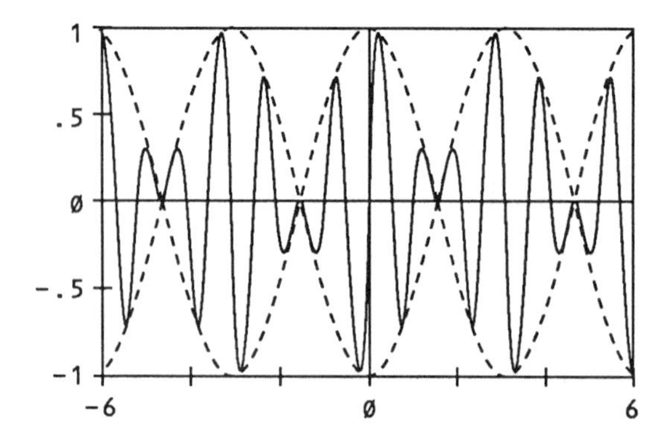

Figure 63 Graph of $y = (\sin 6x)(\cos x)$

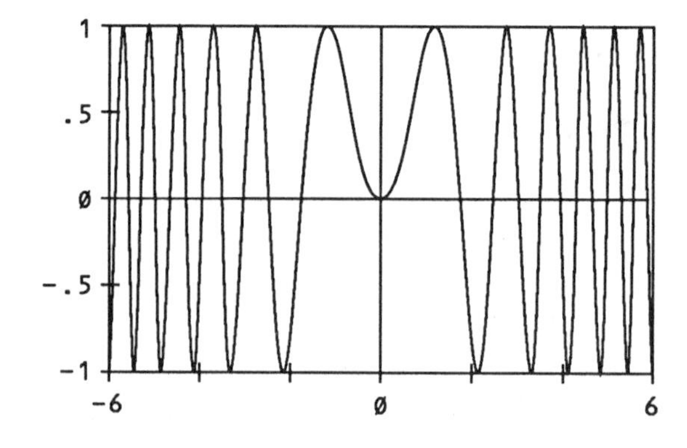

Figure 64 Graph of $y = \sin(x^2)$

$0, \pm 1.772, \pm 2.507, \pm 3.070, \pm 3.545$. Notice that as you move farther from the origin the zeros get closer together so the graph oscillates faster and faster. ■

Exercise Set 10

1. Sketch a graph of each function using the techniques presented in this section. If you are uncertain of a particular graph, examine a computer or calculator graph and then write a few sentences to explain why the graph looks the way it does.

   a. $y = x^2(\sin x)$

   b. $y = 2^x(\sin x)$

   c. $y = x^2 + \sin x$

   d. $y = 2^x + \sin x$

   e. $y = 2^{(\cos x)}$

   f. $y = |x+1| + \cos x$

   g. $y = e^{(\sin x)}$

   h. $y = x(\cos 2x)$

   i. $y = \log(\sin x)$

   j. $y = |\tan x|$

   k. $y = \sin(\frac{1}{x})$

   l. $y = \log|\sin x|$

   m. $y = (\frac{1}{x})\sin x$ (**Hint**: Use your calculator to evaluate the function at $x = 0.1$, $x = 0.01$, and $x = 0.001$.)

   n. $y = \dfrac{\sin x + 1}{x}$

   o. $y = (\sin x)^2$

   p. $y = \tan(x^2 + 1)$

   q. $y = \sin(4x)\cos x$

2. According to the theory of biorhythms, each person has 3 cycles which begin when he or she is born. They are the physical cycle of 23 days, the emotional cycle of 28 days, and the intellectual cycle of 33 days. Assume that each cycle can be represented by a sinusoid and that all have the same amplitude; let $p(x)$ represent the physical cycle, $e(x)$ the emotional cycle, and $i(x)$ the intellectual cycle.

   a. Write equations for $p(x), e(x)$, and $i(x)$.

   b. Sketch graphs of $p(x), e(x)$, and $i(x)$.

   c. Your general sense of well being is said to be determined by $F(x) = p(x) + e(x) + i(x)$. Use a computer or graphing calculator to sketch $F(x)$.

   d. Identify some of the days on which $F(x)$ attains a relative maximum.

   e. How many days elapse until all three of the cycles are the same as they were at birth?

3. After a person dives off a diving board the board slows down and eventually stops vibrating because of friction and air resistance. Sketch the motion of a diving board over time. What function has a graph with the same characteristics as the one you have just sketched?

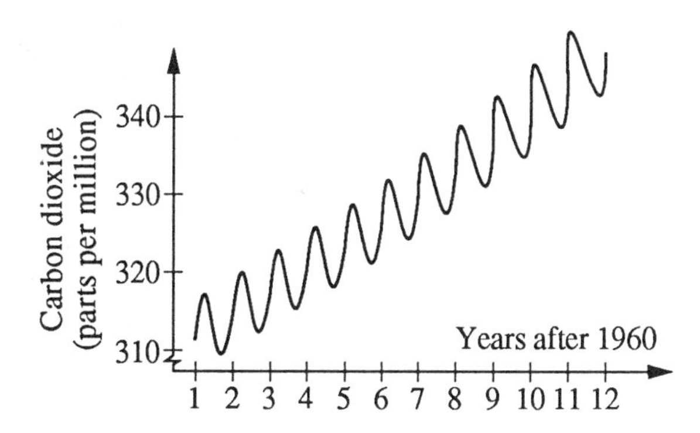

Figure 65 Concentration of Carbon Dioxide

4. The graph in Figure 65 shows the concentration of carbon dioxide in the atmosphere near Mauna Loa in Hawaii.

   a. Write an expression for a function whose graph is similar to the graph in Figure 65.

   b. Write a few sentences to explain the steady increase in carbon dioxide over the years and the peaks and valleys within each year.

11
Trigonometric Identities

Sound travels in waves that are like sine curves. The pitch of the sound depends on the frequency of the sine wave, and the intensity of the sound depends on the amplitude of the sine wave. Physicists often write the equation of a sine wave in the form $y = \sin 2\pi ft$ where t represents time and f is the frequency. Recall that the frequency of a sine wave is the number of cycles completed in one second; frequency is often measured in units called Hertz, abbreviated Hz. One Hertz corresponds to a frequency of one cycle per second. Convince yourself that a sine wave with equation $y = \sin 2\pi ft$ does indeed have period $\frac{1}{f}$ and frequency f.

Suppose two tuning forks that produce sounds with frequencies f_1 and f_2 are struck at the same time. What kind of sound is produced? The sound that results is the sum of two sine waves:

$$y = \sin 2\pi f_1 t + \sin 2\pi f_2 t.$$

Because the two waves do not have the same frequency, they will interfere with each other and the sound level will rise and fall. These regularly spaced changes in sound intensity are called *beats.* Musicians use this fact to tune their instruments. When an instrument is in tune with a standard tone, no beats will be heard.

Let's use a specific example to analyze the interference of sound waves. Suppose two tuning forks with frequencies $f_1 = 105$ Hz and $f_2 = 100$ Hz are struck simultaneously. The individual sound waves can be represented by the equations

$$y = \sin 2\pi 105t = \sin 210\pi t$$

and

$$y = \sin 2\pi 100t = \sin 200\pi t.$$

The combined sound is represented by the equation

$$y = \sin 210\pi t + \sin 200\pi t.$$

Graphs of $y = \sin 210\pi t$ and $y = \sin 200\pi t$ are shown in Figure 66; the graph of $y = \sin 210\pi t + \sin 200\pi t$ is shown in Figure 67.

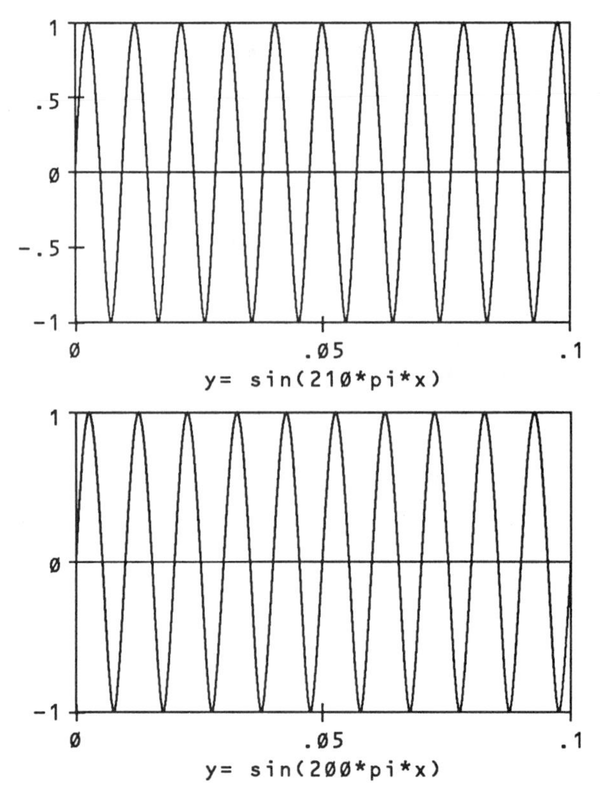

Figure 66 Graphs of $y = \sin 210\pi t$ and $y = \sin 200\pi t$

The sound wave pictured in Figure 67 resembles one sinusoid oscillating within the envelope of another sinusoid. The intensity of the sound rises and falls in a regular pattern. When the envelope reaches a maximum or minimum we hear

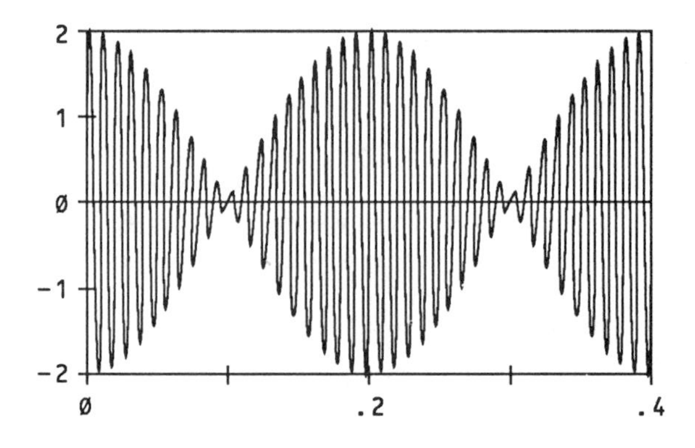

Figure 67 Graph of $y = \sin 210\pi t + \sin 200\pi t$

the most intense sound, and when the envelope reaches zero we hear no sound. In order to understand why these beats are produced we need to study more trigonometry, in particular more about trigonometric identities.

Some of the important identities we have already studied are listed below.

$$\sin^2 x + \cos^2 x = 1$$

$$\sin(\tfrac{\pi}{2} - x) = \cos x$$

$$\cos(-x) = \cos x$$

$$\sin(-x) = -\sin x$$

The first identity on this list is often called a Pythagorean identity since it can be proved using the Pythagorean theorem. Two other identities can easily be derived from $\sin^2 x + \cos^2 x = 1$. Dividing both sides by $\cos^2 x$ yields

$$\tan^2 x + 1 = \sec^2 x.$$

Dividing both sides of the Pythagorean identity by $\sin^2 x$ gives

$$1 + \cot^2 x = \csc^2 x.$$

Another important identity involves the cosine of a difference:

$$\cos(a - b) = \cos a \cos b + \sin a \sin b.$$

This identity may surprise you a bit; you may have expected that $\cos(a - b)$ would be equal to $\cos a - \cos b$. To convince yourself that $\cos(a-b) \neq \cos a - \cos b$ choose $a = \pi$ and $b = \frac{\pi}{2}$. Substituting into $\cos(a - b)$ gives $\cos(\pi - \frac{\pi}{2}) = \cos(\frac{\pi}{2}) = 0$; substituting into $\cos a - \cos b$ gives $\cos \pi - \cos \frac{\pi}{2} = -1 - 0 = -1$.

You will prove the identity for the cosine of a difference in the exercises at the end of this section. This identity is important for many reasons, one of which is that if you know this identity you can easily derive five other identities:

$$\cos(a + b) = \cos a \cos b - \sin a \sin b$$

$$\cos(2a) = \cos^2 a - \sin^2 a$$

$$\sin(a - b) = \sin a \cos b - \cos a \sin b$$

$$\sin(a + b) = \sin a \cos b + \cos a \sin b$$

$$\sin(2a) = 2 \sin a \cos a$$

In pre-calculator days identities were very useful for finding particular values of trigonometric functions. For instance, you can find an exact value for $\sin \frac{\pi}{12}$ by writing it as $\sin(\frac{\pi}{3} - \frac{\pi}{4})$ and then using the identity for $\sin(a - b)$. With access to a calculator the importance of identities has changed. Even if you have a calculator to evaluate $\sin \frac{\pi}{12}$, it is important that you be sufficiently familiar with identities to recognize situations in which they may be useful. One such situation is in solving trigonometric equations.

Example 1

Find all solutions of $2 \cos x + \sin^2 x = 0$.

Solution We will first express each term in this equation in terms of a single trigonometric function. The identity $\sin^2 x + \cos^2 x = 1$ can be written as $\sin^2 x = 1 - \cos^2 x$, so we can substitute $1 - \cos^2 x$ for $\sin^2 x$ in our equation:

$$2 \cos x + 1 - \cos^2 x = 0$$

$$\cos^2 x - 2 \cos x - 1 = 0$$

Now we have a quadratic equation in $\cos x$. Using the quadratic formula yields

$$\cos x = 1 \pm \sqrt{2}.$$

The number $1 + \sqrt{2}$ is greater than 1; since $\cos x$ is never greater than 1 we must reject this value. Therefore

$$\cos x = 1 - \sqrt{2},$$

so

$$x = \cos^{-1}(1 - \sqrt{2}) = \cos^{-1}(-0.4142136) = 1.998.$$

Thus $x = 1.998$ is a number whose cosine is $1-\sqrt{2}$. By the symmetry of the circle, $2\pi - x = 4.285$ is another number whose cosine is $1 - \sqrt{2}$. The complete solution set of the equation is $\{x \mid x = 1.998 + 2k\pi \text{ or } x = 4.285 + 2k\pi, k \text{ any integer}\}$. ∎

Example 2

Solve $\sin 2x = 2\sin x$.

Solution This equation involves only the sine function, but the arguments of the sines are different. The identity $\sin 2x = 2\sin x\cos x$ allows us to substitute $2\sin x\cos x$ for $\sin 2x$ in our equation:

$$2\sin x\cos x = 2\sin x.$$

It is not necessary to express everything in terms of one trigonometric function because we can isolate the sine and the cosine by factoring. Move all the terms to one side of the equal sign and then factor.

$$2\sin x\cos x - 2\sin x = 0$$

$$2\sin x(\cos x - 1) = 0$$

You should not divide both sides of the equation by $\sin x$ since you have no assurance that $\sin x$ is not equal to zero. We now know

$$\sin x = 0 \text{ or } \cos x = 1.$$

The solutions of $\sin x = 0$ are multiples of π; the solutions of $\cos x = 1$ are even multiples of π. Therefore, the solution set is $\{x \mid x = k\pi, k \text{ any integer}\}$. Figure 68 shows that the graphs of $y = \sin 2x$ and $y = 2\sin x$ intersect at points that correspond to the x-values in our solution set. ■

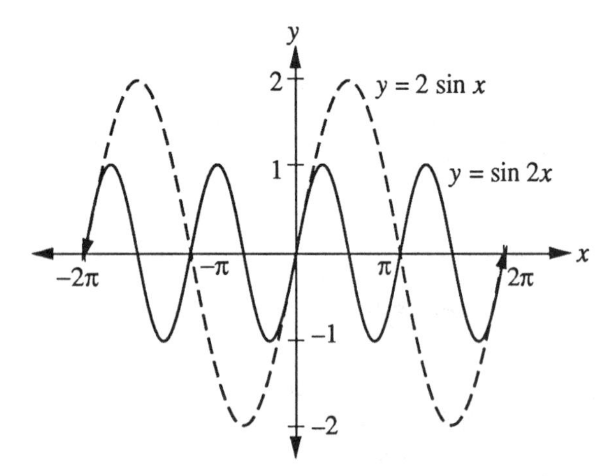

Figure 68 Graphs of $y = \sin 2x$ and $y = 2\sin x$

Example 3

Solve $\frac{\sin x}{1+\cos x} = 1$.

Solution The fraction in this equation will be equal to 1 provided the numerator and denominator are equal:

$$\sin x = 1 + \cos x.$$

The Pythagorean identity allows us to replace $\sin x$ with $\pm\sqrt{1-\cos^2 x}$, which gives

$$\pm\sqrt{1-\cos^2 x} = 1 + \cos x.$$

Now square both sides.

$$\begin{aligned} 1-\cos^2 x &= (1+\cos x)^2 \\ 1-\cos^2 x &= 1 + 2\cos x + \cos^2 x \\ 0 &= 2\cos x + 2\cos^2 x \\ 0 &= 2\cos x(1+\cos x) \end{aligned}$$

We can conclude that $\cos x = 0$ or $\cos x = -1$. However, $1+\cos x$ appears in the denominator of the original equation so we must reject the solutions associated with $\cos x = -1$. The only possible solutions are x-values that satisfy $\cos x = 0$. These are odd multiples of $\frac{\pi}{2}$ so $x = \frac{(2k+1)\pi}{2}$, k any integer. However, because we squared both sides in the solution process we may have introduced extraneous solutions. Checking in the original equation shows that $x = \frac{3\pi}{2}, \frac{7\pi}{2}, \frac{11\pi}{2}$, and so forth are not solutions. Therefore, the solution set is $\{x \mid x = \frac{(4k+1)\pi}{2}, k \text{ any integer }\}$. ■

We will now return to our analysis of sound wave interference. Recall that the equation $f(t) = \sin 210\pi t + \sin 200\pi t$ represents the sound that results when tuning forks of frequencies 105 Hz and 100 Hz are struck simultaneously. The graph of $y = \sin 210\pi t + \sin 200\pi t$ in Figure 67 appears to have an envelope, but this is not obvious from the equation. We want to rewrite the equation in a way that helps us to understand why and how frequently the beats occur. To do this we will derive an identity for $\sin x + \sin y$.

We know that

$$\sin(a+b) = \sin a \cos b + \cos a \sin b$$

and that

$$\sin(a-b) = \sin a \cos b - \cos a \sin b.$$

Adding these two equations yields a new identity:

$$\sin(a+b) + \sin(a-b) = 2 \sin a \cos b.$$

Now let $a + b = x$ and let $a - b = y$. If we solve these equations simultaneously for a and b, we find that $a = \frac{x+y}{2}$ and $b = \frac{x-y}{2}$. We can substitute these expressions for $a + b$, $a - b$, a, and b in the preceding identity to obtain

$$\sin x + \sin y = 2\sin(\frac{x+y}{2})\cos(\frac{x-y}{2}).$$

This identity expresses a sum in terms of a product.

In our sound wave problem, $x = 210\pi t$ and $y = 200\pi t$, so $\frac{x+y}{2} = 205\pi t$ and $\frac{x-y}{2} = 5\pi t$. This means that

$$\sin 210\pi t + \sin 200\pi t = 2\sin(205\pi t)\cos(5\pi t).$$

The graph of $f(t) = \sin 210\pi t + \sin 200\pi t$ is identical to the graph of $g(t) = 2\sin(205\pi t)\cos(5\pi t)$. Since g is the product of $y = \sin(205\pi t)$ and $y = 2\cos(5\pi t)$, the graph of g can be viewed as $y = \sin(205\pi t)$ oscillating between the envelopes $y = 2\cos(5\pi t)$ and $y = -2\cos(5\pi t)$ (see Figure 69).

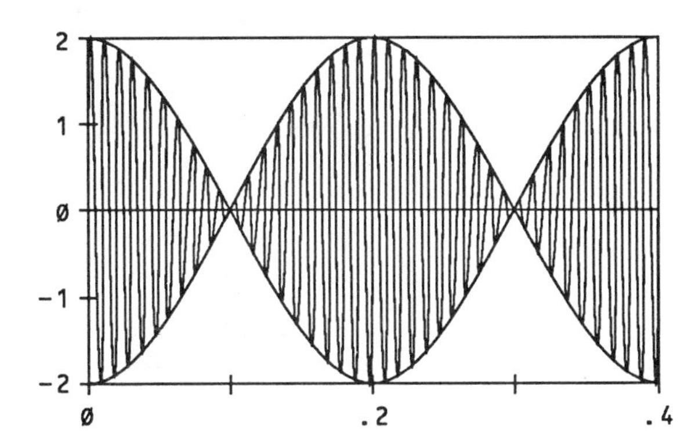

Figure 69 Graph of $y = 2\sin(205\pi t)\cos(5\pi t)$

When the individual sound waves combine, the sound is intensified at some times; at other times there is no sound. This is what causes us to hear a beat. Writing the equation of the combined sound waves in the form $g(t) = 2\sin(205\pi t)\cos(5\pi t)$ makes it easy to analyze the frequency of the beats. A beat will occur whenever the envelope diminishes to zero, and this happens two times within each cycle of the envelope $y = 2\cos(5\pi t)$. The period of the enveloping curve is 0.4 seconds per cycle and the frequency is 2.5 cycles per second. There are two beats in each cycle, so there are 5 beats per second.

Exercise Set 11

1. Suppose you know that $\sin\theta = \frac{1}{4}$ and $\frac{\pi}{2} < \theta < \pi$. Use identities to find the exact value of the following.

a. $\cos\theta$

b. $\sin(\frac{\pi}{2} - \theta)$

c. $\tan(\frac{\pi}{2} + \theta)$

d. $\sin(\theta - \frac{\pi}{4})$

2. Suppose you know that $\tan x = 4$ and $0 < x < \frac{\pi}{2}$. Find the exact value of the following.

a. $\sin x$

b. $\cos 2x$

3. Suppose you know that $\sin a = -\frac{1}{4}$ and $\frac{3\pi}{2} < a < 2\pi$. Find the exact value of the following.

a. $\sin(\frac{\pi}{2} - a)$

b. $\sin(2a)$

c. $\sin(\frac{\pi}{4} + a)$

4. Find all solutions of each equation in the interval $[0, 2\pi]$.

a. $\tan 2x = 1$

b. $\sec^2 x - 1/4 = 0$

c. $\cos x = \cos 2x$

d. $\sin 2x - \sin 4x = 0$

e. $\sec^2 x - \tan x = 1$

f. $(\sin x + \cos x)^2 = 1$

g. $\sin 4x \cos x = \sin x \cos 4x$

h. $\sec^5 \theta = 4 \sec \theta$

5. Solve each equation. Find all real solutions.

a. $2 \cos x - \sin^2 x = 2$

b. $5 \sec^2 x + 2 \tan x - 8 = 0$

c. $\frac{\sin 2x}{x} = 1$

d. $\sin^2 x - \cos^2 x = -\frac{1}{8}$

e. $(\sin x + \cos x)^2 = \frac{(2+\sqrt{3})}{2}$

f. $\sin \frac{x}{2} + \cos x = 1$

6. Find the points of intersection of $y = 2 \cos 2x$ and $y = 2 - 2 \sin x$.

7. Solve $\dfrac{(\cos x)(1 + \cos 2x)}{2 \cos x + \sin 2x} = \dfrac{3}{2}$.

8. Write a few sentences comparing the graphs of $y = \sec^2 x$ and $y = \tan^2 x + 1$.

9. Graph each function.

a. $f(x) = \sin x \cos x$

b. $f(x) = \frac{(\sin x \cos x)}{x}$

c. $f(x) = x\sqrt{1 - \sin^2 x}$

10. Assume that the birth rate for an animal species is given by

$$B = \cos^2 t + 2 \cos t$$

and the death rate is

$$D = -\cos 2t$$

where t is the time in months.
Find the smallest positive value of t for which the birth and death rates are equal.

11. What is the frequency of the beat when sounds of frequency f_1 and f_2 interfere with each other?

12. Suppose two tuning forks are struck simultaneously and one produces a sound with frequency 40 Hz and the other a sound with frequency 50 Hz. How often do the beats occur? Use a computer or calculator to graph each of the sound waves separately; also graph their sum. Write the sum as a product. Does your graph yield the results you would expect?

13. The frequency of the foghorn on a steamboat is 80 Hz when heard by people standing on that steamboat. Suppose that you are on another steamboat that is moving toward the first boat. The pitch of the first boat's foghorn is 10% higher to you, so the frequency you hear is 88 Hz. (This effect of a rise in pitch by an approaching sound source is also commonly heard with trains.) Your steamboat has an identical foghorn on it; you hear its sound at a frequency of 80 Hz. Generate a graph representing the combined sound of the two foghorns.

14. Follow these steps to prove the identity for $\cos(a - b)$. The first unit circle in Figure 70 shows an arc of length a beginning at $(1, 0)$ and ending at P. An arc of length b also begins at $(1, 0)$; it ends at point Q (assume $0 < b < a < 2\pi$). The coordinates of P are thus $(\cos a, \sin a)$ and the coordinates of Q are $(\cos b, \sin b)$. The arc from P to Q has length $a - b$.

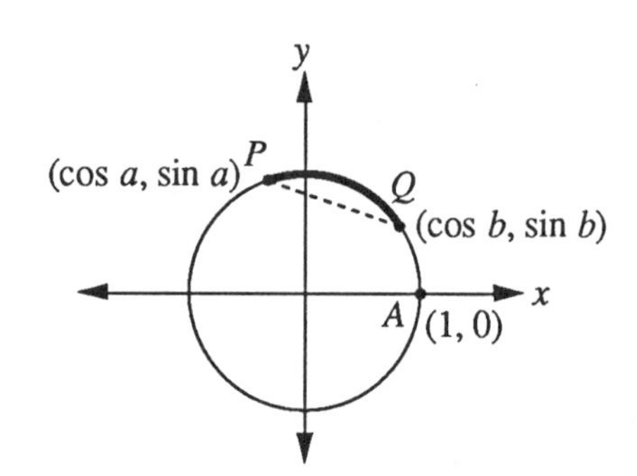

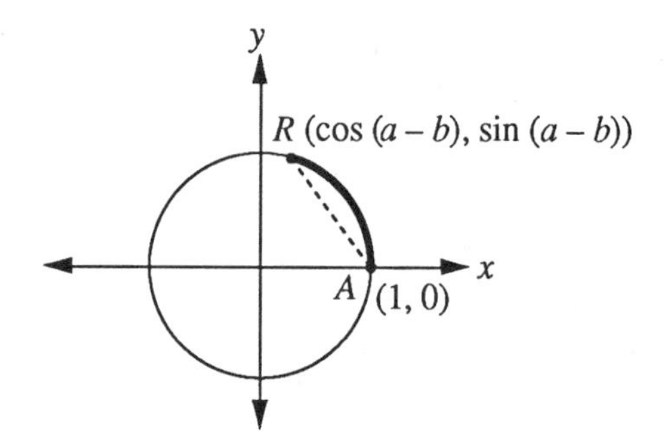

Figure 70 Derivation of Identity for $\cos(a - b)$

The second circle in Figure 70 shows an arc of length $a - b$ beginning at $(1, 0)$ and ending at R. The coordinates of R are thus

$$(\cos(a - b), \sin(a - b)).$$

Note that chord PQ in one circle has the same length as chord AR in the other.

a. Use the distance formula to express the chord lengths PQ and AR in terms of a and b.

b. Set the two expressions equal to each other and simplify. (Recall that $\sin^2 x + \cos^2 x = 1$.)

15. Use the identity $\cos(a - b) = \cos a \cos b + \sin a \sin b$ to derive each of the following identities.

a. $\cos(a + b) = \cos a \cos b - \sin a \sin b$ (**Hint**: $a + b = a - (-b)$.)

b. $\cos 2a = \cos^2 a - \sin^2 a$ (**Hint**: $2a = a + a$.)

c. $\sin(a - b) = \sin a \cos b - \cos a \sin b$ (**Hint**: $\sin x = \cos(x - \frac{\pi}{2})$ so $\sin(a - b) = \cos((a - b) - \frac{\pi}{2}) = \cos(a - (b + \frac{\pi}{2}))$.)

d. $\sin(a + b) = \sin a \cos b + \cos a \sin b$

e. $\sin(2a) = 2 \sin a \cos a$

16. Use the results of the previous exercise to show that $\cos 2a = 1 - 2\sin^2 a$ and that $\cos 2a = 2\cos^2 a - 1$.

12

Solving Triangles with Trigonometry

In Section 5, we provided right triangle definitions of the trigonometric functions. You should recall that for an acute angle θ in a right triangle, we defined the following:

$$\sin \theta = \frac{\text{side opposite } \theta}{\text{hypotenuse}}$$

$$\cos \theta = \frac{\text{side adjacent to } \theta}{\text{hypotenuse}}$$

$$\tan \theta = \frac{\text{side opposite } \theta}{\text{side adjacent to } \theta}$$

These ratios enable us to determine the measures of all parts of a right triangle if we know the measure of at least three parts including the right angle. Recall from your study of geometry that at least one of the known parts must be the measure of a side, since three angles do not determine a unique triangle. The process of determining the measures of all unknown angles and sides is called *solving a triangle.*

Example 1

During an air show, a stunt pilot flies over a crowd of spectators at an altitude of 2500 feet. One minute later the *angle of elevation* of the plane is 15°. The angle of elevation is formed by two rays originating at an observer's eye. One ray extends horizontally; the other is the line of sight from the observer to an elevated object. If the pilot maintains an altitude of 2500 feet and a constant speed, how fast is the plane flying? Ignore the height of the observer for this problem.

Solution To better understand this problem we should make a sketch of the given information. In Figure 71, $\angle CAD$ is the angle of elevation and BC represents the distance traveled by the plane during the minute it was observed. If we find the distance traveled in one minute, we can convert feet per minute to miles per hour to obtain the speed of the plane. Since $\angle BAC$ and $\angle CAD$ are complementary, we know that $\angle BAC = 90° - 15° = 75°$. Using right triangle ABC and the

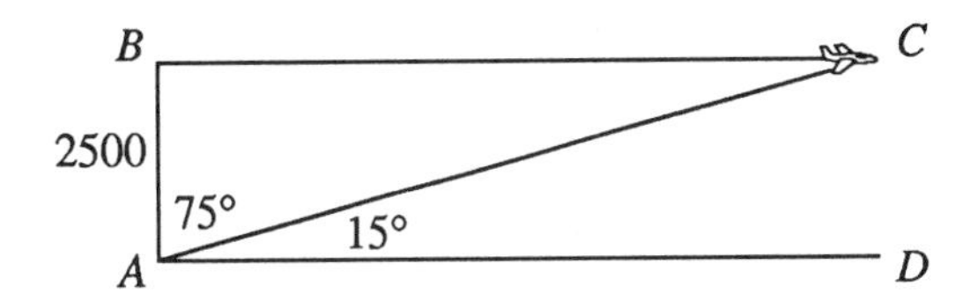

Figure 71 Diagram for Example 1

tangent ratio, we can now solve for side BC as follows:

$$\tan \angle BAC = \frac{BC}{AB}$$

$$\tan 75^\circ = \frac{BC}{2500 \text{ ft}}$$

$$BC = (2500 \text{ ft}) \tan 75^\circ \approx 9330 \text{ ft}.$$

The plane travels about 9330 feet in one minute. We can convert to miles per hour as follows:

$$\frac{9330 \text{ ft}}{1 \text{ min}} = \frac{9330 \text{ ft}}{1 \text{ min}} \cdot \frac{60 \text{ min}}{1 \text{ hr}} \cdot \frac{1 \text{ mile}}{5280 \text{ ft}} \approx 106 \text{ mph}.$$

■

Example 2

Several minutes later another plane, flying at a constant speed, passes over the crowd at an altitude of 2500 feet. When it is directly above the crowd it begins ascending. One minute later the angle of elevation to the plane is 40°, and the plane's altitude is 7200 feet. Find the speed of the plane during its ascent.

Solution The given information is illustrated in Figure 72.

The plane ascends along side AB of $\triangle ABC$, so this is the side whose measure we need. Since $\angle BCD$ and $\angle ACB$ are complementary, we know that $\angle ACB = 50^\circ$. We need at least three parts of a triangle in order to solve it, but we have only one angle and one side of $\triangle ABC$. Using right $\triangle BCD$, we can solve $\sin 40^\circ = \frac{7200}{BC}$ to determine that $BC \approx 11,201$. Now we know two sides and an included angle of $\triangle ABC$. We cannot use right triangle ratios to solve for side AB, however, since $\triangle ABC$ is not a right triangle.

We will complete the solution of this problem after we develop methods for solving triangles that are not right triangles; such triangles are called *oblique*. ■

Law of Cosines

The Law of Cosines can be used to solve some oblique triangles, such as $\triangle ABC$ in Figure 73. By convention, we say that the side opposite $\angle A$ has length a.

Assume that the lengths of sides b and c and the measure of angle A are known. How can we find the length of side a? Draw the altitude h from $\angle B$ to side b. This will divide side b into two segments whose lengths are x and $(b - x)$.

In right triangle BCD the Pythagorean Theorem gives $(b - x)^2 + h^2 = a^2$; in right triangle ABD we have $x^2 + h^2 = c^2$. We can use the second equation to substitute for h^2 in the first equation and obtain $(b - x)^2 + c^2 - x^2 = a^2$. Squaring $(b - x)$, we obtain $b^2 - 2bx + x^2 + c^2 - x^2 = a^2$, or $a^2 = b^2 + c^2 - 2bx$. We would like to express x in terms of known quantities. In right triangle ABD, $\cos A = \frac{x}{c}$ so $x = c \cos A$. Substituting

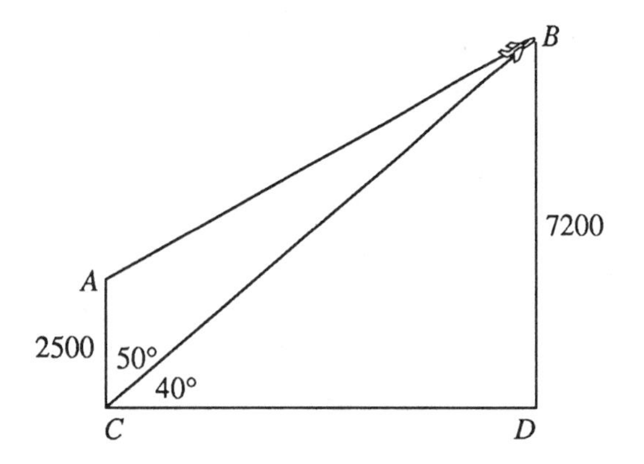

Figure 72 Diagram for Example 2

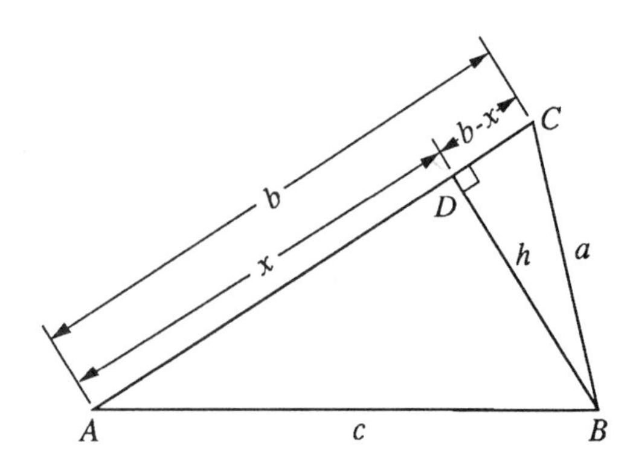

Figure 73 Oblique Triangle ABC

yields $a^2 = b^2 + c^2 - 2bc\cos A$. This is the Law of Cosines.

The Law of Cosines can be represented with cyclic replacements of a, b, and c. Three versions of this law are stated below:

$$a^2 = b^2 + c^2 - 2bc\cos A$$
$$b^2 = a^2 + c^2 - 2ac\cos B$$
$$c^2 = b^2 + a^2 - 2ab\cos C.$$

Example 3

We will now complete the work we began in Example 2 to determine the distance traveled during the first minute of the plane's ascent. Once we have found the distance, we will convert the speed from feet per minute to miles per hour. The diagram of the triangle we must solve is shown in Figure 74.

Solution Using the Law of Cosines we solve for side AB as follows:

$$AB^2 = 2500^2 + 11{,}201^2 - 2(2500)(11{,}201)\cos 50°$$
$$AB \approx 9783 \text{ ft}.$$

We know now that the plane travels about 9783 feet during the first minute of its ascent. We can convert to miles per hour as follows:

$$\frac{9783 \text{ ft}}{1 \text{ min}} = \frac{9783 \text{ ft}}{1 \text{ min}} \cdot \frac{60 \text{ min}}{1 \text{ hr}} \cdot \frac{1 \text{ mile}}{5280 \text{ ft}} \approx 111 \text{ mph}.$$

■

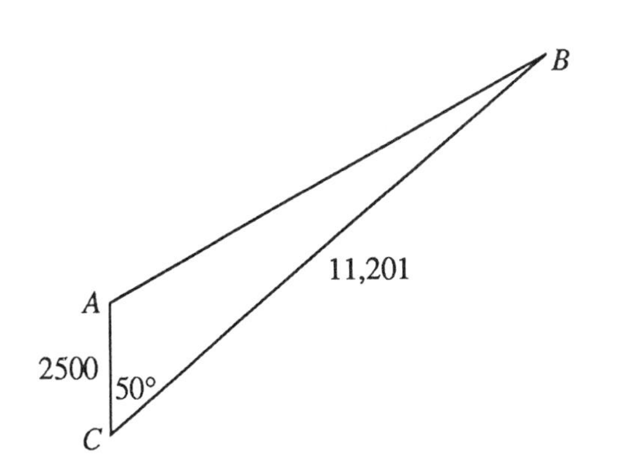

Figure 74 Triangle for Example 3

Law of Sines

The Law of Cosines is useful for solving many oblique triangles, but it does not help in all situations. Suppose, for example, that you know two angles and one side of an oblique triangle. You simply have too many missing pieces to use the Law of Cosines, but the Law of Sines can be used.

Consider the oblique triangle ABC in Figure 75. An altitude h has been drawn from $\angle B$ to side b. In triangle ABD, $\sin A = \frac{h}{c}$; in triangle BCD, $\sin C = \frac{h}{a}$. Rewriting these ratios as $c\sin A = h$ and $a\sin C = h$, we see that $c\sin A = a\sin C$, or $\frac{\sin A}{a} = \frac{\sin C}{c}$.

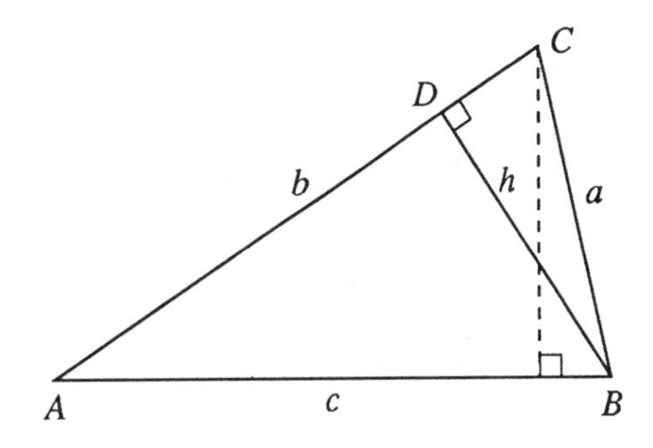

Figure 75 Oblique Triangle ABC

Similarly, if an altitude is drawn from C, it can be shown that $\frac{\sin B}{b} = \frac{\sin A}{a}$. Combining these results we have the Law of Sines:

$$\frac{\sin A}{a} = \frac{\sin B}{b} = \frac{\sin C}{c}.$$

Example 4

Margaret and John want to sail their boat from a marina to an island 15 miles east of the marina. Along the path there are several small islands that they must avoid. When charting the course, they decide to sail first on a heading of 70° and then on a 120° heading. (*Headings* are angle measures rotated clockwise from the north.) What is the total distance they will travel before reaching their destination?

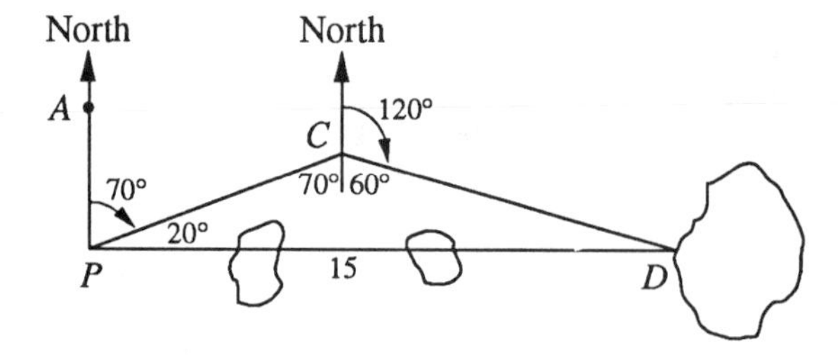

Figure 76 Diagram for Example 4

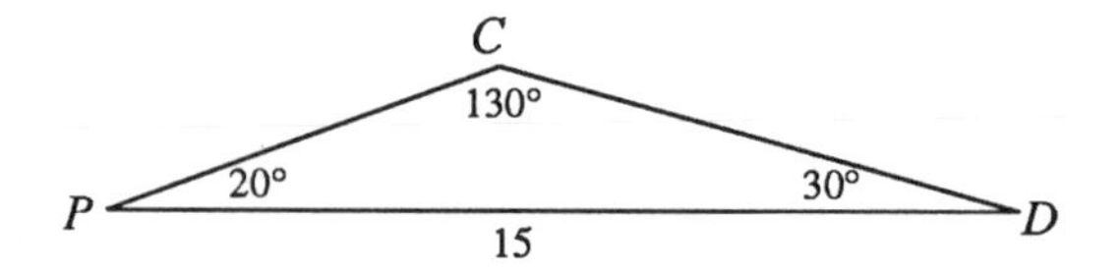

Figure 77 Triangle for Example 4

Solution The drawing shown in Figure 76 displays the information provided in this problem. Since $\angle APC = 70°$ and is complementary to $\angle CPD$, we know that $\angle CPD = 20°$. The measure of $\angle PCD$ can be obtained by summing 70°, obtained from an angle that forms an alternate interior pair with $\angle APC$, and 60°, obtained as the supplement of the 120° angle that gives the second heading. We know that the sum of the angles of the triangle must be 180°, so we can use the fact that $\angle D = 30°$ in our calculations. We want to find the lengths of sides PC and CD in the triangle shown in Figure 77.

We know only one side, so we need to use the Law of Sines to find the other two:

$$\frac{\sin 20°}{CD} = \frac{\sin 130°}{15}$$

$$CD = \frac{15}{\sin 130°} \cdot \sin 20°$$

$$\approx 6.7 \text{ miles}$$

$$\frac{\sin 30°}{PC} = \frac{\sin 130°}{15}$$

$$PC = \frac{15}{\sin 130°} \cdot \sin 30°$$

$$\approx 9.8 \text{ miles}$$

Adding the lengths of these sides, we conclude that they traveled approximately 16.5 miles to reach their destination. ■

Example 5

A triangular piece of land in a park is to be made into a flower bed. Stakes have previously been driven into the ground at the vertices of the triangle, which we will call B, E, and D, but the gardener can locate only the two stakes at B and E. BE measures 6.2 meters, and the gardener recalls that the angle at B is 60° and that the side opposite the 60° angle was to be 5.5 meters in length. Based on this information, where should the gardener search for the missing stake?

Solution Figure 78 shows the information given in this problem. Notice that the conditions given do not determine a unique triangle; there are two triangles that satisfy all of the conditions stated in this problem. This situation can occur when you know two sides and a non-included angle of an oblique triangle. It is often referred to as the *ambiguous case* and is explained in more detail at the end of this example.

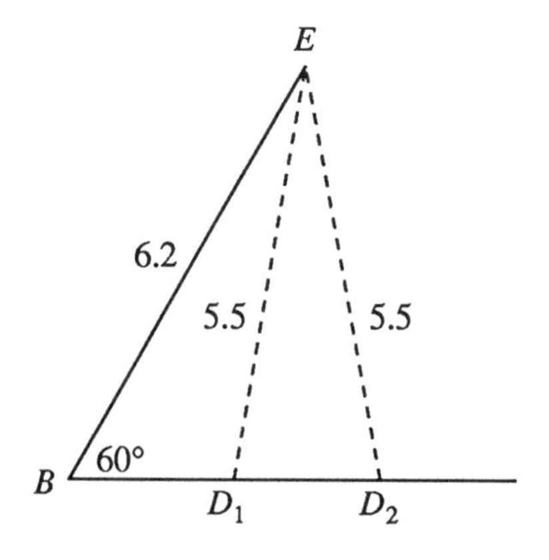

Figure 78 Triangle for Example 5

When we attempt to use the Law of Sines to locate D, we have

$$\frac{\sin 60^\circ}{5.5} = \frac{\sin D}{6.2}$$

$$(6.2)\frac{\sin 60^\circ}{5.5} = \sin D$$

$$\sin D \approx 0.976.$$

We know that there are two angles between 0° and 180° with sine equal 0.976; one is approximately 77.5°, and the other is approximately $180^\circ - 77.5^\circ = 102.5^\circ$. Judging from our diagram, $\angle D_1 \approx 102.5^\circ$ and $\angle D_2 \approx 77.5^\circ$.

We can now examine each triangle to find the possible locations of the third stake. In $\triangle BED_1$ (see Figure 79), $\angle E = 180^\circ - (60^\circ + 102.5^\circ) = 17.5^\circ$ and

$$\frac{BD_1}{\sin 17.5^\circ} = \frac{5.5}{\sin 60^\circ}$$

so $BD_1 \approx 1.9$ meters. In $\triangle BED_2$, $\angle E = 180^\circ - (60^\circ + 77.5^\circ) = 42.5^\circ$ and

$$\frac{BD_2}{\sin 42.5^\circ} = \frac{5.5}{\sin 60^\circ}$$

so $BD_2 \approx 4.3$ meters. ∎

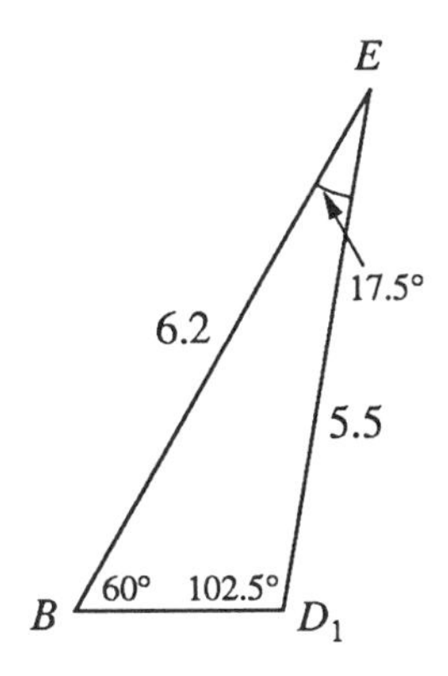

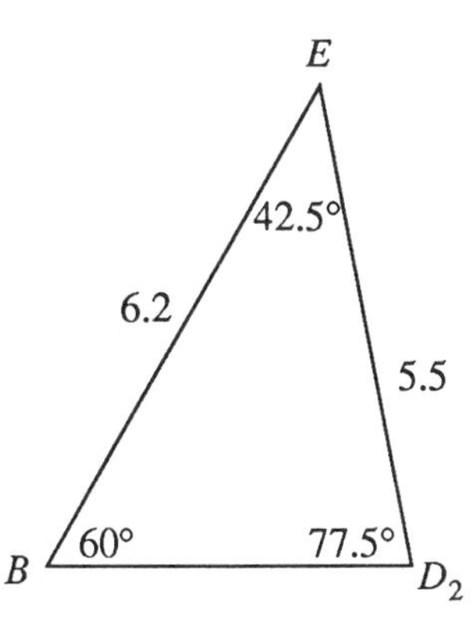

Figure 79 Two Triangles for Example 5

In Example 5 we solved a triangle for which we were given two sides and a non-included angle. We will now explore the SSA case in more detail. What is "ambiguous" about such triangles? When you are given sides a and b and acute angle A, how does side a fit to form the triangle? There are four distinct possibilities; the first three listed are illustrated in Figure 80.

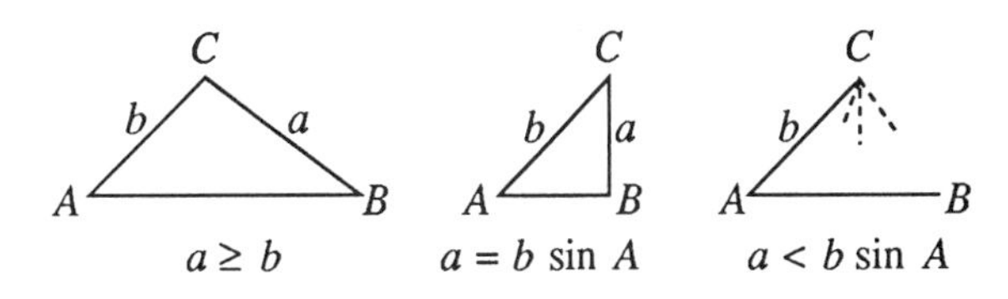

Figure 80 Ambiguous Case Triangles

Case 1: If side a is equal to or longer than side b, then only one triangle can be made. To swing a back to the left would not produce a triangle. So if $a \geq b$, there is exactly one triangle.

Case 2: If a is exactly long enough to touch the opposite side, a right triangle is formed. In this case $\sin A = \frac{a}{b}$, or $a = b \sin A$.

Case 3: Side a may not be long enough to form a triangle. When given SSA, first check to see if $a \geq b$. If it is not, then evaluate $b \sin A$. If $a < b \sin A$, then a is not long enough to form a right triangle (since the shortest distance between a point and a line is the perpendicular distance), so no triangle is formed.

Case 4: Suppose $b \sin A < a < b$. There are two ways for a to form a triangle. In Figure 81 you can

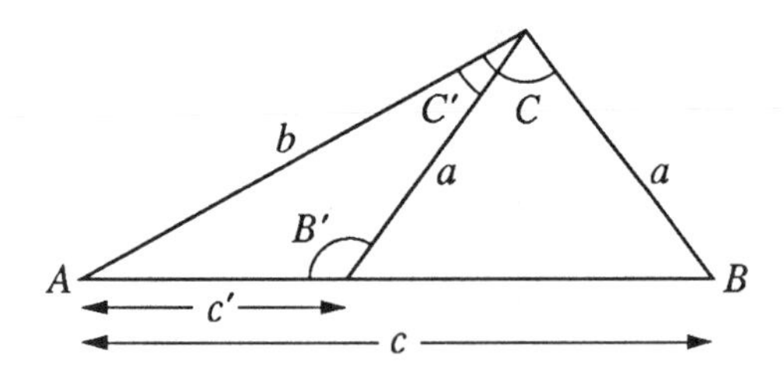

Figure 81 Ambiguous Case Triangle: $b \sin A < a < b$

examine the two triangles that satisfy this case: triangle ABC and triangle $AB'C'$.

When you are given two sides and a non-included angle and asked to solve a triangle, your first task is to determine the number of triangles that can be formed. Then you can use either the Law of Cosines or the Law of Sines to solve the triangle(s). Notice that when the non-included angle A is $90°$ or greater, there will be at most one triangle formed. Side a must be longer than side b since the longest side must be opposite the largest angle.

In Example 5 we knew the measures of two sides and a non-included angle. Since

$$6.2 \sin 60° < 5.5 < 6.2$$

we could have anticipated that there were two possible triangles.

Exercise Set 12

1. Solve the following triangles for the remaining parts.

a. $c = 8,\ B = 50°,\ a = 10$

b. $a = 6,\ b = 12,\ A = 30°$

c. $B = 30°,\ c = 12,\ b = 7$

d. $a = 3,\ b = 5,\ c = 7$

2. A ladder 28 feet long leans against the side of a building. If the angle between the ladder and the building is $20°$, how far does the top of the ladder move down the building if the distance between the building and the ladder increases by 2 feet?

3. An airplane pilot wishes to maintain a true course with heading $250°$ and with a ground speed of 400 mph. If the wind is blowing directly south at 4 mph, approximate the required air speed and compass heading of the plane.

4. Two people come to a fork in a road. One walks down one straight branch at 3 mph and the other walks down the other straight branch at 3.5 mph. If the angle between the roads is $30°$, find the distance between the two people after one hour.

5. A builder wants to build a ramp 30 feet long which rises to a height of 6 feet above level ground. What angle should the ramp make with the horizontal?

6. A triangular plot of land has sides of length 400 feet, 350 feet, and 100 feet. Find the smallest angle between the sides.

7. A jogger running at a constant speed of one mile every ten minutes runs in the direction S30°E for 30 minutes, and then in the direction of N40°E for the next 15 minutes. Approximate, to the nearest tenth of a mile, the distance from the jogger to the starting point. (To obtain the proper angle for S30°E, begin by facing south and rotate $30°$ towards the east.)

8. Suppose you want to install a vertical TV tower on the side of a hill which makes a $25°$ angle with the horizon. The tower is 75 m high and guy wires are to be attached $\frac{2}{3}$ of the way up the tower. If the guy wires are anchored 30 m up the hill and 35 m down the hill from the tower, find the length of the guy wires (see Figure 82).

9. A balloon is sighted from two points on level ground. From point A the angle of elevation is $18°$. From point B the angle of elevation is $12°$. If points A and B are 8.4 miles apart, what is the height of the balloon?

10. Engineers want to measure the distance from P to Q, but the span from P to Q is across the tip of a lake. So they select a point R on land and find that the distance from R to Q is 100

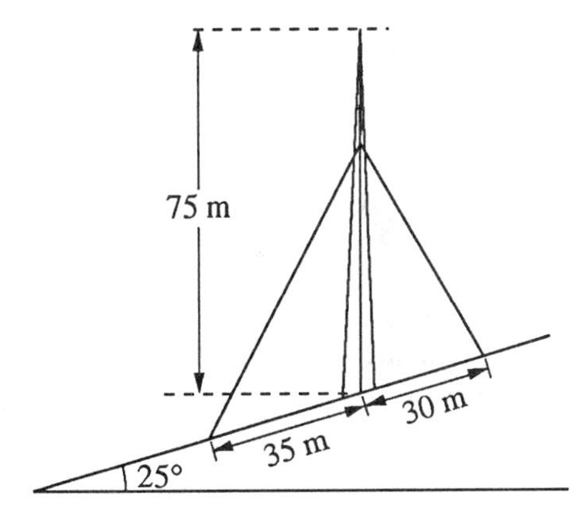

Figure 82 TV Tower with Guy Wires

feet and from R to P is 120 feet. Angle QPR measures 47°. How far is it from P to Q?

11. A golfer takes two putts to get the golf ball into the hole. The first putt rolls the ball 10.2 ft in the northwest direction, and the second putt sends the ball due north 3.7 ft into the hole. How far, and in what direction, should the golfer have aimed the first putt to get the ball into the hole with one stroke? (Assume that the green is level.)

12. If you know two sides and an included angle of a triangle (a, b and $\angle C$), then the area of the triangle is given by $0.5ab\sin C$. Convince yourself that this is true. If you are given values of a and b, what value of $\angle C$ provides the maximum area?

13. Suppose an airplane is traveling toward an observer at 200 mph (293 ft/sec) at an altitude of 3000 feet. Eventually, the plane will pass directly over an observer on the ground.

a. If sound traveled infinitely fast, then the sound that the plane made when its angle of elevation was 20° would be heard immediately. However, since sound travels at 1100 feet/sec, if the observer looks in the direction of 20° upon hearing the sound, the plane will no longer be there. At what angle should the observer look to see the plane?

b. Suppose the observer views the plane at an angle of 25° above the horizon and simultaneously hears the sound of its engines. At what angle was the plane above the horizon when it made the sound the observer hears?

14. Use the Law of Cosines to solve the problem in Example 5.

15. Lines of latitude run east to west around the earth. The equator is the 0° line of latitude. Longitudinal lines run from north to south around the earth through the North and South Poles (see Figure 83). Assume that the earth is a sphere with radius 3960 miles.

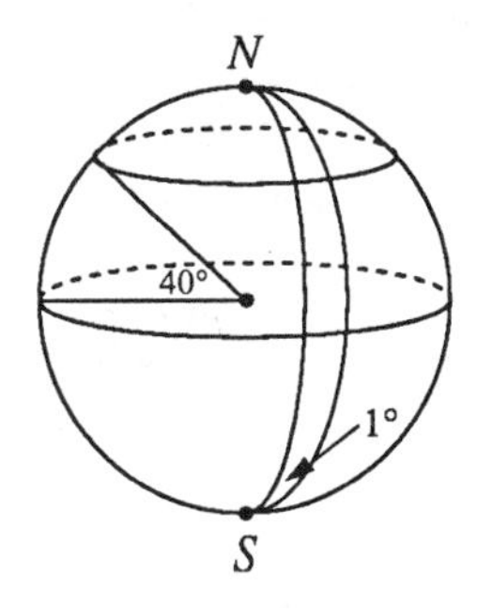

Figure 83 Latitude and Longitude

a. What is the length, in miles, of one degree of longitude at the equator?

b. What is the length, in miles, of one degree of longitude at 36° N?

c. What is the length, in miles, of one degree of longitude at 40° S?

d. At what latitude does one degree of longitude encompass 30 miles?

13

A Trigonometric Model for Geology

Geologists have developed techniques that allow them to determine the composition and depth of layers under the earth's surface without drilling. One technique of this type is called *controlled source seismology*. It consists of creating a sound on the earth's surface and measuring the time required for the sound to reach a second location. When geologists combine this information with knowledge about the speed of sound in different media they can estimate the depth and composition of subsurface layers. In this section we will develop a mathematical model for this geological technique.

The technique is based on the fact that sound travels at different speeds in different media. For instance, sound travels 330 meters per second in air and 1450 meters per second in water. In general, sound travels faster through more compact media and slower through less compact media. Since sound travels at known speeds in different media, finding the speed of the sound in an unknown material allows seismologists to identify the material.

The simplifying assumptions for this model are as follows. Assume that a layer of earth of unknown, uniform thickness lies above a second layer of unknown composition (see Figure 84). The second layer is adjacent to the bottom of the upper surface and d meters below the surface. If a noise source is set up at point A on the earth's surface, sound waves will spread in all directions from A. Some sound waves will travel straight across to the receiver at point B, L units away. Other waves will travel down through material M_1 until they reach the adjacent material M_2. Here the sound waves will be bent, or refracted. Some waves will be refracted so that they travel along the interface between the two materials and then back up to the surface at B.

Suppose the velocity of sound in material M_1 is v_1 and the velocity along the interface of M_2 is v_2; suppose further that $v_2 > v_1$. Two major sound waves will reach point B, one traveling along the top of the upper layer at speed v_1, and the other traveling to the lower medium at speed v_1, along the interface at speed v_2, and back to the surface at speed v_1. Our goal is to determine d, the depth of the unknown layer below the earth's surface, as well as v_2, the velocity of sound in the unknown material. Knowing v_2 allows seismologists to make an educated guess about the composition of material M_2.

The sound wave that travels along the upper layer from A to B travels a distance L at speed v_1. It is easy to determine v_1 since we can measure both L and the travel time t_1:

$$v_1 = \frac{L}{t_1}.$$

The other major sound wave that reaches B travels down through M_1 and is refracted at the interface between M_1 and M_2. This sound travels along three distinct line segments; we can use right triangle trigonometry to write an expression for the total travel time along this path. Suppose sound takes t_2 seconds to travel from A to B along the path shown in Figure 85. The time of travel for each segment is equal to the distance traveled

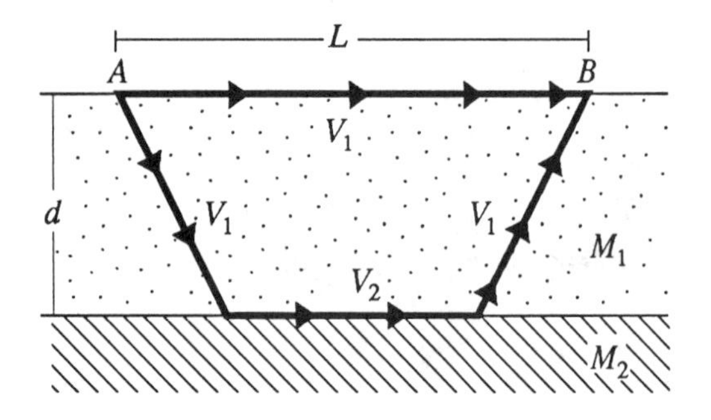

Figure 84 Sound Traveling from A to B

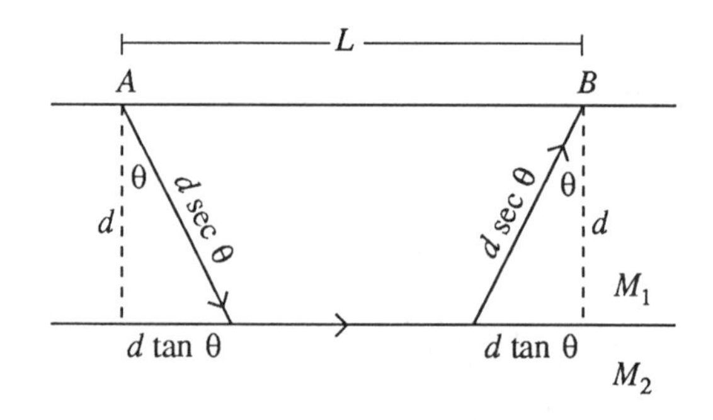

Figure 85 Path from A to B

divided by the velocity in that medium. Thus,

$$t_2 = \frac{d\sec\theta}{v_1} + \frac{L - 2d\tan\theta}{v_2} + \frac{d\sec\theta}{v_1}.$$

Simplifying this equation yields

$$t_2 = \frac{2d\sec\theta}{v_1} + \frac{L - 2d\tan\theta}{v_2}$$

$$t_2 = \frac{L}{v_2} + 2d\left[\frac{\sec\theta}{v_1} - \frac{\tan\theta}{v_2}\right].$$

The preceding equation expresses the relationship between six variables: t_2, d, θ, v_1, L, and v_2. The quantities t_2 and L can be measured directly, and v_1 can be determined by computing the ratio of L and t_1 (the travel time for the sound wave that travels along the surface of material M_1). The variables d and v_2 represent the thickness of M_1 (which is the same as the depth of M_2 below the surface) and the speed of sound in M_2; these are the quantities that allow seismologists to identify the depth and composition of the subsurface layer. What about the variable θ? The Class Practice exercises will explore how θ is related to the other variables in our model.

Class Practice

A sound is made on the earth's surface at point P. Let the horizontal distance from P to Q be 100 meters and the vertical distance be 20 meters (see Figure 86). Furthermore suppose that the speed of sound in material M_1 is v_1 and in material M_2 the speed is v_2 with $v_2 > v_1$. What path from P to Q will minimize the travel time of sound waves? The shortest path is clearly the straight line from P to Q, but this is not the least time path since the sound is traveling at the slower speed v_1.

The total time required for a sound wave to travel from P to Q is the sum of the travel times in the two materials. We can find these travel times by dividing the distance by the velocity in each material. Thus,

$$T = \frac{20\sec\theta}{v_1} + \frac{100 - 20\tan\theta}{v_2}.$$

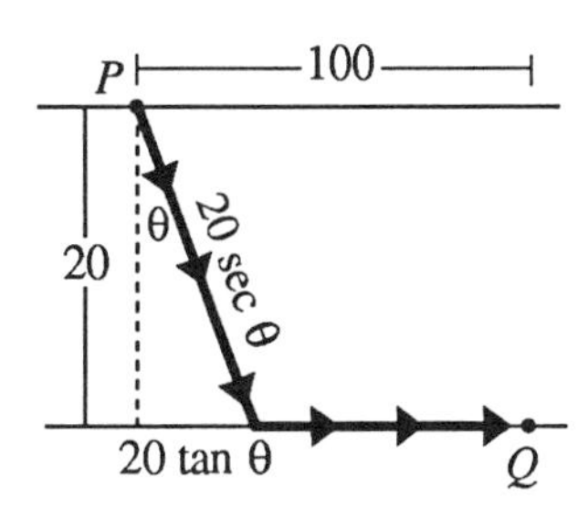

Figure 86 Path from P to Q

For particular values of v_1 and v_2, T is a function of θ.

1. For each pair of values for v_1 and v_2, use a computer or graphing calculator to find the value of θ that minimizes the total travel time T. Then complete the chart.

v_1	v_2	θ	$\frac{v_1}{v_2}$	$v_1 \times v_2$	$\tan\theta$	$\sin\theta$
10	20					
5	10					
15	20					
3	4					
17	20					
7	8					
1	4					
50	55					

2. What appears to be the relationship between v_1, v_2, and θ?

The Class Practice exercises indicate that $\sin\theta = \frac{v_1}{v_2}$; this is a special case of Snell's Law. Now we will use the relationship between v_1, v_2, and θ to express $\sec\theta$ and $\tan\theta$ in terms of v_1 and v_2. The triangle in Figure 87 is based on the relationship $\sin\theta = \frac{v_1}{v_2}$. Using the triangle we can conclude that

$$\sec\theta = \frac{v_2}{\sqrt{(v_2)^2 - (v_1)^2}}$$

and

$$\tan\theta = \frac{v_1}{\sqrt{(v_2)^2 - (v_1)^2}}.$$

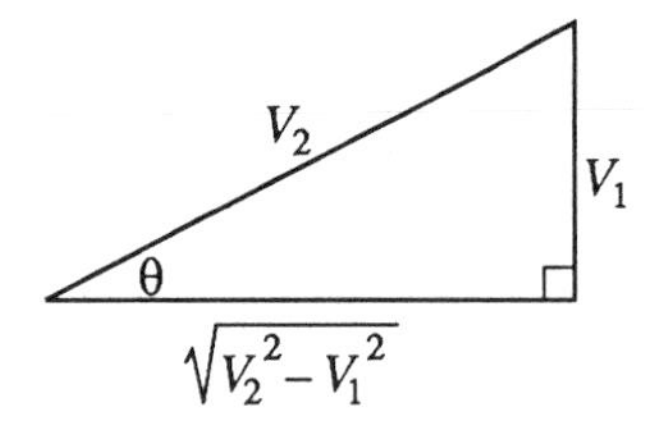

Figure 87 $\sin\theta = \frac{v_1}{v_2}$

Substituting for $\sec\theta$ and $\tan\theta$ in our previous equation for t_2 and then simplifying gives:

$$t_2 = \frac{L}{v_2} + 2d\left[\frac{\sec\theta}{v_1} - \frac{\tan\theta}{v_2}\right]$$

$$t_2 = \frac{1}{v_2}L + 2d\left[\frac{v_2}{v_1\sqrt{(v_2)^2-(v_1)^2}} - \frac{v_1}{v_2\sqrt{(v_2)^2-(v_1)^2}}\right]$$

$$= \frac{1}{v_2}L + 2d\left[\frac{(v_2)^2-(v_1)^2}{v_1v_2\sqrt{(v_2)^2-(v_1)^2}}\right]$$

$$= \frac{1}{v_2}L + 2d\left[\frac{\sqrt{(v_2)^2-(v_1)^2}}{v_1v_2}\right].$$

Notice that t_2 is a linear function of L. The preceding equation is in the form $t_2 = mL + b$ where $m = \frac{1}{v_2}$ and $b = \frac{2d\sqrt{(v_2)^2-(v_1)^2}}{v_1v_2}$. Solving for v_2 and d yields

$$v_2 = \frac{1}{m}$$

and

$$d = \frac{bv_1v_2}{2\sqrt{(v_2)^2-(v_1)^2}}.$$

Let's take stock of where we are. Our goal is to find v_2 and d, which represent the velocity of sound in the unknown layer and the depth of the unknown layer below the earth's surface. We can measure L and t_1 and t_2, we know that $v_1 = \frac{L}{t_1}$, and we have shown algebraically that t_2 is a linear function of L. If we can determine the slope m and y-intercept b of this linear function, then v_2 will be equal to $\frac{1}{m}$ and d will be equal to $\frac{bv_1v_2}{2\sqrt{(v_2)^2-(v_1)^2}}$.

How can we determine the slope and intercept of the linear function? By generating sounds at point A and varying the distance L to the receiver at B, we can measure the times t_1 and t_2 for each L and therefore estimate the value of v_1. Then we can use data analysis techniques to find an equation for a line that fits the ordered pairs (L, t_2).

Seven measurements of time and distance are given in Table 88.

Table 88 Data for Identification of Rock Formations

Distance (L) in meters	Time (t_1) in seconds	Time (t_2) in seconds
4000	2.035	0.941
5000	2.477	1.145
6000	3.033	1.341
7000	3.504	1.558
8000	3.969	1.701
9000	4.461	1.960
10000	5.032	2.203

We know that $v_1 = \frac{L}{t_1}$. Each pair of values for L and t_1 estimates a value for v_1. Due to measurement error the values we calculate for v_1 vary; we will use the average value for v_1 which is 1997 meters per second.

From the data in Table 88, we can generate the least-squares line shown in Figure 89. The slope m of the line through the ordered pairs (L, t_2) is 0.000206286 and the y-intercept b is 0.105857.

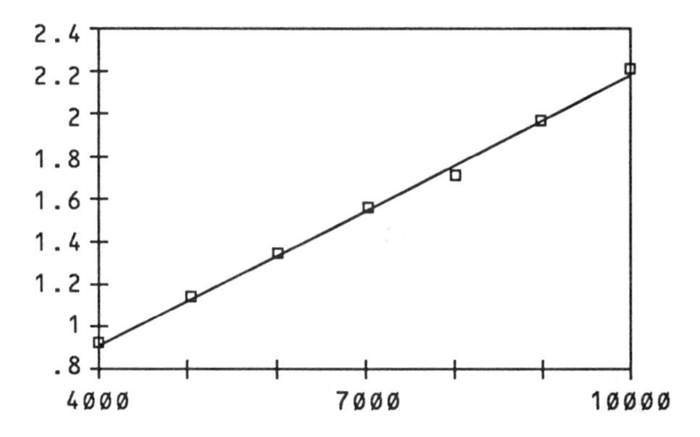

Figure 89 Line Through Ordered Pairs (L, t_2)

Now we can use the slope and intercept of this line to approximate the values of v_2 and d. We know that $v_2 = \frac{1}{m}$, so $v_2 \approx \frac{1}{0.000206286} \approx 4848$ meters per second. Substituting the values $v_1 = 1997$ meters per second, $v_2 = 4848$ meters per second, and $b = 0.105857$ in the equation

$$d = \frac{bv_1v_2}{2\sqrt{(v_2)^2 - (v_1)^2}}$$

gives $d \approx 116.0$ meters.

The fact that $v_2 \approx 4848$ meters per second allows a geologist to make an educated guess about the composition of the material M_2. A collection of known velocities is given in Table 90. The velocity of sound in the same material increases as the depth beneath the surface increases and the material becomes more compressed. For instance, the velocity of sound in shale is about 2300 meters per second near the earth's surface and is 4700 meters per second at a depth of 1800 meters. Also notice that a material that varies in consistency (such as sandstone or limestone) has a wide range of velocities associated with it; a material whose consistency is more uniform (such as granite) has less variation in velocities.

A velocity of about 4848 meters per second could be associated with several different materials. Since the layer is shallow (the depth is only about 116 meters), the material will be less compressed and therefore the velocity given should be at the lower part of the range. What material do you think is below the surface?

Table 90 Velocities Associated with Various Materials

Material	**Velocity in meters/sec**
Sand and Soil	110 to 185
Sandstone	1400 to 4300
Shale and Slate	2300 to 4700
Limestone	2800 to 6400
Granite	4800 to 5600
Basalt	5100 to 5600
Dunite	7400 to 8600

Source: *UMAP Unit 292-293*, COMAP, INC., 1980

Table 91 Data for Exercise 1

Distance (L) in meters	**Time (t_1) in seconds**	**Time (t_2) in seconds**
4000	2.106	2.097
5000	2.501	2.453
6000	3.004	2.786
7000	3.494	3.155
8000	4.132	3.491
9000	4.546	3.850
10000	4.983	4.187

Exercise Set 13

1. A layer of unknown composition lies below the earth's surface. Use the data given in Table 91 and the model we have developed to determine the composition of the medium and how far below the surface the layer is located.
2. Suppose there is a third layer below the first two. In this case three major sounds will be heard at B: one through the uppermost layer, a second refracting off the middle layer, and a third refracting off the lowest layer (see Figure 92). Develop a model to determine the depth and composition of each of the two lower layers.

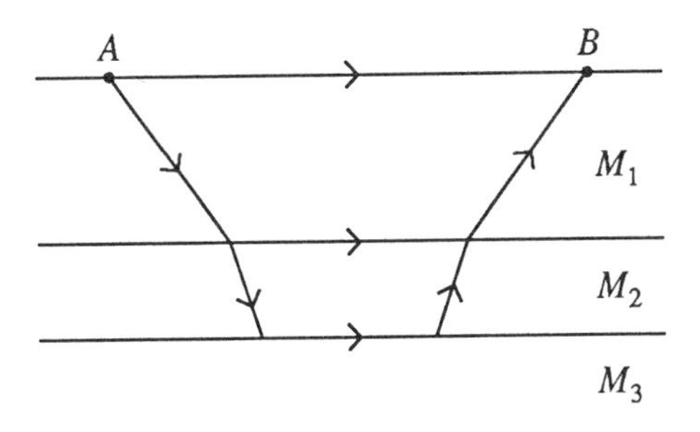

Figure 92 Two Unknown Layers Below the Surface

CHAPTER 8

Matrices

1

Introduction

Three major countries that produce cars for sale in the U.S. are Japan, Germany, and the U.S. itself. When it is time to buy a new car, people will choose a car based in part on the satisfaction they have received from the car they presently own. Suppose that of car buyers who presently own a U.S. car, 55% will purchase another American-made car, 25% will buy a Japanese-made car, 10% will buy a German-made car, and 10% will buy a car made in none of these three countries. Of those who presently own a Japanese car, 60% will buy another Japanese car, whereas 25% will buy American, 10% German, and 5% none of the three. Of those car buyers who own a German car, 40% will again purchase a German car, 35% will switch to American cars, 15% will switch to Japanese cars, and 10% will buy from another country. Of those who presently own a car from a country other than the three major producers, 20% will switch to American, 25% will switch to Japanese, 15% will switch to German, and 40% will continue to buy from another country.

The details of this information are difficult to grasp all at once; however, the following display of the data offers distinct advantages over the verbal description given above.

$$T = \begin{array}{c} \\ \textit{U.S.} \\ \textit{Jap.} \\ \textit{Ger.} \\ \textit{Other} \end{array} \begin{array}{c} \begin{array}{cccc} \textit{U.S.} & \textit{Jap.} & \textit{Ger.} & \textit{Other} \end{array} \\ \begin{pmatrix} 0.55 & 0.25 & 0.10 & 0.10 \\ 0.25 & 0.60 & 0.10 & 0.05 \\ 0.35 & 0.15 & 0.40 & 0.10 \\ 0.20 & 0.25 & 0.15 & 0.40 \end{pmatrix} \end{array}$$

The table of numbers T shown above is an example of a *matrix*, in which the numbers are known as *entries*. The *dimension* of a matrix is given by the number of rows and the number of columns, so that T is considered a 4×4 matrix. If a matrix has m rows and n columns, then it is said to be an $m \times n$ matrix. If $m = n$, as in T, then the matrix is called *square*. Individual entries in a matrix are identified by row number and column number, in that order. For example, the number 0.05 is the entry in row 2 and column 4 of T, abbreviated as T_{24}. We also have $T_{32} = 0.15$ and $T_{13} = 0.10$ as two other entries in the matrix T.

Each entry in T has a specific, unique meaning; therefore, the dimension of a matrix cannot be reduced without losing essential information. In T, the rows represent the country of origin of the presently owned car, whereas the columns represent the origin of the next car. T is an example of a *transition matrix*, for it contains information concerning the owner's transition from the present car to the new car. The concept of a transition matrix is essential to the later section on Markov chains.

When mathematics is used to analyze real-world phenomena, the interaction of mathematical concepts and the real world is often based upon data. We apply mathematics to information that is gathered through measurement and observation. The discipline of mathematics includes many concepts that aid us in analyzing and interpreting data, so that data can be used and expressed in summary form. Various data representations exist that allow us to discern trends, to form generalizations, and to make predictions, all on the basis of data. For example, an economist may use a linear function to forecast the growth of an industry. A demographer may use an exponential function to predict when the world population will exceed 10 billion. These two examples are characterized by using mathematics to describe the behavior of a given situation, a process called *mathematical modeling*. Most of the mathematical models studied in school involve real-world phenomena approximated by continuous functions, which have graphs that can be drawn without lifting the pencil from the paper, such as lines and parabolas. A principal advantage of using continuous functions is the availability of a large body of well-developed theoretical results, such as those studied in high school algebra and precalculus courses.

A matrix, on the other hand, can be used with a collection of data that does not lend itself to the use of a continuous model. *Matrices* (the plural of matrix) are important in a larger branch of mathematics called *discrete mathematics*. The term discrete refers to the fact that these techniques of mathematical modeling deal with finite sets of noncontinuous data rather than continuous functions or continuous sets of data. The advantage of using a matrix to organize a set of data is illustrated by the preceding example of purchasing cars. Just as a functional rule like $f(x) = x^2$ is a good continuous model for certain real-world phenomena (the trajectory of a falling object, for example), a matrix is a mathematical tool used to handle noncontinuous data sets in summary form. Because matrices operate with discrete data, they possess dimension; that is, a matrix cannot be reduced to a single number such as the value of a function at a point. A matrix can be thought of as a single entity for the sake of simplicity; however, it is a single entity that contains many data values. In addition, just as special algebraic rules exist for functions, matrices have a special algebra associated with them. Although matrix algebra seems rather peculiar at first, it involves operations defined in ways that allow matrices to be used in many mathematical models. As a complement to the repertoire of continuous techniques emphasized in secondary school mathematics, the following sections introduce methods for using matrices to model discrete data.

2

Matrix Addition and Scalar Multiplication

We begin this section with a problem that illustrates the operations of matrix addition and scalar multiplication.

Example 1

The Hobby-Shop Problem

Suppose that you have a small woodworking shop in your garage and you make toys for children as a hobby. Lately you have begun selling your toys at the local flea market. You make four different kinds of toys, namely a train (t), an airplane (a), a dragon (d), and a nameplate (n). Each of these can be made very plainly out of pine (p), or with greater detail and ornamentation out of oak (o). Let matrices O, N, and D represent your sales for October, November, and December, as given below.

$$O = \begin{matrix} & t & a & d & n \\ p & 3 & 5 & 0 & 2 \\ o & 1 & 2 & 1 & 3 \end{matrix}$$

$$N = \begin{matrix} & t & a & d & n \\ p & 4 & 2 & 1 & 3 \\ o & 1 & 0 & 2 & 4 \end{matrix}$$

$$D = \begin{array}{c} \\ p \\ o \end{array} \begin{array}{c} \begin{array}{cccc} t & a & d & n \end{array} \\ \begin{pmatrix} 4 & 8 & 5 & 3 \\ 4 & 2 & 1 & 6 \end{pmatrix} \end{array}$$

Notice that, as with the transition matrix of the previous example, each entry in matrices O, N, and D has a specific meaning. For example, the number of pine dragons made in December is given by entry D_{13}, which is 5. How many oak trains were made in November? What does O_{24} represent?

We will now examine the following question: how many of each item were sold for the entire three month period?

Solution The total number of pine trains sold is $3+4+4=11$, which is the sum of the numbers in the upper-left corner of each matrix. The total number of oak dragons is $1+2+1=4$, which is the sum of the entries in row 2 and column 3, or

$$O_{23} + N_{23} + D_{23} = 4.$$

We can construct a matrix S that has entries representing the total number of each item sold during the three months. From the pattern established above, we see that each entry of S is the sum of the corresponding entries of O, N, and D. In symbols, we have

$$S_{ij} = O_{ij} + N_{ij} + D_{ij},$$

and S contains the entries shown below:

$$S = \begin{array}{c} \\ p \\ o \end{array} \begin{array}{c} \begin{array}{cccc} t & a & d & n \end{array} \\ \begin{pmatrix} 11 & 15 & 6 & 8 \\ 6 & 4 & 4 & 13 \end{pmatrix} \end{array}$$

■

The fact that each entry of S equals the sum of the corresponding entries in O, N, and D is represented by the matrix sum

$$S = O + N + D.$$

Addition of matrices is a common-sense operation. Since each entry in a matrix has a meaning based on its position within the matrix, the operation of matrix addition is performed by adding the corresponding entries in each matrix. Adding matrices provides a concise structure for organizing an otherwise complicated operation. For example, in the Hobby-Shop Problem, the single matrix sum $S = O + N + D$ represents 8 sums of 3 numbers each.

Example 2

Using the data from Example 1, suppose that in the following year you sell the same number of each item in October, double the number sold of each item in November, and triple the number sold in December. How many of each item do you sell during the three months?

Solution The matrices O', N', and D' that we want to add are shown below:

$$O' = \begin{array}{c} \\ p \\ o \end{array} \begin{array}{c} \begin{array}{cccc} t & a & d & n \end{array} \\ \begin{pmatrix} 3 & 5 & 0 & 2 \\ 1 & 2 & 1 & 3 \end{pmatrix} \end{array}$$

$$N' = \begin{array}{c} \\ p \\ o \end{array} \begin{array}{c} \begin{array}{cccc} t & a & d & n \end{array} \\ \begin{pmatrix} 8 & 4 & 2 & 6 \\ 2 & 0 & 4 & 8 \end{pmatrix} \end{array}$$

$$D' = \begin{array}{c} \\ p \\ o \end{array} \begin{array}{c} \begin{array}{cccc} t & a & d & n \end{array} \\ \begin{pmatrix} 12 & 24 & 15 & 9 \\ 12 & 6 & 3 & 18 \end{pmatrix} \end{array}$$

The entries of O' are identical to O. Each entry of N' is twice the corresponding entry of N, which we can represent by the notation

$$N' = 2N.$$

The number 2, which is not a matrix, is a dimensionless quantity called a *scalar*. The result of a matrix multiplied by a scalar, called *scalar multiplication*, is a matrix derived by multiplying each entry of the original matrix by the scalar. Thus, we also have

$$D' = 3D.$$

Using addition and scalar multiplication, we can represent S', the number sold during the three months, by

$$S' = O' + N' + D',$$

or, equivalently,

$$S' = O + 2N + 3D,$$

which gives

$$S' = \begin{matrix} \\ p \\ o \end{matrix} \begin{pmatrix} t & a & d & n \\ 23 & 33 & 17 & 17 \\ 15 & 8 & 8 & 29 \end{pmatrix}$$

■

Matrix addition and scalar multiplication have reasonable definitions; likewise, subtraction of matrices is defined just as you would expect—finding the difference of corresponding entries.

Example 3

Suppose that in the original Hobby-Shop Problem, we know that sales for the entire year are given by the matrix Y below:

$$Y = \begin{matrix} \\ p \\ o \end{matrix} \begin{pmatrix} t & a & d & n \\ 26 & 25 & 14 & 16 \\ 13 & 7 & 8 & 28 \end{pmatrix}$$

How many of each item did you sell during the months other than October, November, and December? The answer is given by the matrix R defined as

$$\begin{aligned} R &= Y - (O + N + D) \\ &= Y - S. \end{aligned}$$

The result is

$$R = \begin{matrix} \\ p \\ o \end{matrix} \begin{pmatrix} t & a & d & n \\ 15 & 10 & 8 & 8 \\ 7 & 3 & 4 & 15 \end{pmatrix}$$

■

Addition and subtraction of matrices require that the matrices have the same dimension, meaning that they must have the same number of rows and the same number of columns. Furthermore, the corresponding rows and columns must have identical interpretations. Each row and column of a matrix has a specific meaning; therefore, trying to add or subtract matrices with different row or column labels is an attempt to combine incompatible quantities.

Exercise Set 2

1. The Campus Bookstore's inventory of books consists of the following quantities.

Hardcover: textbooks—5280; fiction—1680; nonfiction—2320; reference—1890.

Paperback: textbooks—1940; fiction—2810; nonfiction—1490; reference—2070.

The College Bookstore's inventory of books consists of the following quantities.

Hardcover: textbooks—6340; fiction—2220; nonfiction—1790; reference—1980.

Paperback: textbooks—2050; fiction—3100; nonfiction—1720; reference—2710.

a. Represent the inventory of the Campus Bookstore as a matrix.

b. Represent the inventory of the College Bookstore as a matrix.

c. Use matrix algebra to determine the total inventory of a new company formed by the merger of the College Bookstore and the Campus Bookstore.

2. The Lucrative Bank has three branches in Durham: Northgate (N), Downtown (D), and South Square (S). Matrix A shows the number of accounts of each type—checking (c), savings (s), and market (m)—at each branch office on January 1.

$$A = \begin{matrix} \\ N \\ D \\ S \end{matrix} \begin{pmatrix} c & s & m \\ 40039 & 10135 & 512 \\ 15231 & 8751 & 105 \\ 25612 & 12187 & 97 \end{pmatrix}$$

Matrix B shows the number of accounts of each type at each branch that were opened during the first quarter, and matrix C shows the number of accounts closed during the first quarter.

$$B = \begin{matrix} \\ N \\ D \\ S \end{matrix} \begin{pmatrix} c & s & m \\ 5209 & 2506 & 48 \\ 1224 & 405 & 17 \\ 2055 & 771 & 21 \end{pmatrix}$$

$$C = \begin{matrix} & c & s & m \\ N & 2780 & 1100 & 32 \\ D & 565 & 189 & 25 \\ S & 824 & 235 & 14 \end{matrix}$$

a. Calculate the matrix representing the number of accounts of each type at each location at the end of the first quarter.

b. The sudden closing of a large textile plant has led bank analysts to estimate that all accounts will decline in number by 7% during the second quarter. Construct a matrix that represents the anticipated number of each type of account at each branch at the end of the second quarter. Assume that fractions of accounts are rounded to integer values.

c. The bank president announces that the Lucrative Bank will merge with the Me. D. Okra Bank, which has branches in the same locations as those of the Lucrative Bank. The accounts at each branch of the Me. D. Okra Bank on January 1 are:

$$\begin{matrix} & c & s & m \\ N & 1345 & 2531 & 52 \\ D & 783 & 1987 & 137 \\ S & 2106 & 3765 & 813 \end{matrix}$$

Find the total number of accounts of each type at each branch of the bank formed by the merger of the two banks. Use the January 1 figures and assume that the accounts stay at their current branch offices.

3

A Common-Sense Approach to Matrix Multiplication

In the previous section, we saw that matrices can be used to organize information that is otherwise more difficult to grasp. We also saw that matrix addition and scalar multiplication are defined just as one would expect, given the meaning attached to the data in a matrix. Is it likewise possible to define matrix multiplication in a common-sense way that relates to real-life situations? The multiplication of integers can be thought of as repeated additions. For example, 5 times 3 can be thought of as $5+5+5$. This interpretation can be applied to scalar multiplication, since $3A$ is equal to the sum $A+A+A$. The product AB of two matrices A and B does not fit the same interpretation: A cannot be added to itself B times, since B is a matrix, not a number. The following example provides motivation for matrix multiplication.

Example 1

The Cutting-Board Problem

You and a friend decide to go into a partnership making cutting boards and selling them at the local flea market. Suppose that each of you makes 3 different types of cutting boards.

Style 1: made of alternating oak and walnut strips

Style 2: made of oak, walnut, and cherry strips

Style 3: made in a checkerboard pattern of walnut and cherry

You and your partner plan to make the number of cutting boards of each style shown in matrix A.

$$A = \begin{matrix} & \mathit{1} & \mathit{2} & \mathit{3} \\ \mathit{You} & 8 & 4 & 6 \\ \mathit{Part} & 6 & 6 & 8 \end{matrix}$$

Each cutting board is made by gluing together one-inch strips of wood of the appropriate type in the desired pattern. Matrix B describes the number of strips of oak (o), walnut (w), and cherry (c) needed for each style.

$$B = \begin{matrix} & o & w & c \\ \mathit{1} & 10 & 10 & 0 \\ \mathit{2} & 8 & 6 & 6 \\ \mathit{3} & 0 & 10 & 10 \end{matrix}$$

To determine how much of each type of wood you need to allocate to produce the cutting boards listed in matrix A, we will examine the following questions.

a. How much oak will you use to make the cutting boards?

You will make 8 boards of Style 1, each of which uses 10 oak strips; 4 boards of Style 2, each of which uses 8 oak strips; and 6 boards of Style 3, which uses no oak. The total number of oak strips you will use is expressed by the sum of the products

$$8(10) + 4(8) + 6(0) = 112,$$

so you need a total of 112 oak strips.

b. How much oak will your partner use?

Your partner will use an amount of oak given by the sum

$$6(10) + 6(8) + 8(0) = 108,$$

so your partner requires a total of 108 oak strips.

c. How much cherry will you use?

The amount of cherry you will use is

$$8(0) + 4(6) + 6(10) = 84$$

for a total of 84 cherry strips. ■

For each of the questions above, we found the amount of a particular type of wood you or your partner need through addition of the products obtained from multiplying the number of each style to be made by the corresponding amount of the wood needed to make one cutting board. In each of these products, the first factor is a number from a row of A, whereas the second factor is from a column of B. We can summarize the amount of wood you and your partner will use with the following matrix C, in which the entries are the numbers of wood strips:

$$C = \begin{matrix} & o & w & c \\ \textit{You} & 112 & 164 & 84 \\ \textit{Part} & 108 & 176 & 116 \end{matrix}$$

Observe that the entry in the first row and first column of C is obtained by lining up the first row of A and the first column of B, then multiplying the corresponding entries and adding the products together. Row 1 of A and column 1 of B are

$$\begin{matrix} & \mathit{1} & \mathit{2} & \mathit{3} \\ \textit{You} & (8 & 4 & 6) \end{matrix} \quad \begin{matrix} & o \\ \mathit{1} & 10 \\ \mathit{2} & 8 \\ \mathit{3} & 0 \end{matrix}$$

Multiplying pairwise term by term gives

$$8(10) + 4(8) + 6(0) = 112;$$

this sum is entry C_{11}. Likewise, entry C_{23} is obtained by multiplying pairwise term by term the second row of A by the third column of B, as shown below.

$$\begin{matrix} & \mathit{1} & \mathit{2} & \mathit{3} \\ \textit{Part} & (6 & 6 & 8) \end{matrix} \quad \begin{matrix} & c \\ \mathit{1} & 0 \\ \mathit{2} & 6 \\ \mathit{3} & 10 \end{matrix}$$

$$= 6(0) + 6(6) + 8(10) = 116$$

What row of A multiplied by what column of B gives entry C_{12}? The entry in the first row and second column of C is found by multiplying the first row of A by the second column of B.

All of the entries in C can be found using the method illustrated above. This way of combining entries in two matrices to yield a third matrix is called *matrix multiplication*; matrix C is defined as the product of matrices A and B. The operation can be written in the form shown below.

$$C = AB$$

$$= \begin{matrix} & \mathit{1} & \mathit{2} & \mathit{3} \\ \textit{You} & 8 & 4 & 6 \\ \textit{Part} & 6 & 6 & 8 \end{matrix} \quad \begin{matrix} & o & w & c \\ \mathit{1} & 10 & 10 & 0 \\ \mathit{2} & 8 & 6 & 6 \\ \mathit{3} & 0 & 10 & 10 \end{matrix}$$

$$= \begin{matrix} & o & w & c \\ \textit{You} & 112 & 164 & 84 \\ \textit{Part} & 108 & 176 & 116 \end{matrix}$$

In general, the matrix multiplication $C = AB$ is defined as follows: Each entry C_{ij} is obtained by multiplying pairwise term by term the ith row of the left-hand matrix A by the jth column of the right-hand matrix B. In symbols, this definition means that

$$C_{ij} = A_{i1}B_{1j} + A_{i2}B_{2j} + A_{i3}B_{3j} + \cdots + A_{in}B_{nj}.$$

Example 2

Let matrix D represent the cost in dollars per strip for each type of wood in the Cutting-Board Problem.

$$D = \begin{array}{c} \\ o \\ w \\ c \end{array} \begin{array}{c} \textit{Cost} \\ \begin{pmatrix} 0.18 \\ 0.22 \\ 0.20 \end{pmatrix} \end{array}$$

a. You would like to determine a selling price for the cutting boards. What is the total cost of the wood for one cutting board of each style?

The cost of the wood for a cutting board of Style 1 is equal to the number of strips of each type of wood multiplied by the cost per strip, or

$$10(0.18) + 10(0.22) + 0(0.20) = 4.00.$$

The cost of the wood for a cutting board of Style 1 is \$4.00. This number was calculated by multiplying pairwise term by term the first row of B by the column in D. Using similar reasoning for Styles 2 and 3, we see that the matrix product BD gives the required information for each style.

$$BD = \begin{array}{c} \\ 1 \\ 2 \\ 3 \end{array} \begin{array}{c} \begin{array}{ccc} o & w & c \end{array} \\ \begin{pmatrix} 10 & 10 & 0 \\ 8 & 6 & 6 \\ 0 & 10 & 10 \end{pmatrix} \end{array} \begin{array}{c} \\ o \\ w \\ c \end{array} \begin{array}{c} \textit{Cost} \\ \begin{pmatrix} 0.18 \\ 0.22 \\ 0.20 \end{pmatrix} \end{array}$$

$$= \begin{array}{c} \\ 1 \\ 2 \\ 3 \end{array} \begin{array}{c} \textit{Cost} \\ \begin{pmatrix} 4.00 \\ 3.96 \\ 4.20 \end{pmatrix} \end{array}$$

The cost of the wood for a cutting board of Style 1 is \$4.00, the cost for Style 2 is \$3.96, and the cost for Style 3 is \$4.20.

b. You and your partner would like to know how much money to budget for purchasing the wood for the cutting boards. What are the total costs for you and your partner to produce the number of cutting boards listed in matrix A of Example 1?

The product AB from Example 1 gives the number of strips of wood used by you and your partner. Multiplying AB by D results in a matrix containing the costs for you and your partner, as shown below.

$(AB)(D)$

$$= \begin{array}{c} \\ \textit{You} \\ \textit{Part} \end{array} \begin{array}{c} \begin{array}{ccc} o & w & c \end{array} \\ \begin{pmatrix} 112 & 164 & 84 \\ 108 & 176 & 116 \end{pmatrix} \end{array} \begin{array}{c} \\ o \\ w \\ c \end{array} \begin{array}{c} \textit{Cost} \\ \begin{pmatrix} 0.18 \\ 0.22 \\ 0.20 \end{pmatrix} \end{array}$$

$$= \begin{array}{c} \\ \textit{You} \\ \textit{Part} \end{array} \begin{array}{c} \textit{Cost} \\ \begin{pmatrix} 73.04 \\ 81.36 \end{pmatrix} \end{array}$$

The total cost for you is \$73.04; the total cost for your partner is \$81.36. ■

The rows and columns of data in the matrices in Examples 1 and 2 are described by labels. In matrix A, the row labels are names (you and your partner) and the column labels are styles of cutting board (1, 2, and 3), so that matrix A classifies data according to name and style; we refer to A as a name-by-style matrix. Consistent with this notation, matrix B is a style-by-wood matrix. The row and column labels of matrices are especially helpful in interpreting the results of matrix multiplication.

Observe that in Example 1 we multiplied a name-by-style matrix (A) by a style-by-wood matrix (B) to get a name-by-wood matrix (C). In Example 2, we multiplied a style-by-wood matrix (B) by a wood-by-cost matrix (D) to get a style-by-cost matrix. We also found that the product of a name-by-wood matrix (AB) and a wood-by-cost matrix (D) is a name-by-cost matrix. In each example, matrix multiplication eliminated the labels of the first factor's columns and the second factor's rows, leaving a product matrix with exactly the row and column labels we desired in our answer. Matrix multiplication, which at first glance may seem very strange, actually is designed to give us the information we want in a straightforward manner.

Two matrices are multiplied by multiplying the elements of a row of the left-hand matrix by the corresponding elements of a column of the right-hand matrix and then adding the products. If matrix S is multiplied by matrix T, the number of columns of S must equal the number of rows of T. If $ST = U$, the product matrix U has the same number of rows as S and the same number of columns as T. In symbols,

$$S_{m\times n}T_{n\times p} = U_{m\times p}.$$

In order for the product ST to be meaningful, the column labels of S must be the same as the row labels of T. If $ST = U$, then U has the row labels of S and the column labels of T.

A special type of matrix that appeared in Example 2 is a *vector*, a matrix that consists of either one column, called a column vector, or one row, called a row vector. In other branches of mathematics, a vector with n entries represents a point in n-dimensional space. For example, a vector with 3 entries can represent the x-, y-, and z-components of a geometric vector in 3-space.

The *identity matrix* $I_{n\times n}$ is a square matrix with the property that multiplying a matrix A by I returns A as the product. If A is a 2×3 matrix and $I\cdot A = A$, then I must be a 2×2 matrix. Can you explain why? If A is a 2×3 matrix and $A\cdot I = A$, then I must be a 3×3 matrix.

The elements of an identity matrix are 1's on the main diagonal (going from upper left to lower right) and 0's elsewhere. Several examples of identity matrices are shown below.

$$I_{2\times 2} = \begin{pmatrix} 1 & 0 \\ 0 & 1 \end{pmatrix} \qquad I_{3\times 3} = \begin{pmatrix} 1 & 0 & 0 \\ 0 & 1 & 0 \\ 0 & 0 & 1 \end{pmatrix}$$

$$I_{n\times n} = \begin{pmatrix} 1 & 0 & \cdots & 0 & 0 \\ 0 & 1 & \cdots & 0 & 0 \\ \vdots & \vdots & \vdots & \vdots & \vdots \\ 0 & 0 & \cdots & 1 & 0 \\ 0 & 0 & \cdots & 0 & 1 \end{pmatrix}$$

When you try multiplying a matrix by the appropriate identity matrix, the reason for the 0's and 1's will be evident.

Exercise Set 3

1. The following is a set of abstract matrices (without row and column labels).

$$M = \begin{pmatrix} 1 & -1 \\ 2 & 0 \end{pmatrix} \quad N = \begin{pmatrix} 2 & 4 & 1 \\ 0 & -1 & 3 \\ 1 & 0 & 2 \end{pmatrix}$$

$$O = \begin{pmatrix} 6 \\ -1 \end{pmatrix} \quad P = \begin{pmatrix} 0 & 1/2 \\ -1 & 1/2 \end{pmatrix}$$

$$Q = \begin{pmatrix} 4 \\ 1 \\ 3 \end{pmatrix} \quad R = \begin{pmatrix} 3 & 1 \\ -1 & 0 \end{pmatrix}$$

$$S = \begin{pmatrix} 3 & 1 \\ 1 & 0 \\ 0 & 2 \\ -1 & 1 \end{pmatrix} \quad T = \begin{pmatrix} 1 \\ 2 \\ -3 \\ 4 \end{pmatrix}$$

$$U = \begin{pmatrix} 4 & 2 & 6 & -1 \\ 5 & 3 & 1 & 0 \\ 0 & 2 & -1 & 1 \end{pmatrix}$$

List all orders of pairs of matrices from this set for which the product is defined. State the dimension of each product.

2. Using the matrices M and P from Exercise 1, find the matrix products MP and PM. What property do you notice about these matrices?

3. Is matrix multiplication associative? In other words, is it always true that $A(BC) = (AB)C$, assuming these matrix products are defined? Use some of the matrices from Exercise 1 to test your conjecture.

4. What is the result of multiplying a vector times a vector, assuming the multiplication is defined?

5. Is matrix multiplication commutative? In other words, is it always true that $AB = BA$? Is multiplication by the identity matrix commutative? Use some matrices from Exercise 1 to test your conjecture. Tell why it is necessary to use the terms *left-multiply* and *right-multiply* when referring to matrix multiplication. (Specifically, in the product AB, matrix A is right-multiplied by B, and B is left-multiplied by A.)

6. A company has investments in three states—North Carolina, North Dakota, and New Mexico. Its deposits in each state are divided among bonds, mortgages, and consumer loans. The amount of money (in millions of dollars) invested in each category on June 1 is displayed in the table below.

	NC	ND	NM
Bonds	13	25	22
Mort.	6	9	4
Loans	29	17	13

The current yields on these investments are 7.5% for bonds, 11.25% for mortgages, and 6% for consumer loans. Use matrix multiplication to find the total earnings for each state.

7. Several years ago an investor purchased growth stocks, with the expectation that they would increase in value over time. The purchase included 100 shares of stock A, 200 shares of stock B, and 150 shares of stock C. At the end of each year the value of each stock is recorded. The table below shows the price per share (in dollars) of stocks A, B, and C at the end of the years 1984, 1985, and 1986.

	1984	1985	1986
Stock A	68.00	72.00	75.00
Stock B	55.00	60.00	67.50
Stock C	82.50	84.00	87.00

Calculate the total value of the stock purchases at the end of each year.

8. A virus hits campus. Students are either sick, well, or carriers of the virus. The following percentages of people are in each category, depending on whether they are a junior or a senior:

	Junior	Senior
Well	15%	25%
Sick	35%	40%
Carrier	50%	35%

The student population is distributed by class and sex as follows:

	Males	Females
Junior	104	80
Senior	107	103

How many sick males are there? How many well females? How many female carriers?

9. The Sound Company produces stereos. Their inventory includes four models—the Budget, the Economy, the Executive, and the President models. The Budget model needs 50 transistors, 30 capacitors, 7 connectors, and 3 dials. The Economy model needs 65 transistors, 50 capacitors, 9 connectors, and 4 dials. The Executive model needs 85 transistors, 42 capacitors, 10 connectors, and 6 dials. The President model needs 85 transistors, 42 capacitors, 10 connectors, and 12 dials. The daily manufacturing goal in a normal quarter is 10 Budget, 12 Economy, 11 Executive, and 7 President stereos.

a. How many transistors are needed each day? capacitors? connectors? dials?

b. During August and September, production is increased by 40%. How many Budget, Economy, Executive, and President models are produced daily during these months?

c. It takes 5 person-hours to produce the Budget model, 7 person-hours to produce the Economy model, 6 person-hours for the Executive model, and 7 person-hours for the President model. Determine the number of employees needed to maintain the normal production schedule, assuming everyone works an average of 7 hours each day. How many employees are needed in August and September?

10. The parabola $y = -4x^2+16x-12$ contains the points (1,0), (2,4), and (3,0). Find the new ordered pair (x', y') that is produced from each pair (x, y) using the matrix multiplication below when $\theta = 30°$.

$$\begin{pmatrix} \cos\theta & -\sin\theta \\ \sin\theta & \cos\theta \end{pmatrix} \begin{pmatrix} x \\ y \end{pmatrix} = \begin{pmatrix} x' \\ y' \end{pmatrix}$$

Accurately plot the new points in the same coordinate system as the three given points of the original parabola. Sketch the remainder of the parabola as you think it would appear if every point on the original curve were transformed by the matrix multiplication above. Compare this sketch with a sketch of the original parabola. What happened? Try $\theta = 90°$ to test your conjecture.

11. The president of the Lucrative Bank is hoping for a 21% increase in checking accounts, a 35% increase in savings accounts, and a 52% increase in market accounts. The current statistics on the number of accounts at each branch are as follows:

$$\begin{array}{l} \\ Northgate \\ Downtown \\ S.\ Square \end{array} \begin{array}{c} \begin{array}{ccc} Check. & Sav. & Mark. \end{array} \\ \begin{pmatrix} 40039 & 10135 & 512 \\ 15231 & 8751 & 105 \\ 25612 & 12187 & 97 \end{pmatrix} \end{array}$$

What is the goal for each branch in each type of account? (**Hint:** multiply by a 3×3 matrix with certain nonzero entries on the diagonal and zero entries elsewhere.) What will be the total number of accounts at each branch?

12. Winners at a science fair are determined by a scoring system based on five items with different weights attached to each item. The items and associated weights are the summary of background research — weight 3; experimental procedure — weight 5; research paper — weight 6; project display — weight 8; and creativity of idea — weight 4. Each project is judged by grading each of the five items on a scale from 0 to 10, with 10 highest. The total score for a project is derived by adding the products of the corresponding weights and points for each item.

a. What is the maximum total score possible for a project?

b. Calculate the score for a student who earns 8 points on background research, 9 points on experimental procedure, 7 points on the research paper, 8 points on the project display, and 6 points on creativity.

c. The table shown below contains the points for the finalists in the biology division. Calculate the total scores to determine the first, second, and third place entries.

	Peter	Shelia	Arvind	Kathy	Nia	Chris	Maurice
Backgr. research	9	8	10	7	8	9	10
Exper. procedure	10	9	9	10	10	9	10
Research paper	7	9	8	9	7	8	8
Project display	9	10	9	8	10	8	9
Creativity of idea	8	7	8	10	6	8	7

13. The Metropolitan Opera is planning its last cross-country tour. It plans to perform *Carmen* and *La Traviata* in Atlanta in May. The person in charge of logistics wants to make plane reservations for the two troupes. *Carmen* has 2 stars, 25 other adults, 5 children, and 5 staff members. *La Traviata* has 3 stars, 15 other adults, and 4 staff members. There are 3 airlines to choose from. Redwing charges round-trip fares to Atlanta of $630 for first class, $420 for coach, and $250 for youth. Southeastern charges $650 for first class, $350 for coach, and $275 for youth. Air Atlanta charges $700 first class, $370 coach, and $150 youth. Assume stars travel first class, other adults and staff travel coach, and children travel for the youth fare.

a. Find the total cost for each opera troupe with each airline.

b. If each airline will give a 30% discount to the *Carmen* troupe because they will stay over Saturday, what is the total cost on each airline?

c. Suppose instead that each airline will give a discount to both troupes just to get their business. Redwing gives a 30% discount, Southeastern 20%, and Air Atlanta 25%. Use matrix multiplication to find a new 3×3 cost matrix. (See the hint for Exercise 11.)

14. On the Sunday before the 1986 NCAA basketball finals, a national survey was taken of people's choices to win the game, along with their income. The following information was collected:

435 for Duke making over $30,000 per year

105 for Louisville making over $30,000 per year

115 with no choice making over \$30,000 per year

125 for Duke making under \$30,000 per year

205 for Louisville making under \$30,000 per year

231 with no choice making under \$30,000 per year.

A survey was done in a restaurant in New York on the night of the game to determine the incomes of the people eating there. The following information was collected:

302 making over \$30,000 per year

276 making under \$30,000 per year.

Using matrix operations, estimate the number of Duke fans, the number of Louisville fans, and the number of fans with no choice in the restaurant, based on the survey from Sunday. These numbers are found using probabilities based on the Sunday survey and the data collected on the night of the game. Before you attempt to find an answer, the information from each survey should be converted to ratios and displayed in matrices.

15. A company that produces and markets stuffed animals has three plants—one on the East Coast, one on the West Coast, and one in the central part of the country. Among other items, each plant manufactures stuffed pandas, kangaroos, and rabbits. Personnel are needed to cut fabric, sew appropriate parts together, and provide finish work for each animal. Matrix A gives the time (in hours) of each type of labor required to make each type of stuffed animal; matrix B gives the daily production capacity at each plant; matrix C provides hourly wages of the different workers at each plant; and matrix D contains the total orders received by the company in October and November.

$$A = \begin{array}{c} \\ \textit{Pan.} \\ \textit{Kang.} \\ \textit{Rab.} \end{array} \begin{array}{c} \begin{array}{ccc} \textit{Cut} & \textit{Sew} & \textit{Finish} \end{array} \\ \begin{pmatrix} 0.5 & 0.8 & 0.6 \\ 0.8 & 1.0 & 0.4 \\ 0.4 & 0.5 & 0.5 \end{pmatrix} \end{array}$$

$$B = \begin{array}{c} \\ \textit{East} \\ \textit{Cent.} \\ \textit{West} \end{array} \begin{array}{c} \begin{array}{ccc} \textit{Pan.} & \textit{Kang.} & \textit{Rab.} \end{array} \\ \begin{pmatrix} 25 & 15 & 12 \\ 10 & 20 & 15 \\ 20 & 15 & 15 \end{pmatrix} \end{array}$$

$$C = \begin{array}{c} \\ \textit{East} \\ \textit{Cent.} \\ \textit{West} \end{array} \begin{array}{c} \begin{array}{ccc} \textit{Cut} & \textit{Sew} & \textit{Finish} \end{array} \\ \begin{pmatrix} 7.50 & 9.00 & 8.40 \\ 7.00 & 8.00 & 7.60 \\ 8.40 & 10.50 & 10.00 \end{pmatrix} \end{array}$$

$$D = \begin{array}{c} \\ \textit{Pan.} \\ \textit{Kang.} \\ \textit{Rab.} \end{array} \begin{array}{c} \begin{array}{cc} \textit{Oct.} & \textit{Nov.} \end{array} \\ \begin{pmatrix} 1000 & 1100 \\ 600 & 850 \\ 800 & 725 \end{pmatrix} \end{array}$$

Use the matrices above to compute the following quantities:

a. the hours of each type of labor needed each month (October, November) to fill all orders

b. the production cost per item at each plant

c. the cost of filling all October orders at the East Coast plant

d. the daily hours of each type of labor needed at each plant if production levels are at capacity

e. the daily amount each plant will pay its personnel when producing at capacity

---

4

The Leontief Input-Output Model and the Inverse of a Matrix

We have seen that matrices are useful for organizing and manipulating data, and that the arithmetic of matrices makes sense in light of their applications. Another example of the use of matrices is in the Leontief Input-Output Model of an economy. In the April 1965 issue of *Scientific American*, the economist Wassily Leontief explained his input-output system using the 1958 American economy. He divided the economy into

81 sectors grouped into 6 families, viewing the economy as a large 81×81 matrix. For his work, Leontief won the 1973 Nobel Prize for economics, and his model is now used world-wide. We will use a simplified version of his model to demonstrate the real-world importance of matrices.

Leontief divided the economy into 81 sectors (transportation, manufacturing, steel, utilities, etc.), each of which relies on input resources taken from the output of other sectors. For example, the steel industry uses the output of the utilities, heavy manufacturing, and transportation sectors, and even some steel, as inputs in its production. Therefore, not all steel is available to meet consumer demand. To develop the details of the Leontief model, we will examine a simplified economy that has only three sectors — agriculture, manufacturing, and transportation. The model hinges on the fact that some of the output from each sector is used in the production process, so that not all output is available to meet consumer demand.

The following technology information has been gathered by a research team:

— Production of a unit of output of agriculture requires inputs consisting of 1/10 of a unit of agriculture, 1/5 of a unit of manufacturing, and 1/5 of a unit of transportation. (A unit here refers to the value of a unit of output from the agriculture sector; a unit can have some given monetary equivalence.)

— Production of a unit of output of manufacturing requires inputs consisting of 1/15 of a unit of agriculture, 1/4 of a unit of manufacturing, and 1/5 of a unit of transportation. (A unit here refers to the value of a unit of output from the manufacturing sector.)

— Production of a unit of output of transportation requires inputs consisting of no agriculture, 1/4 of a unit of manufacturing, and 1/6 of a unit of transportation. (A unit here refers to the value of a unit of output from the transportation sector.)

The diagram in Figure 1, called a *state diagram*, illustrates the flow of resources from one

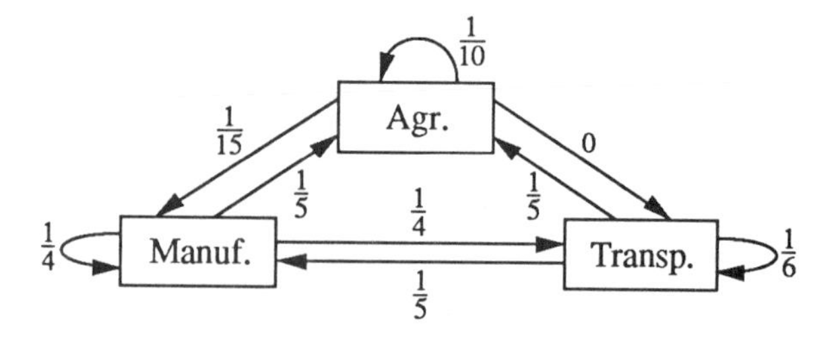

Figure 1 State Diagram for a Three-Sector Economy

sector to another. For example, the arc from manufacturing to transportation is assigned the number 1/4, indicating that the production of one unit of output of transportation requires the input of 1/4 of a unit of manufacturing. The arc from agriculture to agriculture is assigned the number 1/10, indicating that each unit of output of agriculture requires the input of 1/10 of a unit of agriculture. In general, an arc from sector i to sector j assigns a number indicating the units of input required from sector i to produce one unit of output from sector j.

The matrix T reflecting this information, called the *technology matrix*, is shown below.

$$T = \begin{array}{r} \\ Ag. \\ Manu. \\ Tran. \end{array} \begin{array}{c} \begin{array}{ccc} Ag. & Manu. & Tran. \end{array} \\ \begin{pmatrix} 1/10 & 1/15 & 0 \\ 1/5 & 1/4 & 1/4 \\ 1/5 & 1/5 & 1/6 \end{pmatrix} \end{array}$$

For the sake of uniformity, the units of output for each sector are usually measured in dollars rather than physical units such as tons. Assuming a unit equals one dollar, from the matrix T we observe that to produce one dollar's worth of manufactured goods requires about 7 cents worth (1/15 dollar) of agriculture, 25 cents worth of manufacturing, and 20 cents worth of transportation. In general, a column of the technology matrix gives the fraction of one dollar's worth of input from each sector needed to produce one dollar's worth of output in the sector represented by that column.

It is conventional, but not essential, for the column headings to represent output and for the row labels to represent input, as in matrix T.

The interpretations of rows and columns could be exchanged, but the matrix operations that we are about to describe would need to be modified in order to achieve correct results.

Suppose we know that the economy produces 100 million dollars worth of agriculture, 120 million dollars worth of manufacturing, and 120 million dollars worth of transportation. This information is displayed in the production matrix P shown below. Note that in the production matrix each entry represents millions of dollars.

$$P = \begin{array}{r} \\ \textit{Ag.} \\ \textit{Manu.} \\ \textit{Tran.} \end{array} \begin{array}{c} \textit{Prod.} \\ \begin{pmatrix} 100 \\ 120 \\ 120 \end{pmatrix} \end{array}$$

Since the output of each sector requires input produced in the other sectors, the total production is not available for the demands of consumers. Some of this production is consumed internally by the economy. How much of each sector's output must be used to achieve the production levels given in matrix P? By determining what resources are used in production, we will be able also to determine what amount remains for consumers. We will answer this question for each of the three sectors.

First, how many units of agriculture are used in the production of all three sectors? Each unit of agricultural output requires 0.1 units of agriculture as input; therefore, the 100 units of agriculture specified in P require $(1/10)(100)$ units of agriculture as input. Likewise, 120 units of manufacturing output require $(1/15)(120)$ units of agricultural input. The 120 units of transportation output require $(0)(120)$ units of agricultural input. The total input, in millions of dollars, of agriculture necessary to meet the production levels given in P is

$$(\tfrac{1}{10})(100) + (\tfrac{1}{15})(120) + (0)(120) = 18$$

How many units of manufacturing are used to produce at a level given by P? By the same reasoning as above, the total input, in millions of dollars, of manufacturing required is

$$(\tfrac{1}{5})(100) + (\tfrac{1}{4})(120) + (\tfrac{1}{4})(120) = 80$$

Finally, how many units, in millions of dollars, of transportation are needed to achieve the production matrix P? Using the information in the technology matrix, we find that we require

$$(\tfrac{1}{5})(100) + (\tfrac{1}{5})(120) + (\tfrac{1}{6})(120) = 64$$

In the calculations above, notice that the second factors in the products are the numbers found in P. The first factors in the expression for agriculture come from the agriculture row in the technology matrix. Likewise, the first factors in the expression for manufacturing come from the manufacturing row, and the first factors for transportation come from the transportation row. Clearly, the concept of matrix multiplication is at work in these calculations. In fact, the matrix product TP has entries corresponding to the calculations above, specifically,

$$TP = \begin{array}{r} \\ \textit{Ag.} \\ \textit{Manu.} \\ \textit{Tran.} \end{array} \begin{array}{c} \begin{array}{ccc} \textit{Ag.} & \textit{Manu.} & \textit{Tran.} \end{array} \\ \begin{pmatrix} 1/10 & 1/15 & 0 \\ 1/5 & 1/4 & 1/4 \\ 1/5 & 1/5 & 1/6 \end{pmatrix} \end{array}$$

$$\times \begin{array}{r} \\ \textit{Ag.} \\ \textit{Manu.} \\ \textit{Tran.} \end{array} \begin{array}{c} \textit{Prod.} \\ \begin{pmatrix} 100 \\ 120 \\ 120 \end{pmatrix} \end{array}$$

$$= \begin{array}{r} \\ \textit{Ag.} \\ \textit{Manu.} \\ \textit{Tran.} \end{array} \begin{array}{c} \textit{Int. Cons.} \\ \begin{pmatrix} 18 \\ 80 \\ 64 \end{pmatrix} \end{array}.$$

In general, if T is a technology matrix and P is a production matrix, then TP is a matrix that represents the amount of the output consumed by the system internally. The matrix that results from the product TP is called the *internal consumption matrix*. We observe that TP gives us the amount of each sector's output needed as input by the other sectors to meet production goals.

How much output is left to meet the demands of the consumers after the input requirements of each sector are met? This can be represented by the *demand matrix* D that is the difference between production and internal consumption. In symbols, D is defined as

$$D = P - TP.$$

Using the numbers provided above gives

$$D = P - TP$$

$$= \begin{matrix} & \textit{Prod.} \\ \begin{matrix} \textit{Ag.} \\ \textit{Manu.} \\ \textit{Tran.} \end{matrix} & \begin{pmatrix} 100 \\ 120 \\ 120 \end{pmatrix} \end{matrix} - \begin{matrix} & \textit{Int. Cons.} \\ \begin{matrix} \textit{Ag.} \\ \textit{Manu.} \\ \textit{Tran.} \end{matrix} & \begin{pmatrix} 18 \\ 80 \\ 64 \end{pmatrix} \end{matrix}$$

$$= \begin{matrix} & \textit{Demand} \\ \begin{matrix} \textit{Ag.} \\ \textit{Manu.} \\ \textit{Tran.} \end{matrix} & \begin{pmatrix} 82 \\ 40 \\ 56 \end{pmatrix} \end{matrix}$$

The amounts left for distribution to consumers are 82 units of agriculture, 40 units of manufacturing, and 56 units of transportation, where a unit represents one million dollars.

Viewing this problem from a different perspective, we see that to have 82 units of agriculture available for consumer demand requires the production of 100 units of agriculture. Furthermore, to have 40 units of manufacturing available for demand, 120 units of manufacturing must be produced. To meet a demand of 56 units of transportation requires production of 120 units of transportation. Suppose the consumer demand for agriculture is not 82 units but is actually 100 units.

We have demonstrated a method for determining the quantity of resources available for consumer demand given a certain level of production. Generally, however, professional analysts will estimate society's demand for specific goods and services. The question usually investigated is the following: given a certain demand by society, how much should each sector of the economy produce in order to meet this demand? To produce less than the estimated demand may cause shortages and hardship; to produce more than the estimated demand leads to waste and inefficiency.

From the discussion above, we have $D = P - TP$. Since $P = IP$, where I is an identity matrix of the same dimension as T, this equation can be rewritten as

$$D = (I - T)P.$$

Because matrix multiplication is not commutative, P must be factored out to the right, since it is to the right of T. In this equation, we know the entries in the matrices D and $I - T$, and we wish to determine the matrix P.

The analogous situation with numbers is the equation

$$ax = b,$$

in which the values of a and b are known. To determine the value of x, we multiply both sides of the equation by the *multiplicative inverse* of a, namely $1/a$. This strategy transforms the equation as follows:

$$\begin{aligned} \frac{1}{a} \cdot a \cdot x &= \frac{1}{a} \cdot b \\ \left(\frac{1}{a} \cdot a\right) \cdot x &= \frac{1}{a} \cdot b \\ 1 \cdot x &= b/a \\ x &= b/a. \end{aligned}$$

The strategy for solving a matrix equation is similar. To isolate P on one side of the equation $D = (I - T)P$, we need to multiply both sides of the equation by the *multiplicative inverse* of $(I - T)$. The inverse of $(I - T)$ is a matrix that when multiplied by the matrix $I - T$ yields an identity matrix. After left-multiplying both sides of the equation $D = (I - T)P$ by the inverse of $(I - T)$, the right side of the equation is the product of an identity matrix and P, or simply P. This process therefore isolates P so that economic production can be determined for a given amount of consumer demand.

Before going further with the problem of determining P, we need a method for finding the inverse of a matrix, which the following section explains.

Finding the Inverse of a Matrix

The inverse of a square matrix R is a matrix S such that the product of R and S is an identity matrix. For example, if

$$R = \begin{pmatrix} 1 & -1 \\ 2 & 0 \end{pmatrix} \quad \text{and} \quad S = \begin{pmatrix} 0 & 1/2 \\ -1 & 1/2 \end{pmatrix},$$

then

$$\begin{pmatrix} 1 & -1 \\ 2 & 0 \end{pmatrix} \begin{pmatrix} 0 & 1/2 \\ -1 & 1/2 \end{pmatrix} = \begin{pmatrix} 1 & 0 \\ 0 & 1 \end{pmatrix},$$

or

$$RS = I.$$

It is also true that $SR = I$. Since $RS = SR = I$, the matrices R and S are inverses of each other. The inverse of R is symbolized by R^{-1}, so that

$$R^{-1} = S,$$

and

$$S^{-1} = R.$$

Given the matrix A shown below, what is the inverse of A?

$$A = \begin{pmatrix} a_{11} & a_{12} & a_{13} \\ a_{21} & a_{22} & a_{23} \\ a_{31} & a_{32} & a_{33} \end{pmatrix}$$

We wish to find a matrix such that when it is multiplied with A, the product is an identity matrix. In other words, what is the matrix A^{-1} such that $A^{-1}A = I$?

In approaching this problem, think of the entries in A as the coefficients of a system of linear equations. Specifically, the system of equations

$$\begin{aligned} a_{11}x_1 + a_{12}x_2 + a_{13}x_3 &= c_1 \\ a_{21}x_1 + a_{22}x_2 + a_{23}x_3 &= c_2 \\ a_{31}x_1 + a_{32}x_2 + a_{33}x_3 &= c_3 \end{aligned}$$

can be expressed as the matrix equation

$$\begin{pmatrix} a_{11} & a_{12} & a_{13} \\ a_{21} & a_{22} & a_{23} \\ a_{31} & a_{32} & a_{33} \end{pmatrix} \begin{pmatrix} x_1 \\ x_2 \\ x_3 \end{pmatrix} = \begin{pmatrix} c_1 \\ c_2 \\ c_3 \end{pmatrix}$$

If we let

$$X = \begin{pmatrix} x_1 \\ x_2 \\ x_3 \end{pmatrix} \quad \text{and} \quad C = \begin{pmatrix} c_1 \\ c_2 \\ c_3 \end{pmatrix}$$

then the matrix equation above is equivalent to

$$AX = C.$$

Left-multiplying both sides of the equation above by the inverse of A provides the equivalent equation

$$A^{-1}AX = A^{-1}C.$$

Recall that since matrix multiplication is not commutative, both sides must be left-multiplied. This equation can be simplified to

$$IX = A^{-1}C,$$

so the solution for vector X is

$$X = A^{-1}C.$$

When a system of equations is solved without introducing matrix notation, a common procedure for finding the solution involves transforming the system with the following operations:

1. Exchange the positions of two equations.
2. Multiply an equation by a nonzero constant.
3. Replace an equation with the sum of the equation and a multiple of another equation.

These operations each produce a system with a solution identical to that of the original system. Eventually a system results whose solution can be determined by inspection. This same strategy can be applied to the problem of finding the inverse of a matrix.

We can rewrite the matrix equation $AX = C$ in the equivalent form

$$AX = IC,$$

where I is an identity matrix of the same dimension as A. Suppose operations are performed on A and I that are equivalent to the manipulations performed on a system of equations. In other words, we manipulate the matrices A and I as we would the coefficients of a system of equations, performing the same operations on both sides of the matrix equation. If these operations eventually transform A into an identity matrix, then the left side of the matrix equation will be IX, or just X. Since X is on the left side, the right side of the equation will be $A^{-1}C$. To summarize, if we can transform A into I, then the same operations will transform I into A^{-1}.

There are matrix operations analogous to the three transformations for systems of equations. These can be used to find the inverse of a matrix and are called *elementary row operations* (EROs). The three EROs are as follows:

1. Exchange two rows.
2. Multiply one row by a nonzero constant.
3. Replace a row with the sum of the row and a multiple of another row.

The use of EROs to determine the inverse of a matrix is demonstrated with the matrix A shown below.

$$A = \begin{pmatrix} 2 & 1 & 0 \\ 1 & 0 & 2 \\ 0 & 1 & -1 \end{pmatrix}$$

The goal is to transform A using EROs until we reach a 3×3 identity matrix. The inverse of A results from performing the same EROs on $I_{3\times3}$, so we write $I_{3\times3}$ alongside A. This form is called an *augmented matrix* and is shown below.

$$\left(\begin{array}{ccc|ccc} 2 & 1 & 0 & 1 & 0 & 0 \\ 1 & 0 & 2 & 0 & 1 & 0 \\ 0 & 1 & -1 & 0 & 0 & 1 \end{array}\right)$$

We begin the sequence of EROs by interchanging rows 1 and 2 so that a 1 is in the upper left corner, which results in the new augmented matrix

$$\left(\begin{array}{ccc|ccc} 1 & 0 & 2 & 0 & 1 & 0 \\ 2 & 1 & 0 & 1 & 0 & 0 \\ 0 & 1 & -1 & 0 & 0 & 1 \end{array}\right)$$

Now add -2 times the first row to the second row and replace row 2 with the sum.

$$\left(\begin{array}{ccc|ccc} 1 & 0 & 2 & 0 & 1 & 0 \\ 0 & 1 & -4 & 1 & -2 & 0 \\ 0 & 1 & -1 & 0 & 0 & 1 \end{array}\right)$$

Already we are finished transforming the first column of A—it is identical to the first column of $I_{3\times3}$. Next add -1 times the second row to the third row, which finishes the second column, as shown below.

$$\left(\begin{array}{ccc|ccc} 1 & 0 & 2 & 0 & 1 & 0 \\ 0 & 1 & -4 & 1 & -2 & 0 \\ 0 & 0 & 3 & -1 & 2 & 1 \end{array}\right)$$

To transform the third column, multiply the third row by 1/3, yielding

$$\left(\begin{array}{ccc|ccc} 1 & 0 & 2 & 0 & 1 & 0 \\ 0 & 1 & -4 & 1 & -2 & 0 \\ 0 & 0 & 1 & -1/3 & 2/3 & 1/3 \end{array}\right)$$

Add 4 times row 3 to row 2, and add -2 times row 3 to row 1, leaving the matrix

$$\left(\begin{array}{ccc|ccc} 1 & 0 & 0 & 2/3 & -1/3 & -2/3 \\ 0 & 1 & 0 & -1/3 & 2/3 & 4/3 \\ 0 & 0 & 1 & -1/3 & 2/3 & 1/3 \end{array}\right)$$

The transformation process is now complete, and the inverse of A is

$$A^{-1} = \begin{pmatrix} 2/3 & -1/3 & -2/3 \\ -1/3 & 2/3 & 4/3 \\ -1/3 & 2/3 & 1/3 \end{pmatrix}$$

Calculate the product of A^{-1} and A to verify that $A^{-1}A = I$.

Class Practice

The EROs performed above to find the inverse of A can also be accomplished through matrix multiplication, as the following exercises demonstrate.

1. The first ERO performed on A interchanged rows 1 and 2. Verify that the matrix

$$\begin{pmatrix} 0 & 1 & 0 \\ 1 & 0 & 0 \\ 0 & 0 & 1 \end{pmatrix}$$

   interchanges rows 1 and 2 when A is left-multiplied by this matrix. What matrix will interchange rows 1 and 3? Rows 2 and 3?

2. Multiplying a row by a nonzero constant is a second ERO used to find the inverse of a

matrix. Verify that left-multiplying A by the matrix

$$\begin{pmatrix} k & 0 & 0 \\ 0 & 1 & 0 \\ 0 & 0 & 1 \end{pmatrix}$$

multiplies the first row of A by k. What matrix will multiply row 2 by k? Row 3?

3. The third ERO involves multiplying a row by a constant k and adding it to another row. Verify that left-multiplying A by the matrix

$$\begin{pmatrix} 1 & k & 0 \\ 0 & 1 & 0 \\ 0 & 0 & 1 \end{pmatrix}$$

adds k times the second row of A to the first row of A. What matrix will add k times the third row to the first row? k times row 2 to row 3?

4. List the 6 matrices that correspond to the 6 EROs used to find the inverse of A.

5. Find the product of the 6 matrices in Exercise 4, left-multiplying in order with the first ERO matrix being right-most, the second ERO matrix immediately to the left of the first, and so on. Notice that the result is A^{-1}.

6. Why must the ERO matrices be left-multiplied in Exercise 5 above? Is it possible to use right-multiplication?

Do All Matrices Have Inverses?

If B is the inverse of A, then A is the inverse of B, which means that

$$A \cdot B = B \cdot A = I.$$

Inverses can be multiplied in either of the two orders and the result will be an identity matrix. An implication of this property is that only square matrices have inverses.

If a matrix has an inverse, it is said to be *invertible*. Are all square matrices invertible? If not, what are the general conditions for a square matrix A to have an inverse? Our method for finding an inverse involves using EROs to reduce matrix A to an identity matrix. If this is possible, then the same sequence of EROs will transform the identity matrix into A^{-1}. Recall that EROs were introduced as analogies of operations used to solve a system of linear equations. Suppose A is the *coefficient matrix* for a system of equations, so that the matrix equation $AX = C$ represents the system. If A is invertible, then the system can be solved with EROs; therefore the existence of A^{-1} is equivalent to the existence of a solution for the system $AX = C$. Asking the question of the invertibility of a matrix is equivalent to examining the question of the existence of a solution for a linear system.

Example 1

The equations in the 2-by-2 system

$$2x_1 + x_2 = 3$$
$$4x_1 + 2x_2 = 8$$

describe parallel lines because the coefficients of the second equation are twice the coefficients of the first equation, but the constant terms are not in a 2-to-1 ratio. This system has no solution. A system of linear equations that has no solution is called *inconsistent.*

The equations in the 2-by-2 system

$$3x_1 - x_2 = 5$$
$$9x_1 - 3x_2 = 15$$

both describe the same line because the coefficients and the constant term of the second equation are 3 times the corresponding numbers in the first equation. The system has an infinite number of solutions. A system of linear equations that has an infinite number of solutions is called *dependent.* ■

If a linear system does not have a unique solution, then the matrix equation $AX = C$ that represents the system cannot be transformed to $X = A^{-1}C$; therefore, the coefficient matrix of an inconsistent or dependent system does not have an inverse. On the other hand, if a linear system represented by $AX = C$ does have a unique solution, the system is called *consistent* and A is invertible.

To find the inverse of a matrix, elementary row operations are used to transform the coefficient matrix A. If at any point in the process two rows of A are found to be equal, then A must not be invertible. What this means is that if any row of a matrix can be expressed as a sum of multiples of the other rows, then that matrix does not have an inverse. Adding multiples of rows is called *linear combination*. The property of a matrix that determines if it is invertible can be stated as follows: *A square matrix is invertible if and only if no row can be expressed as a linear combination of the other rows.*

Example 2

The first system from Example 1 above has a coefficient matrix

$$\begin{pmatrix} 2 & 1 \\ 4 & 2 \end{pmatrix}$$

This matrix is not invertible because the second row is twice the first row.

In the matrix

$$\begin{pmatrix} 5 & -3 & 0 \\ 1 & 2 & -2 \\ 9 & -8 & 2 \end{pmatrix}$$

note that the third row is equal to twice the first row minus the second row, so that this matrix is not invertible. ■

Inverses That Are Difficult to Calculate

If matrix B is not invertible and the entries in a matrix A are close to the entries in a matrix B, then the inverse of A can be difficult to find using a computer or calculator. Computers and calculators have limited precision, meaning only a certain number of digits are stored in any series of calculations. The fact that these machines round off calculations due to the limits of their internal precision leads to a phenomenon called *round-off error*. In sequences of calculations such as those involved in inverting a matrix, round-off errors tend to increase as more calculations are completed. As we attempt to invert matrix A, within a few operations the computer or calculator will probably find A indistinguishable, or nearly so, from B, which does not have an inverse. In such a situation the computer or calculator would not be able to successfully find the inverse of A, or it would find an incorrect inverse. If the computer or calculator is used to compute the solution of a linear system, then the solution may need to be checked in the original system. A system of linear equations with a coefficient matrix that is close to a matrix that is not invertible is an example of what is called an *ill-conditioned* system. In general, an ill-conditioned system is one that is extremely sensitive to small changes in the values of coefficients of the system. A large change in the answer can result from a small change in a coefficient.

Example 3

This example illustrates the difficulties associated with an ill-conditioned system of linear equations. The system

$$\begin{aligned} 2x_1 + x_2 &= 3 \\ 1.9876x_1 + x_2 &= 6 \end{aligned}$$

has solutions $x_1 \approx -241.94$ and $x_2 \approx 486.87$. If the coefficients of this system are rounded to two decimal places, then the new system

$$\begin{aligned} 2x_1 + x_2 &= 3 \\ 1.99x_1 + x_2 &= 6 \end{aligned}$$

has solutions $x_1 = -300$ and $x_2 = 603$. A change in a coefficient by an amount 0.0014 causes a wild fluctuation in the solution. Rounding off 1.9876 to 1.99 leads to a solution quite different from the actual solution to the original system.

The system is ill-conditioned because it is close to the inconsistent system

$$\begin{aligned} 2x_1 + x_2 &= 3 \\ 2x_1 + x_2 &= 6, \end{aligned}$$

which is a system whose coefficient matrix is not invertible. Although computers and calculators store many more digits than are presented in this example, they still produce approximate values which give rise to the difficulties of ill-conditioned systems. ■

Leontief's Model Revisited

Recall that Leontief's Input-Output Model gives the following relationship between consumer demand (D), the economy's production (P), and the input-output interaction between sectors represented by the technology matrix (T):

$$D = P - TP,$$

or

$$D = (I - T)P.$$

Each entry in the technology matrix T represents the output of one sector utilized as input by another sector. More specifically, the entry in row i and column j of T is the relative amount of goods and services needed as input from sector i for the production of the output resources of sector j. Although the technology matrix changes over time, the entries can be determined through research for a certain time period. The technology matrix from the beginning of this section is rewritten below with decimals.

$$T = \begin{matrix} \\ \textit{Ag.} \\ \textit{Manu.} \\ \textit{Tran.} \end{matrix} \begin{matrix} \begin{matrix} \textit{Ag.} & \textit{Manu.} & \textit{Tran.} \end{matrix} \\ \begin{pmatrix} 0.1 & 0.067 & 0 \\ 0.2 & 0.25 & 0.25 \\ 0.2 & 0.2 & 0.167 \end{pmatrix} \end{matrix}$$

The entries of the demand matrix D are often estimated through market research. We wish to determine a matrix P such that $P - TP$ will yield the given entries of D. To solve for the matrix P, left-multiply both sides of the equation $D = (I - T)P$ by the inverse of $(I - T)$ to get

$$(I - T)^{-1}D = (I - T)^{-1}(I - T)P,$$

which simplifies to

$$(I - T)^{-1}D = IP,$$

or

$$(I - T)^{-1}D = P.$$

Once again, it is important to maintain the correct order of multiplication on both sides of the equation.

Suppose the consumer demand is for 100 units of agriculture, 120 units of manufacturing, and 90 units of transportation, where each unit represents one million dollars worth of a sector's goods. This is displayed in a demand matrix as

$$D = \begin{matrix} \\ \textit{Ag.} \\ \textit{Manu.} \\ \textit{Tran.} \end{matrix} \begin{matrix} \textit{Dem.} \\ \begin{pmatrix} 100 \\ 120 \\ 90 \end{pmatrix} \end{matrix}$$

Given this level of consumer demand, we wish to determine the level of production necessary to satisfy this demand as well as the amount of internal consumption given by the product TP.

The first step in calculating the matrix P is to determine $(I - T)^{-1}$. We start with

$$\begin{aligned} I - T &= \begin{pmatrix} 1 & 0 & 0 \\ 0 & 1 & 0 \\ 0 & 0 & 1 \end{pmatrix} - \begin{pmatrix} 0.1 & 0.067 & 0 \\ 0.2 & 0.25 & 0.25 \\ 0.2 & 0.2 & 0.167 \end{pmatrix} \\ &= \begin{pmatrix} 0.9 & -0.067 & 0 \\ -0.2 & 0.75 & -0.25 \\ -0.2 & -0.2 & 0.833 \end{pmatrix} \end{aligned}$$

Using EROs and aided by computer software, we find that

$$(I - T)^{-1} \approx \begin{pmatrix} 1.14 & 0.11 & 0.03 \\ 0.43 & 1.49 & 0.45 \\ 0.38 & 0.38 & 1.32 \end{pmatrix}$$

Now we can calculate P as follows:

$$P = (I - T)^{-1}D$$
$$\approx \begin{pmatrix} 1.14 & 0.11 & 0.03 \\ 0.43 & 1.49 & 0.45 \\ 0.38 & 0.38 & 1.32 \end{pmatrix} \begin{pmatrix} 100 \\ 120 \\ 90 \end{pmatrix}$$

$$\approx \begin{matrix} \textit{Ag.} \\ \textit{Manu.} \\ \textit{Tran.} \end{matrix} \begin{pmatrix} 129.9 \\ 262.3 \\ 202.4 \end{pmatrix}$$

The economy must produce 129.9 million dollars worth of agriculture, 262.3 million dollars worth of manufacturing, and 202.4 million dollars worth of transportation to satisfy consumer demand after the requirements of internal consumption are met.

Exercise Set 4

1. Consider a 4-sector economic system consisting of petroleum, textiles, transportation, and chemicals. The production of 1 unit of petroleum requires 0.2 units of transportation, 0.4 units of chemicals, and 0.1 unit of itself. The production of 1 unit of textiles requires 0.4 units of petroleum, 0.1 unit of textiles, 0.15 units of transportation, and 0.3 units of chemicals. The production of 1 unit of transportation requires 0.6 units of petroleum, 0.1 unit of itself, and 0.25 units of chemicals. Finally, the production of 1 unit of chemicals requires 0.2 units of petroleum, 0.1 unit of textiles, 0.3 units of transportation, and 0.2 units of chemicals.

   a. Write a technology matrix to represent this information.

   b. On what sector is petroleum most dependent? Least dependent?

   c. If the textiles sector has an output of \$4 million, what is the input in dollars from petroleum?

   d. Suppose the production matrix is

$$P = \begin{matrix} \textit{Petr.} \\ \textit{Text.} \\ \textit{Tran.} \\ \textit{Chem.} \end{matrix} \begin{pmatrix} 800 \\ 200 \\ 700 \\ 750 \end{pmatrix}$$

   What is the internal consumption matrix? How much petroleum is left over for external use?

   e. Suppose the demand matrix, in millions of dollars, is

$$D = \begin{matrix} \textit{Petr.} \\ \textit{Text.} \\ \textit{Tran.} \\ \textit{Chem.} \end{matrix} \begin{pmatrix} 25 \\ 14 \\ 30 \\ 42 \end{pmatrix}$$

   How much of each sector must be produced?

2. Suppose the demand matrix given in Exercise 1e is doubled. What is the new production matrix? How does the new production matrix compare with the original production matrix?

3. An economy with the four sectors manufacturing, petroleum, transportation, and hydroelectric power has the following technology matrix:

$$T = \begin{matrix} \\ \textit{Manu.} \\ \textit{Petr.} \\ \textit{Tran.} \\ \textit{HP} \end{matrix} \begin{matrix} \begin{matrix} \textit{Manu.} & \textit{Petr.} & \textit{Tran.} & \textit{HP} \end{matrix} \\ \begin{pmatrix} 0.15 & 0.18 & 0.3 & 0.1 \\ 0.22 & 0.12 & 0.37 & 0 \\ 0.09 & 0.3 & 0.11 & 0 \\ 0.27 & 0.05 & 0.07 & 0.1 \end{pmatrix} \end{matrix}$$

   Find the production matrix if all the entries of the demand matrix are 200.

4. Consider a 3-sector system consisting of steel, coal, and transportation. The production of one unit of coal requires 0.23 units of transportation, 0.19 units of itself, and 0.2 units of steel. The production of 1 unit of steel requires 0.2 units of itself, 0.3 units of coal, and 0.15 units of transportation. Finally, producing 1 unit of transportation requires 0.1 units of itself, 0.15 units of coal, and 0.35 units of steel.

   a. Write a technology matrix to represent this information.

   b. On what sector does coal rely most? Rely least?

c. Which sector depends most on steel?

d. Suppose the production matrix is

$$P = \begin{matrix} \textit{Steel} \\ \textit{Coal} \\ \textit{Tran.} \end{matrix} \begin{pmatrix} 18 \\ 23 \\ 15 \end{pmatrix}.$$

What is the surplus available beyond internal consumption?

e. Suppose the demand matrix is

$$D = \begin{matrix} \textit{Steel} \\ \textit{Coal} \\ \textit{Tran.} \end{matrix} \begin{pmatrix} 24 \\ 19 \\ 12 \end{pmatrix}.$$

Find the production matrix.

5

Additional Applications of the Inverse of a Matrix

This section includes an example and exercises to illustrate other applications of inverses.

Example 1

McDougal's Restaurant sponsors special funding for three projects: scholarships for employees, special public service projects, and beautification of the exteriors of the restaurants. Each of the three locations of McDougal's in Durham, Hillsborough, Northgate, and Boulevard, made requests for funds, with the relative amounts requested by each location for the three projects distributed as shown in Table 2. Headquarters decided to allocate \$100,000 for these projects to the Durham area. The money was to be distributed with 43% to scholarships, 28% to public service projects, and the remaining 29% to beautification. How much will each of the three locations receive?

Solution This problem can be approached by letting the variables x, y, and z stand for the amounts that each of the three locations will receive. A 3-by-3 system of equations based on the

Table 2 Distribution of Funding Requests

Project	Location Hills.	North.	Blvd.
Scholarships	50%	30%	40%
Public service	20%	30%	40%
Beautification	30%	40%	20%

information in this problem can be written, and the system can then be solved by using matrices. Make the following variable assignments:

x = the amount of money for Hillsborough
y = the amount of money for Northgate
z = the amount of money for Boulevard

Each of the three projects leads to an equation in terms of x, y, and z, resulting in the following system of linear equations:

$$\begin{aligned} &\text{Scholarships:} && 0.5x + 0.3y + 0.4z = \$43{,}000 \\ &\text{Public service:} && 0.2x + 0.3y + 0.4z = \$28{,}000 \\ &\text{Beautification:} && 0.3x + 0.4y + 0.2z = \$29{,}000 \end{aligned}$$

A matrix equation for this system is shown below.

$$\begin{matrix} \\ \textit{Schol.} \\ \textit{Pub. Ser.} \\ \textit{Beaut.} \end{matrix} \begin{matrix} \textit{Hills.} & \textit{North.} & \textit{Blvd.} \\ \left(\begin{matrix} 0.5 \\ 0.2 \\ 0.3 \end{matrix}\right. & \begin{matrix} 0.3 \\ 0.3 \\ 0.4 \end{matrix} & \left.\begin{matrix} 0.4 \\ 0.4 \\ 0.2 \end{matrix}\right) \end{matrix} \begin{matrix} \\ H \\ N \\ B \end{matrix} \begin{matrix} \textit{Amt.} \\ \begin{pmatrix} x \\ y \\ z \end{pmatrix} \end{matrix}$$

$$= \begin{matrix} \textit{Schol.} \\ \textit{Pub. Ser.} \\ \textit{Beaut.} \end{matrix} \overset{\textit{Amt.}}{\begin{pmatrix} \$43,000 \\ \$28,000 \\ \$29,000 \end{pmatrix}}$$

This matrix equation is of the form $AX = B$, so we can solve for X by finding A^{-1} and left-multiplying both sides, giving $X = A^{-1}B$. Substituting the entries of the matrices and using the computer or calculator to find the inverse of the coefficient matrix give

$$\begin{pmatrix} x \\ y \\ z \end{pmatrix} \approx \begin{pmatrix} 3.333 & -3.333 & 0 \\ -2.667 & 0.667 & 4 \\ 0.333 & 3.667 & -3 \end{pmatrix} \begin{pmatrix} \$43,000 \\ \$28,000 \\ \$29,000 \end{pmatrix}.$$

After performing this multiplication, we find that

$$\begin{pmatrix} x \\ y \\ z \end{pmatrix} = \begin{pmatrix} \$50,000 \\ \$20,000 \\ \$30,000 \end{pmatrix}$$

in which the entries have been rounded to four significant digits. The $100,000 should be distributed so that $50,000 goes to the Hillsborough location, $20,000 goes to the Northgate location, and $30,000 goes to the Boulevard location. ∎

For the preceding example, we did not actually need to know the entries in A^{-1}. All we are really interested in is the result of the multiplication $A^{-1}B$. Recall that we can find A^{-1} by performing EROs on the augmented matrix $(A|I)$. If we perform the same EROs on B, the result will be $A^{-1}B$; therefore, we can actually find the product $A^{-1}B$ by performing the same EROs on $(A|B)$ that are performed on $(A|I)$ in finding A^{-1}. In the example above, this is accomplished by first forming the augmented matrix

$$\left(\begin{array}{ccc|c} 0.5 & 0.3 & 0.4 & 43,000 \\ 0.2 & 0.3 & 0.4 & 28,000 \\ 0.3 & 0.4 & 0.2 & 29,000 \end{array}\right)$$

then using EROs to transform the left three columns into an identity matrix. The numbers in the right-hand column will be the solution to the system of equations found in the previous example.

Exercise Set 5

1. Explain why a matrix that contains a column of all zeros cannot have an inverse.

2. Solve the following systems using matrix algebra.

a. $2x - 5y + 7z = 4$
$3x + y - 12z = -8$
$5x + 2y - 4z = 3$

b. $2x - 5y + 7z = 4$
$3x + y - 12z = -8$
$5x - 4y - 5z = -4$

c. $2x - 5y + 7z = 4$
$3x + y - 12z = -8$
$7x - 9y + 2z = 1$

3. Find the inverses of the following matrices.

a.
$$\begin{pmatrix} 2 & -7 & 5 \\ 1 & -3 & -10 \\ 3 & 4 & -5 \end{pmatrix}$$

b.
$$\begin{pmatrix} 1 & 2 & 4 & 8 \\ 1 & 3 & 9 & 27 \\ 1 & 4 & 16 & 64 \\ 1 & 5 & 25 & 125 \end{pmatrix}$$

4. Suppose that the Hillsborough location of McDougal's changes its request to by asking for 43% for scholarships, 28% for public service, and 29% for beautification, using the same allocation scheme as the headquarters did. The other stores keep their original requests. How much money will each store receive?

5. The snack bar makes two types of sandwiches. The chicken sandwich takes 6 minutes to cook and 2 minutes to put on a bun with the lettuce, tomato, and mayonnaise. The hamburger takes 4 minutes to cook and 3 minutes to put on a bun with the tomato, lettuce, and mustard.

a. How many hamburgers and chicken sandwiches can be produced if 40 minutes is spent on cooking and 20 minutes on preparing the sandwiches?

b. How many hamburgers and chicken sandwiches can be produced with 35 minutes spent on cooking and 25 on preparing the sandwiches?

c. How many with 45 minutes spent on cooking and 15 minutes spent preparing the sandwiches?

6. In Exercise 10 in section 3, we found that the matrix below rotates a point (x, y) in the plane through an angle θ about the origin. Find the inverse of this matrix. What transformation does the inverse perform?

$$\begin{pmatrix} \cos\theta & -\sin\theta \\ \sin\theta & \cos\theta \end{pmatrix}$$

7. A total of \$30,000 is available in a school for student groups to spend on their projects. The Student Council, the Beta Club, and the 4-H Club were asked to submit proposals describing how they would spend their portion of the \$30,000. The principal accepted the following proposals for how each group would spend its allocation of the money:

	Activities	Commun. Service	Club Exp.
Student Council	40%	40%	20%
Beta Club	20%	50%	30%
4-H Club	20%	30%	50%

a. The principal wants to allocate 30% of the funds to activities, 50% to community service, and 20% to club expenses. How much money would each club receive under these conditions? What problem results from this allocation?

b. Assuming the principal's allocation scheme in part a, how much would each club receive if the Beta Club changed its proposal so that it would spend 30% on activities, 50% on community services, and 20% on club expenses?

c. The principal actually decides to allocate 25% of the funds to activities, 40% to community service, and 35% to club expenses. On the basis of the original proposals from each group, how much money does the principal allocate to each group?

6

The Leslie Matrix Model

Population growth is a significant phenomenon for which many mathematical models have been developed. A frequently used model is the exponential function

$$P(t) = P_0 e^{kt},$$

in which $P(t)$ is the size of a population growing without limits. Constrained growth of a population can be modeled by the logistic growth function. The logistic equation is

$$P(t) = \frac{QM}{Q + e^{-kt}},$$

in which M is the maximum sustainable population and

$$Q = \frac{P_0}{M - P_0}$$

is the ratio of the initial population to the room for growth. Both of these models are macromodels, meaning that the models consider the population as a whole. In this section, a micromodel is developed that allows us to investigate questions about the different age groups within an entire population.

Modeling Age-Specific Population Growth

The future of Social Security, the future of veteran's benefits, and the changing school population in different regions of the country are current issues in public policy. The principal question arising in each of these discussions is how many people will be of a certain age after a period of time. The total population can be modeled with the equations above, but these macromodels provide little help in answering age-specific growth questions. We would like to be able to examine the growth and decline of future populations within various age groups. The model developed in this section will enable us to make these age-specific projections.

A fundamental assumption we will use in our model is that the proportion of males in the population is the same as the proportion of females, an assumption justified for most species. Consider a female population of small woodland mammals; Table 3 gives the populations for 3-month age groups. The total population in each age group is assumed to be twice the female population. The life span of this mammal is assumed to be 15–18 months, so none advance beyond the final column in Table 3. Our primary task with this data is to derive a mathematical model that will allow us to predict the number of animals in each age group after some number of years. To proceed, we first need to know something about the birth rate and death rate for each age group, rates that vary with age for most animal populations.

Table 3 A Population of Small Woodland Mammals

Age (months)	0–3	3–6	6–9	9–12	12–15	15–18
Number of Females	14	8	12	4	0	0

The birth rate depends on a variety of factors, such as the probability that an animal will become pregnant, the number of pregnancies that can occur in any age group, and the average number of newborns in a litter. For animals that bear only one young and require a long gestation period, the birth rate is low; however, for insects, fish, and other species that bear thousands of young at a time, the birth rate is high.

Generally, the birth rate is given as a proportion of the total population. For example, if an animal population of 100 has a birth rate of 0.4, then it is understood that the 100 animals will produce 40 newborns, or 0.4 of the population. If we consider only the female population, the 50 females of the 100 animals (half of the total) will be reproducing at a rate of 0.8 of their population, since the 50 females will produce 40 young. Only 20 of these 40 young are expected to be female, however, so the birth rate of females would be 0.4 of the female population. In our model, the birth rate will represent the average number of daughters born to each female in the population during a specified time interval. Under the assumption of equal female and male proportions, this definition of birth rate is equivalent to viewing birth rate as a proportion of the total population.

For the species with the age distribution given in Table 3, the birth rate and death rate by age are listed in Table 4.

Table 4 Birth Rate and Death Rate for Each Age Group

Age (months)	0–3	3–6	6–9	9–12	12–15	15–18
Birth Rate	0	0.3	0.8	0.7	0.4	0
Death Rate	0.4	0.1	0.1	0.2	0.4	1

To investigate the mammal population in each age group over time, we begin by considering the following question: After 3 months, how many females will there be in each age group?

The populations of age groups 3–6, 6–9, 9–12, and 12–15 are easily found. The calculations can be simplified by introducing a quantity called the survival rate (SR), which is equal to one minus the death rate (DR). In symbols, the relationship is

$$SR = 1 - DR.$$

Whereas the death rate indicates the proportion of a population group that dies in a 3-month interval, the survival rate is the proportion of a population group that survives a 3-month period. The survival rate of the 0–3 age group is 0.6 ($= 1 - 0.4$), and the number of mammals from the original 14 that advance to the 3–6 age group after 3 months is

$$(0.6)(14) = 8.4.$$

The survival rate of the 3–6 age group is 0.9 ($= 1 - 0.1$), and of the original 8 mammals in this group, the number that advance to the 6–9 age group after 3 months is

$$(0.9)(8) = 7.2.$$

The calculations for the number of mammals moving up to the next age group after 3 months are summarized in Table 5.

The numbers in Table 5 may seem strange because the populations are not rounded to integers but contain fractional parts. These fractions of animals should remain in our analysis. The birth rates and survival rates used in the model are probabilistic quantities. They represent the

Table 5 Movement of Females Up through Age Groups

Age	DR	SR	Number	Number Moving Up
0–3	0.4	0.6	14	$(0.6)(14) \Rightarrow 8.4$ move up to the 3–6 age group
3–6	0.1	0.9	8	$(0.9)(8) \Rightarrow 7.2$ move up to the 6–9 age group
6–9	0.1	0.9	12	$(0.9)(12) \Rightarrow 10.8$ move up to the 9–12 age group
9–12	0.2	0.8	4	$(0.8)(4) \Rightarrow 3.2$ move up to the 12–15 age group
12–15	0.4	0.6	0	$(0.6)(0) \Rightarrow 0$ move up to the 15–18 age group
15–18	1	0	0	no animal advances beyond the 15–18 group

probable rates for a given time, perhaps found by averaging data on a species for a number of years. We would not expect these rates to be exact at any one time, but we expect that over the long run they accurately reflect age-specific population growth. The fractional parts can make a significant difference in calculations over time, so they must be retained. To obtain an estimate of the population at a definite time, or a "snapshot" of the process, one would of course round off numbers to the nearest integer.

A question we have left unanswered is how many females enter the 0–3 age group. This is the sum of all the births in each of the other age groups. To find the births in each age group, multiply the birth rate times the population in that age group. Using the data from Table 4, this sum is

$$14(0) + 8(0.3) + 12(0.8) + 4(0.7) + 0(0.4) + 0(0)$$
$$= 0 + 2.4 + 9.6 + 2.8 + 0 + 0 = 14.8.$$

Consolidating this number with the information in Table 5 gives us the female population in each age group after 3 months, as presented in Table 6. Notice that the population has grown from 38 to about 44 animals after 3 months.

After another 3-month period, how many females will be in each group? Sixty percent of the 14.8 females in the 0–3 age group, or 8.88 females, will survive to move into the 3–6 age group. Of the 8.4 females in the 3–6 age group, ninety percent (7.56) will survive to move into the 6–9 age bracket. Ninety percent of the 7.2 females in the 6–9 age bracket, or 6.48 females, will survive to move into the 9–12 age group, while eighty percent (8.64) of the 10.8 females in the 9–12 age group will survive. Of the 3.2 females in the 12–15 age group, sixty percent (1.92) will move into the 15–18 age group. The 0–3 age group will be populated by newborns from each group. The total moving into the 0–3 age group will be

$$14.8(0) + 8.4(0.3) + 7.2(0.8) + 10.8(0.7)$$
$$+3.2(0.4) + 0(0) = 17.12.$$

Table 7 shows the number of animals in the female population after 6 months. The total population of females is now about 51, which implies, by the equal-proportions assumption, that the total animal population has reached about 102.

Table 6 Female Population after Three Months

Age	0–3	3–6	6–9	9–12	12–15	15–18
Number	14.8	8.4	7.2	10.8	3.2	0

Table 7 Female Population after Six Months

Age	0–3	3–6	6–9	9–12	12–15	15–18
Number	17.12	8.88	7.56	6.48	8.64	1.92

Class Practice

1. Find the total population of females after 9 months.

2. Find the total population of females after 12 months.

The Leslie Matrix

In the previous example, we saw an increase in the population from 38 to about 44 during the initial 3-month interval, followed by an increase to about 51 after 6 months. Further calculations reveal that the population remains around 51–52 for the next two cycles, 9 and 12 months. During subsequent intervals will the population remain stable around 51, begin to grow again, or start to die out? We can answer this question by performing the same sequence of operations to calculate the population every 3 months; however, the matrix data structure provides an easier way to determine future age-specific population distributions.

The form of the calculations in the preceding section provides clues for developing a matrix model of age-specific population growth. The sums of products that were used in the calculations can each be represented as a row of one matrix multiplied by a column of another matrix. We will develop a matrix representation for the data that determine the age distribution of a population from one interval to the next, namely the birth rates and survival rates. This matrix is called the *Leslie matrix*, and will be symbolized by L. The survival rate for the kth age group is denoted by S_k. Similarly, denote the birth rate for the initial age group (0–3 months) by B_1, the next (3–6 months) by B_2, and, in general, the birth rate of the kth age group by B_k.

The number of females in each age group can be represented by a column vector called an *age distribution vector*. In our example, we have

$$X_0 = \begin{pmatrix} 14 \\ 8 \\ 12 \\ 4 \\ 0 \\ 0 \end{pmatrix} \quad \text{and} \quad X_1 = \begin{pmatrix} 14.8 \\ 8.4 \\ 7.2 \\ 10.8 \\ 3.2 \\ 0 \end{pmatrix}$$

where the subscript of X signifies the number of 3-month intervals that have elapsed. X_0 was given and X_1 was calculated based on X_0 and the birth rates and survival rates.

The first entry in X_1 was obtained by multiplying the values in X_0 times the birth rates for the n different age groups. If the first row of L is

$$\begin{pmatrix} B_1 & B_2 & B_3 & B_4 & \cdots & B_n \end{pmatrix},$$

then multiplying the first row of L with the column vector X_0 will yield the first entry in X_1. The second entry in X_1 is the product of the survival rate for the first age group and the number of females in the first age group from the previous cycle, which is the first entry in X_0. If the second row of L is

$$\begin{pmatrix} S_1 & 0 & 0 & 0 & \cdots & 0 \end{pmatrix},$$

then multiplying this by X_0 will provide the second entry in X_1. The third entry in X_1 is the product of the survival rate for the second age group and the number of females in the second age group from the previous cycle, which is the second entry in X_0. If the third row of L is

$$\begin{pmatrix} 0 & S_2 & 0 & 0 & \cdots & 0 \end{pmatrix},$$

then the result of multiplying this row by X_0 is the third entry in X_1.

This pattern is continued for the remaining rows of L; therefore, the Leslie matrix L is defined as

$$L = \begin{pmatrix} B_1 & B_2 & B_3 & B_4 & \cdots & B_{n-1} & B_n \\ S_1 & 0 & 0 & 0 & \cdots & 0 & 0 \\ 0 & S_2 & 0 & 0 & \cdots & 0 & 0 \\ 0 & 0 & S_3 & 0 & \cdots & 0 & 0 \\ \vdots & \vdots & \vdots & \vdots & & \vdots & \vdots \\ 0 & 0 & 0 & 0 & \cdots & S_{n-1} & 0 \end{pmatrix}$$

For the woodland mammal example we have

$$L = \begin{pmatrix} 0 & 0.3 & 0.8 & 0.7 & 0.4 & 0 \\ 0.6 & 0 & 0 & 0 & 0 & 0 \\ 0 & 0.9 & 0 & 0 & 0 & 0 \\ 0 & 0 & 0.9 & 0 & 0 & 0 \\ 0 & 0 & 0 & 0.8 & 0 & 0 \\ 0 & 0 & 0 & 0 & 0.6 & 0 \end{pmatrix}$$

Why have we chosen the definition of the Leslie matrix explained above? To see the rationale for the definition, examine the product LX_0 shown below.

$$LX_0 = \begin{pmatrix} 0 & 0.3 & 0.8 & 0.7 & 0.4 & 0 \\ 0.6 & 0 & 0 & 0 & 0 & 0 \\ 0 & 0.9 & 0 & 0 & 0 & 0 \\ 0 & 0 & 0.9 & 0 & 0 & 0 \\ 0 & 0 & 0 & 0.8 & 0 & 0 \\ 0 & 0 & 0 & 0 & 0.6 & 0 \end{pmatrix} \begin{pmatrix} 14 \\ 8 \\ 12 \\ 4 \\ 0 \\ 0 \end{pmatrix}$$

$$= \begin{pmatrix} 0(14)+0.3(8)+0.8(12)+0.7(4)+0.4(0)+0(0) \\ 0.6(14)+0(8)+0(12)+0(4)+0(0)+0(0) \\ 0(14)+0.9(8)+0(12)+0(4)+0(0)+0(0) \\ 0(14)+0(8)+0.9(12)+0(4)+0(0)+0(0) \\ 0(14)+0(8)+0(12)+0.8(4)+0(0)+0(0) \\ 0(14)+0(8)+0(12)+0(4)+0.6(0)+0(0) \end{pmatrix} = \begin{pmatrix} 14.8 \\ 8.4 \\ 7.2 \\ 10.8 \\ 3.2 \\ 0 \end{pmatrix}$$

The matrix above is just X_1; thus, the form of the Leslie matrix leads to the matrix equation $LX_0 = X_1$.

Class Practice

1. Verify that $LX_k = X_{k+1}$ for $k = 1$, 2, and 3 by comparing these products with the values shown in the text and the preceding Class Practice exercises.
2. Verify that $L^k X_0 = X_k$ for $k = 1$, 2, 3, and 4 by comparing these products with the X_k vectors evaluated previously. L^k is defined as the matrix L raised to the kth power, or in other words, L taken as a factor k times.

The Leslie matrix has been defined so that left-multiplication by it will generate the age distribution vectors in the sequence $X_0, X_1, \ldots, X_{k-1}, X_k$, as shown in the first Class Practice exercise above. Multiplication by L determines the next successive age distribution vector from the current vector; however, we need not go through all of the $k-1$ preceding vectors just to find the kth vector. Since we always left-multiply X_{k-1} by L to find X_k, the associative property of matrix multiplication implies that we can simply multiply L by itself k times (which is L^k) and then right-multiply the result by the initial age distribution vector X_0. This property was demonstrated in the second Class Practice exercise above and is shown in general in the calculations below.

We have observed that

$$X_1 = LX_0 \quad \text{and} \quad X_2 = LX_1.$$

By substitution of the first expression for X_1 into the second equation, we find that

$$X_2 = L(LX_0) = L^2 X_0.$$

Similarly, substitution of the expression for X_2 into the equation

$$X_3 = LX_2$$

yields

$$X_3 = L(L^2 X_0) = L^3 X_0.$$

In general, we have the following sequence of equivalent expressions:

$$\begin{aligned} X_k &= LX_{k-1} = L(LX_{k-2}) \\ &= L(L(L \cdots (LX_0))) \\ &= \overbrace{(L(L(L \cdots L) \cdots))}^{k \text{ times}} X_0 \\ &= L^k X_0. \end{aligned}$$

An interesting pattern emerges in the long-term behavior of a process modeled with the Leslie matrix. The total female population of the woodland mammals is shown below for the first 5 cycles (a total elapsed time of 15 months).

Cycle	0	1	2	3	4	5
Female Pop.	38	44.4	50.6	52.14	51.33	53.76

No obvious pattern is apparent from examining the total female population during the first 5 cycles; however, a pattern does emerge further along in the process. The total female population during cycles 10–13 increases by a gradually larger percent as shown in the table below.

Cycle	10	11	12	13
Female Pop.	62.7874	64.5985	66.5059	68.5738
Percent Growth		2.88	2.95	3.11

Does the percentage growth of the population during each cycle continue to increase? Moving even further along in the process sheds light on this question. The results of calculations for cycles 20–23 are shown below.

Cycle	20	21	22	23
Female Pop.	84.5867	87.1639	89.8168	92.5485
Percent Growth		3.05	3.04	3.04

The growth rate appears to be stabilizing at about 3.04% per 3-month cycle. After 30 cycles, the total female population is 114.158, and after 31 cycles, it is 117.632 — an increase of 3.04%. Additional calculations confirm that the growth rate of the population converges to about 3.04%. This is called the *long-term growth rate* of the total population. The growth rate of the total population after a small number of cycles, called the *short-term growth rate*, is variable and does not reveal a pattern. After a large number of cycles, however, a stable long-term growth rate appears.

Examination of the age distribution vector X_k for large k also leads to some interesting results. The age distributions of the population after 20 and 21 cycles are

$$X_{20} = \begin{pmatrix} 27.46 \\ 15.99 \\ 13.97 \\ 12.20 \\ 9.47 \\ 5.51 \end{pmatrix} \quad \text{and} \quad X_{21} = \begin{pmatrix} 28.29 \\ 16.47 \\ 14.39 \\ 12.57 \\ 9.76 \\ 5.68 \end{pmatrix}.$$

The total female population after 20 cycles is 84.60, and after 21 cycles is 87.16. If the entries in an age distribution vector are each divided by the current total female population, then the resulting entries are the proportions of the total population found in each age group at that time. If each entry in X_{20} is divided by the total female population 84.60 and each entry in X_{21} is divided by the total female population 87.16, then the proportions found are

$$X_{20}/84.60 = \begin{pmatrix} 0.3246 \\ 0.1890 \\ 0.1651 \\ 0.1442 \\ 0.1119 \\ 0.0651 \end{pmatrix}$$

and

$$X_{21}/87.16 = \begin{pmatrix} 0.3246 \\ 0.1890 \\ 0.1651 \\ 0.1442 \\ 0.1120 \\ 0.0652 \end{pmatrix}.$$

The proportion of the population in each age group appears to reach a stable distribution. Examining the data after cycles 30 and 31, we find that the differences in the proportions from X_{30} to X_{31} are less than 10^{-6}. In the long run, therefore, the woodland mammal population tends to the following approximate distribution: 32% are 0–3 months old, 19% are 3–6 months old, 17% are 6–9 months old, 14% are 9–12 months old, 11% are 12–15 months old, and 7% are 15–18 months old.

A final property of the Leslie matrix model concerns the growth rates of the different age groups. Since the long-term proportions in each age group are fixed, and the growth rate of the total population eventually remains at 3.04%, then the growth rate of each age group must also eventually converge to 3.04%. This implies that for a large enough value of k, the age distribution vector X_k is equal to the scalar 1.0304 times the previous age distribution vector X_{k-1}. In symbols, we have that eventually, for large k,

$$X_k = 1.0304 X_{k-1}.$$

In the calculations above, we have seen evidence of the following general characteristics of the long-term behavior of the Leslie matrix model:

1. The growth rate of the total population converges to a constant percentage growth rate, called the long-term growth rate.
2. The proportions of the population in each age group eventually approach fixed values, and the growth rate of each age group stabilizes at the same value as the long-term growth rate of the total population.
3. For large k, successive age distribution vectors are related by
$$X_k = (1 + r) X_{k-1},$$
where r is the long-term growth rate.

Exercise Set 6

1. Use the data from Table 5, page 269 , to determine how many mammals will be in the 6–9 month age group in 5 years and in 10 years.

2. How long will it take for the total number of mammals to exceed 500? **Note**: Instead of adding up the entries in an age distribution vector by hand to find the total population, matrix software or some calculators will conveniently provide the total population if the age distribution vector is left-multiplied by the row vector

$$\begin{pmatrix} 1 & 1 & \cdots & 1 \end{pmatrix}.$$

3. For each initial age distribution vector given below, use the Leslie matrix from the woodland mammal example to determine the length of time before the total population reaches 500.

   a. $$X_0 = \begin{pmatrix} 20 \\ 10 \\ 8 \\ 0 \\ 0 \\ 0 \end{pmatrix}$$

   b. $$X_0 = \begin{pmatrix} 38 \\ 0 \\ 0 \\ 0 \\ 0 \\ 0 \end{pmatrix}$$

   c. $$X_0 = \begin{pmatrix} 6 \\ 6 \\ 6 \\ 6 \\ 6 \\ 8 \end{pmatrix}$$

4. For each of the initial age distributions given in Exercise 3, determine the long-term growth rate of the total population. Use the same Leslie matrix. How does the initial age distribution appear to be related to the long-term growth rate?

5. Using the Leslie matrix from the woodland mammal example, compare the long-term proportions of the population in each age group for the initial age distribution vectors

$$X_0 = \begin{pmatrix} 20 \\ 15 \\ 10 \\ 0 \\ 0 \\ 0 \end{pmatrix} \quad \text{and} \quad X_0 = \begin{pmatrix} 22 \\ 16 \\ 12 \\ 9 \\ 8 \\ 10 \end{pmatrix}.$$

   Form a conjecture based on your results.

6. Suppose an animal population has the characteristics described in the table below.

Age Group (years)	0–5	5–10	10–15	15–20
Birth Rate	0	0	1.2	0.8
Death Rate	0.5	0.2	0.1	0.1

Age Group (years)	20–25	25–30	30–35
Birth Rate	0.7	0.2	0
Death Rate	0.3	0.5	1

   a. What is the expected life span of this animal?

   b. Construct the Leslie matrix for this animal.

   c. For the initial female population given in the table shown below, find the female age distribution and the total female population after 300 years.

Age Group	0–5	5–10	10–15	15–20
Number	30	30	26	28

Age Group	20–25	25–30	30–35
Number	32	15	10

   d. Determine the long-term growth rate for this population.

   e. If the maximum sustainable population for this animal in its native habitat is 700, when will the maximum population be reached?

7

Markov Chains

This section begins with an example that illustrates many of the essential features of constructing a class of matrix models called Markov chains.

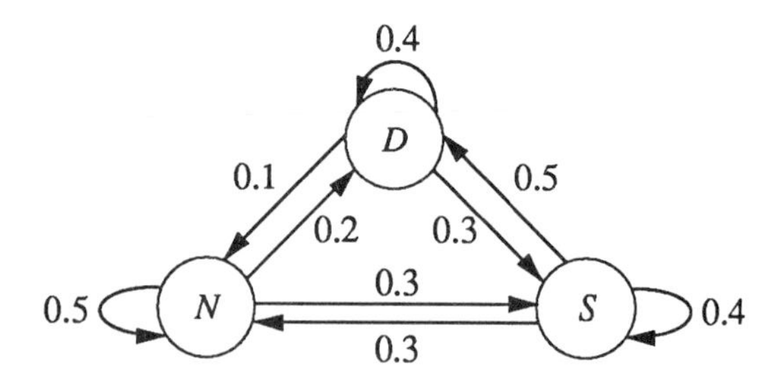

Figure 8 State Diagram for the Taxi Problem

Example 1

The Taxi Problem

A taxi company has divided the city into three regions — Northside, Downtown, and Southside. By keeping track of pickups and deliveries, the company has found that of the fares picked up in Northside, 50% stay in that region, 20% are taken Downtown, and 30% go to Southside. Of the fares picked up Downtown, only 10% go to Northside, 40% stay Downtown, and 50% go to Southside. Of the fares picked up in Southside, 30% go to each of Northside and Downtown, while 40% stay in Southside.

We would like to know what the distribution of taxis will be over time as they pick up and drop off successive fares. This is a difficult analysis that we will approach first through a simpler question in this example. Examples later in this section will gradually work toward the analysis of the long-term distribution of taxis throughout the city. The question we will examine for now is the following: If a taxi starts off Downtown, what is the probability that it will be Downtown after letting off its third fare?

Solution The information in this example can be represented with a state diagram which includes (a) three states D, N, and S corresponding to the three regions of the city; and (b) the probabilities of moving from one region to another. The state diagram for the Taxi Problem is shown in Figure 8. In general, the movement from one state to another is called a *transition*. In this example, a transition corresponds to a customer being picked up in a region and dropped off in a region. Transitions to all three regions are feasible from every region; therefore, each state in the state diagram is connected to all other states, including itself.

A taxi that starts off Downtown and ends up Downtown after three fares can follow several different paths through the state diagram. It could go Southside after one fare, then Northside after two fares, and end up Downtown after three fares; or, it could go Northside, Downtown, and Downtown on its three fares. This reasoning shows that we must consider all possible combinations of fares such that the third fare ends up Downtown. The picture in Figure 9, called a *tree diagram*, shows all possible paths starting Downtown and picking up three fares, with the third fare ending up Downtown. With a taxi picking up a fare Downtown, there is a probability of 0.1 that the taxi will go to Northside, a probability of 0.4 of dropping off the fare Downtown, and a probability of 0.5 of heading to Southside. These probabilities are indicated in the tree diagram on the lines, called *branches*, that represent the possible destinations of the first fare. The first set of branches start at D and end at N, D, and S. The branches representing the second fare likewise have the associated probabilities shown in the tree diagram. Each path through the tree from the starting D to the ending D represents a possible sequence of three fares in which the first fare starts Downtown and the third fare ends Downtown. Every such path has a probability of occurring that is determined by multiplying the probabilities for each branch in the path. Our objective is to determine the probabilities associated with following each of the possible paths through the tree diagram.

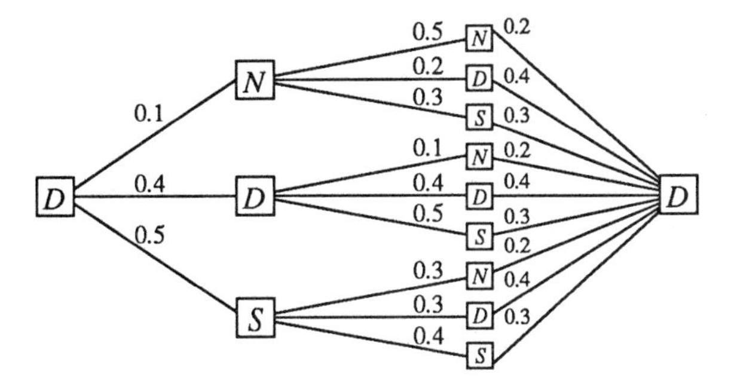

Figure 9 Tree Diagram for the Taxi Problem

After one fare, the probabilities of being Northside (N), Downtown (D), and Southside (S) are

$$P(N_1) = 0.1, \qquad P(D_1) = 0.4, \qquad P(S_1) = 0.5,$$

where the subscripts refer to the number of fares.

Referring to the tree diagram, we observe that after two fares, the probability of being Northside can be expressed in symbols as

$$\begin{aligned} P(N_2) = {} & P(N_1)P(NN) + P(D_1)P(DN) \\ & + P(S_1)P(SN), \end{aligned}$$

where $P(NN)$ is the probability of going from Northside to Northside, $P(DN)$ is the probability of going from Downtown to Northside, and $P(SN)$ is the probability of going from Southside to Northside. Likewise, the probabilities of being Downtown and Southside after two fares are

$$\begin{aligned} P(D_2) = {} & P(N_1)P(ND) + P(D_1)P(DD) \\ & + P(S_1)P(SD) \end{aligned}$$

and

$$\begin{aligned} P(S_2) = {} & P(N_1)P(NS) + P(D_1)P(DS) \\ & + P(S_1)P(SS). \end{aligned}$$

Each of the terms in the sums above is a product of two probabilities — of being in a certain first region after one fare and of going from the first to the second region on the second fare. For example, $P(N_1)P(NS)$ is the probability of going to Northside on the first fare times the probability of going from Northside to Southside on the second fare. Substituting the appropriate values from the tree diagram in the sums gives the following:

$$\begin{aligned} P(N_2) &= (0.1)(0.5) + (0.4)(0.1) + (0.5)(0.3) \\ &= 0.24 \\ P(D_2) &= (0.1)(0.2) + (0.4)(0.4) + (0.5)(0.3) \\ &= 0.33 \\ P(S_2) &= (0.1)(0.3) + (0.4)(0.5) + (0.5)(0.4) \\ &= 0.43 \end{aligned}$$

Notice that $P(N_2)+P(D_2)+P(S_2) = 1$, meaning that the taxi must be in one of the three regions after two fares, as expected.

The probability of being Downtown after three fares is

$$\begin{aligned} P(D_3) &= P(N_2)P(ND) + P(D_2)P(DD) \\ &\quad + P(S_2)P(SD) \\ &= (0.24)(0.2) + (0.33)(0.4) + (0.43)(0.3) \\ &= 0.309. \end{aligned}$$

A taxi starting Downtown has a probability of 0.309 of being Downtown after three fares. ■

The preceding example describes mathematical modeling techniques for analyzing the behavior of a system that includes a finite number of states — the three regions of the city — and probabilities of transitions from each state to every other possible state. The probabilities are constant and independent of previous behavior. We assume that a transition — picking up and dropping off a fare — occurs each time the system is observed, and that observations occur at regular intervals. Systems with these characteristics are called *Markov chains* or *Markov processes.*

Suppose we wish to take the Taxi Problem further and determine the probability of a taxi being Downtown after five fares if it started Downtown. At this point, finding the probability of being Downtown after five fares appears to be a tedious endeavor. Let us return to the calculations we used to determine the probability of being Downtown after three fares. From our previous experiences with matrices, we might guess that the calculations above can be simplified considerably with the use of matrix multiplication. The probabilities of being in each region after two fares, represented by $P(N_2)$, $P(D_2)$, and $P(S_2)$, were calculated by finding sums of products. The first product in each sum has a factor of 0.1; the second product in each sum has a factor of 0.4; and the third product in each sum has a factor of 0.5. These observations lead us to deduce that the probabilities of being in each region after two fares are given by the row vector resulting from the matrix multiplication shown below:

$$\begin{pmatrix} 0.1 & 0.4 & 0.5 \end{pmatrix} \begin{pmatrix} 0.5 & 0.2 & 0.3 \\ 0.1 & 0.4 & 0.5 \\ 0.3 & 0.3 & 0.4 \end{pmatrix} = \begin{pmatrix} 0.24 & 0.33 & 0.43 \end{pmatrix}.$$

The matrix T defined by

$$T = \begin{matrix} & \begin{matrix} N & D & S \end{matrix} \\ \begin{matrix} N \\ D \\ S \end{matrix} & \begin{pmatrix} 0.5 & 0.2 & 0.3 \\ 0.1 & 0.4 & 0.5 \\ 0.3 & 0.3 & 0.4 \end{pmatrix} \end{matrix}$$

is crucial to our calculations. T is called a *transition matrix* because it contains entries such that T_{ij} is the probability of a transition from region i to region j. For example, T_{13} is the probability of a fare that originates in Northside going to Southside. The probability of a fare that originates in Southside going to Downtown is T_{32}. Continuing with our analysis of the previous calculations, observe that the probability of being Downtown after three fares is equal to the product of the row vector after two fares and the column vector composing the second column of T, as shown below:

$$\begin{pmatrix} 0.24 & 0.33 & 0.43 \end{pmatrix} \begin{pmatrix} 0.2 \\ 0.4 \\ 0.3 \end{pmatrix} = 0.309$$

Substitution of the matrix product that produced the leftmost matrix in the equation above yields

$$\begin{pmatrix} 0.1 & 0.4 & 0.5 \end{pmatrix} \begin{pmatrix} 0.5 & 0.2 & 0.3 \\ 0.1 & 0.4 & 0.5 \\ 0.3 & 0.3 & 0.4 \end{pmatrix} \times \begin{pmatrix} 0.2 \\ 0.4 \\ 0.3 \end{pmatrix} = 0.309$$

Instead of using only the second column of T as the rightmost matrix in the product above, substitute the entire matrix T as follows:

$$\begin{pmatrix} 0.1 & 0.4 & 0.5 \end{pmatrix} \begin{pmatrix} 0.5 & 0.2 & 0.3 \\ 0.1 & 0.4 & 0.5 \\ 0.3 & 0.3 & 0.4 \end{pmatrix} \times \begin{pmatrix} 0.5 & 0.2 & 0.3 \\ 0.1 & 0.4 & 0.5 \\ 0.3 & 0.3 & 0.4 \end{pmatrix} = \begin{pmatrix} 0.282 & 0.309 & 0.409 \end{pmatrix}$$

The leftmost matrix above is merely the second row of T. It can be found by left-multiplying T by the vector (0 1 0); therefore, the calculations through three fares can be written as

$$\begin{pmatrix} 0 & 1 & 0 \end{pmatrix} \begin{pmatrix} 0.5 & 0.2 & 0.3 \\ 0.1 & 0.4 & 0.5 \\ 0.3 & 0.3 & 0.4 \end{pmatrix} \times \begin{pmatrix} 0.5 & 0.2 & 0.3 \\ 0.1 & 0.4 & 0.5 \\ 0.3 & 0.3 & 0.4 \end{pmatrix} \begin{pmatrix} 0.5 & 0.2 & 0.3 \\ 0.1 & 0.4 & 0.5 \\ 0.3 & 0.3 & 0.4 \end{pmatrix}$$

which is equivalent to

$$\begin{pmatrix} 0 & 1 & 0 \end{pmatrix} T^3 = \begin{pmatrix} 0.282 & 0.309 & 0.409 \end{pmatrix}.$$

The probability of being Downtown after three fares when starting out Downtown is 0.309, the middle entry in the row vector shown above. What is the meaning of the other entries in this row vector? The result of multiplying T^3 by (0 1 0) is the second row of T^3, specifically (0.282 0.309 0.409). Rather than focus only on the second row of T^3, let us examine the following more general question: What meaning should we attach to all of the entries of T^3?

Before answering this question, consider the entries in T^2 shown below.

$$T^2 = \begin{matrix} & \begin{matrix} N & D & S \end{matrix} \\ \begin{matrix} N \\ D \\ S \end{matrix} & \begin{pmatrix} 0.5 & 0.2 & 0.3 \\ 0.1 & 0.4 & 0.5 \\ 0.3 & 0.3 & 0.4 \end{pmatrix} \end{matrix} \begin{matrix} & \begin{matrix} N & D & S \end{matrix} \\ \begin{matrix} N \\ D \\ S \end{matrix} & \begin{pmatrix} 0.5 & 0.2 & 0.3 \\ 0.1 & 0.4 & 0.5 \\ 0.3 & 0.3 & 0.4 \end{pmatrix} \end{matrix}$$

$$= \begin{matrix} & \begin{matrix} N & D & S \end{matrix} \\ \begin{matrix} N \\ D \\ S \end{matrix} & \begin{pmatrix} 0.36 & 0.27 & 0.37 \\ 0.24 & 0.33 & 0.43 \\ 0.3 & 0.3 & 0.4 \end{pmatrix} \end{matrix}$$

The entry in row 1, column 2 of T^2 is derived by multiplying row 1 of T by column 2 of T as follows:

$$(0.5)(0.2) + (0.2)(0.4) + (0.3)(0.3) = 0.27.$$

Row 1 of T contains the probabilities for the transitions from N to each of the regions, whereas column 2 of T contains the probabilities for the transitions from each of the regions to D. In symbols, the product of row 1 times column 2 is

$$P(NN)P(ND) + P(ND)P(DD) \\ +P(NS)P(SD),$$

which is the probability of going from region N to region D after 2 transitions. In the matrix T^2, the entry in row 1, column 2 has row label N and column label D, and it represents the probability of going from N to D in 2 transitions. Using similar reasoning, it can be shown that the entry in row 1, column 1 of T^2 gives the probability of starting in N and ending in N after 2 transitions. In general, the ij entry of T^2 gives the probability of starting in region i and ending in region j after 2 transitions.

What do the entries in T^3 represent?

$$T^3 = \begin{matrix} & N & D & S \\ N & 0.318 & 0.291 & 0.391 \\ D & 0.282 & 0.309 & 0.409 \\ S & 0.3 & 0.3 & 0.4 \end{matrix}$$

The row and column labels were useful in interpreting T^2, and they are included with T^3 as well. The probability of being in region D three transitions after starting in region D, which we found to be 0.309, is the entry with row label D and column label D. By reasoning similar to that used with T^2, we can deduce that the entry in row D, column S is the probability of ending in region S three transitions after starting in region D. In general, entry ij of T^3 gives the probability of starting in region i and ending in region j after three transitions.

We now return to the question under consideration: If a taxi starts off Downtown, what is the probability that it will be Downtown after letting off its fifth fare? The answer to this question is found in the entry in row D and column D of T^5, which is 0.30081, as shown in the matrix below calculated with computer software.

$$T^5 = \begin{matrix} & N & D & S \\ N & 0.30162 & 0.29919 & 0.39919 \\ D & 0.29838 & 0.30081 & 0.40081 \\ S & 0.3 & 0.3 & 0.4 \end{matrix}$$

A taxi that starts off Northside must be either Northside, Downtown or Southside after letting off its fifth fare. Therefore, the sum of these probabilities must be 1. Notice the sum of the elements of the first row (.30162 .29919 .39919) is 1. In order to maintain accuracy of the elements of these matrices, digits created in the process of multiplication must be kept. A very important check of your work in Markov Chains is to sum the appropriate probabilities in each row to see if you get 1.

Example 2

In the Taxi Problem, where should a taxi start to have the best chance of being Northside after three fares?

Solution The probability of being in each region for each possible starting region is given by the matrix T^3 shown below.

$$T^3 = \begin{matrix} & N & D & S \\ N & 0.318 & 0.291 & 0.391 \\ D & 0.282 & 0.309 & 0.409 \\ S & 0.3 & 0.3 & 0.4 \end{matrix}$$

The probability of ending up in Northside for each starting place is found in column 1, the N column, of T^3. The Northside entry is the largest in this column, so starting in Northside offers the best chance of being Northside after three fares — a probability of 0.318. ■

On the basis of the work in the first two examples, the following general observations can be made about a transition matrix T for a Markov chain:

1. A transition matrix is square. This characteristic is obvious because the number of rows

and the number of columns are both the same as the number of states.

2. All entries are between 0 and 1 inclusive. This follows from the fact that the entries correspond to transition probabilities from one state to another.

3. The sum of the entries in any row must be 1. The sum of the entries in an entire row is the sum of the transition probabilities from one state to all other states. Since a transition is sure to take place, this sum must be 1.

4. The ij entry in the matrix T^n gives the probability of being in state j after n transitions, with state i as the initial state.

5. The entries in the transition matrix are constant. A Markov chain model depends upon the assumption that the transition matrix does not change throughout the process. This implies that to determine the state of the system after any transition, it is necessary to know only the immediately preceding state of the system. Knowledge of the prior behavior of the system is not needed provided that the immediately preceding state is known.

Example 3

A bag contains 3 red and 4 green jelly beans. Suppose you take out 3 beans, one at a time, and eat them. What is the probability that the third bean chosen is green?

Solution This problem is an example of a probabilistic situation in which all of the previous behavior of the system must be examined; therefore, it is not modeled as a Markov chain. The set of beans chosen in all previous selections will affect the probability of choosing each color in the present selection. The outcomes that have a green jelly bean chosen third are RRG, RGG, GRG, and GGG, where R stands for choosing a red jelly bean and G stands for choosing a green jelly bean. Taking into account the beans remaining after each selection, each outcome has the probabilities listed below.

$$P(RRG) = \left(\frac{3}{7}\right)\left(\frac{2}{6}\right)\left(\frac{4}{5}\right)$$

$$P(RGG) = \left(\frac{3}{7}\right)\left(\frac{4}{6}\right)\left(\frac{3}{5}\right)$$

$$P(GRG) = \left(\frac{4}{7}\right)\left(\frac{3}{6}\right)\left(\frac{3}{5}\right)$$

$$P(GGG) = \left(\frac{4}{7}\right)\left(\frac{3}{6}\right)\left(\frac{2}{5}\right)$$

The answer to this problem is the sum of the 4 products above, which is

$$\frac{24 + 36 + 36 + 24}{7 \cdot 6 \cdot 5} = \frac{120}{210} = \frac{4}{7}.$$

The probability that the third jelly bean drawn from the bag is green is 4/7. ■

The problem in Example 3 is a simple example of a *stochastic process*. In a generalized stochastic process, the transition probabilities are not necessarily constant or independent of the previous behavior of the system. Notice that in Example 3, the probability of drawing a red jelly bean will vary depending on the number of beans of each color that have been chosen previously. A Markov chain is a special case of a stochastic process in which the transition probabilities are constant and independent of the previous behavior of the system. In a Markov chain, the next state is determined solely by the unchanging transition probabilities and the current state of the system. The route that is followed to arrive at the current state does not affect the transition matrix; only the current state of the system is relevant.

Example 4

In the Taxi Problem, the cab company initially places 25% of the cars Northside, 40% of the cars Downtown, and 35% of the cars Southside. What will be the distribution of cars after each has made three pickups?

Solution Represent the initial distribution of cars with the row vector

$$X_0 = \begin{matrix} N & D & S \\ (0.25 & 0.40 & 0.35) \end{matrix}.$$

To find the percentage of cars Northside after one pickup, multiply the percentage of the cars in each region by the probability of going from that region to Northside and add the resulting products:

$$(0.25)(0.5) + (0.40)(0.1) + (0.35)(0.3) = 0.27.$$

This sum is just the product of X_0 and the first column of the transition matrix T. Likewise, the percentage of cars Downtown after one pickup is equal to the product of X_0 and the second column, the D column, of T. The percentage of cars Southside after one pickup is equal to the product of X_0 and the third column of T. The distribution X_1 of cars after one pickup is therefore given by

$$\begin{aligned} X_1 &= X_0T \\ &= \begin{matrix} N & D & S \\ (0.25 & 0.40 & 0.35) \end{matrix} \begin{matrix} \\ N \\ D \\ S \end{matrix} \begin{matrix} \quad N \quad\; D \quad\; S \\ \begin{pmatrix} 0.5 & 0.2 & 0.3 \\ 0.1 & 0.4 & 0.5 \\ 0.3 & 0.3 & 0.4 \end{pmatrix} \end{matrix} \\ &= \begin{matrix} N & D & S \\ (0.27 & 0.315 & 0.415) \end{matrix}. \end{aligned}$$

Continuing this line of reasoning, we see that X_2, the distribution of cars after two fares, is given by

$$X_2 = X_1T = X_0T^2.$$

The distribution of cars after three pickups is

$$X_3 = X_2T = X_0T^3,$$

which is

$$\begin{aligned} &\begin{matrix} N & D & S \\ (0.25 & 0.40 & 0.35) \end{matrix} \begin{matrix} \\ N \\ D \\ S \end{matrix} \begin{matrix} \quad N \quad\quad D \quad\quad S \\ \begin{pmatrix} 0.318 & 0.291 & 0.391 \\ 0.282 & 0.309 & 0.409 \\ 0.3 & 0.3 & 0.4 \end{pmatrix} \end{matrix} \\ &= \begin{matrix} N & D & S \\ (0.2973 & 0.30135 & 0.40135) \end{matrix}. \end{aligned}$$

After three fares, about 30% of the taxis are Northside, about 30% are Downtown, and about 40% are Southside. ■

The vector $X_k = X_0T^k$ is called the *state vector* for a Markov chain after k transitions with an initial distribution X_0. The jth entry of X_k is the probability of being in state j after k transitions.

Example 5

For the Taxi Problem with the initial state vector given in Example 4, what is the long-term distribution of cars?

Solution We found X_3 in Example 4. After five pickups, we have

$$\begin{aligned} X_5 &= \begin{matrix} N & D & S \\ (0.25 & 0.40 & 0.35) \end{matrix} T^5 \\ &= \begin{matrix} N & D & S \\ (0.299757 & 0.300121 & 0.400121) \end{matrix}. \end{aligned}$$

Moving along even further, the state vector after ten pickups is

$$X_{10} = X_0T^{10} = \begin{matrix} N & D & S \\ (0.299999 & 0.3 & 0.4) \end{matrix},$$

and after fifteen pickups,

$$X_{15} = X_0T^{15} = \begin{matrix} N & D & S \\ (0.3 & 0.3 & 0.4) \end{matrix}.$$

A trend is clear in the state vector X_k—it converges to the vector $(0.3 \;\; 0.3 \;\; 0.4)$. Further transitions after the fifteenth pickup will not change this distribution. ■

Evidently, the system represented by the Markov process in the Taxi Problem stabilizes; that is, after a certain number of transitions the distribution of taxis is not changed by additional transitions. In other words, the state vector eventually reaches a stable distribution even after repeated multiplication by the transition matrix. In symbols, stability means that for large enough k, we find that $X_kT = X_k$. From another point of view, since $X_{k+1} = X_kT$, we eventually reach a point beyond which $X_k = X_{k+1}$. This is called the *stable state*. The *stable state vector* of a Markov process is a vector X such that $X = XT$, where T is the transition matrix. The entries in the stable state vector were found in Example 5 by multiplying by T a large number of times, calculations made easy by the use of computer software.

Class Practice

1. Let the following matrix represent a transition matrix for a Markov chain.

$$T = \begin{matrix} & \begin{matrix} 1 & 2 & 3 \end{matrix} \\ \begin{matrix} 1 \\ 2 \\ 3 \end{matrix} & \begin{pmatrix} 0.2 & 0.4 & 0.4 \\ 0.4 & 0.2 & 0.4 \\ 0 & 0.3 & 0.7 \end{pmatrix} \end{matrix}$$

   a. What is the probability of moving from state 1 to state 3? From state 3 to state 1?

   b. If the system is in state 2, what is the probability of staying there on the next transition?

2. What is the stable state vector of the Taxi Problem? Does the stable state vector depend upon the initial distribution? What does changing the initial distribution have to do with reaching a stable state?

We have examined the behavior of the distribution X_k for a particular initial distribution X_0. As X_k approaches the stable state vector, what is happening to the entries of T^k? The calculation of T^5, T^{10}, and T^{15}, shown below, reveals that each row of T^k converges to the stable state vector.

$$T^5 = \begin{matrix} & \begin{matrix} N & D & S \end{matrix} \\ \begin{matrix} N \\ D \\ S \end{matrix} & \begin{pmatrix} 0.30162 & 0.29919 & 0.39919 \\ 0.29838 & 0.30081 & 0.40081 \\ 0.3 & 0.3 & 0.4 \end{pmatrix} \end{matrix}$$

$$T^{10} = \begin{matrix} & \begin{matrix} N & D & S \end{matrix} \\ \begin{matrix} N \\ D \\ S \end{matrix} & \begin{pmatrix} 0.300004 & 0.299998 & 0.399998 \\ 0.299996 & 0.300002 & 0.400002 \\ 0.3 & 0.3 & 0.4 \end{pmatrix} \end{matrix}$$

$$T^{15} = \begin{matrix} & \begin{matrix} N & D & S \end{matrix} \\ \begin{matrix} N \\ D \\ S \end{matrix} & \begin{pmatrix} 0.3 & 0.3 & 0.4 \\ 0.3 & 0.3 & 0.4 \\ 0.3 & 0.3 & 0.4 \end{pmatrix} \end{matrix}$$

What is the implication of this phenomenon? Recall that row i of T^k contains the probabilities of ending in each state after k transitions from a starting state i. Since each row converges to the stable state vector, we can infer that the stable state vector is independent of the starting state. No matter what initial distribution is specified, the state vector will converge to the same stable state vector, which is equal to each row of T^k for large k. Care must be taken, however, when T is raised to a large power using computer software. Round-off errors could actually cause T^k to diverge for sufficiently large k. One strategy for dealing with the effects of round-off errors is to examine the behavior of T^k at various points in the process, and not just at one large value of k.

Not all Markov chains have a stable state vector. Consider the transition matrix A shown below.

$$A = \begin{pmatrix} 0.3 & 0.1 & 0.4 & 0 & 0.2 \\ 0 & 1 & 0 & 0 & 0 \\ 0.2 & 0.3 & 0.1 & 0.1 & 0.3 \\ 0 & 0 & 0 & 1 & 0 \\ 0.2 & 0.2 & 0.1 & 0.5 & 0 \end{pmatrix}$$

For large powers of k, we find that A^k converges to

$$\begin{pmatrix} 0 & 0.587097 & 0 & 0.412903 & 0 \\ 0 & 1 & 0 & 0 & 0 \\ 0 & 0.589247 & 0 & 0.410753 & 0 \\ 0 & 0 & 0 & 1 & 0 \\ 0 & 0.376344 & 0 & 0.623656 & 0 \end{pmatrix}.$$

The diversity of the rows of A^k for large k assures us that no stable state vector exists for this system. The probability of ending in a certain state after stability is reached clearly depends on the starting state.

Observe that in transition matrix A, once the system enters state 2 or state 4, it can never leave that state. Zeros in rows 2 and 4 indicate a probability of zero that the system will change to another state once it is in state 2 or state 4. These states are called *absorbing states.* The row corresponding to an absorbing state contains a 1 on the diagonal of the matrix and 0s for all of the other entries. An *absorbing-state Markov chain* is a Markov chain such that (a) it has at least one absorbing state and (b) it is possible to move from any nonabsorbing state to an absorbing state. A process modeled with an absorbing-state Markov chain will eventually end up in one of the absorbing states, as can be observed with matrix A above. The only nonzero entries in a large power of A are found in the absorbing-state columns.

How do we know if a Markov chain has a stable state vector? A sufficient condition, which is stated without proof, for a Markov chain to have a stable state vector is that some power of the transition matrix has only nonzero entries. A Markov chain satisfying this criterion is called *regular.* For example, a Markov chain with transition matrix

$$B = \begin{pmatrix} 0.2 & 0.8 \\ 1 & 0 \end{pmatrix}$$

is regular since

$$B^2 = \begin{pmatrix} 0.84 & 0.16 \\ 0.2 & 0.8 \end{pmatrix}.$$

For regular Markov chains, we can find the stable state vector by raising the transition matrix to a large power. As the power increases, the rows of the transition matrix each approach the stable state vector. A stable state vector may not exist in an absorbing-state Markov chain; however, the transition matrix raised to a large power may have rows that converge to some vector, but not necessarily the same vector for each row.

Exercise Set 7

1. A rat is placed in the maze shown in Figure 10. During a fixed time interval, the rat randomly chooses one of the doors available to it (depending upon which room it is in) and moves through that door to the next room—it does not remain in the room it occupies. Each movement of the rat is taken as a transition in a Markov chain in which a state is identified with the room the rat is in. The first row of the transition matrix is

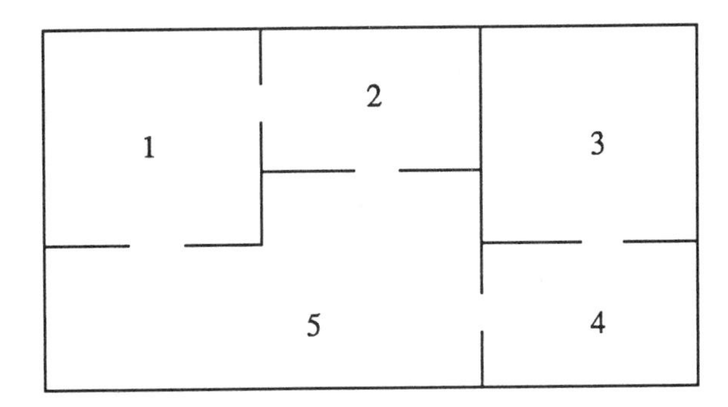

Figure 10 Maze for Problem 1

$$\begin{array}{cccccc} & 1 & 2 & 3 & 4 & 5 \\ 1 & (0 & \frac{1}{2} & 0 & 0 & \frac{1}{2}). \end{array}$$

a. Construct the entire transition matrix for this process.

b. If the rat starts in room 1, what is the probability that it is in room 3 after two transitions? After three transitions?

c. Determine the stable state vector.

d. After a large number of transitions, what is the probability that the rat is in room 4?

e. In the long run, what percentage of the time will the rat spend in rooms 2 or 3?

2. Suppose jar A contains 3 beads and jar B contains 4 beads. Of the 7 beads, 3 are red and 4 are black. We start a Markov process by picking at random one bead from each jar and interchanging them (that is, the bead from jar A is placed in jar B, and the bead from jar B is placed in jar A). This process is continued. Let the state of the process be identified by the number of red beads in jar A. State 0 represents 0 red beads in jar A, state 1 represents 1 red bead in jar A, state 2 represents 2 red beads in jar A, and state 3 represents 3 red beads in jar A.

a. Find the transition matrix for this process, focusing on the number of red beads in jar A.

b. If we start with the 3 red beads in A and the 4 black beads in B, what is the probability that there will be 2 red beads in jar A after three transitions?

c. In the long run, what is the probability that there will be 2 red beads in jar A?

3. Wade, Donald, and Andrea are playing frisbee. Wade always throws to Donald, Donald always throws to Andrea, but Andrea is equally likely to throw to Wade or Donald.

a. Represent this information as a transition matrix of a Markov chain.

b. Notice that this transition matrix has zero entries in several places. Compare the values of the transition matrix if raised to the second, fourth, sixth, and tenth powers. Are the zero entries still there? Can you explain?

4. The snack bar at school sells three items that students especially like: onion rings, french fries, and chocolate chip cookies. The manager noticed that what each student ordered depended on what he or she ordered on the last previous visit. She ran a survey during the first two weeks of school and found out that 50% of those who ordered onion rings on their last snack break ordered them again this time, while 35% switched to french fries, and 15% switched to chocolate chip cookies. Of those who ordered french fries on their last visit 40% did so the next time, but 30% switched to onion rings, and another 30% switched to chocolate chip cookies. Of the students who ordered chocolate chip cookies on their last visit, 20% switched to onion rings, and 55% switched to french fries.

a. Set up the transition matrix for this Markov process.

b. On Monday, 30 students buy french fries, 40 buy onion rings, and 25 buy chocolate chip cookies. If these same students come in on Tuesday and each buys one of these items, how many orders of french fries should the manager expect to sell?

c. Suppose the students in part 4b continue buying from the snack bar every day for two weeks. How many orders of onion rings, french fries, and cookies should the manager expect to sell on the third Monday?

d. If these same people come all year, how many orders of onion rings, french fries, and cookies should the manager expect to sell to them each day?

e. In the long run, what percent of the orders for these three items will be onion rings? French fries? Chocolate chip cookies?

5. The manager of the snack bar in the previous problem decided to add soft custard ice-cream cones to the menu. Lots of students tried the new item, but few of them liked it. (The machine didn't work right and the ice cream came out lumpy.) A two-week survey gave the results in the transition matrix below.

$$\begin{array}{c} \\ Rings \\ Fries \\ Cookies \\ Cones \end{array} \begin{array}{c} \begin{array}{cccc} Rings & Fries & Cookies & Cones \end{array} \\ \begin{pmatrix} 0.4 & 0.3 & 0.1 & 0.2 \\ 0.25 & 0.35 & 0.2 & 0.2 \\ 0.15 & 0.4 & 0.2 & 0.25 \\ 0.3 & 0.35 & 0.3 & 0.05 \end{pmatrix} \end{array}$$

a. Does this system reach a stable state?

b. If everyone really dislikes the ice cream, why does the stable state matrix show that many students still buy it?

c. Why is a Markov chain not a good model for this system? Are people likely to forget that they did not like something only two days after they ate it? (Recall the principal assumption of a Markov process.) Why would this model work if the customers liked the ice cream?

6. A mouse is in the maze shown in Figure 11. Doors are shown by openings between rooms. Arrows indicate one-way doors and the direction of passage through the one-way doors. The mouse does not have to change rooms at each transition, but can stay in a room. Notice some of the rooms are impossible to leave once they are entered. During each transition, the mouse has an equal chance of leaving a room by a particular door or staying in the room. For example, during a single transition, a mouse in room 2 has a $1/6$ chance of moving to room 1, a $1/6$ chance of staying

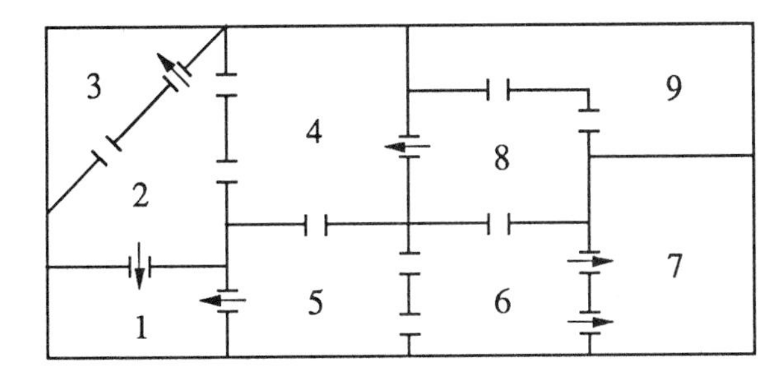

Figure 11 Maze for Problem 6

in room 2, a 1/3 chance of moving to room 3, and a 1/3 chance of moving to room 4.

a. Find the transition matrix which describes the movement of the mouse.

b. If the mouse starts in room 4, what is the probability that it will eventually be trapped in room 1?

c. In which room besides 7 should the mouse be started to have the best chance of being trapped in room 7? Besides 1, the worst chance?

7. A research article [A. W. Marshall and H. Goldhamer, "An Application of Markov Processes to the Study of the Epidemiology of Mental Diseases," *American Statistical Association Journal*, March 1955, pp. 99–129] on the application of Markov chains to mental illness suggests that we consider a person to be in one of four states:

State I — severely insane and hospitalized
State II — dead, with death occurring while unhospitalized
State III — sane
State IV — insane and unhospitalized

For this model, assume that states I and II are absorbing states. Suppose that of the people in state III, after one year, 1.994% will be in state II, 98% will be in state III, and 0.006% will be in state IV. Also suppose that of the people in state IV, after one year, 2% will be in state I, 3% will be in state II, and 95% will be in state IV.

a. Set up the transition matrix that describes this model.

b. Find the stable state matrix.

c. Determine the probability that a person who is currently well eventually will be severely insane and hospitalized.

8. Go to the library and determine whether the presidents elected during the twentieth century, starting with Theodore Roosevelt, were Democrats or Republicans. Note that Gerald Ford was not elected. The election process can be viewed as a Markov chain — each new election is a transition and the states of the process are Democrat and Republican. Design a transition matrix based on the probability of electing Democratic and Republican presidents in the twentieth century. In the transition matrix, count each president only once — a re-election is not considered a transition. According to your matrix, will a Republican or Democrat be elected in the next presidential election? When the year 2001 comes around, to what party will our president belong? What is the main weakness with this model of presidential elections?

9. A dreaded strain of flu is studied by research biologists. Statistics are taken each week in an effort to describe the probabilities after exposure of staying well, getting ill, becoming immune, and dying. A person becomes immune to this flu by having a mild case of it. Any well person once exposed to this flu has a 20% chance of getting the illness. Once a person becomes ill there is a 55% chance of remaining ill for more than a week, a 40% chance of being permanently immune after a mild illness, and a 5% chance of dying from the illness.

a. Construct a transition matrix to represent the information above.

b. Compute the probability of becoming immune after ten weeks for groups that begin as follows:

i. 100% well and exposed

ii. 50% well and 50% sick

iii. 80% well and 20% immune

iv. 100% sick

c. Does this Markov process have a stable state?

10. An ant walks from one corner to another of a square $ABCD$. Assume that between successive observations of the process, the ant has moved from one corner to another corner. The transition probabilities between corners are shown in Figure 12. Construct a Markov chain model for this situation.

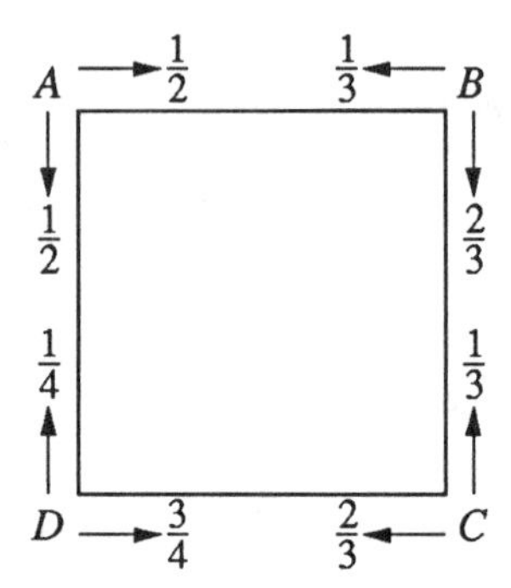

Figure 12 Ant Walking around a Square

a. Investigate the behavior of successive powers of the transition matrix for this Markov chain. In successive powers, notice the oscillation of the transition probabilities between states represented by the rows of the powers of the transition matrix.

b. What percentage of the time will the ant spend at each of the corners?

ANSWERS

CHAPTER 1
Data Analysis One

Exercise Set 3

1. The relationship between rebounds and assists is a weak one. Many points are near the horizontal at 50 assists. Five points appear to be removed from the others.

3. **a.** The slope indicates the change in waist measure for each one-inch increase in forearm circumference. The y-intercept gives a constant correction to the forearm circumference that estimates the waist measure. Since it is not possible to have a zero value for forearm circumference, the y-intercept provides no additional information about the relationship.

   b. A person whose data point is above the line has a large waist relative to forearm length. This person may need to lose weight, or may simply have shorter than average arms.

Exercise Set 4

1. Based on the summary points $(4, 10)$, $(37, 198)$, and $(90.5, 332)$, the equation of the median-median line is

$$y = 3.72x + 16.86.$$

   Using the median-median line, y-values are calculated for several x-values given in Table 1.

3. The equation of the median-median line is $y = 1.72x + 35.00$. The slope indicates that the winning speed is increasing about 1.72 mph each year. The model is satisfactory. The data points are fairly well scattered about the line and are reasonably close to the line.

Table 1 Predicted and Actual Points

Fouls	**Predicted Points (y-value from line)**	**Actual Points**
6	39	12
55	221	253
66	262	211
79	311	274
102	396	349

5. The median-median line has equation $y = -.33x + 883.07$.
   The slope indicates that times are decreasing at a rate of about 1/3 second per year.
   The model predicts in 2000, a time of 223.1 seconds.
   The record will be 3 minutes (180 seconds) in the year 2131.
   The record will be 1 minute (60 seconds) in the year 2494.
   The model looks fairly good over the data that we have, although the cluster of points above the line from about 1910 to 1950 causes some concern. There will be some time in future years after which the model will diverge from reality, since the record time will level off at some physical limit.

Exercise Set 5

1. Speeds calculated from the median-median line $y = 1.72x + 35.00$ are provided in Table 2.
   The residuals are not randomly scattered, since the residuals on the ends are negative and the residuals in the middle (with only one exception) are positive. This gives cause for concern and suggests that the model possibly could be improved.

Table 2 Predicted Speeds and Residuals

Year	Actual Speed (in mph)	Calculated Speed	Residual
1961	139.1	139.9	−.8
1962	140.3	141.6	−1.3
1963	143.1	143.4	−.3
1964	147.4	145.1	2.3
1965	151.4	146.8	4.6
1966	144.3	148.5	−4.2
1967	151.2	150.2	1.0
1968	152.9	152.0	.9
1969	156.9	153.7	3.2
1970	155.7	155.4	.3
1971	157.7	157.1	.6
1972	163.5	158.8	4.7
1973	159.0	160.6	−1.6
1974	158.6	162.3	−3.7

Exercise Set 6

1. **a.** $d = 493.08t^2 + 2.25$

   b. $\frac{d}{t} = 492.17t + 5.36$ or $d = 492.17t^2 + 5.36t$

   c. Each of the three models provides a good fit to linearized data. The model obtained from the re-expression $(t, \sqrt{d})$ includes constant, linear, and quadratic terms in t. The models in a and b are each missing one of the terms of the general quadratic function. In terms of the phenomenon we are trying to model, the equation obtained from the re-expression $(t, \frac{d}{t})$ may be best, since its graph contains the point $(0, 0)$. Though this point is not included in the data set, it is clear that $d = 0$ when $t = 0$.

4. **a.** The scatter plot curves with downward concavity.

   b. $T = .024L + .425$.

   c. The residuals at both ends are rather large and negative. Residuals in the middle are positive. This somewhat parabolic pattern in the residuals suggests that a better model is needed.

   d. Transform the data by using the square root of the L values. The resulting scatter plot appears linear. Similar results are obtained by squaring the T values.

   e. $T = .224\sqrt{L} - .0805$ or $T^2 = .0468L - .0823$, which is equivalent to

   $$T = \sqrt{.0468L - .0823}.$$

   The residuals confirm that both transformations linearize the data.

   f. Both models seem to fit fairly well. The residual on the far right does not cause concern, since the relative error is small (the residual divided by the actual T value).

CHAPTER 2

Functions

Exercise Set 1

1. **a.**

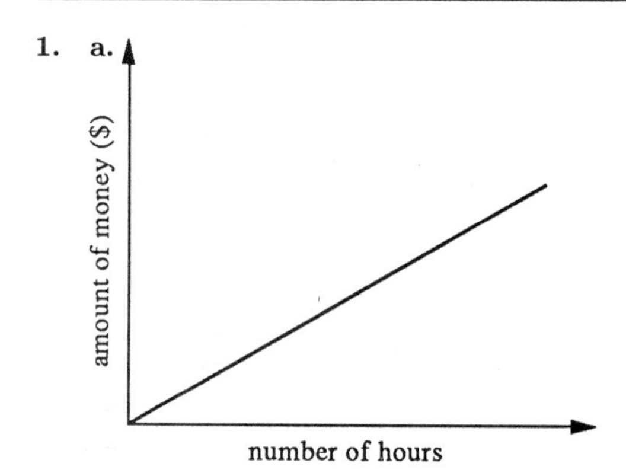

As the number of hours worked increases, the amount of money earned increases.

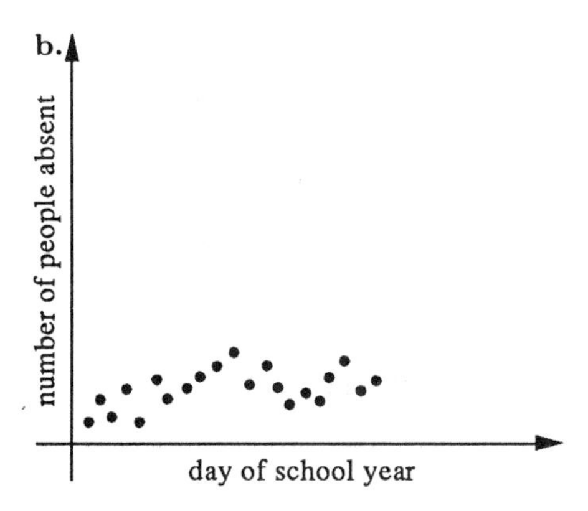

The number of people absent changes daily depending on several factors — health, weather, holidays, etc.; not every day is shown.

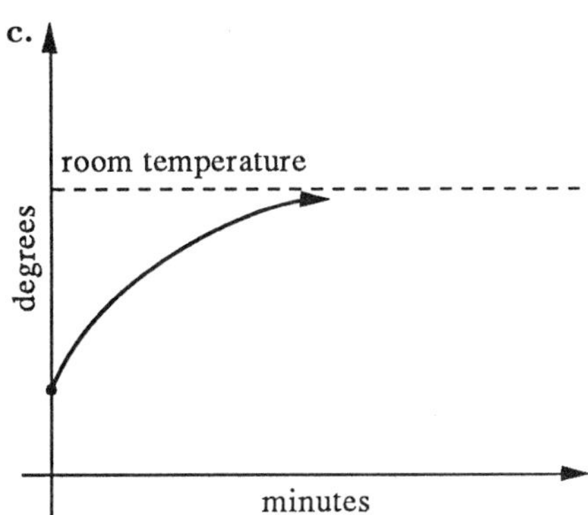

The temperature rises until it reaches room temperature.

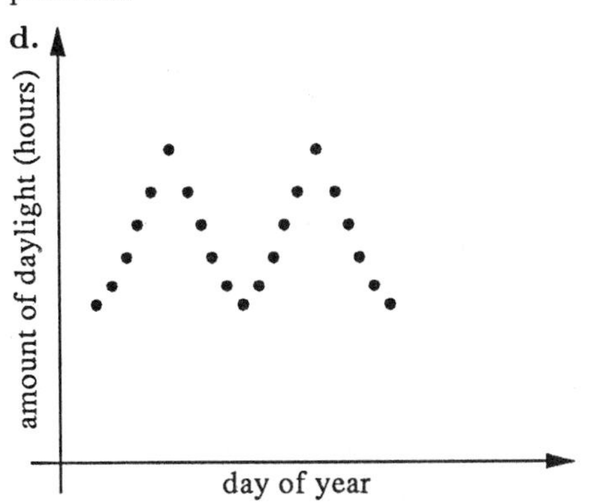

The number of hours of daylight varies from the shortest day in winter to the longest day in summer; not every day is shown.

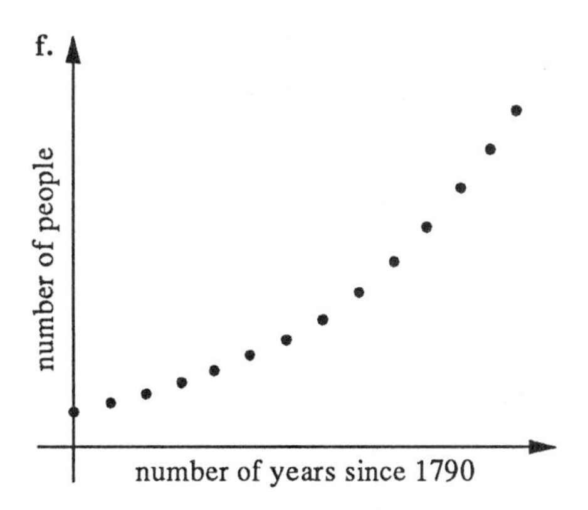

The census is taken every ten years; the population has increased over time.

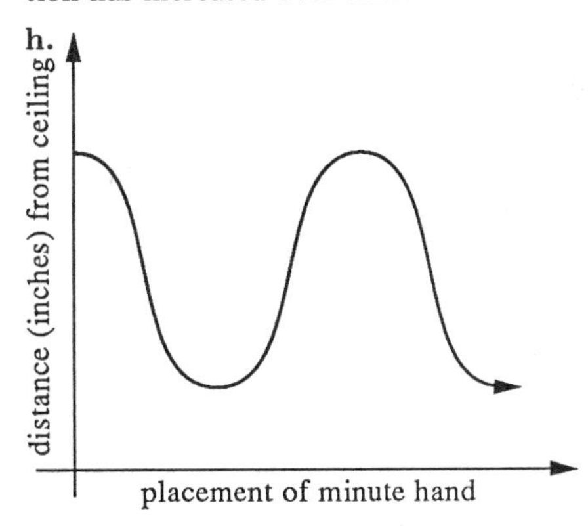

The tip of the minute hand is closest to the ceiling at 12 o'clock. The cycle repeats every hour.

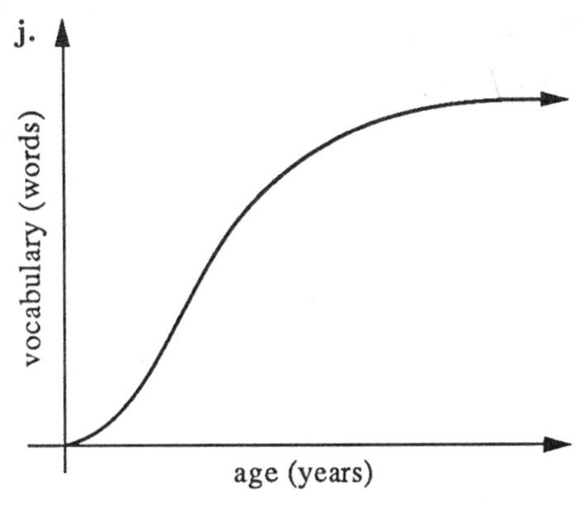

As a person ages, his vocabulary increases quickly at first, then more slowly.

2. **a.**

1.4 1.2 1.0 .8 .6 .4 .2
rate ($)
2 4 6 8 10 12 14 16
weight (oz.)

Independent variable: weight
Dependent variable: postage rate
The postage rate is constant over each interval.

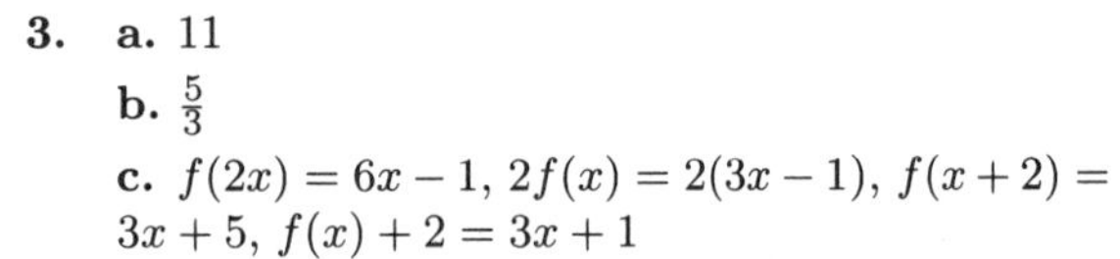

3. **a.** 11

b. $\frac{5}{3}$

c. $f(2x) = 6x - 1$, $2f(x) = 2(3x - 1)$, $f(x+2) = 3x + 5$, $f(x) + 2 = 3x + 1$

5. **a.** $x^2 + 3x + 2$

c. $0.25x^2 + 0.5x$

e. $x^2 + |x|$

6. **a.** $y = f(x - 4)$

c. $y = f(x) + 5$

d. $y = g(x - 1)$

e. $y = 2 + 7f(x)$

Exercise Set 2

1. $y = x^4$

$y = x^5$

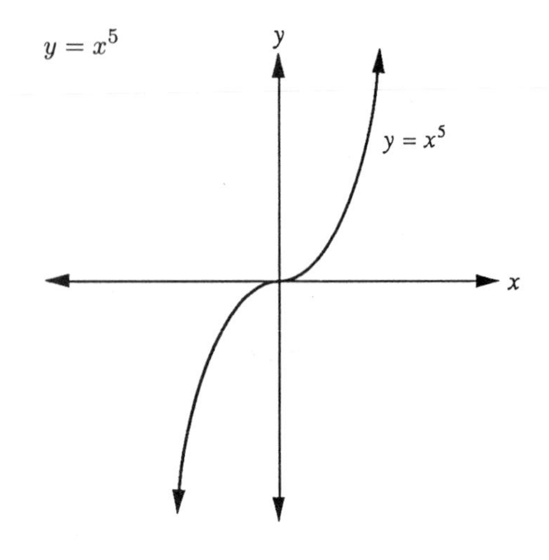

$y = x^6$

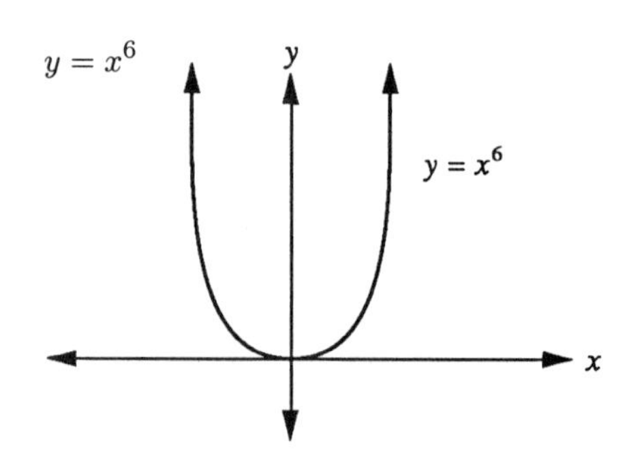

$y = x^7$

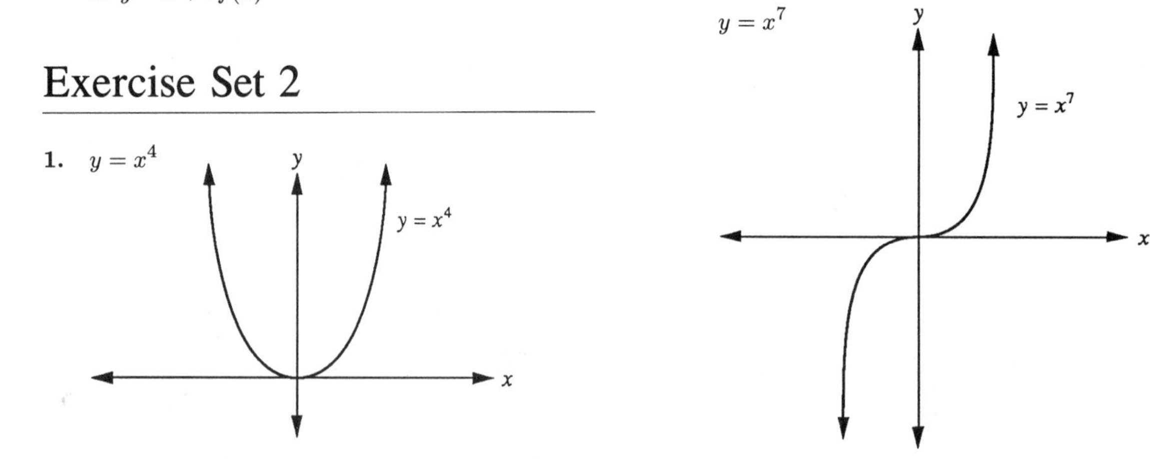

When n is even, $y = x^n$ is always in the 1st and 2nd quadrants, always concave up, decreases for $x < 0$, increases for $x > 0$, contains points $(-1, 1)$, $(1, 1)$, and $(0, 0)$.

When n is odd, $y = x^n$ is always in the 1st and 3rd quadrants, concave down for $x < 0$, concave up for $x > 0$, always increasing, contains points $(-1, -1)$, $(1, 1)$, and $(0, 0)$.

2. **a.**

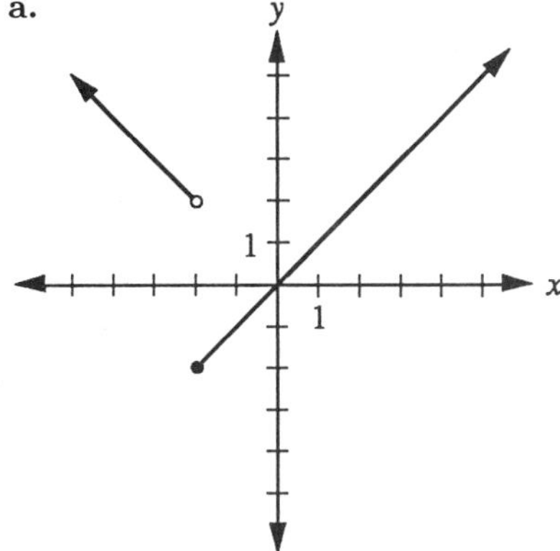

c.

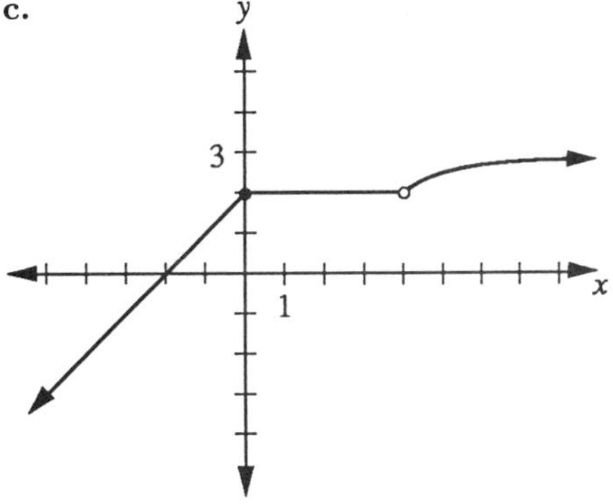

3. $f(x) = \frac{3}{2}x + 4$

4. **a.** $y = 1 + q(x)$

c. $y = \frac{1}{2}q(x - 1)$ or $q(2x - 2)$

e. $y = 3 - p(x)$

5. On $0 < x < 1$, $y = x$ is the highest, $y = x^2$ is next, and $y = x^3$ is the lowest. This is because the cube and the square of a number between 0 and 1 are less than the number.

Exercise Set 3

1. **a.** The graph shows a straight, horizontal line, very close to the x-axis.

c. At the left edge of the screen (x's close to -15) the y-values are large but the graph ends so it is difficult to tell what is happening there.

3. The graph is a square root function that opens to the left from the point $(20, 0)$. All y-values are positive. There are no y-values for $x > 20$.

4. **a.** Move g up 2 units vertically.

c. Move g 4 units to the left.

e. f is g together with g's reflection across the y-axis.

g. Stretch g vertically by a factor of 3.

Exercise Set 4

1.

Function	Domain	Range
$y = c$	$\mathbf{R}$	$y = c$
$y = x$	$\mathbf{R}$	$\mathbf{R}$
$y = x^2$	$\mathbf{R}$	$x \geq 0$
$y = x^3$	$\mathbf{R}$	$\mathbf{R}$
$y = \sqrt{x}$	$x \geq 0$	$y \geq 0$
$y = \lvert x \rvert$	$\mathbf{R}$	$y \geq 0$
$y = \frac{1}{x}$	$x \in \mathbf{R}, x \neq 0$	$y \in \mathbf{R}$, $y \neq 0$
$y = \sin x$	$\mathbf{R}$	$-1 \leq y \leq 1$

2. **a.** $x \in \mathbf{R}, x \neq 11$

c. $x \leq 3$

e. $x \in \mathbf{R}, x \neq 3$

g. $x \geq \frac{7}{2}$

i. $x < -2$ or $x > 2$

k. $x \geq 3$ or $x \leq -3$

m. $x < -1$ or $x \geq \frac{1}{2}$

n. $x \in \mathbf{R}, x \neq n\pi$, where n is an integer

3. **a.** $y \geq -1$

c. $1 \leq y \leq 3$

e. $y \geq 0$

g. $y \geq 3$

h. $-3 \leq y \leq 3$

5. **a.** 0

b. 0.5

c. $x = n + 0.75$, where n is an integer

d. domain: **R**
range: $0 \le y < 1$

7. **a.** yes ($x = \frac{1}{2}$)
b. yes ($x = -\frac{13}{18}$)
c. yes ($x = 6$)
d. no

9. $C = \begin{cases} 5, & \text{if } 0 < x \le 5 \\ 5 + 0.1[x - 5], & \text{if } x > 5 \end{cases}$
domain: $0 < x \le 50$
range: $C = 5, 5.1, 5.2, \ldots, 9.5$

Exercise Set 5

1. The graph of $y = x^n$ is symmetric about the y-axis if n is even and is symmetric about the origin if n is odd.

3. No, because $f(-x) = f(x)$ and $f(-x) = -f(x)$ cannot both be true (except for the trivial case where $f(x) = 0$).

4. **b.** symmetric about the x-axis, y-axis, origin, and the line $y = x$

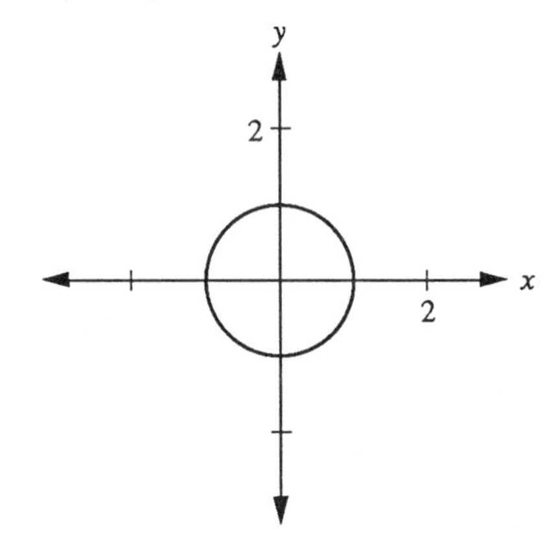

e. odd

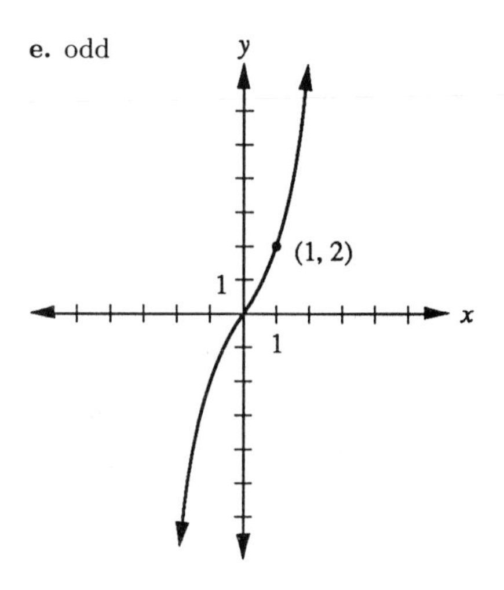

g. even

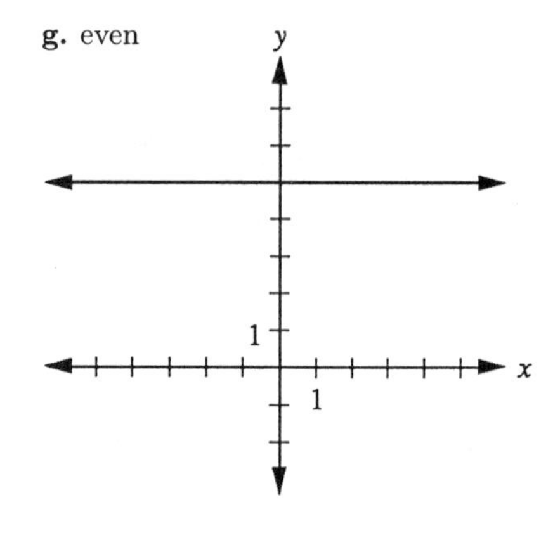

i. odd

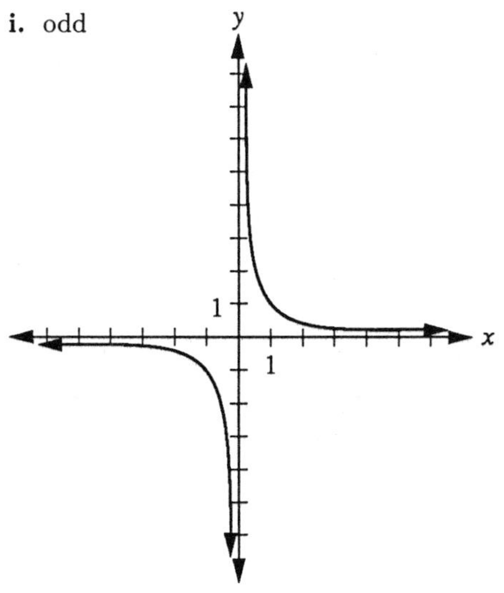

5. **a.** $f(x) = -3x - 2$

b. $f(x) = -3x + 2$

c. $f(x) = \frac{1}{3}x + \frac{2}{3}$

d. $f(x) = -3x + 4$

6. **a.**

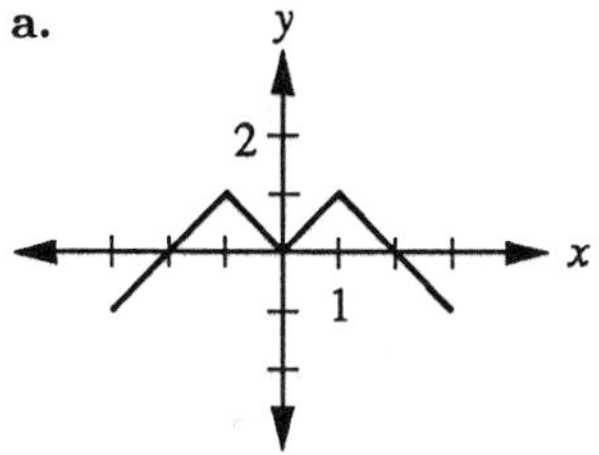

Exercise Set 6

1. **a.** Absolute value moved up 5 units.

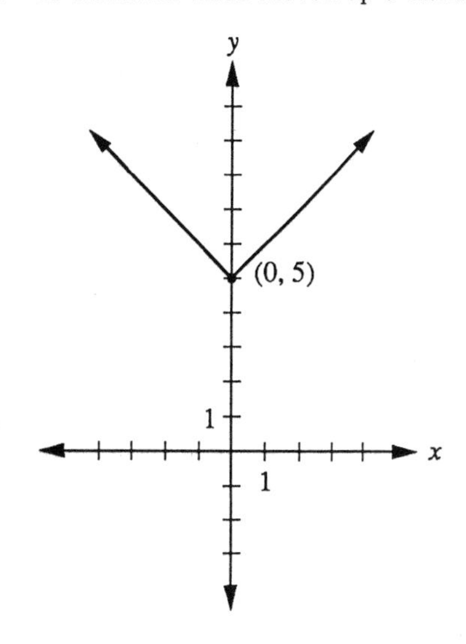

c. Reciprocal moved to the left 2 units.

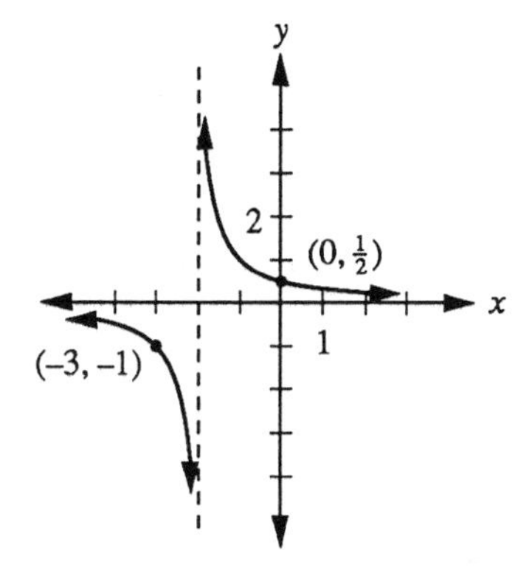

e. Absolute value moved to the left 3 units and down 3 units.

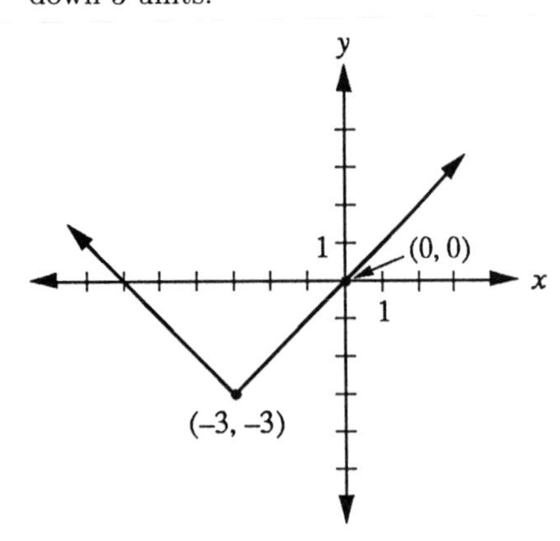

g. Square root compressed horizontally by a factor of $\frac{1}{9}$.

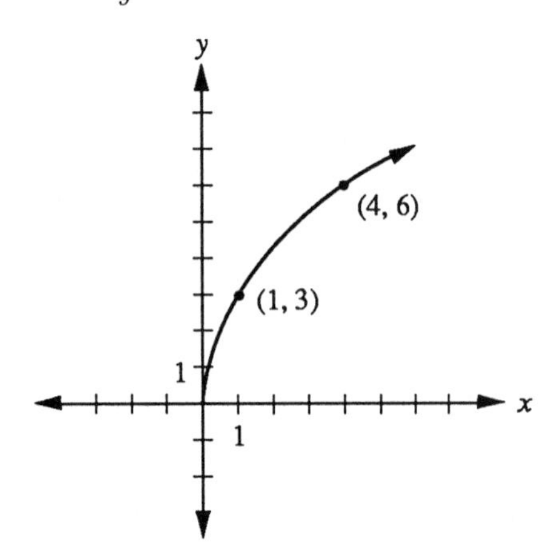

i. Reciprocal stretched vertically by a factor of 2 and moved up 4 units.

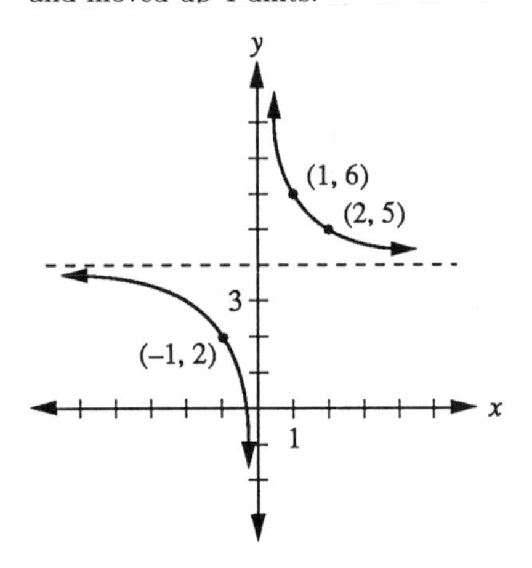

2. Compress g horizontally by a factor of $\frac{1}{4}$ to obtain $\sqrt{4x}$. Stretch g vertically by a factor of 2 to obtain $2\sqrt{x}$. For the function g a horizontal compression of $\frac{1}{k}$ is equivalent to a vertical stretch by $\sqrt{k}$.

3. **a.** domain: $-3 \leq x \leq 4$
range: $-5 \leq y \leq -1$

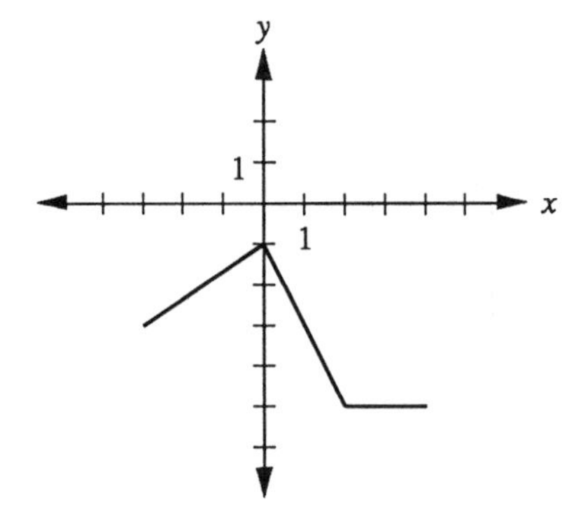

c. domain: $-9 \le x \le 12$
range: $-2 \le y \le 2$

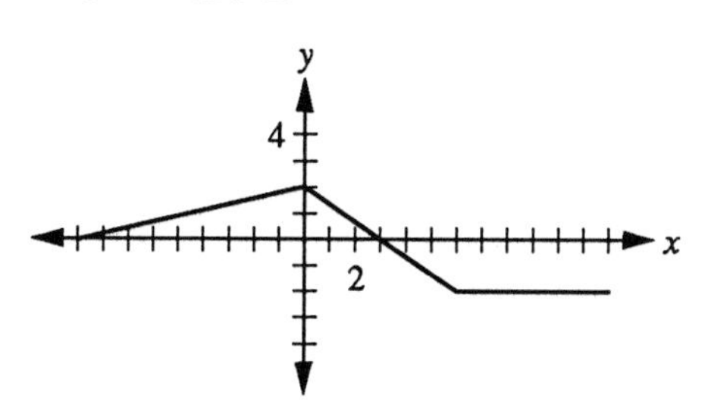

4. **a.**

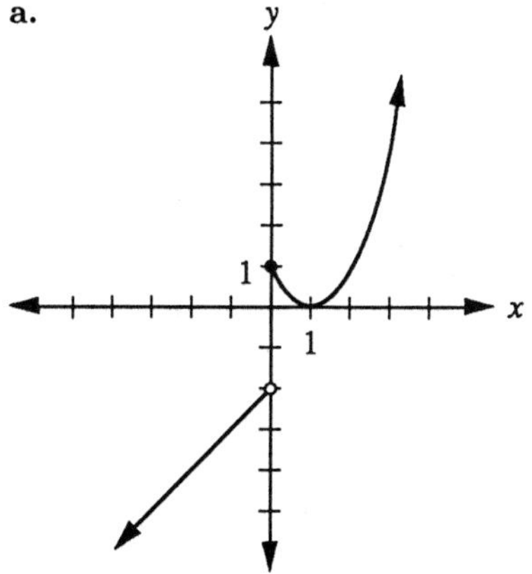

c.

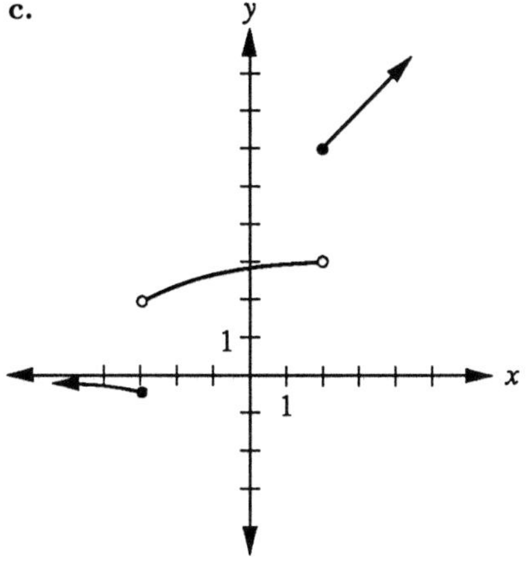

Exercise Set 7

1. **a.** domain: $-6 \le x \le 1$
range: $-2 \le y \le 2$

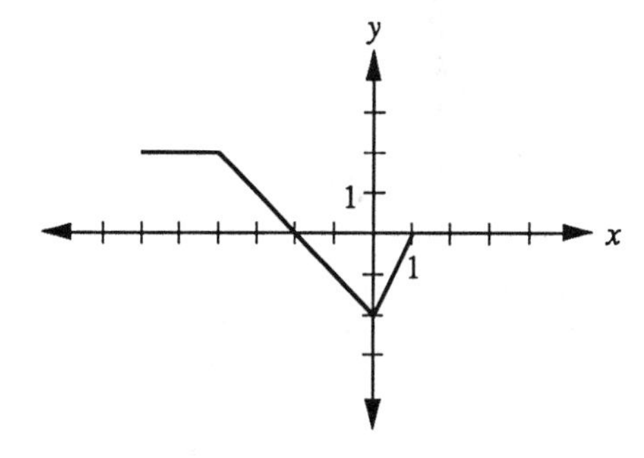

c. domain: $-2 \le x \le \frac{3}{2}$
range: $-2 \le y \le 2$

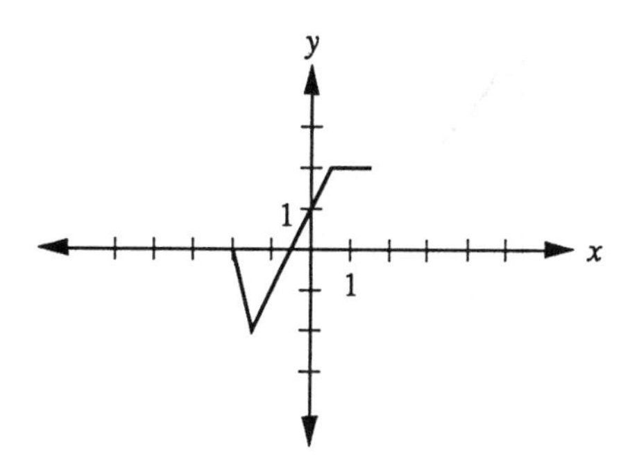

e. domain: $-3 \le x \le 4$
range: $-2 \le y \le 2$

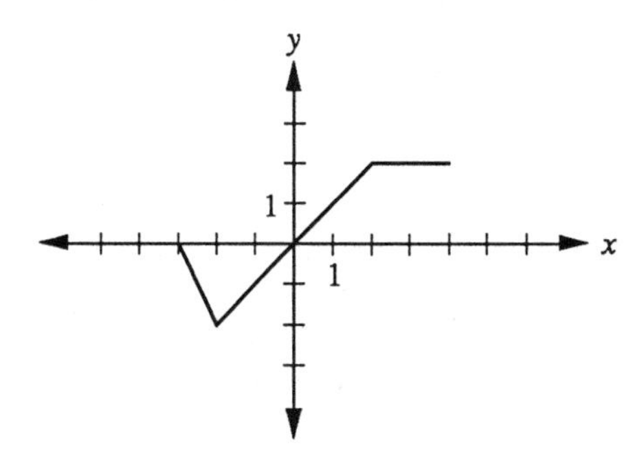

g. domain: $-\frac{1}{2} \leq x \leq 3$
range: $-2 \leq y \leq 2$

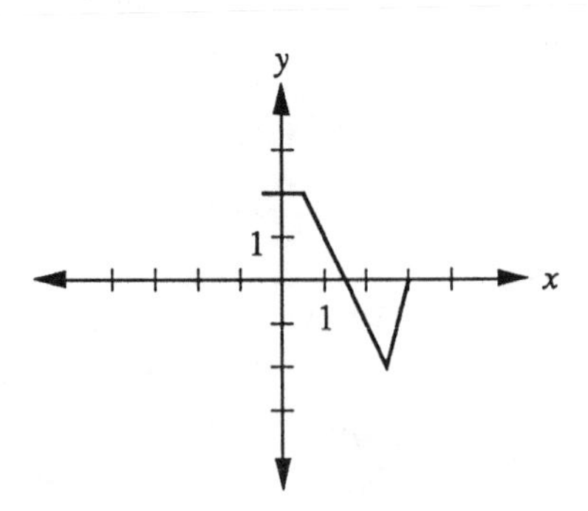

2. a.

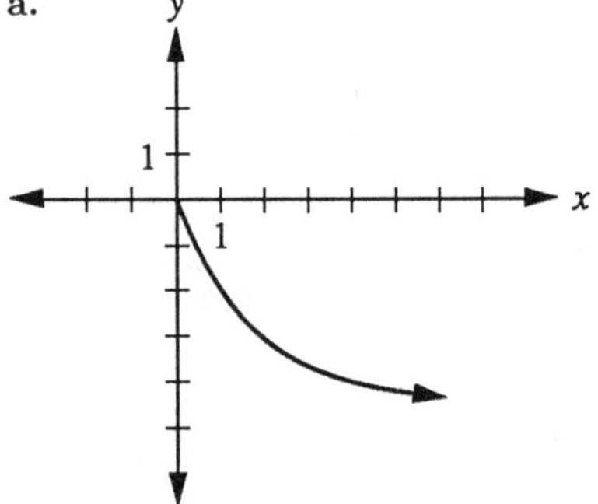

c.

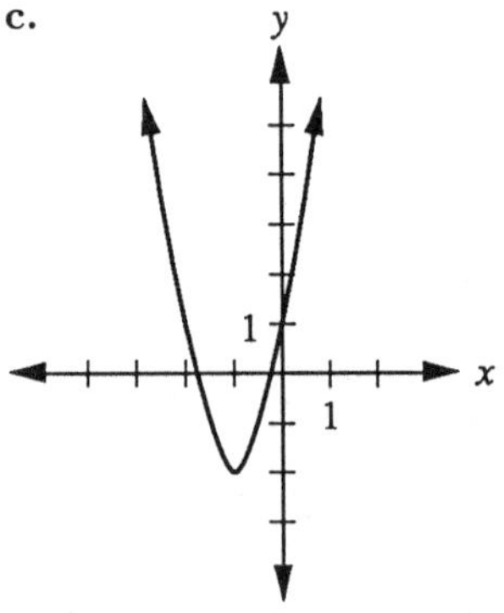

e.

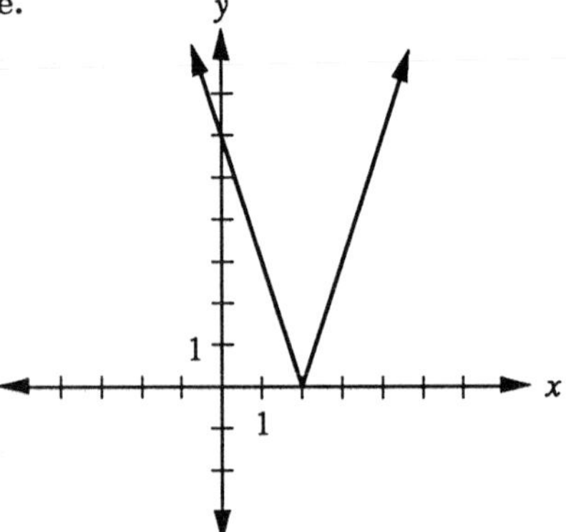

g.

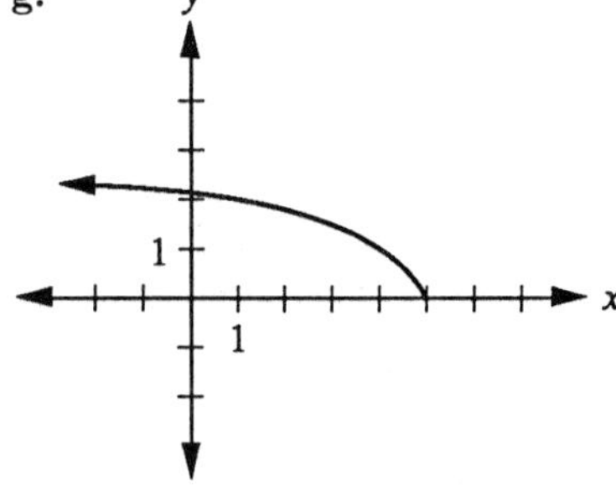

h.

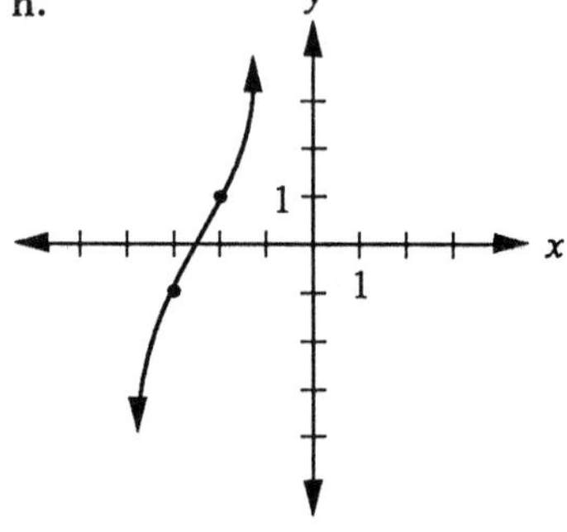

k.

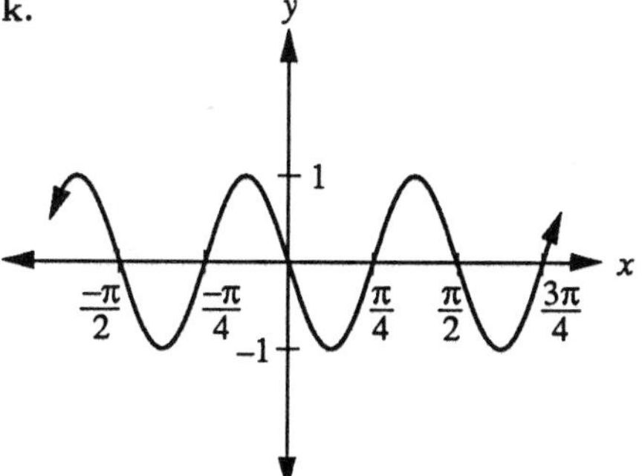

m.

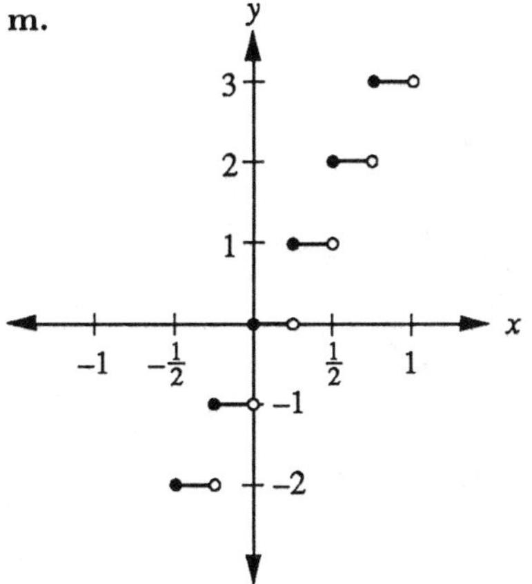

n.

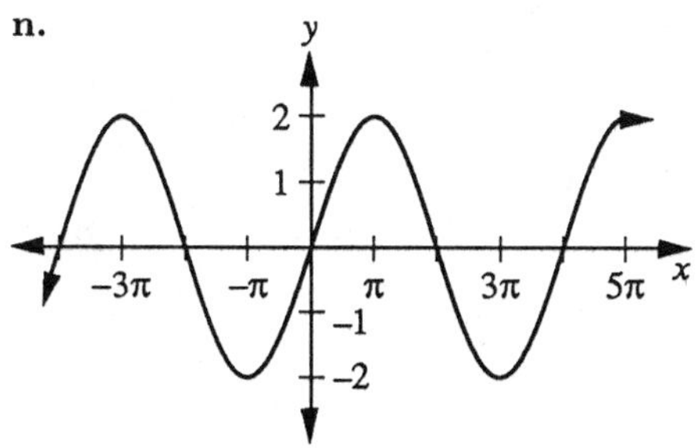

q.

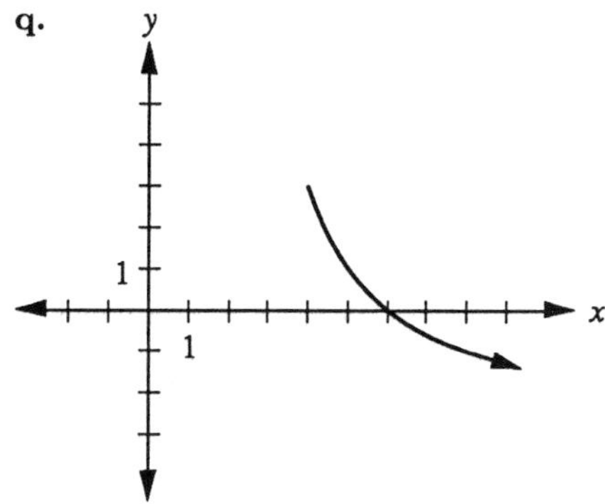

3. **a.** $y = x^2 - 2x - 24 = (x-1)^2 - 25$

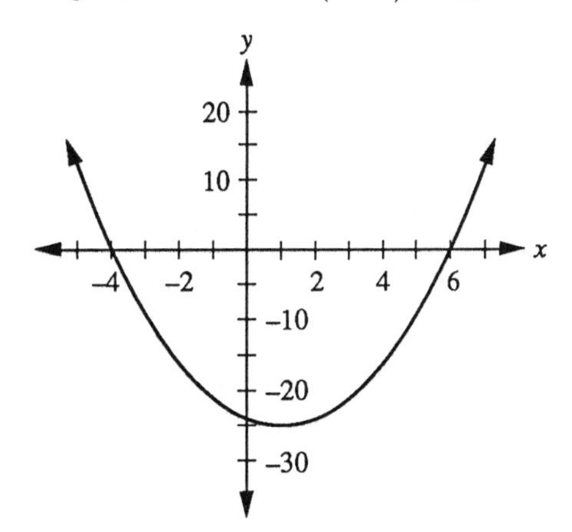

d. $y = \frac{1}{2}x^2 + 4x + 7 = \frac{1}{2}(x+4)^2 - 1$

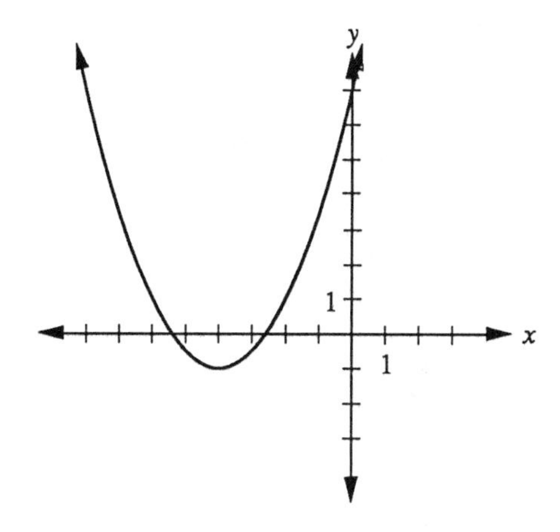

4. **a.**

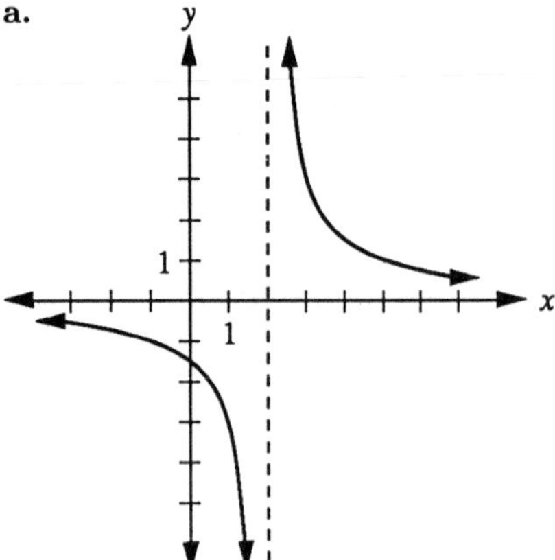

c.

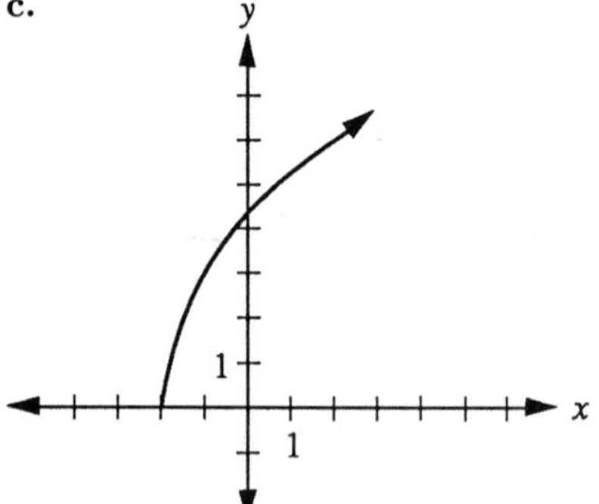

e.

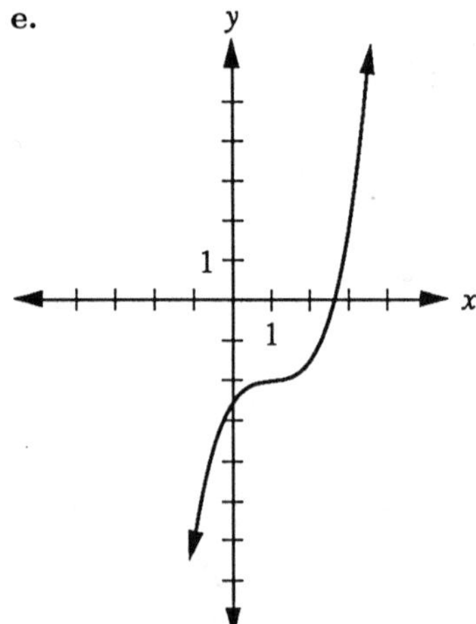

g.

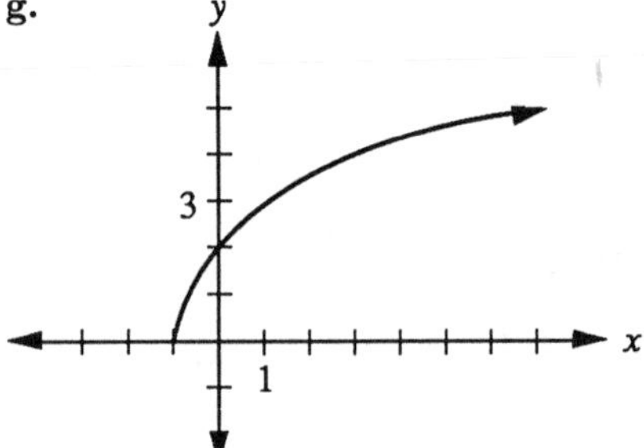

i.

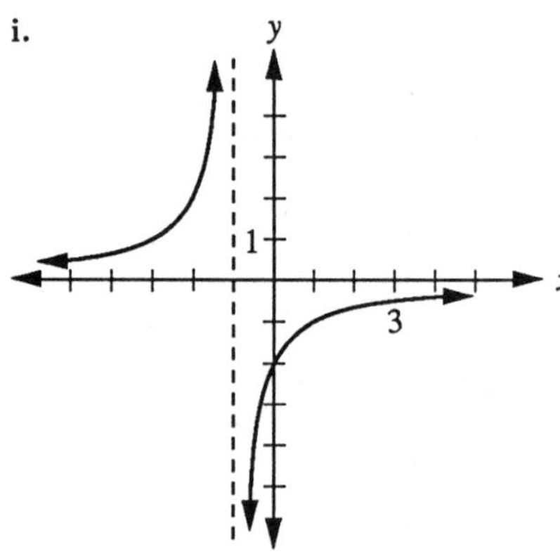

k.

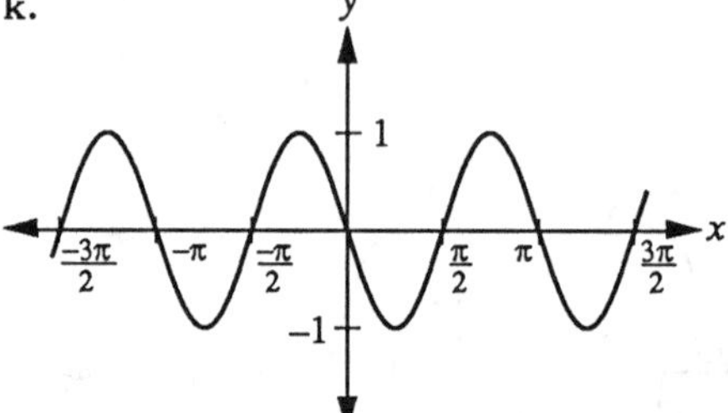

m.

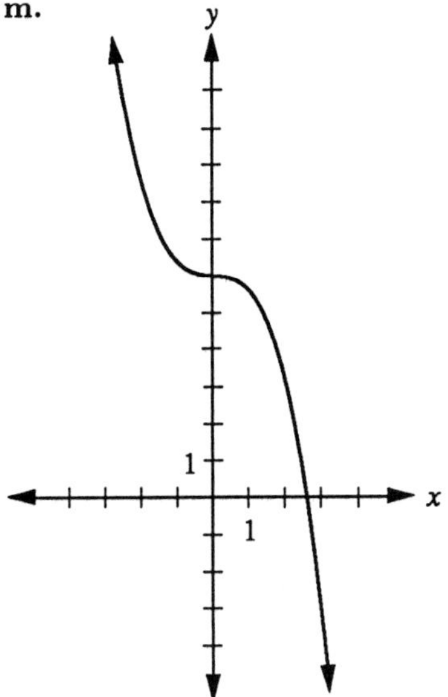

p.

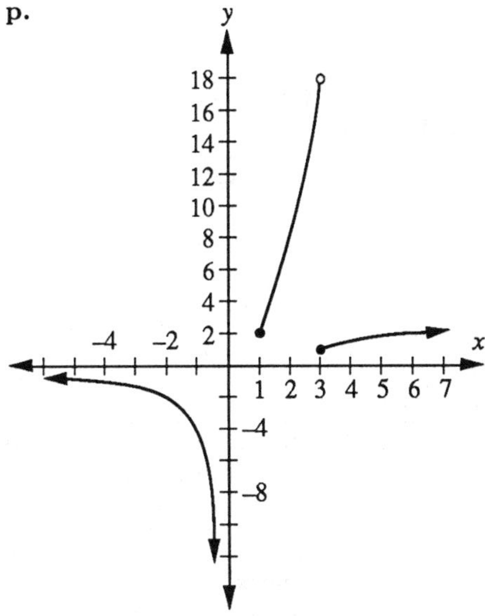

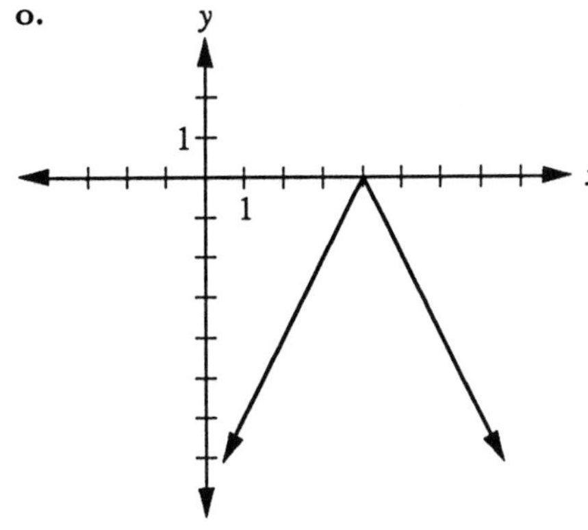

Exercise Set 8

1. **a.** domain: $0 \leq t \leq 7.35$
range: $0 \leq h \leq 218.9$

b. about 219 feet

2. zeros: -1.88, 0.35, 1.53
turning points: $(-1, 3)$ $(1, -1)$
as $x \to \infty$, $f(x) \to \infty$
as $x \to -\infty$, $f(x) \to -\infty$
f appears to have point symmetry about (0, 1).

4. **a.** $0 < y \leq 1$

c. $y \leq 0$ or $y > 1$

e. $0 < y \leq 0.25$

5. **a.** ± 0.71

c. 3

e. $\ldots, -2.64, -1.07, 0.5, 2.07, \ldots$

6. **a.** zeros: $x = 0$, $x = 3$, $x = -4$
turning points: $x \approx -2.36$, $x \approx 1.69$

c. zeros: $x \approx 0.618$, $x \approx -1.618$
turning points: $x = 0.00$, $x = 2.00$

e. zeros: no zeros
turning points: $x = 0$

g. zeros: $x = 1$, $x = -5$
turning points: none

7. **a.** an infinite number of zeros

b. $x \approx 0.644$, $x \approx 2.498$

9. $d = \sqrt{(x-1)^2 + (x^2+1.5)^2}$ has its minimum value at $x = 0.23$.

11. $R = (15+x)(50,000 - 2500x)$ has its maximum value at $x = 2.5$.

13. **a.** $n = 1 \quad P = \$10,100$
$n = 12 \quad P = \$888.49$
$n = 120 \quad P = \$143.47$
$n = 360 \quad P = \$102.86$

b. domain: $x \geq 1$
range: $100 < y \leq 10,100$

c. A payment of \$100 would pay the monthly interest but wouldn't reduce the principal at all. The outstanding balance would never be paid off.

15. For x-values close to 50, small changes in x correspond to drastic changes in f. For large x-values, f changes very little even when x changes significantly. Close up photographs are difficult to focus because slight differences in distance between lens and object (x-values) cause large differences in distance between lens and film (f-values).

Exercise Set 9

1. $r < 0.009$

2. Graph f and look for the places it is above the x-axis.

3. **a.** $x > 3$ or $x < -2$

c. $x > \frac{-1+\sqrt{17}}{2}$ or $\frac{-1-\sqrt{17}}{2} < x < 1$

e. $0 + k\pi < x < \frac{\pi}{2} + k\pi$, k any integer

g. $-3 < x < 3$

i. $\frac{5}{8} < x < \frac{9}{2}$

k. $x \leq -4$ or $0 \leq x \leq 2$

m. $-2 < x < -\frac{1}{2}$ or $x > 3$

o. $-2 < x < \dfrac{-1+\sqrt{5}}{2}$

p. $0 + 2k\pi < x < \pi + 2k\pi$, k any integer

q. $x \leq \frac{5-\sqrt{7}}{2}$ or $3 < x \leq \frac{5+\sqrt{7}}{2}$

s. $x < -\sqrt{5}$ or $-\sqrt{3} < x < \sqrt{3}$ or $x > \sqrt{5}$

u. $-2 < x < -\frac{3}{2}$

5. $x > 10\sqrt{21} - 10 \approx 36$ mph

Exercise Set 10

1. domain: $x > 0$
range: $\mathbf{R}$

2. **a.** $f \circ g = \frac{x+1}{-x}$, domain: $\mathbf{R}, x \neq -1, 0$
$g \circ f = \frac{x-1}{x}$, domain: $\mathbf{R}, x \neq 1, 0$

c. $f \circ g = \sqrt{\frac{1}{x^2} - 5} = \sqrt{\frac{1-5x^2}{x^2}}$,
domain: $-\frac{1}{\sqrt{5}} \leq x \leq \frac{1}{\sqrt{5}}$, $x \neq 0$
$g \circ f = \frac{1}{x-5}$, domain: $x > 5$

e. $f \circ g = \sqrt{x-1} + 1$, domain: $x \geq 1$
$g \circ f = \sqrt{|x+1| - 1}$, domain: $x \leq -2$ or $x \geq 0$

g. $f \circ g = \frac{\sqrt{x+2}}{\sqrt{x+2}-1}$, domain: $x \geq -2$, $x \neq -1$
$g \circ f = \sqrt{\frac{3x-2}{x-1}}$, domain: $x > 1$ or $x \leq \frac{2}{3}$

i. $f \circ g = \sqrt{\sin x}$, domain: $0 + 2k\pi \leq x \leq \pi + 2k\pi$, k an integer
$g \circ f = \sin\sqrt{x}$, domain: $x \geq 0$

3. **a.**

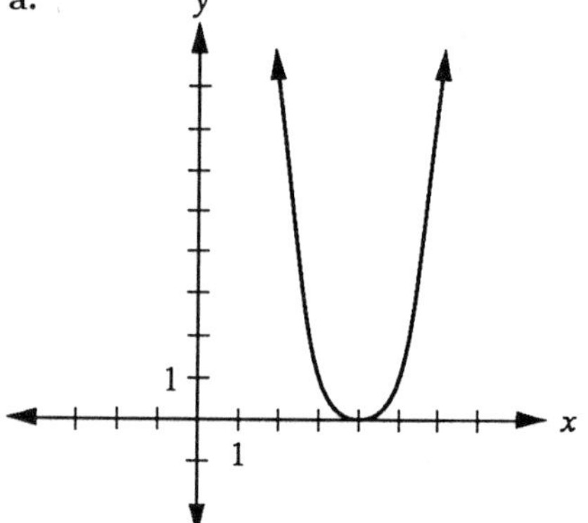

c.

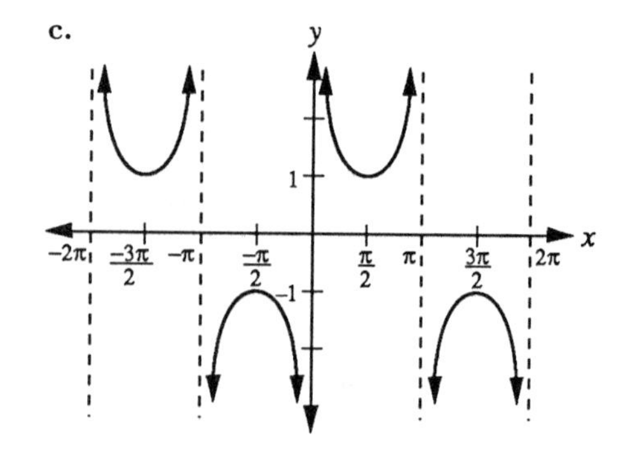

e.

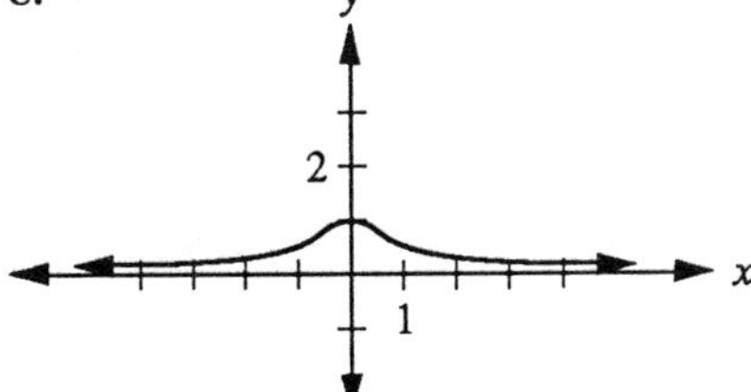
y
2
1
x

g.

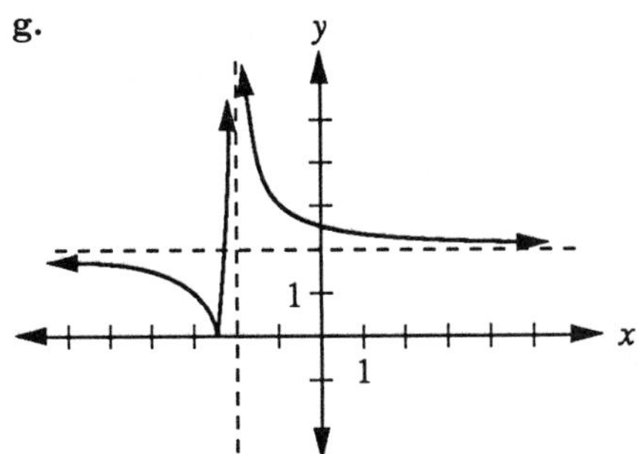
y
1
1
x

i.

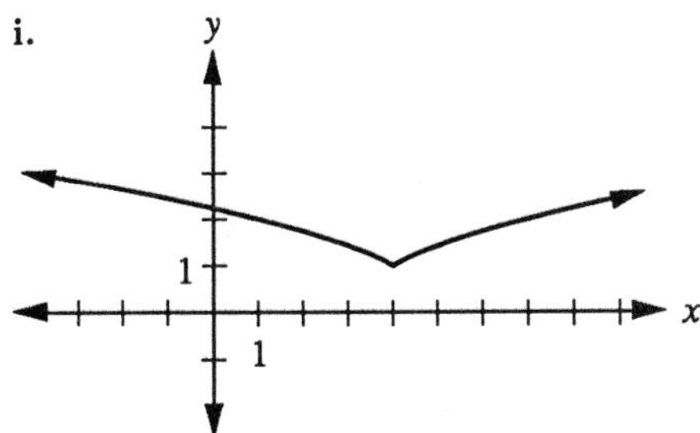
y
1
1
x

k.

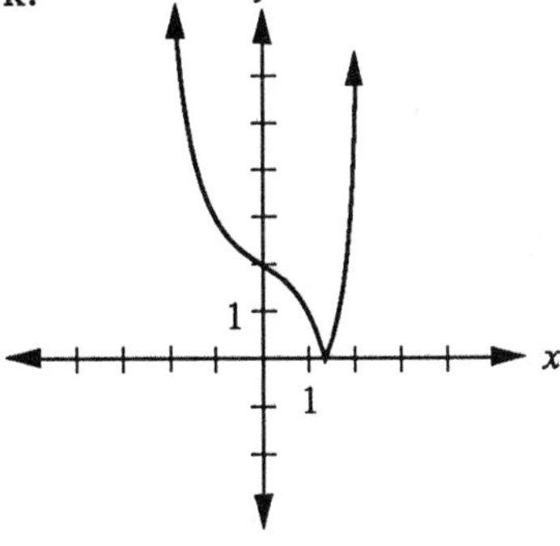
y
1
1
x

m.

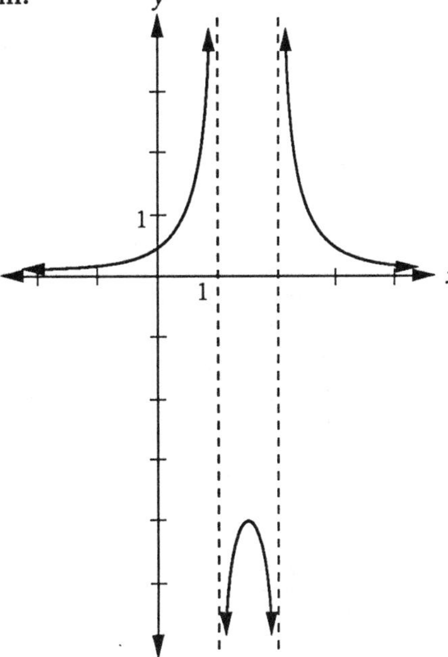
y
1
1
x

o.

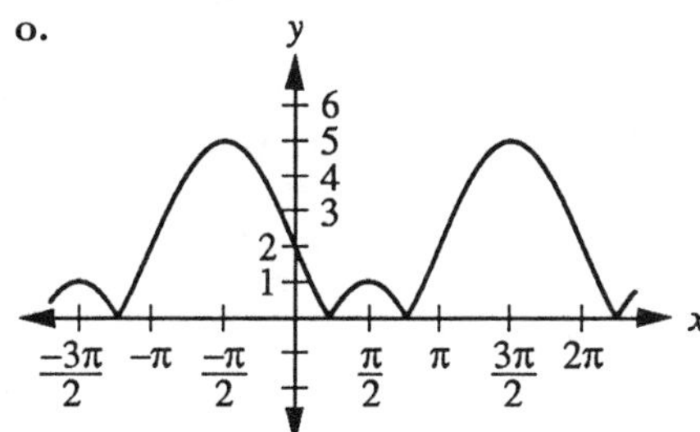
y
6
5
4
3
2
1
x
$\frac{-3\pi}{2}$ $-\pi$ $\frac{-\pi}{2}$ $\frac{\pi}{2}$ π $\frac{3\pi}{2}$ 2π

q.

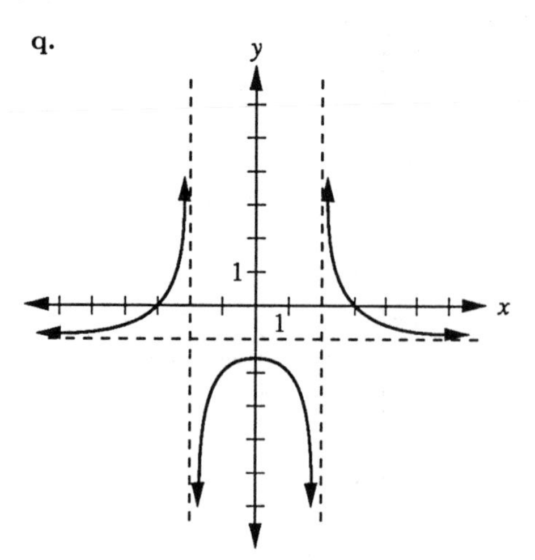

4. c.

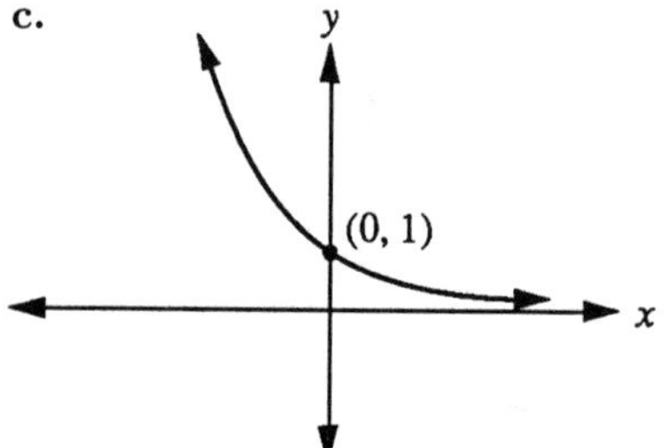

d.

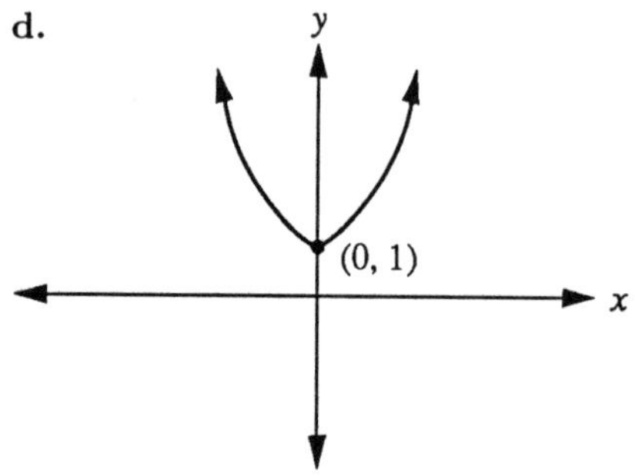

5. Graph $g(x) = x - 5$. Then take the reciprocal of the y-values.

7. Graph $q(x) = x + 3$. Points where $y < 0$ should be reflected across the x-axis.

10. The domain is all real numbers. The radicand is a perfect square. The result of taking the square root is linear: $\sqrt{a^2} = |a|$.

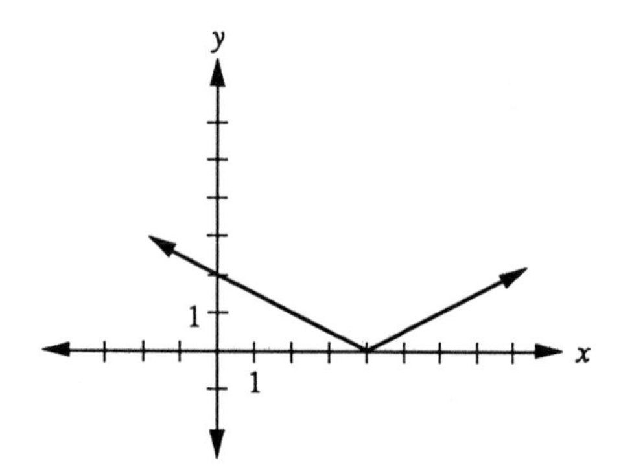

11. a. 10, 10, 10,10

b. 10, 11, 13, 16

13.

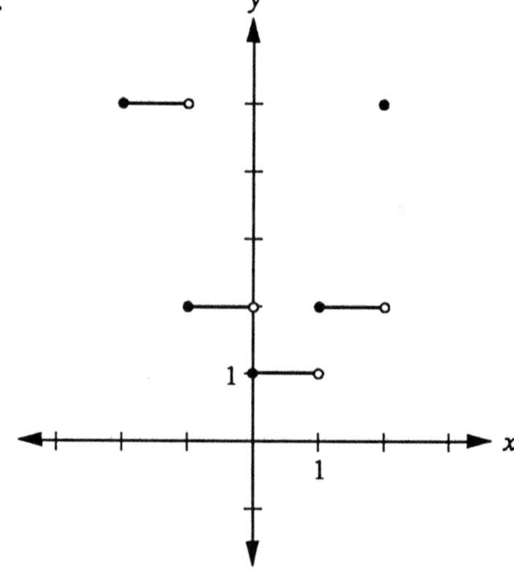

15. $g(g(1)) = \frac{5}{2}$, $g(g(g(1))) = \frac{11}{4}$

17. 17

18. a. $f(102) = 92$

c. $f(97) = 91$

Exercise Set 11

1. a. $f^{-1}(x) = \sqrt[3]{x-2}+1$
   c. $f^{-1}(x) = 5 - x^2,\ x \geq 0$
   e. $f^{-1}(x) = \sqrt{9-x^2},\ x \geq 0$
   g. $f^{-1}(x) = 3 - \sqrt{9-x}$
   h. $f^{-1}(x) = x^2 - 1,\ x \leq 0$

2. a.

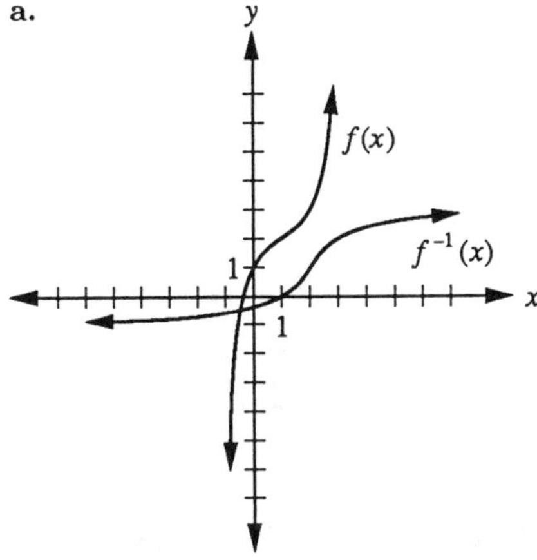

c.

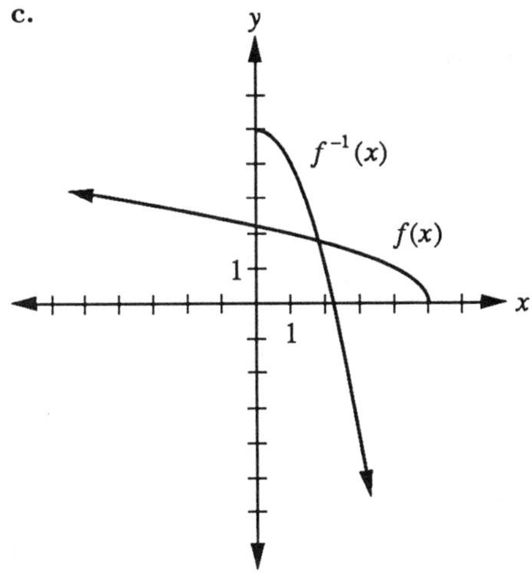

e.

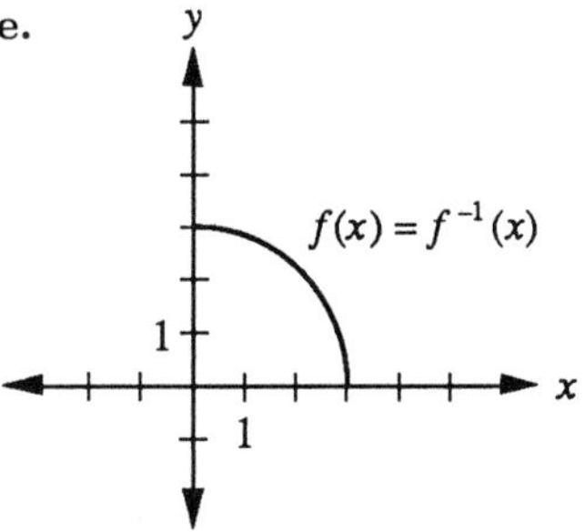

3.

	Domain of f	Domain of f^{-1}
a.	**R**	**R**
b.	**R**, $x \neq -1$	**R**, $x \neq 0$
c.	$x \leq 5$	$x \geq 0$
d.	**R**	**R**
e.	$0 \leq x \leq 3$	$0 \leq x \leq 3$
f.	**R**, $x \neq 1$	**R**, $x \neq 1$
g.	$x \leq 3$	$x \leq 9$
h.	$x \geq -1$	$x \leq 0$

4. a. $x = \frac{1}{3}$

5. a. $h^{-1}(x) = (1-x^3)^{\frac{1}{3}}$
   b. $h^{-1}(x) = \sqrt{1-x^2}, x \geq 0$
   c. $h^{-1}(x) = \frac{3x-1}{2x-3}$
   $h(x)$ is symmetric with respect to $y = x$ in each case.

7. Answers will vary because of different domain restrictions.
   a. $f^{-1}(x) = 1 + \sqrt{x+1}$ if domain of $f(x)$ is restricted to $x \geq 1$.
   b. $f^{-1}(x) = x$ if domain of $f(x)$ is restricted to $x \geq 1$.
   c. $f^{-1}(x) = \sqrt{1-x^2}$ if the domain of $f(x)$ is restricted to $0 \leq x \leq 1$.

9. **a.** $p^{-1}(x) = x^2 - 3,\ x \leq 0$

b.

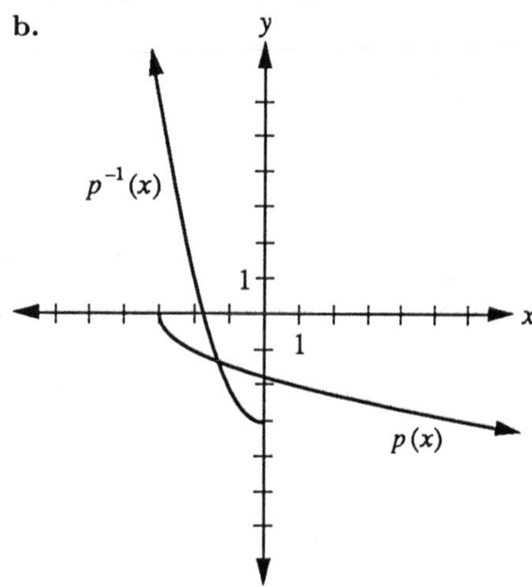

c. domain of p: $x \geq -3$
domain of p^{-1}: $x \leq 0$

d. $x \leq 0$

e. $x \geq -3$

f.

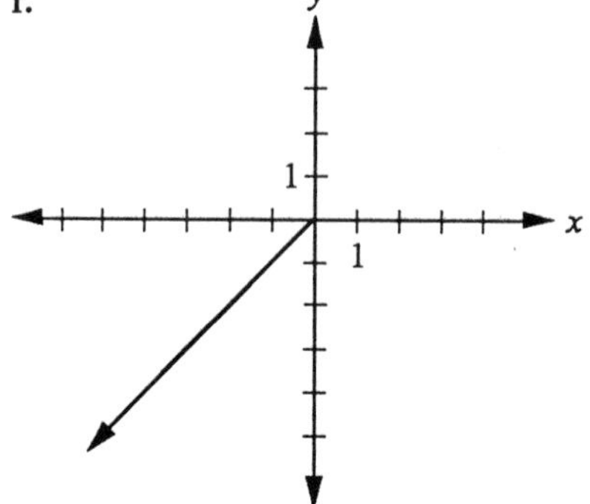

g.

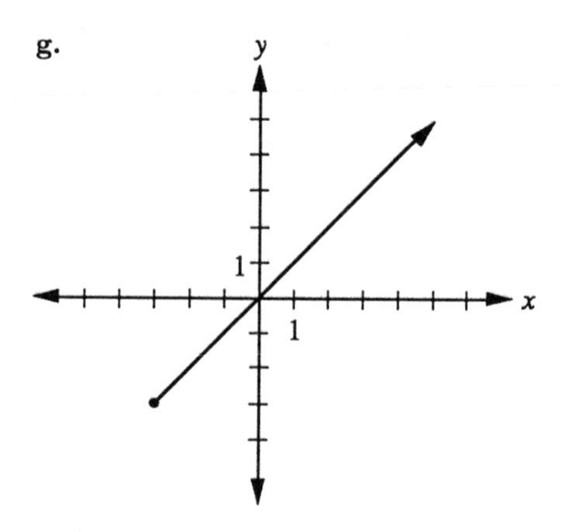

10. **a.** $g^{-1}(x) = \begin{cases} x + 2 & \text{for } x \leq 0 \\ 0.5x + 2 & \text{for } x > 0 \end{cases}$

CHAPTER 3

Polynomials, Rational Functions, and Algorithms

Exercise Set 1

1. zeros: $x = 0,\ -1$

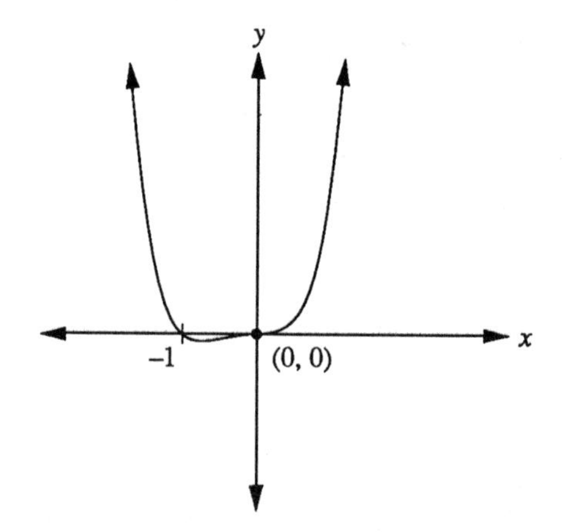

2. **a.**

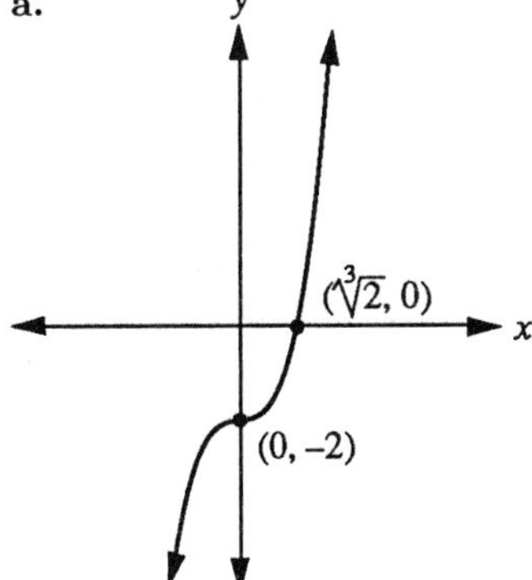

c.

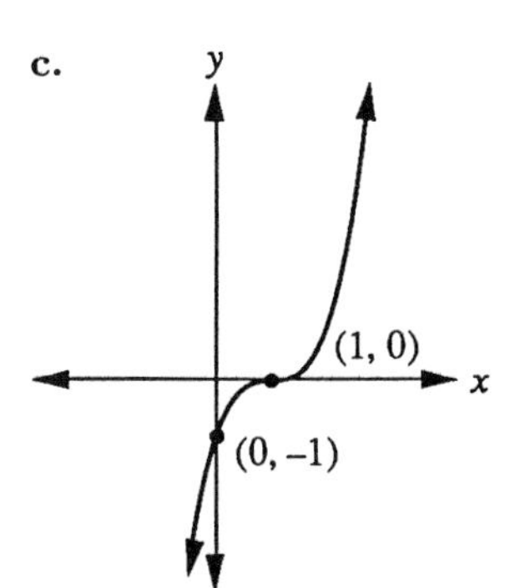

e.

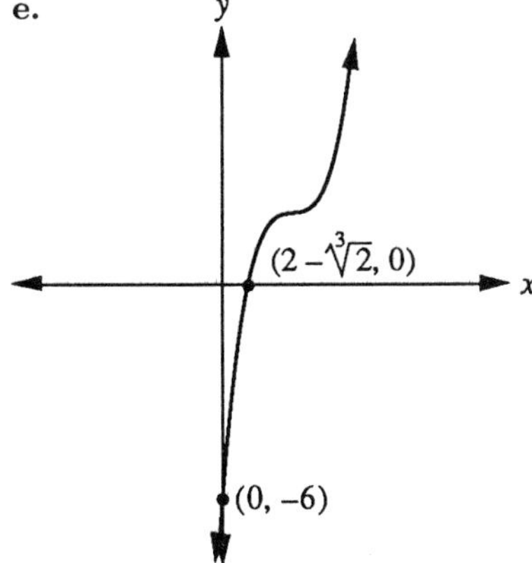

g.

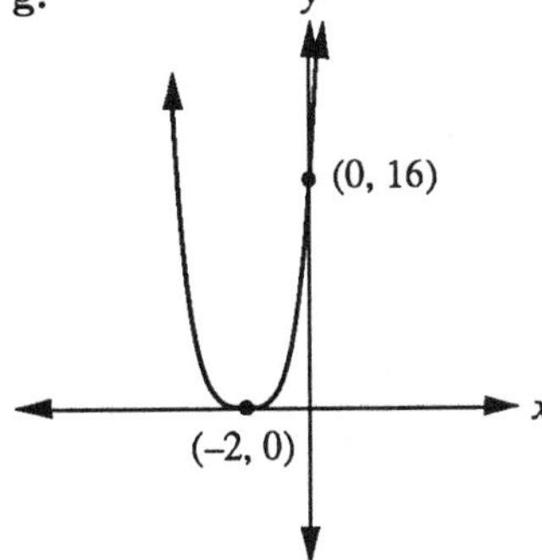

i.

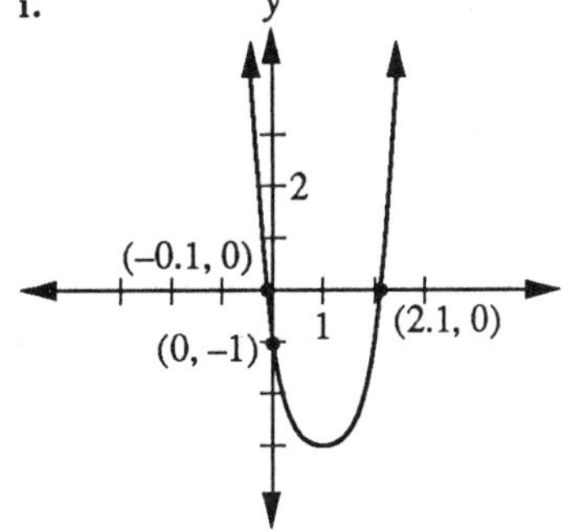

3. A 17th degree polynomial is similar in shape to a cubic (when viewed globally) but much steeper. It can have at most 16 turning points and 17 zeros.

5. Even-degree polynomials have either a lower limit or an upper limit to their range. Odd-degree polynomials have all reals as their range.

7. **a.** $f(x) = (x+2)(x-3)(x-5)$

b. $f(x) = (x+3)(x+1)(x-2)(x-4)$

c. $f(x) = (x+2.1)(x-1.4)(x-3.5)$

d. $f(x) = (x-4)(2x-3)^2$

Exercise Set 2

1. To have three distinct zeros, a function must have three distinct linear factors. If each factor has multiplicity one, the polynomial will have degree 3. The degree will be higher if any factor has multiplicity greater than one.

3. $f(x) = -(x-2)(x+3)(x+4)(x-1)$

Exercise Set 3

1. $y = x\frac{1}{2}(60 - 3x)(24 - 2x)$
   height $\approx$ 4.9 in
   width $\approx$ 14.2 in
   length $\approx$ 22.65 in
3. $y = (300 + 3.5x)(95 - x)$; \$28,575.00
5. max. at $(-1.73, 5.39)$
   min. at $(1.73, -15.39)$

Exercise Set 4

1. **a.** $5.0, -2.0$

   b. -2.3

   c. $0.8, 3.0$
4. A function may have a vertical asymptote and may be positive on one side of the asymptote and negative on the other side. If $f(x) = \frac{1}{x-2}$ then $f(0) < 0$ and $f(3) > 0$ but $f(x)$ is never equal to zero.

Exercise Set 6

1. 18
3. **b.**

Year	0	1	2	3	4
Loan Balance	4000	3070	2065.60	980.85	−190.68

   c. \$1207.68
5. zero at 2.24; 10 iterations needed for bisection and 5 for secant.
7. 1.476

Exercise Set 7

1. 24 iterations
2. $x \approx -2.3$; we know there is a zero in the interval $-2.25 > x > -2.3125$ and the endpoints of this interval agree when rounded to tenths but do not agree when rounded to hundredths.
3. **a.** 3

   b. 5

   c. 7

   d. 7

Exercise Set 8

1. Use bisection or secant algorithm, computer or calculator to locate zeros; x-intercepts occur where the numerator equals 0 and vertical asymptotes occur at zeros of the denominator.
3. **a.** $x \to \infty$, $f(x) \to \infty$
   $x \to -\infty$, $f(x) \to -\infty$
   Global behavior similar to $y = x - 1$

   b. $x \to \infty$, $f(x) \to 0^+$
   $x \to -\infty$, $f(x) \to 0^-$
   global behavior similar to $y = \frac{1}{x}$

   c. $x \to \infty$, $f(x) \to 2^+$
   $x \to -\infty$, $f(x) \to 2^-$
   global behavior similar to $y = 2$
4. **a.**

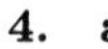

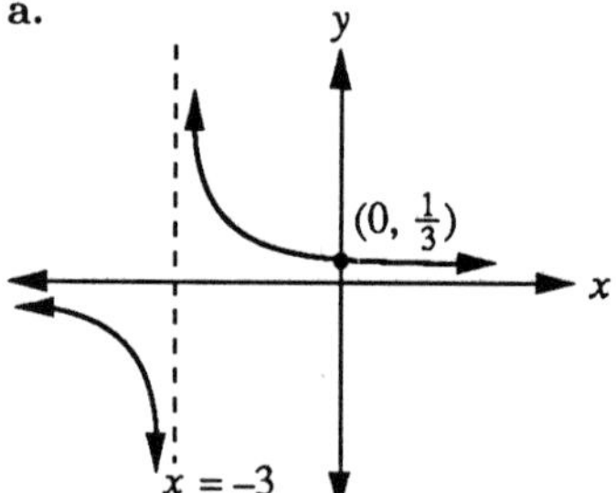

   c.

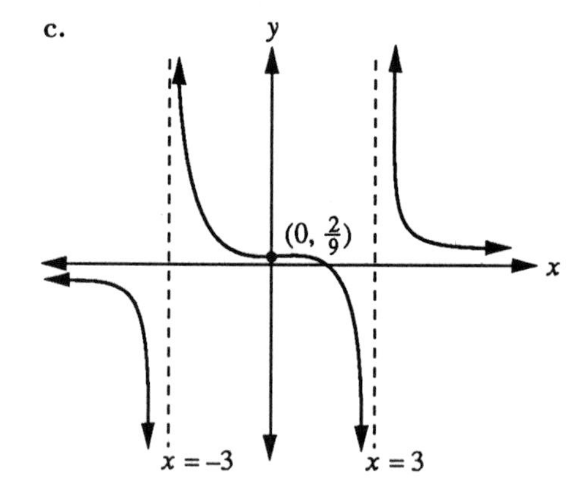

e.

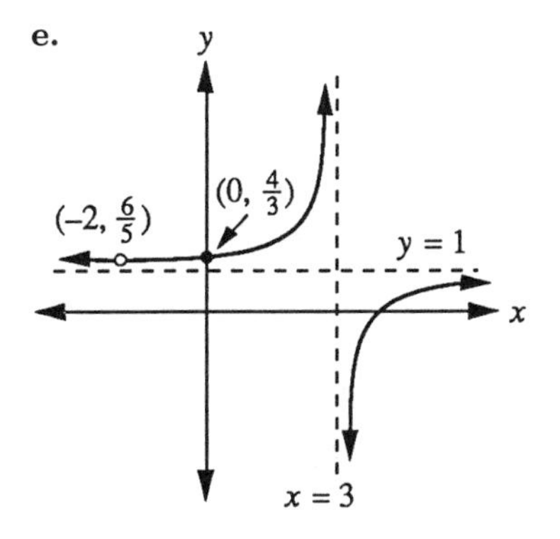

g.

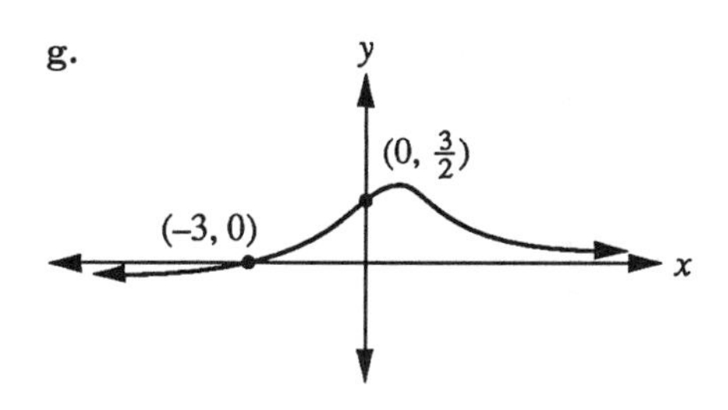

i.

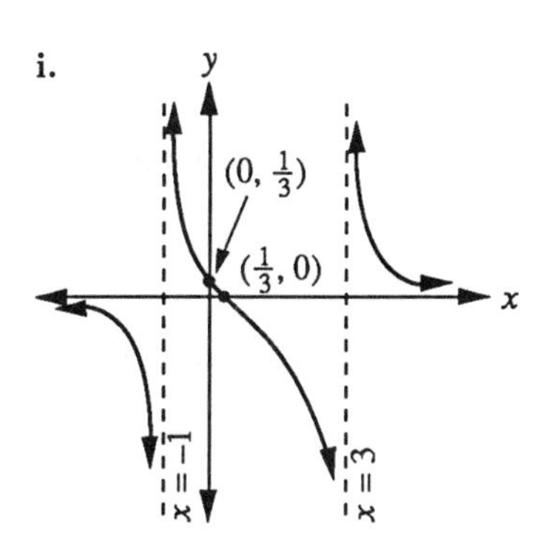

k.

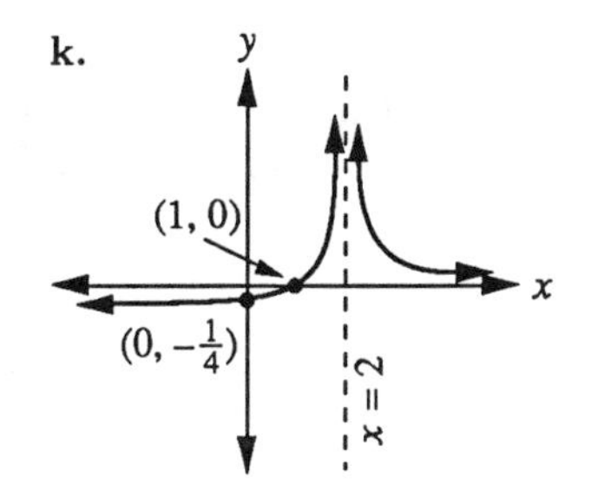

m.

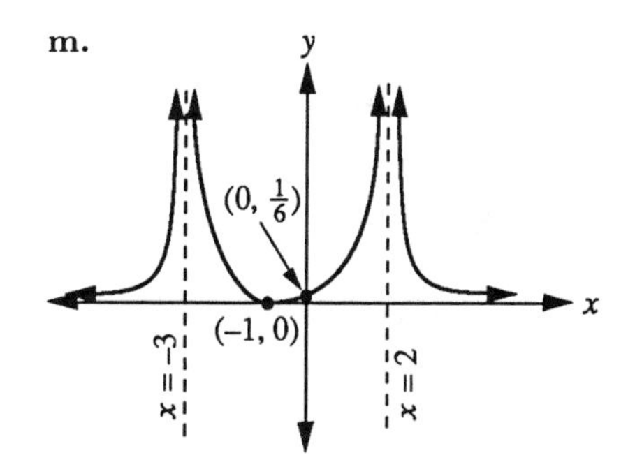

5. **a.** 3.82, 1.18

b. $-2 < x < 3$ or $x > 8$

c. $x < -1$ or $x > \frac{1}{2}$

7. **a.**

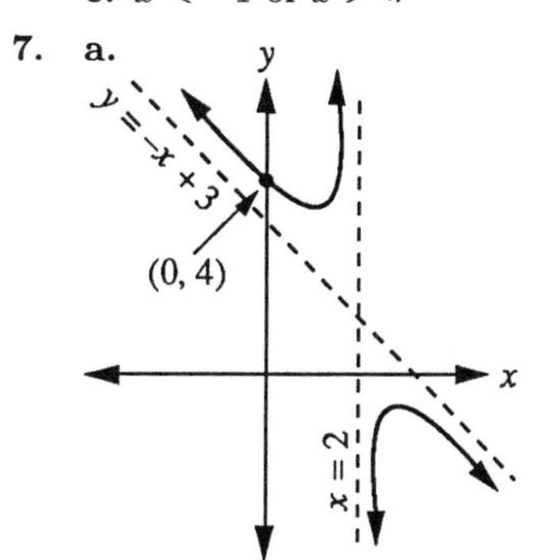

b. $y = \dfrac{-x^2 + 5x - 8}{x - 2}$

Exercise Set 9

1. 164.9 lbs, 1636 mi

3. 6.934 in × 6.934 in × 5.198 in

4. $r = 3.83$

5. **a.** (100, 2.5), (300, 3.6), (500, 4.3), (700, 4.8), (900, 5.2), (1100, 5.6), (1300, 5.9), (1500, 6.2) Scatter plot is curved and concave down.

b. $r = 0.542v^{\frac{1}{3}}$

c. when $v = 2000$ $r = 6.829$

d. $d = h$; manufacturers may not produce cans with d and h equal since these proportions are not pleasing to the eye.

CHAPTER 4

Exponential and Logarithmic Functions

Exercise Set 1

1. \$159.91 during the tenth year compared to \$80.00 and \$86.40.
2. 10 years for balance to reach \$2,000.
3. **a.** \$4,660.96; closed form is easier.

   b. \$5,033.83; recursive form is easier given answer in part a.

Exercise Set 2

1. **c.** range: $y > 0$

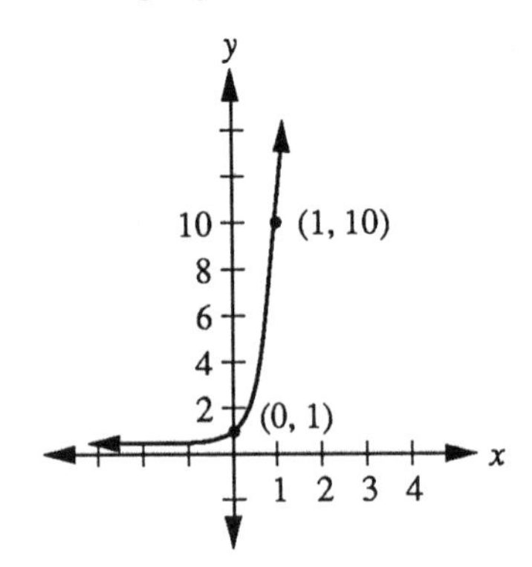

d. range: $y > 0$

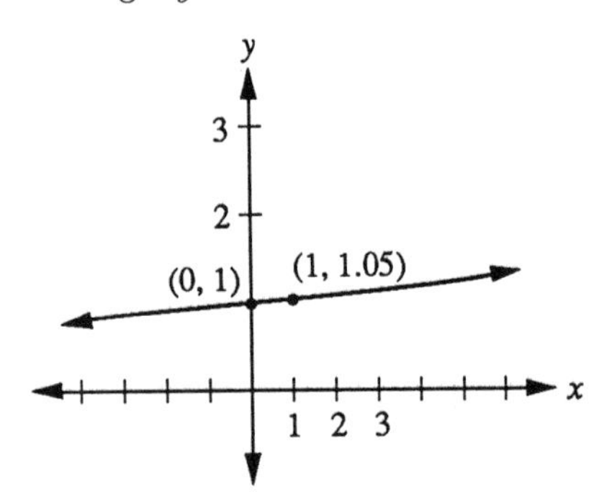

2. **b.** y-values are multiplied by a factor of 0.25.

   c. y-values are multiplied by a factor of 10.

5.

a	0	1	2	3	4	5
C	13,000	10,400	8,320	6,656	5,324	4,259

6. \$1.22
9. $2^{50}(0.003) \approx 3.38(10^{12})$ inches. This is over 50 million miles, farther than the moon and half-way to the sun.
10. **a.** 7

    c. $-\frac{1}{3}$
11. $a = 5$, $b = \frac{5}{6}$

Exercise Set 3

2. **a.**

   b. The graph of $y = e^x$ is between the graphs of $y = 2^x$ and $y = 3^x$.
3. **a.** \$2164.86

   b.
   Continuous compounding approximation, \$2000 $(e^{.08}) =$ \$2166.57, is \$1.71 greater.
4. **a.** \$290.06

   c. \$290.43

   e. \$290.46
5. **a.** The greater future value results from the 8% deposit (\$1469.33 compared to \$1467.84). The number of years does not affect which deposit has greater future value.
7. **a.** \$108.24

   b. 8.24%

   c. 8.03%
9. **a.** \$6810.21

   b. 6.17%
10. Present value of first payment = \$200,000. Present value of second payment $\approx$ \$183,486 since $183,486(1.09)^1 \approx 200,000$.

Present value of third payment $\approx \$168,336$ since $168,336(1.09)^2 \approx 200,000$.
Present value of fourth payment $\approx \$154,437$, and present value of fifth payment $\approx \$141,685$.
So present value is sum of these $\approx \$847,944$.

13. **a.** \$2089.68

b. \$2094.59

c. Since the present value in this situation represents the amount of money which must be invested to achieve a specified future value, the investor would opt for the lower present value.

14. \$3807.26

15. **b.** After 19.4 years.

17. **a.** Choose the certificate of deposit; its value is \$82,436.06 (compared to \$10.24).

b. Choose doubling fund; value is $\$1.0995 \cdot 10^{10}$ (compared to \$369,452.80).

c. After approximately twelve years.

Exercise Set 4

1. **a.** $y = e^{2x} = (e^2)^x \approx (7.4)^x$
$y = e^{2x}$: horizontal compression by factor of $\frac{1}{2}$; $(0, 1)$, $(\frac{1}{2}, e)$.
$y = 7.4^x$: exponential function base 7.4; $(0, 1)$, $(1, 7.4)$.
horizontal asymptote: $y = 0$

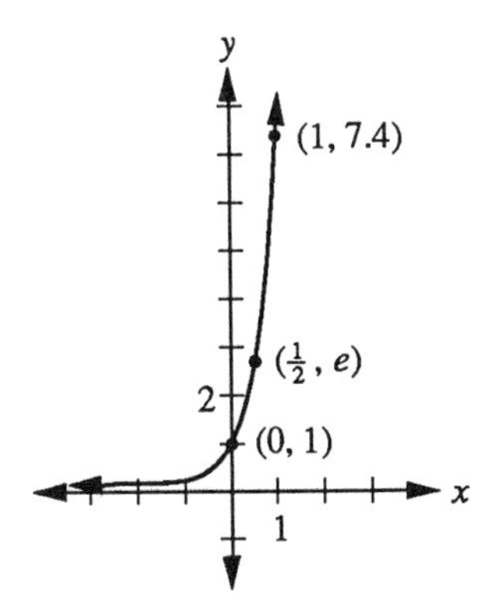

b. $y = 4 \cdot 2^{x-1} = 2^2 2^{x-1} = 2^{x+1}$
$y = 4 \cdot 2^{x-1}$: horizontal shift one unit to right, vertical stretch by factor of 4; $(1, 4)$, $(2, 8)$.

$y = 2^{x+1}$: horizontal shift one unit to left; $(-1, 1)$, $(0, 2)$.
horizontal asymptote: $y = 0$

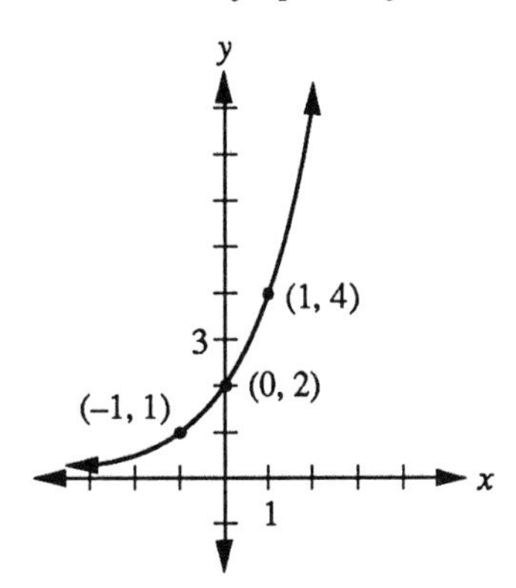

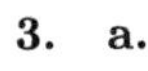

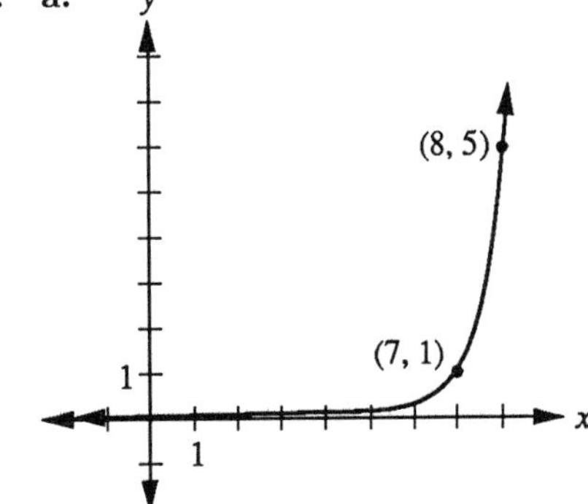

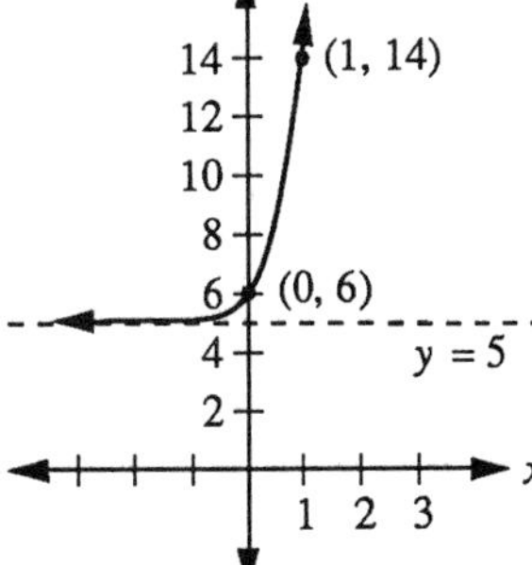

e.

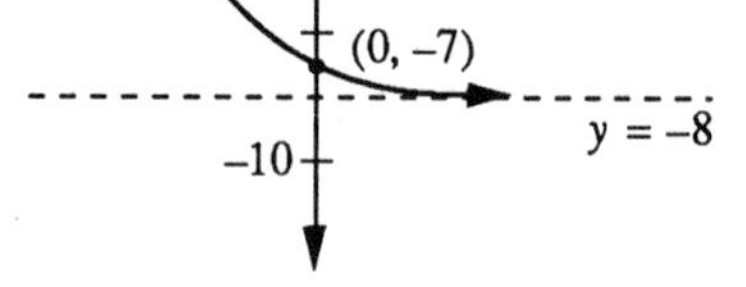

g.

i.

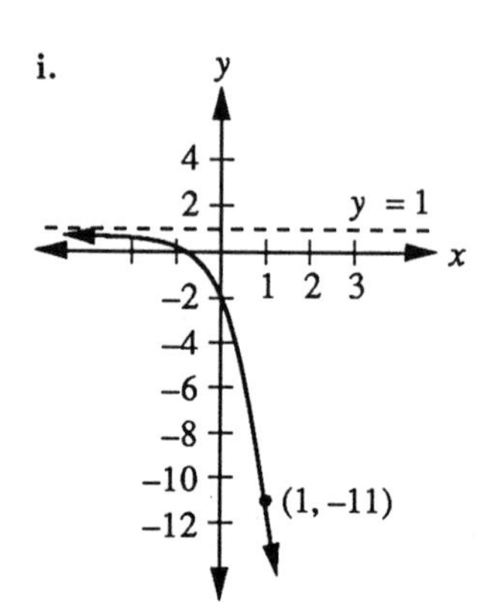

4. a.

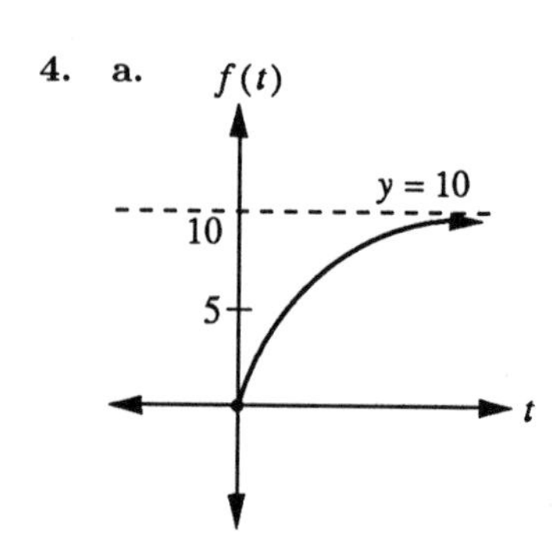

b. 0

c. 8.6 cakes

d. 10 cakes

5. a.

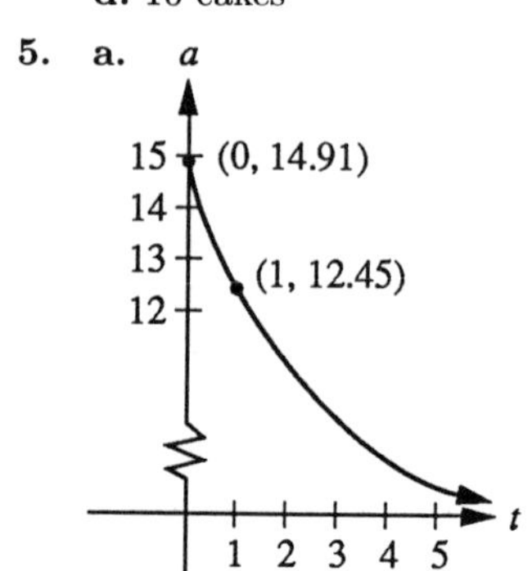

b. 14.91 mg per cc

c. 7.26 mg per cc

d. slightly over 6 hours

Exercise Set 5

1. **a.** $4^3 = 64$

   c. $10^4 = 10,000$

2. **a.** $1 = \log_e e$ or $1 = \ln e$

   c. $\frac{1}{2} = \log_{16} 4$

3. **a.** -1

   d. $\frac{3}{5}$

4. **a.** 1.1

   c. 3.1

   e. 3.3

5. **b.** $f^{-1}(x) = 1 + \log_3(x)$
   domain: $x > 0$

   d. $f^{-1}(x) = 3 + 2^x$
   domain: $\mathbf{R}$

Exercise Set 6

1. $f(x) = \log_b(x)$ increases for $b > 1$ and decreases for $0 < b < 1$.

2. **b. i.** $\log(x)$ has negative values for $0 < x < 1$, is zero when $x = 1$, and has positive values for $x > 1$.

iii. Values of $\log(x)$ are between 0 and 1 when x is between 1 and 10; $\log(10) = 1$; $\log(x) > 1$ for $x > 10$.

3. $g(x) = \ln(x)$ grows at a slower rate than the function $f(x) = \sqrt{x}$.

5. **a.** For $n = 10$, T_C is fastest and T_A is slowest. For $n = 300$, T_A is fastest and T_C is slowest.

b. For large values of n, T_A is fastest and T_C is slowest. This is reasonable, since we know that the logarithmic function increases very slowly and the exponential function increases rapidly.

6. **b.** domain: $x > 0$
range: **R**

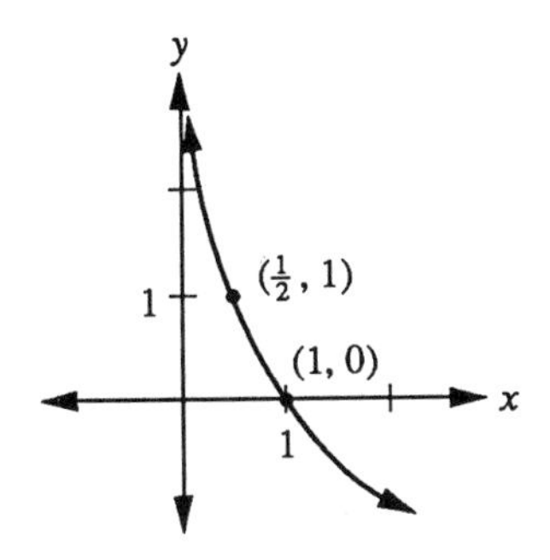

d. domain: $x > 0$
range: **R**

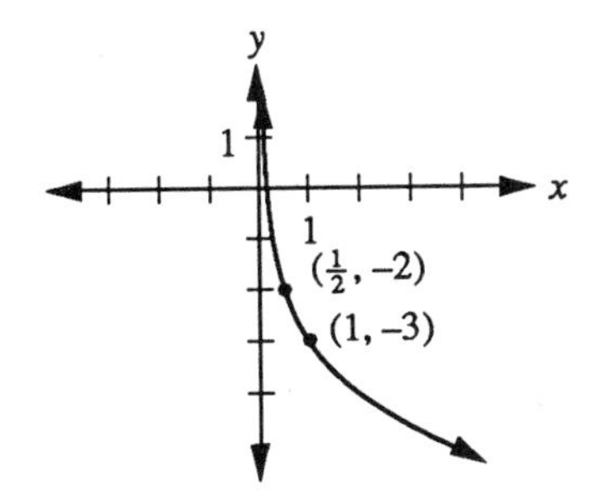

f. domain: $x > 0$
range: **R**

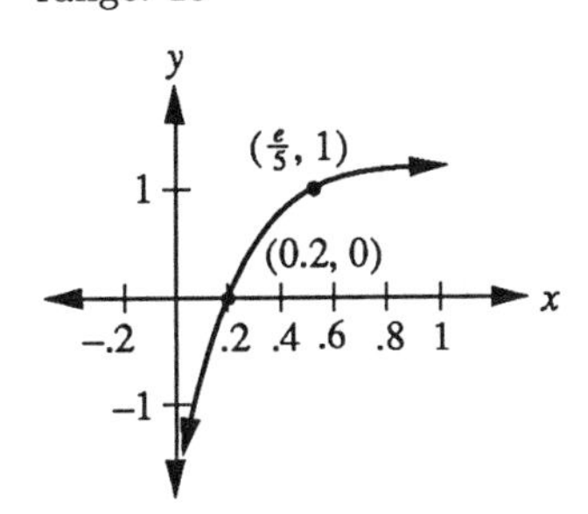

g. domain: $x < 0$
range: **R**

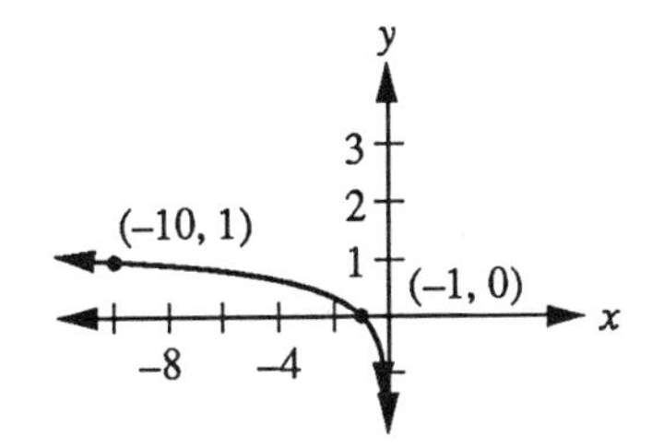

h. domain: $x > 0$
range: $y \geq 0$

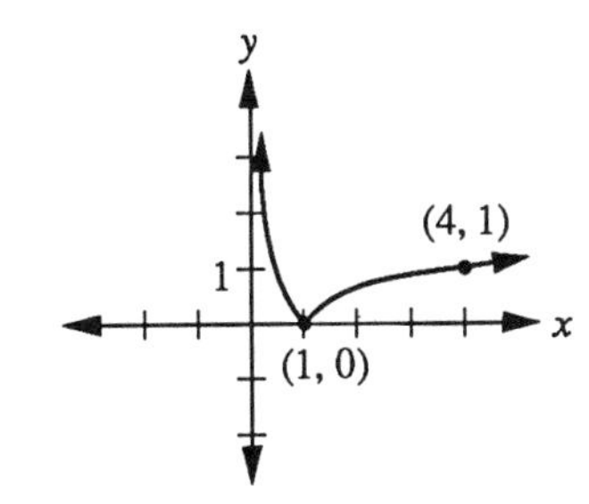

i. domain: $x > \frac{3}{4}$
range: **R**

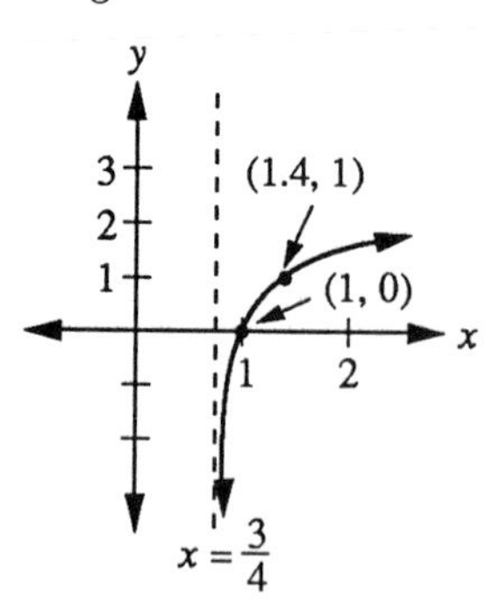

k. domain: $x > 0$
range: **R**

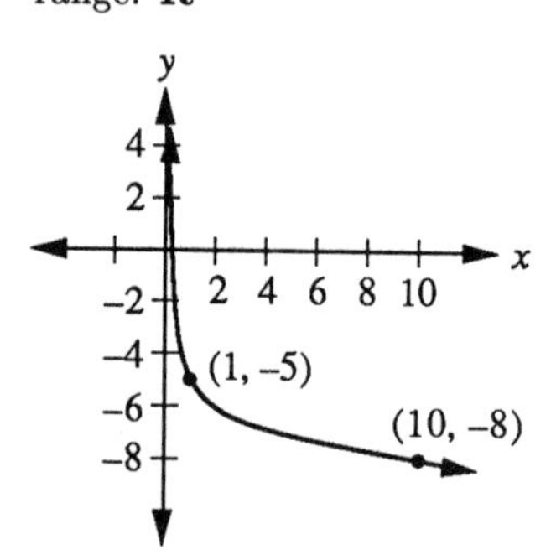

7. **a.** 5

b. $a + 3$, $a > -3$

d. undefined

9. **b.** $\ln \sqrt[3]{x} = \ln(x^{\frac{1}{3}}) = \frac{1}{3} \ln x$

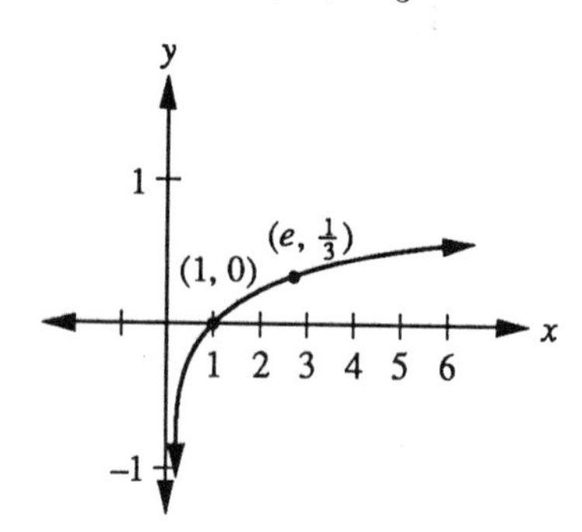

10. $1.03 \cdot 10^{375}$

Exercise Set 7

1. $g(x) = \frac{1}{2}^x = 10^{-0.301x}$

2. **a.** $3^5 = e^{5.493} = 10^{2.386}$

3. $F = 5000(1 + \frac{.09}{12})^{12N} =$
$5000(2^{0.01078})^{12N} = 5000(e^{0.0075})^{12N}$

5. **b.** 1.56

d. -3.32

6. $\log x = \dfrac{\ln x}{\ln 10} = 0.434 \ln x$

8. **a.** $\frac{1}{3}$

c. 36

e. -6 or 1

g. 2 or -1

i. -1

k. 0.86 or 0.68

l. $\emptyset$

m. $e^e - 6 \approx 9.15$

p. $\frac{1}{100}$ or 1 or 100

9. **b.** $c = -\frac{1}{Ax} \ln(\frac{A-y}{yB})$

10. **a.** $f^{-1}(x) = -1 + \log_2(x + 3)$
domain: $x > -3$

c. $f^{-1}(x) = \log_3(\dfrac{x + \sqrt{x^2 - 4}}{2})$
domain: $x \geq 2$

11. $y = 2e^{\frac{1}{3}(\ln 5)x} \approx 2e^{0.536x}$

13. 9.88%

14. Since the values of $f(x)$ range from -1.43×10^{17} to 6.53×10^{16} on the interval $[58, 59]$, the selection of a small value of epsilon for the stopping criterion $|f(x)| < \epsilon$ would be impossible. Use either $|x_1 - x_2| < \epsilon$ or $\dfrac{|x_1 - x_2|}{1 + |x_2|} < \epsilon$.

Exercise Set 8

1. $\ln(1.001) \approx 0.001$; $\ln(1.05) \approx 0.05$; $\ln(0.93) \approx -0.07$

3. $2^{-10}A$

4. approximately 23 years; 80 million

6. 19.5 minutes

7. **a.** $N = \dfrac{\ln 2}{k \ln(1 + \frac{r}{k})}$

8. **a.** 8.625 years

   b. Increasing the numerator will slightly increase the doubling time. This is appropriate since doubling time is longer for less frequent compounding.

10. $r \approx 0.087$, or 8.7%

11. approximately 0.04 square miles; 194 years ago

13. $r \approx 0.896$; approximately 94,000 cases

15. no, too little carbon-14 would remain in objects this old.

Exercise Set 9

1. $1.56 \cdot 10^{-9}$ W/m^2

3. 25.9%; 3 years

5. $1.58 \cdot 10^5$

7. **a.** 1362

   b. 7.8

   c. 7.3

9. 54 dB

11. 1.2

Exercise Set 10

2. **b.** The steepness of $y = x^3$ takes us away from the root, so we must use the inverse, $\sqrt[3]{x}$.

3. **a.** 2

   c. 0.57

   e. 0.68

Exercise Set 11

2.

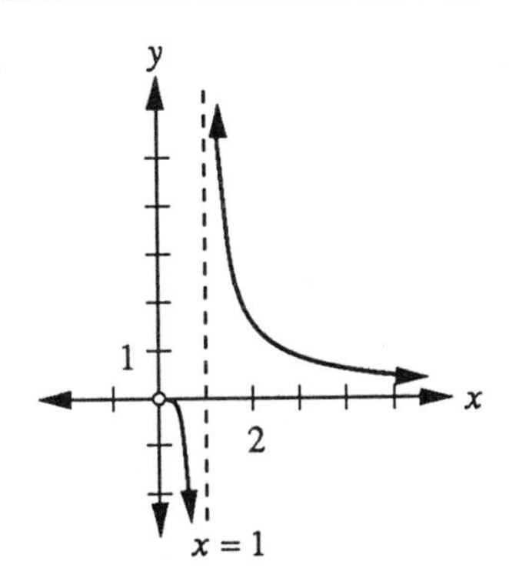

3.

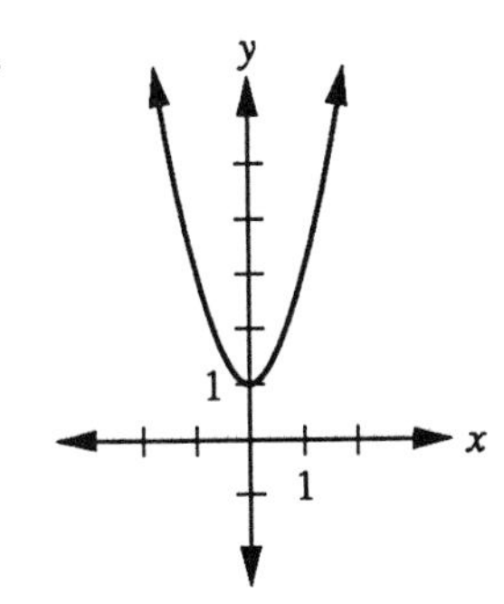

6.

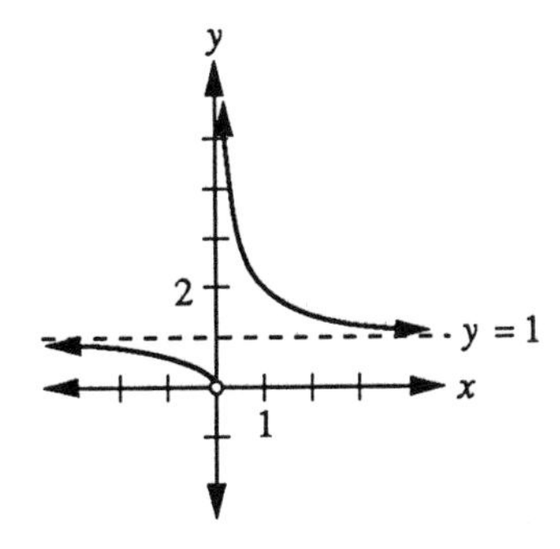

7.

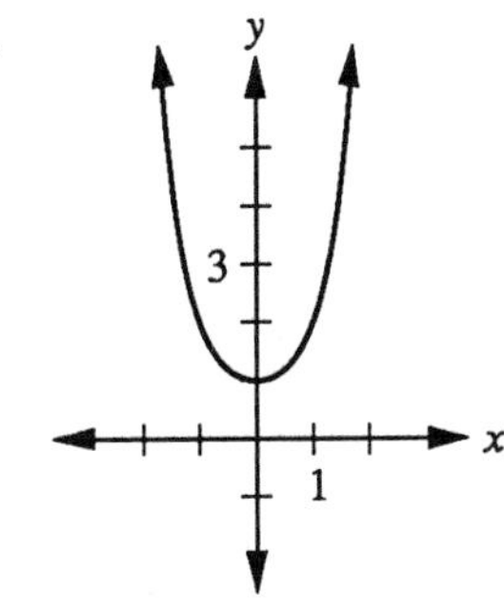

8.

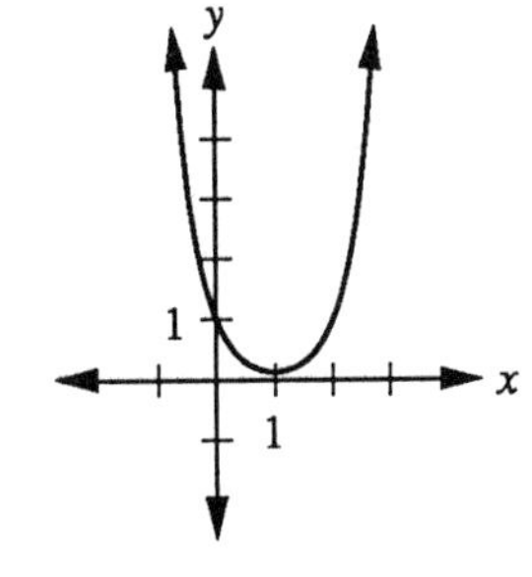

9.

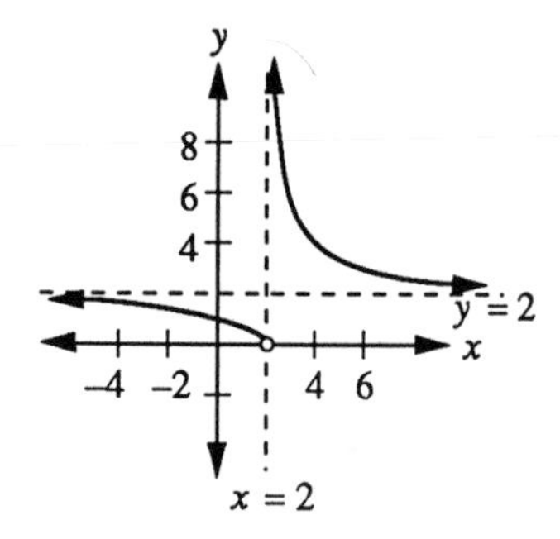

10.

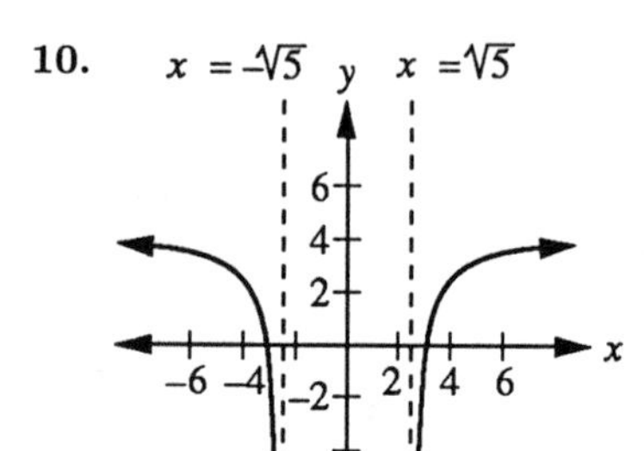

13.

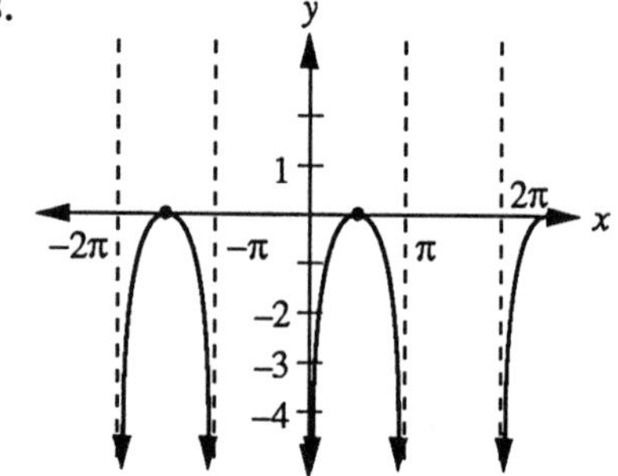

15.

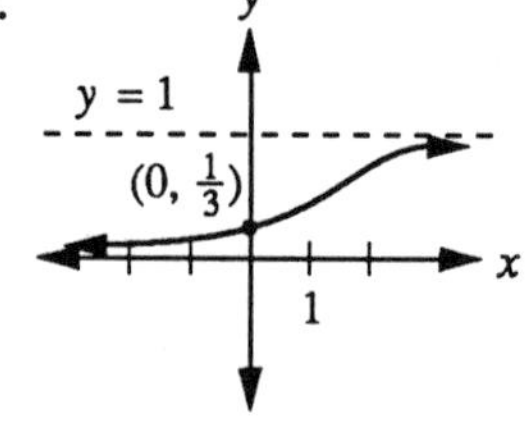

16.

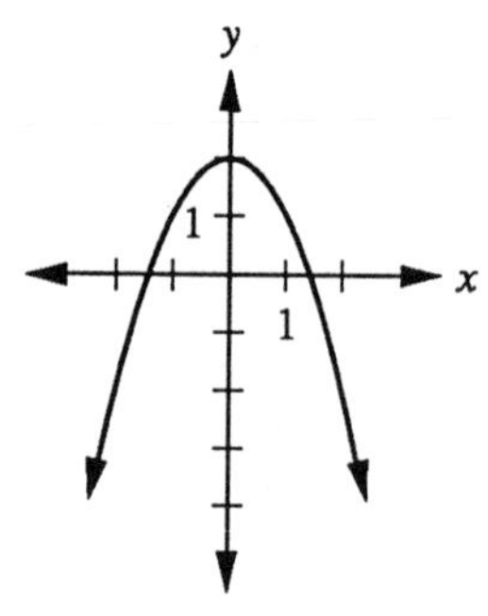

CHAPTER 5

Data Analysis Two

Exercise Set 2

1. Semi-log re-expression should pull the data points too far down so that the scatter plot of transformed data is concave down. This transformation is too severe, indicating that the exponential grows more quickly than the data.

3. The point (0, 0) creates a problem because log(0) is undefined. This data point must be removed, or all points must be shifted up and to the right. Removing the data point would not create a problem, since power functions (which result from log-log re-expressions) are forced through the point (0, 0).

5. **a.** $y = 11.49x^2 + 0.88x + 0.02$

b. $y = 11.94x^2 + 0.33$

c. $y = 11.50x^2 + 0.93x$

d. $y = 12.61x^{1.90}$

e. The re-expression techniques used for parts a, b, and c assume a quadratic model, and the power function obtained in d tends to confirm that assumption. Coefficients of the quadratic term are reasonably close in all models, but other terms are quite different.

An examination of the models superimposed on the original data and the corresponding residual plots reveals that the quadratic model in part a fits the data best. All other residual plots show that residuals are small in magnitude, but there is a pattern of positive and negative residuals in each. In terms of the phenomenon we are attempting to model, the models in c and d both pass through the point (0, 0), and the models from parts a and c both have a linear term that allows for a nonzero initial velocity of the object.

7. Semi-log re-expression and median-median fit produce the model $y = 14.91e^{-0.18x}$. On the basis of this model, 2.46 mg will remain ten hours after the 15 mg reading.

9. **a.** Functions of the form $y = \frac{1}{x^n} + 45$ for $n > 0$ or $y = e^{kx} + 45$ for $k < 0$ might be good models, since both have horizontal asymptotes at $y = 45$.

b. Answers to this question will vary but should be close to 72° after 100 minutes and close to 45° after three days.

Exercise Set 3

2. **a.** The equation of the least squares line is $y = 3.73x - 9.11$. The slope is misleading, since one extremely high point pulls the line up.

b. The equation of the median-median line is $y = x$. This line conveys more accurate information about the general nature of the data. It is a little misleading, however, since the point (9, 50) has no influence.

3. **a.** Since the summary points do not change, this point has no effect on the median-median line.

b. The point (323, 946.1) changes the least squares line to $y = 2.92x - 31.57$. As expected, the point pulls the line up, increasing the slope and decreasing the y-intercept.

c. Since this point is to the right of the middle of the data set, it will pull the line down slightly, resulting in a smaller slope and a greater y-intercept.

5. **a.** $t = 0.24c + 37.26$

b. The slope indicates that a temperature increase of 0.24° is associated with an increase of one chirp per minute. If the model is appropriate for low numbers of chirps per minute, the y- or t-intercept is an estimate of the temperature when there are no chirps. This interpretation is probably not reasonable, however, since $c = 0$ is much smaller than observed c-values.

c. Since $t \approx \frac{1}{4}c + 37$ where c represents the number of chirps per minute, it follows that $\frac{1}{4}c$ represents the number of chirps per one-fourth minute (15 seconds). By counting chirps for 15 seconds and adding 37, one can quickly estimate the temperature.

d. In terms of the relationship, if one variable is dependent on the other, then it must be the case that cricket chirps are dependent on temperature. This is not the same variable that was treated as the dependent variable in a. The new least squares line through ordered pairs of the form (temperature, chirps per minute) is $y = 4.15x - 153.70$. The slope implies that for each one-degree increase in temperature, cricket chirps increase about four chirps per minute. The y-intercept is not meaningful, since there would be no crickets chirping at 0° F.

Exercise Set 4

1. **a.** The least squares line through the ordered pairs $(x, \ln y)$ is $\ln y = 0.140x - 1.250$. Converting to the original variables, the model is $y = 0.287e^{0.140x}$.

b. The size of the tumor increases at the rate of about 14% per day.

2. **a.** The least squares line through the ordered pairs $(\ln x, y)$ is $y = 0.299 \ln x + 2.849$ where y represents walking speed.

b. When $y = 0$, $x = 0.000073$ thousand, or 0.073; there would be less than one person in the "city" when the walking speed is zero.

c. According to this model, walking speed increases as population increases. The increase is slow, however, since the walking speed is a logarithmic function of population. There is no upper bound to walking speed according to this model.

d. The model implies that in large urban areas the pace of life is faster. Though this situation may create more stress, one must be careful not to infer that a causal relationship necessarily exists.

CHAPTER 6

Modeling

Exercise Set 2

3. **a.** All darts hit the board, and the darts are randomly thrown.

b. 0.068

c. 2.27

5. 0.64

7. $k = 4$; $k \geq 8$

Exercise Set 3

1. 20 shots

Exercise Set 5

1. $t = 0.004c^2 + 2.60c + 0.40$. The coefficients in this model are very close to those for the models developed in the discussion.

CHAPTER 7

Trigonometry

Exercise Set 1

1. **a.** -1

c. 0

e. 0

g. 1

2.

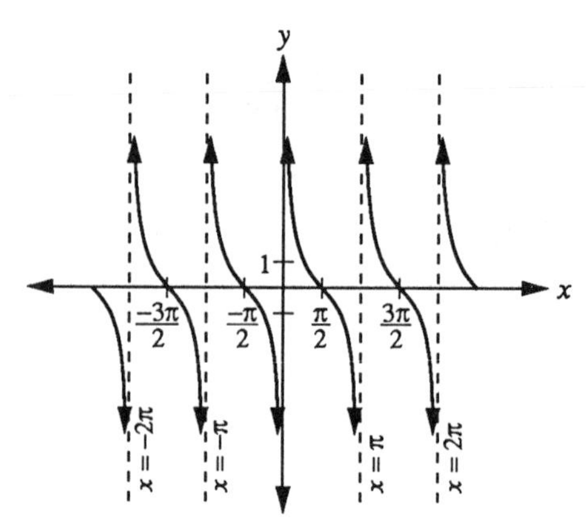

domain: **R** except $k\pi$, k any integer
range: **R**
x-intercepts: $\frac{\pi}{2} + k\pi$, k any integer
asymptotes: $x = k\pi$, k any integer
period: π

3.

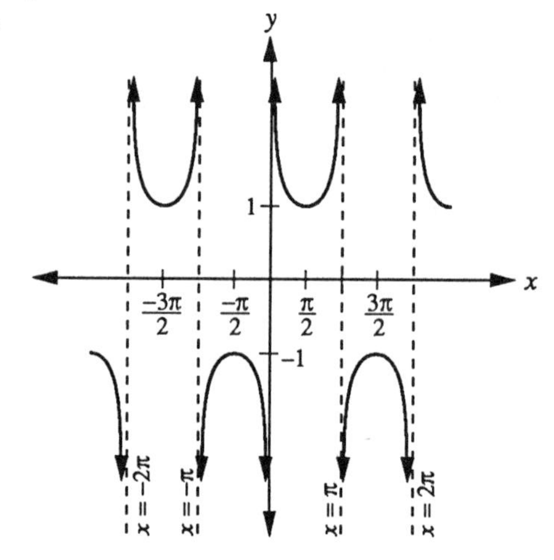

domain: **R** except $k\pi$, k any integer
range: $y \leq -1$ or $y \geq 1$
x-intercepts: none
asymptotes: $x = k\pi$, k any integer
period: 2π

5. **a.**

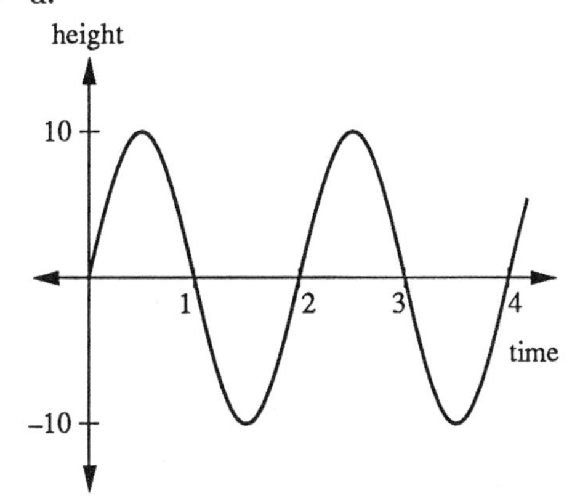

c.

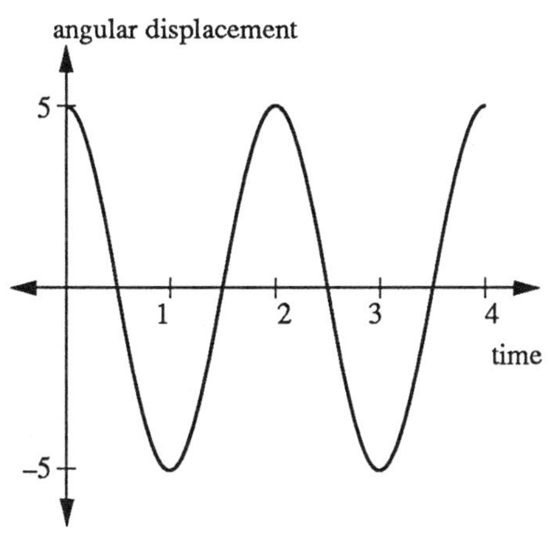

7. Shift $y = \sec x$ to the right $\frac{\pi}{2}$ units to obtain the graph of $y = \csc x$: $\csc x = \sec(x - \frac{\pi}{2})$.

Exercise Set 2

1. $y = 1.3 \cos \pi(x - 1)$

3. **a.** period: 4π frequency: $\frac{1}{4\pi}$

c. period: $\frac{\pi}{2}$ frequency: $\frac{2}{\pi}$

e. period: 1 frequency: 1

4. period: $\frac{2\pi}{B}$ frequency: $\frac{B}{2\pi}$

5. **a.**

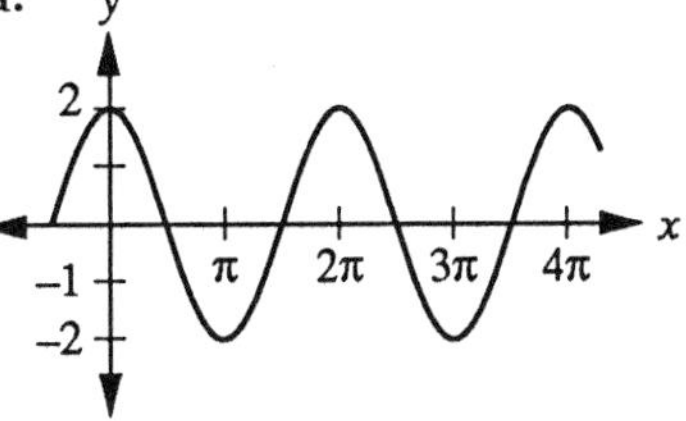

c.

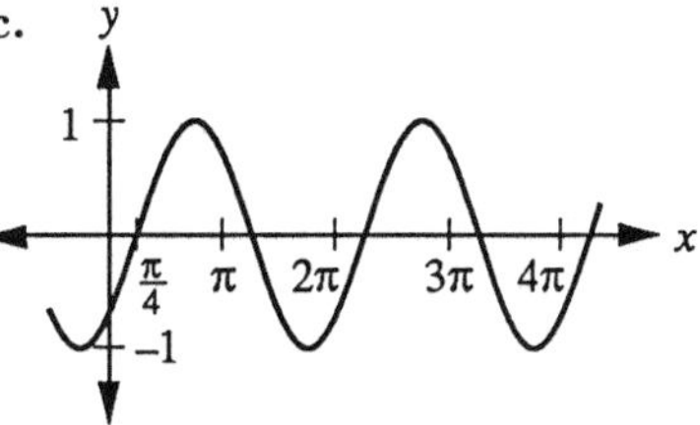

e.

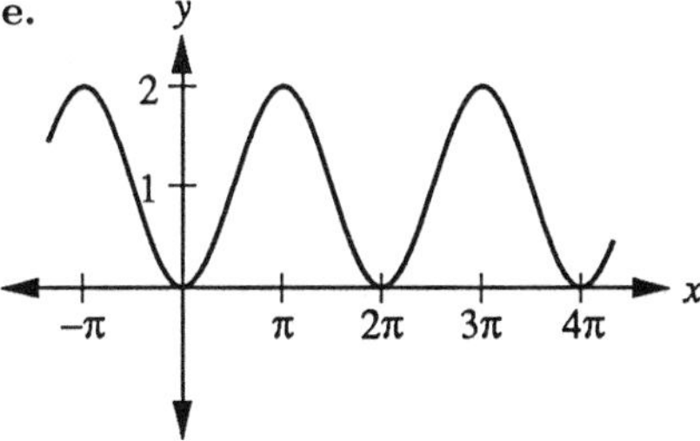

g.

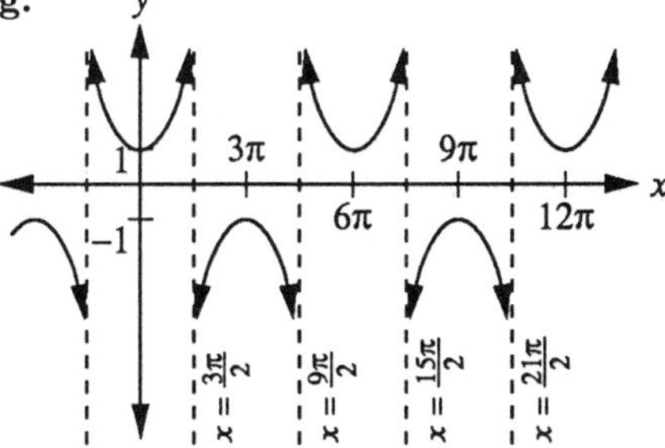

i.

j.

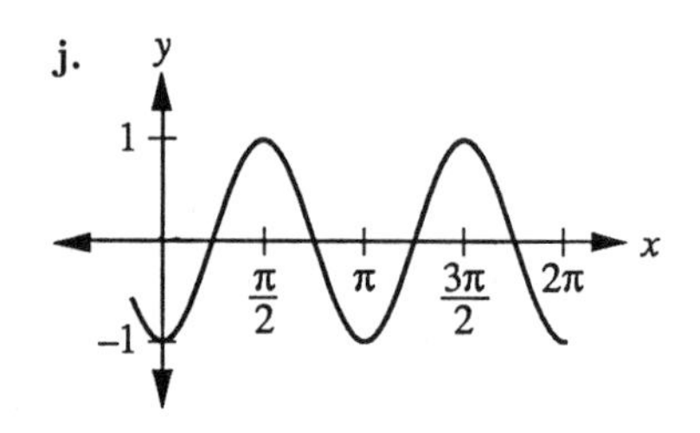

l.

n.

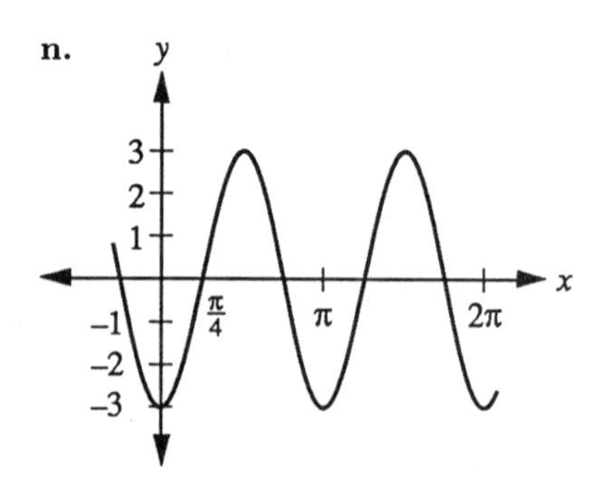

8. **a.** $y = 4\sin x$

b. $y = 0.1\sin 6(x + \frac{\pi}{4})$

c. $y = \frac{1}{3}\sin 2(x - 1)$

d. $y = 1\sin \pi(x + \pi)$

9. $y = -33.2\cos(\frac{2\pi}{15}x) + 37.2$

10. **a.**

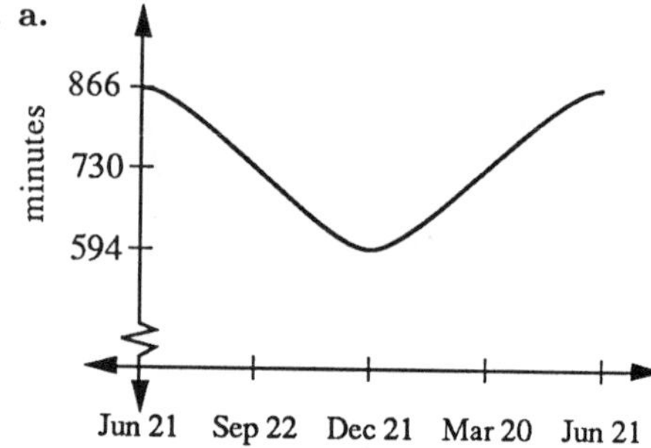

b. $d = 136\cos(\frac{\pi}{6}x) + 730$

11. $y = 1.3\cos(2\pi x) + 23.1$

Exercise Set 3

1. **a.** 1

c. 1

e. 0

g. 0

i. undefined

k. -1

3. 3rd quadrant; $(-0.416, -0.909)$

4. **a.** $\frac{1}{2}$, $\frac{1}{2}$, 0, 0, $\frac{1}{2}$, $-\frac{1}{2}$

b.

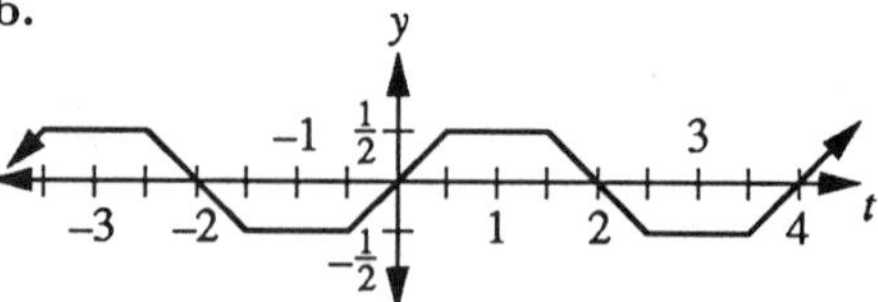

c. yes; 4

e. yes;
square-cosine(x) = square-sine$(x + 1)$

Exercise Set 4

1. $\sin\frac{\pi}{6}=\frac{1}{2}$, $\cos\frac{\pi}{6}=\frac{\sqrt{3}}{2}$
3. a. $\frac{\sqrt{2}}{2}$
   c. $\frac{-\sqrt{3}}{2}$
   d. $\frac{-\sqrt{3}}{3}$
   e. $\frac{\sqrt{2}}{2}$
   g. $\frac{\sqrt{3}}{3}$
   i. $-\sqrt{2}$
5. a. 0.36
   b. 0.9330
   c. 0.3859
   d. −0.3859
7. $x=3+2\pi k$ or $x=(\pi-3)+2\pi k$, k any integer
9. arc $(2\pi-5)$: $(0.28, 0.96)$, arc $(5+\pi)$: $(-0.28, 0.96)$, arc $(3\pi-5)$: $(-0.28,-0.96)$
10. $\sin(a)=0.8980$, $\cos(a+\pi)=-0.44$, $\sin(2\pi-a)=-0.90$
13. a. $-\sqrt{1-w^2}$
    b. $\dfrac{-\sqrt{1-w^2}}{w}$
    c. $-\sqrt{1-w^2}$
    d. $\frac{1}{w}$
    e. $-\frac{1}{w}$
    f. $\dfrac{w}{\sqrt{1-w^2}}$
    g. w

Exercise Set 5

1. $\frac{7\sqrt{53}}{53}=0.9615$
3. $\frac{6\sqrt{61}}{61}=0.7682$
5. $-\frac{44\sqrt{257}}{1285}=-0.5489$
7. $a=3.9847$
   $b=2.0303$
8. 54.74°
9. $(-1.684, 2.483)$, $(-1.684, -2.483)$

Exercise Set 6

1. a. 300°
   c. 286.48°
   e. 500°
2. a. 17.45
   c. 12.57
   e. 0.017
3. $\frac{2\pi}{7}+2k\pi$, $-\frac{2\pi}{7}+2k\pi$, k any integer
5. 4.48 cm
8. 12.73 turns
10. a. $g(0)=0$, $g(\pi)=2$, $g(2\pi)=0$
    b. domain: **R**
    c. range: $-2\le y\le 2$
    d. period: 4π
    e.

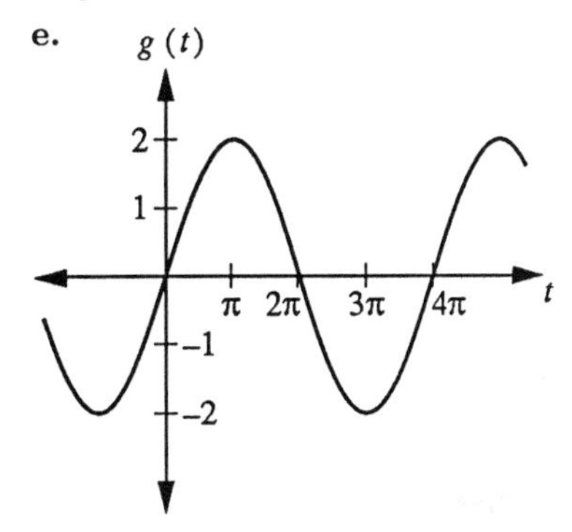

    f. $g(t)=2\sin(\frac{1}{2}t)$
11. $f(t)=r\cos(\frac{1}{r}t)$, $g(t)=r\sin(\frac{1}{r}t)$
    The r before the sine or cosine indicates that the ant is r units from the center of the circle; hence both x- and y-coordinates vary from $-r$ to r. The r inside the parentheses indicates that the period is $2\pi r$, which is the circumference of the circle.

Exercise Set 7

1. a. $\pm 0.72+2k\pi$, k any integer
   c. $1.197+2k\pi$ or $3.52+2k\pi$, k any integer
   e. $-2.77+2k\pi<x<1.20+2k\pi$, k any integer
   g. $0.845+k\pi$ or $0.202+k\pi$, k any integer
2. 2 a.m.; 5a.m. $+6k$, k any integer
3. a. $\pm\frac{\pi}{6}+k\pi$, k any integer

c. $-0.623 + k\pi$ or $-1.23 + k\pi$, k any integer

e. ± 1.18

g. $k\pi$, k any integer

i. ± 0.95

j. 0.88

k. 0

m. $2k\pi \leq x \leq (2k+1)\pi$, k any integer

5. a.

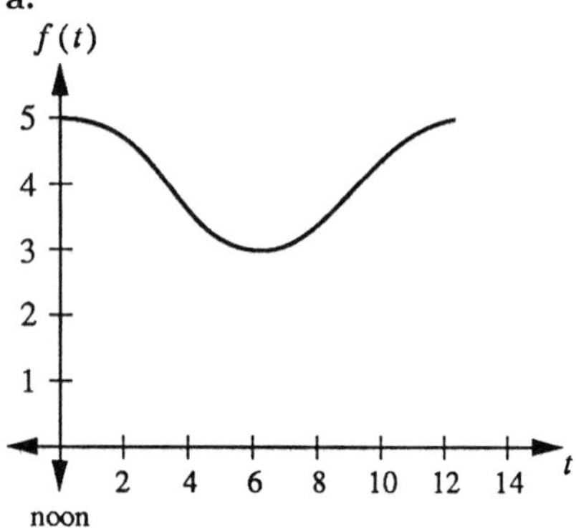

$f(t) = \cos(\frac{\pi}{6.2})t + 4$

b. 8:54 a.m.; 3:06 p.m.

c. 7:52 a.m.; 4:08 p.m.

d. 3:53 p.m.

Exercise Set 8

1. a. 1.25

c. 1.55

e. 2.31

g. 1.56

i. 0.79 or $\frac{\pi}{4}$

k. 0.84

m. 2.09 or $\frac{2\pi}{3}$

o. -0.21

2. b. 1.14; the inverse sine is between $-\frac{\pi}{2}$ and $\frac{\pi}{2}$

c. -0.39; the cosine of the inverse cosine is -0.39

3. a.

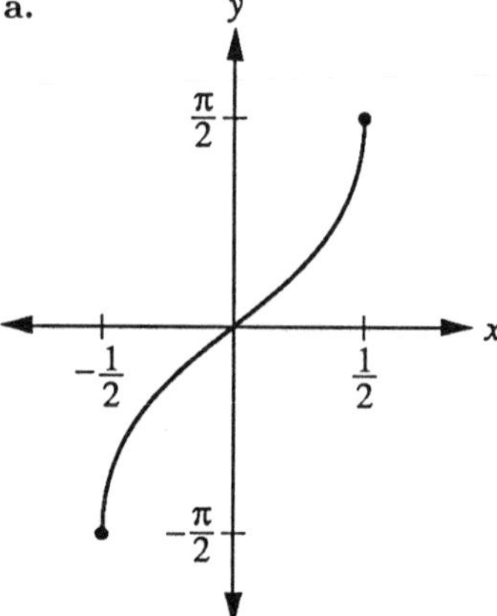

c.

y
1 + π
4
3
1 + π/2
2
1
-1
1
x

4. a. $f^{-1}(x) = \frac{1}{2}\sin x$

c. $h^{-1}(x) = \tan\frac{x}{2}$

5. a. domain: $-\frac{\pi}{2} \leq x \leq \frac{\pi}{2}$
range: $-\frac{1}{2} \leq y \leq \frac{1}{2}$

c. domain: $-\pi < x < \pi$
range: $\mathbf{R}$

7.

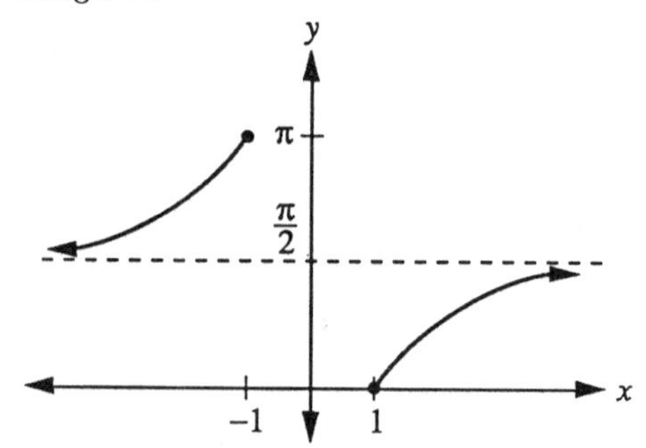

9. a. not continuous at $x = 0$

b. if $x < 0$ add π to the result

Exercise Set 9

1. **a.** $\frac{\sqrt{15}}{4}$

c. $\dfrac{1}{\sqrt{1+4x^2}}$

e. $\pm\sqrt{x^2-2x}$, $x \le 0$ or $x \ge 2$

2. **a.**

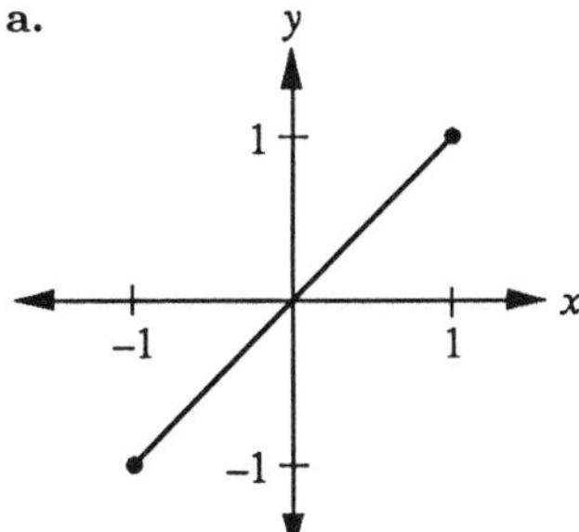

b.

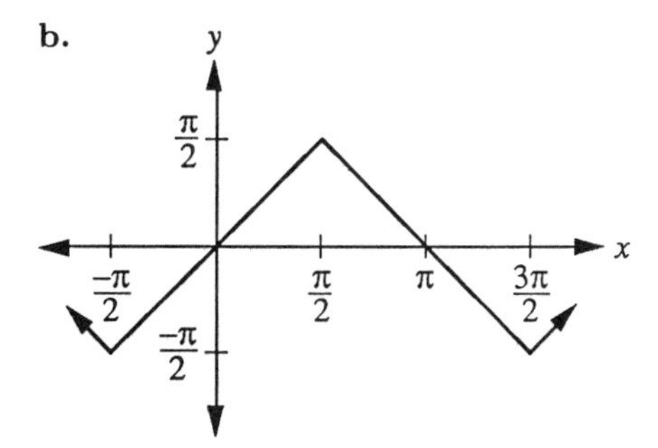

3. **a.** $\frac{1}{\sqrt{5}}$

b. 0.6

c. no solution

5. $\theta = \tan^{-1}(\frac{70}{100-x}) + \tan^{-1}(\frac{115}{x})$; 36.9 meters from the taller building

Exercise Set 10

1. **a.**

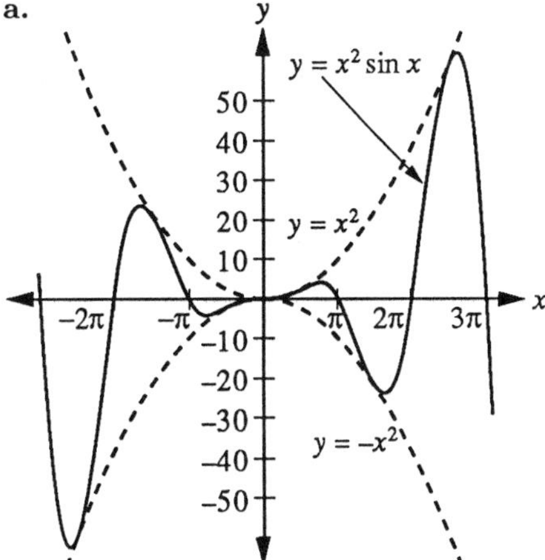

c.

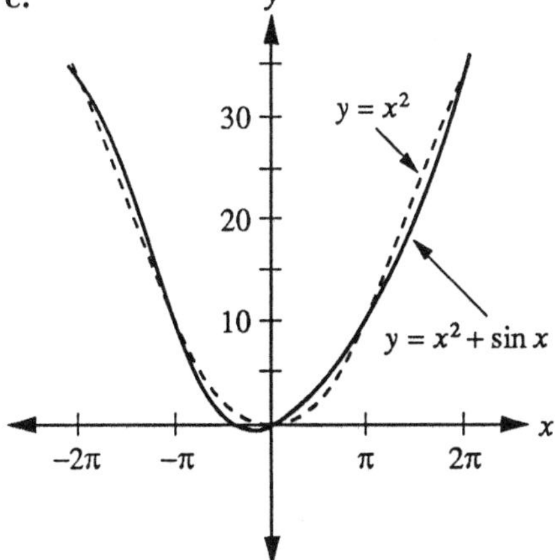

e.

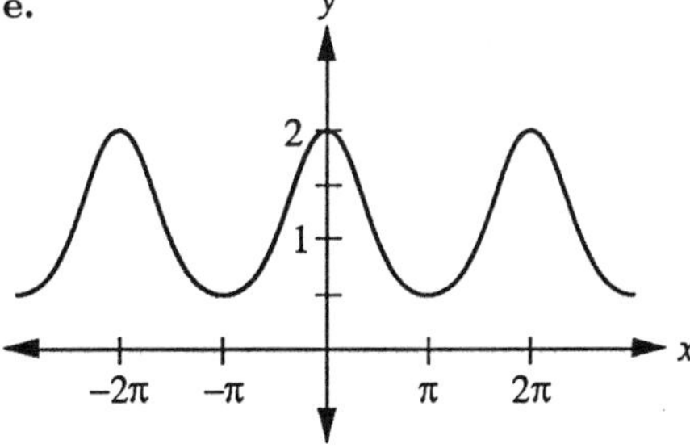

f.

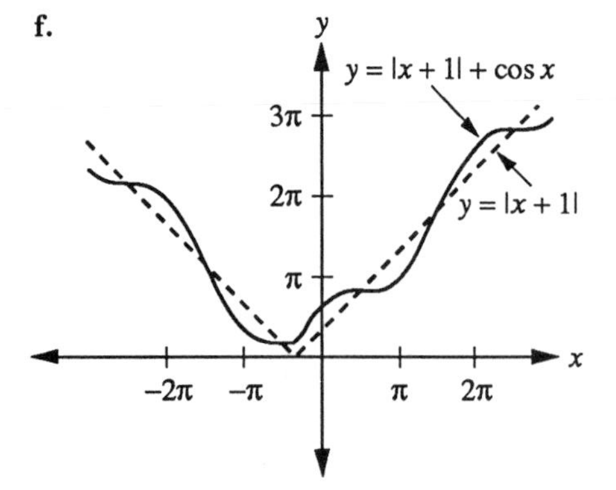

h.

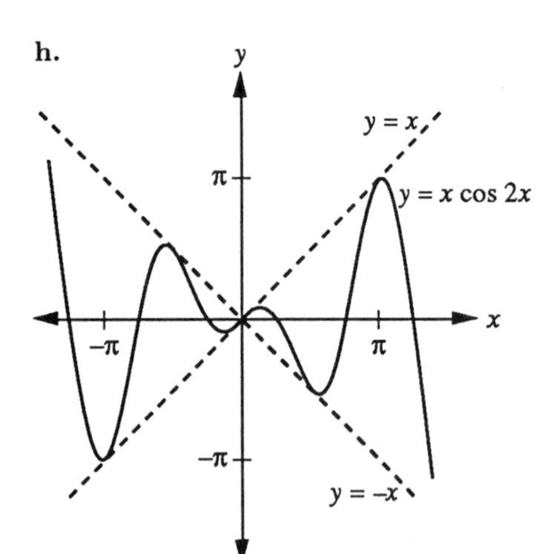

i.

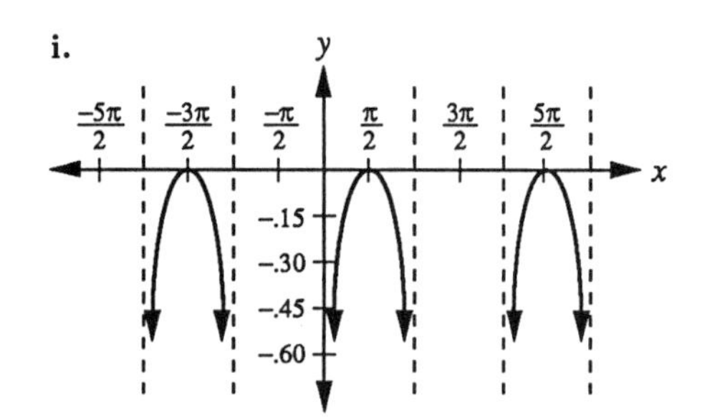

k.

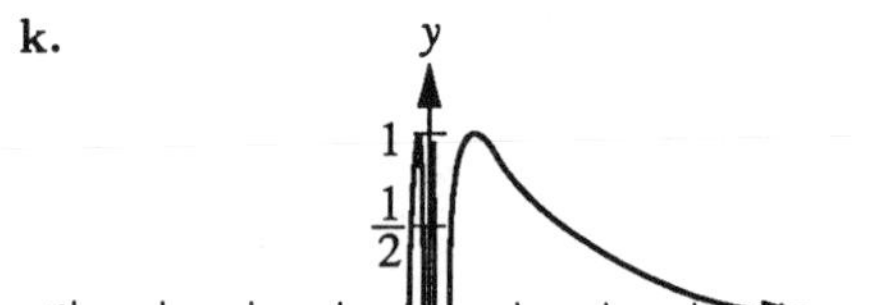
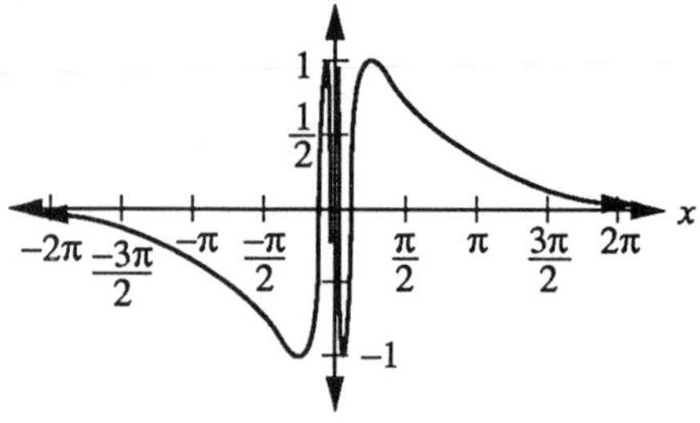

n.

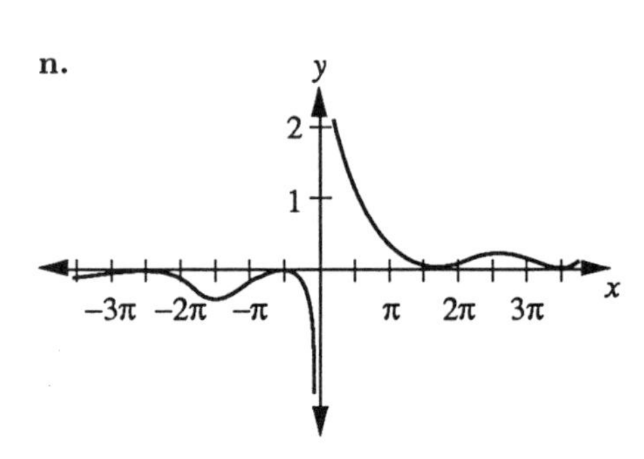

o.

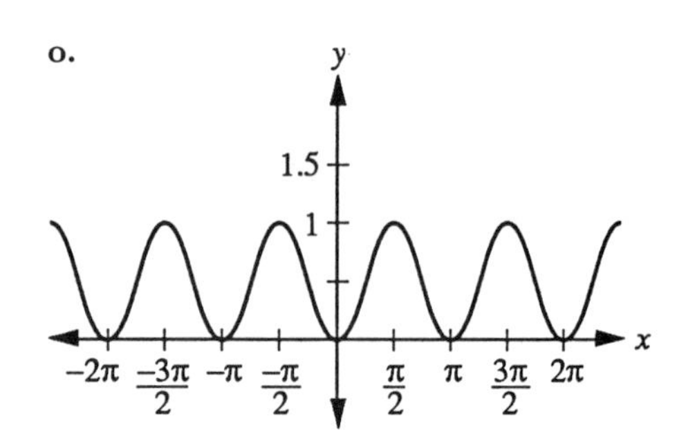

q.

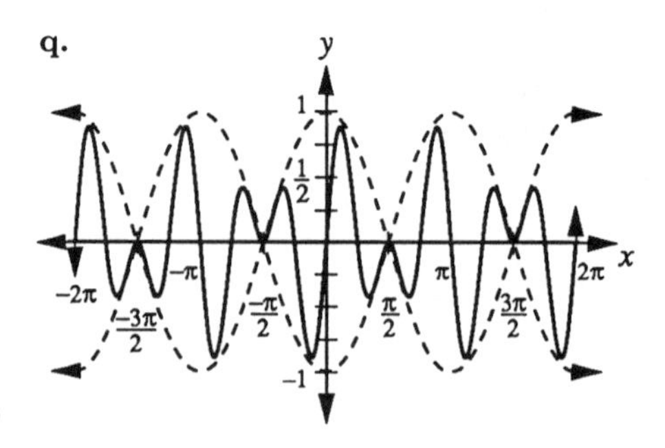

2. **a.** $p(x) = \sin(\frac{2\pi}{23}x)$

$e(x) = \sin(\frac{2\pi}{28}x)$

$i(x) = \sin(\frac{2\pi}{33}x)$

d. $F(x)$ has relative maxima at $x \approx 7$ and $x \approx 372$.

e. 21,252 days

3.

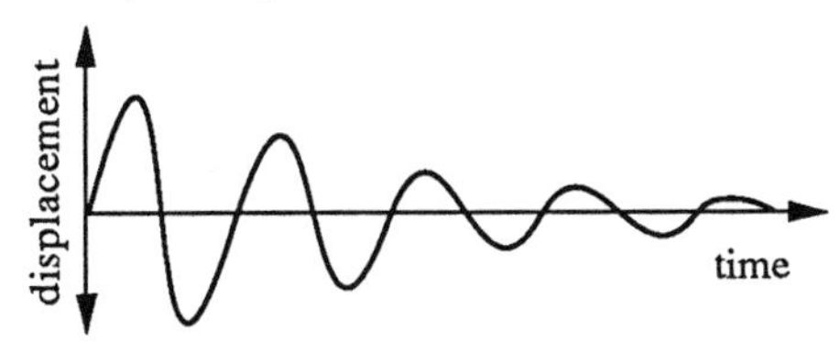

$f(x) = e^{-ax} \sin bx$

Exercise Set 11

1. a. $-\frac{\sqrt{15}}{4}$

b. $-\frac{\sqrt{15}}{4}$

c. $\sqrt{15}$

d. $\frac{\sqrt{2}}{8}(1 + \sqrt{15})$

3. a. $\frac{\sqrt{15}}{4}$

b. $-\frac{\sqrt{15}}{8}$

c. $\frac{\sqrt{2}}{8}(\sqrt{15} - 1)$

4. a. $\frac{\pi}{8}, \frac{5\pi}{8}, \frac{9\pi}{8}, \frac{13\pi}{8}$

c. $0, \frac{2\pi}{3}, \frac{4\pi}{3}, 2\pi$

e. $0, \frac{\pi}{4}, \pi, \frac{5\pi}{4}, 2\pi$

g. $0, \frac{\pi}{3}, \frac{2\pi}{3}, \pi, \frac{4\pi}{3}, \frac{5\pi}{3}, 2\pi$

5. a. $x = 2k\pi$, k any integer

c. $x = \pm 0.947747$

e. $x = \frac{\pi}{6} + k\pi$ or $\frac{\pi}{3} + k\pi$, k any integer

6. $x = k\pi$ or $\frac{\pi}{6} + 2k\pi$ or $\frac{5\pi}{6} + 2k\pi$, k any integer

7. no solution

9. a.

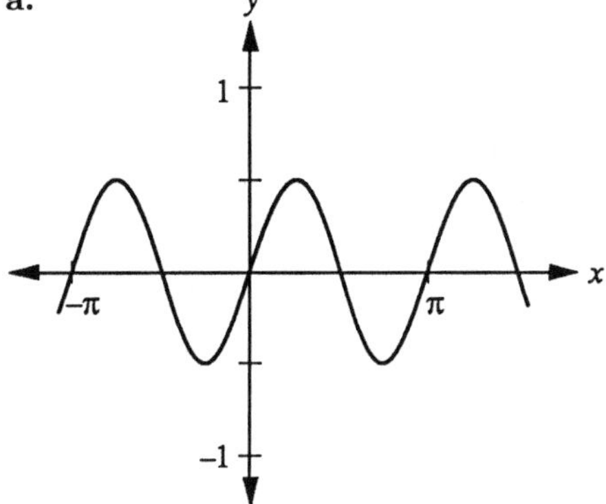

c.

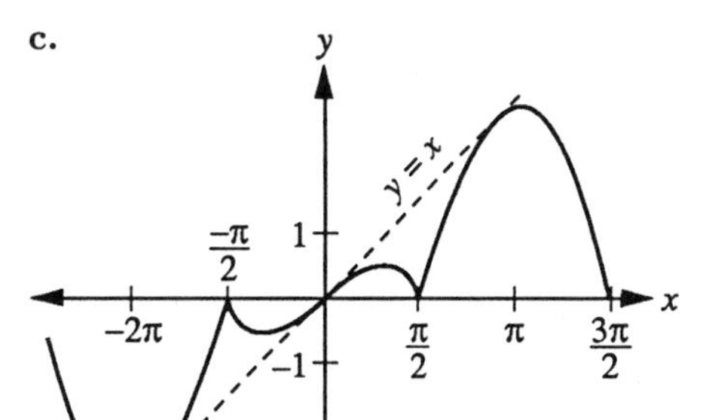

11. $|f_1 - f_2|$ beats per unit time

13. $y = 2\sin(168\pi t)\cos(8\pi t)$

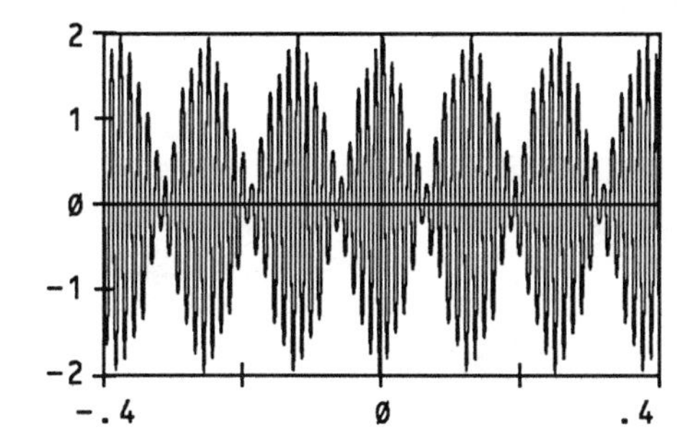

14. a. $PQ = \sqrt{(\cos a - \cos b)^2 + (\sin a - \sin b)^2}$

$AR = \sqrt{(\cos(a - b) - 1)^2 + (\sin(a - b) - 0)^2}$

Exercise Set 12

1. **a.** $b = 7.8$, $C = 51.8°$, $A = 78.2°$

b. $B = 90°$, $C = 60°$, $c = 10.4$

c. $C = 59°$, $A = 91°$, $a = 14$ and $C = 121°$, $A = 29°$, $a = 6.8$

d. $A = 21.8°$, $B = 38.2°$, $C = 120°$

3. air speed ≈ 398.6 mi/hr.
heading $= 250.54°$

5. 11.5°

7. 2.9 mi.

9. 1.08 miles

11. 13 ft., 33.4° west of north

13. **a.** 27° elevation

b. 18.5° elevation

15. **a.** 69 mi.

d. 64.3° latitude

Exercise Set 13

1. Average $v_1 = 1974$
$L = my_2 + b$ where $m = 0.000349$ and $b = 0.7033221$
$v_2 \approx 2865$ meters per second
$d \approx 957$ meters
Material located about 957 m below surface is probably sandstone.

CHAPTER 8

Matrices

Exercise Set 2

1. **a.**

$$\begin{array}{c} \\ H \\ P \end{array} \begin{array}{c} \begin{array}{cccc} T & F & N & R \end{array} \\ \begin{pmatrix} 5280 & 1680 & 2320 & 1890 \\ 1940 & 2810 & 1490 & 2070 \end{pmatrix} \end{array}$$

b.

$$\begin{array}{c} \\ H \\ P \end{array} \begin{array}{c} \begin{array}{cccc} T & F & N & R \end{array} \\ \begin{pmatrix} 6340 & 2220 & 1790 & 1980 \\ 2050 & 3100 & 1720 & 2710 \end{pmatrix} \end{array}$$

c.

$$\begin{array}{c} \\ H \\ P \end{array} \begin{array}{c} \begin{array}{cccc} T & F & N & R \end{array} \\ \begin{pmatrix} 11{,}620 & 3900 & 4110 & 3870 \\ 3990 & 5910 & 3210 & 4780 \end{pmatrix} \end{array}$$

Exercise Set 3

1. MO: 2×1, MP: 2×2, PM: 2×2, MR: 2×2, RM: 2×2, NQ: 3×1, NU: 3×4, PO: 2×1, PR: 2×2, RP: 2×2, RO: 2×1, SM: 4×2, SO: 4×2, SP: 4×2, SR: 4×2, US: 3×2, UT: 3×1

3. yes, matrix multiplication is associative: $M(RP) = (MR)P$

5. Matrix multiplication is not commutative; AB may not be equal to BA.

7.

$$\begin{array}{ccc} 1984 & 1985 & 1986 \end{array}$$
$$\begin{pmatrix} 30{,}175 & 31{,}800 & 34{,}050 \end{pmatrix}$$

9. **a.**

$$\begin{array}{cccc} t & ca & co & d \end{array}$$
$$\begin{pmatrix} 2810 & 1656 & 358 & 228 \end{pmatrix}$$

b.

$$\begin{array}{cccc} B & Ec & Ex & P \end{array}$$
$$\begin{pmatrix} 14 & 16.8 & 15.4 & 9.8 \end{pmatrix},$$

which should be rounded to integer quantities.

c. With 7-hour workdays, the number of employees needed is $249/7 = 35.6$, which implies that 36 employees are needed to maintain full production. For August and September, we want $348.6/7$, which rounds to 50.

11. The goal for each branch in each type of account is:

$$\begin{array}{c} \\ N \\ D \\ S \end{array} \begin{array}{c} \begin{array}{ccc} c & s & m \end{array} \\ \begin{pmatrix} 48{,}448 & 13{,}683 & 779 \\ 18{,}430 & 11{,}814 & 160 \\ 30{,}991 & 16{,}453 & 148 \end{pmatrix} \end{array}$$

The total number of accounts at each branch is:

$$\begin{array}{c} \\ N \\ D \\ S \end{array} \begin{array}{c} Total \\ \begin{pmatrix} 62{,}910 \\ 30{,}404 \\ 47{,}592 \end{pmatrix} \end{array}$$

13. **a.**

$$\begin{array}{c} \\ P \\ E \\ AA \end{array} \begin{array}{c} \begin{array}{cc} C & LT \end{array} \\ \begin{pmatrix} 15{,}110 & 9870 \\ 13{,}175 & 8600 \\ 13{,}250 & 9130 \end{pmatrix} \end{array}$$

b.

$$\begin{array}{c} \\ P \\ E \\ AA \end{array} \begin{array}{c} Cost \\ \begin{pmatrix} 20,447 \\ 17,822.5 \\ 18,405 \end{pmatrix} \end{array}$$

c.

$$\begin{array}{c} \\ P \\ E \\ AA \end{array} \begin{array}{c} \begin{array}{ccc} 1st & C & Y \end{array} \\ \begin{pmatrix} 441 & 294 & 175 \\ 520 & 280 & 220 \\ 525 & 277.5 & 112.5 \end{pmatrix} \end{array}$$

15. a.

$$\begin{array}{c} \\ Cutting \\ Sewing \\ Finishing \end{array} \begin{array}{c} \begin{array}{cc} Oct & Nov \end{array} \\ \begin{pmatrix} 1300 & 1520 \\ 1800 & 2092.5 \\ 1240 & 1362.5 \end{pmatrix} \end{array}$$

b.

$$\begin{array}{c} \\ Panda \\ Kangaroo \\ Rabbit \end{array} \begin{array}{c} \begin{array}{ccc} East & Central & West \end{array} \\ \begin{pmatrix} 15.99 & 14.46 & 18.60 \\ 18.36 & 16.64 & 21.22 \\ 11.70 & 10.60 & 13.61 \end{pmatrix} \end{array}$$

c.

$$\begin{array}{c} \\ Oct \\ Nov \end{array} \begin{array}{c} \begin{array}{ccc} East & Central & West \end{array} \\ \begin{pmatrix} 36,366 & 32,924 & 42,220 \\ 41,677.5 & 37,735 & 48,364.3 \end{pmatrix} \end{array}$$

d.

$$\begin{array}{c} \\ East \\ Central \\ West \end{array} \begin{array}{c} \begin{array}{ccc} Cutting & Sewing & Finishing \end{array} \\ \begin{pmatrix} 29.3 & 41 & 27 \\ 27 & 35.5 & 21.5 \\ 28 & 38.5 & 25.5 \end{pmatrix} \end{array}$$

e.

$$\begin{array}{c} \\ East \\ Central \\ West \end{array} \begin{array}{c} \begin{array}{ccc} East & Central & West \end{array} \\ \begin{pmatrix} 815.55 & 738.30 & 946.62 \\ 702.60 & 636.40 & 814.55 \\ 770.70 & 697.80 & 894.45 \end{pmatrix} \end{array}$$

Exercise Set 4

1. a.

$$\begin{array}{c} \\ Petr. \\ Text. \\ Tran. \\ Chem. \end{array} \begin{array}{c} \begin{array}{cccc} Petr. & Text. & Tran. & Chem. \end{array} \\ \begin{pmatrix} 0.1 & 0.4 & 0.6 & 0.2 \\ 0 & 0.1 & 0 & 0.1 \\ 0.2 & 0.15 & 0.1 & 0.3 \\ 0.4 & 0.3 & 0.25 & 0.2 \end{pmatrix} \end{array}$$

b. chemicals, textiles

c. $1,600,000

d.

$$\begin{array}{c} Petr. \\ Text. \\ Tran. \\ Chem. \end{array} \begin{pmatrix} 730 \\ 95 \\ 485 \\ 705 \end{pmatrix}$$

70 units of petroleum

e.

$$\begin{array}{c} Petr. \\ Text. \\ Tran. \\ Chem. \end{array} \begin{pmatrix} 195.6 \\ 39.3 \\ 154.4 \\ 213.3 \end{pmatrix}$$

3.

$$\begin{array}{c} \\ Manu. \\ Petr. \\ Tran. \\ HP \end{array} \begin{array}{c} P \\ \begin{pmatrix} 579.25 \\ 572.31 \\ 476.21 \\ 464.83 \end{pmatrix} \end{array}$$

Exercise Set 5

1. There are no elementary row operations which can create a one in a column with all zeros; thus, such a matrix could never be transformed into an identity matrix.

2. **a.** $x = 1$, $y = 1$, $z = 1$.

b. The system is dependent and has an infinite number of solutions.

c. The system is inconsistent and has no solutions.

3. **a.**

$$\begin{pmatrix} 0.16 & -0.04 & 0.24 \\ -0.07 & -0.07 & 0.07 \\ 0.04 & -0.08 & 0 \end{pmatrix}$$

5. **a.** 4 hamburgers and 4 chicken sandwiches

b. 1/2 hamburger (which really means no hamburgers) and 8 chicken sandwiches

c. 7.5 hamburgers (so just 7 complete hamburgers) and 0 chicken sandwiches

7. **a.** The Student Council would receive $15,000, and the Beta Club would receive $22,500, but the 4-H Club would have to pay $7,500 — an obvious problem with the solution.

b. The Beta Club would receive $30,000 and the other groups would receive nothing.

c. The Student Council receives $7,500, the Beta Club receives $11,250, and the 4-H Club receives $11,250.

Exercise Set 6

1. 27.9304, after 5 years, 50.866 after 10 years

3. **a.** There are 243.934 females after 56 quarters and 251.357 females after 57 quarters.

   b. There are 249.012 females after 66 quarters and 256.589 females after 67 quarters.

   c. There are 246.026 females after 69 quarters and 253.513 females after 70 quarters.

5.

$$\begin{array}{c} \\ 0\text{-}3 \\ 3\text{-}6 \\ 6\text{-}9 \\ 9\text{-}12 \\ 12\text{-}15 \\ 15\text{-}18 \end{array} \begin{array}{c} \textit{Prop.} \\ \begin{pmatrix} 0.324597 \\ 0.189007 \\ 0.165082 \\ 0.144186 \\ 0.111942 \\ 0.065200 \end{pmatrix} \end{array}$$

for the first distribution, and

$$\begin{array}{c} \\ 0\text{-}3 \\ 3\text{-}6 \\ 6\text{-}9 \\ 9\text{-}12 \\ 12\text{-}15 \\ 15\text{-}18 \end{array} \begin{array}{c} \textit{Prop.} \\ \begin{pmatrix} 0.324598 \\ 0.189007 \\ 0.165083 \\ 0.144187 \\ 0.111943 \\ 0.065200 \end{pmatrix} \end{array}$$

for the second distribution. The long-term proportions of the population in each age group are independent of the initial distribution.

Exercise Set 7

1. **a.**

$$\begin{array}{c} \\ 1 \\ 2 \\ 3 \\ 4 \\ 5 \end{array} \begin{array}{c} \begin{array}{ccccc} 1 & 2 & 3 & 4 & 5 \end{array} \\ \begin{pmatrix} 0 & \frac{1}{2} & 0 & 0 & \frac{1}{2} \\ \frac{1}{2} & 0 & 0 & 0 & \frac{1}{2} \\ 0 & 0 & 0 & 1 & 0 \\ 0 & 0 & \frac{1}{2} & 0 & \frac{1}{2} \\ \frac{1}{3} & \frac{1}{3} & 0 & \frac{1}{3} & 0 \end{pmatrix} \end{array}$$

   b. 0; 0.083

   c. The stable state vector is

$$\begin{array}{c} \begin{array}{ccccc} 1 & 2 & 3 & 4 & 5 \end{array} \\ \begin{pmatrix} 0.2 & 0.2 & 0.1 & 0.2 & 0.3 \end{pmatrix}. \end{array}$$

   d. 0.2

   e. In the long run, the rat will spend 30% of the time in rooms 2 or 3.

3. **a.**

$$\begin{array}{c} \\ W \\ D \\ A \end{array} \begin{array}{c} \begin{array}{ccc} W & D & A \end{array} \\ \begin{pmatrix} 0 & 1 & 0 \\ 0 & 0 & 1 \\ 0.5 & 0.5 & 0 \end{pmatrix} \end{array}$$

   b.

$$P^2 = \begin{pmatrix} 0 & 0 & 1 \\ 0.5 & 0.5 & 0 \\ 0 & 0.5 & 0.5 \end{pmatrix}$$

$$P^4 = \begin{pmatrix} 0 & 0.5 & 0.5 \\ 0.25 & 0.25 & 0.5 \\ 0.25 & 0.5 & 0.25 \end{pmatrix}$$

$$P^6 = \begin{pmatrix} 0.25 & 0.5 & 0.25 \\ 0.125 & 0.375 & 0.5 \\ 0.25 & 0.375 & 0.375 \end{pmatrix}$$

$$P^{10} = \begin{pmatrix} 0.1875 & 0.375 & 0.4375 \\ 0.21875 & 0.40625 & 0.375 \\ 0.1875 & 0.40625 & 0.40625 \end{pmatrix}$$

   The zero entries vanish because by the sixth throw everyone has a chance to catch the frisbee no matter who has it initially.

5. **a.** yes; the stable state vector is

$$\begin{pmatrix} 0.2825 & 0.3455 & 0.1900 & 0.1822 \end{pmatrix}$$

   b. The stable state vector shows that 18.22% of the students buy ice cream. A Markov chain model assumes that each step in the process depends only on what happened in the previous step, so it is assumed that people who had ice cream more than 1 day before forgot that they did not like it.

   c. A Markov chain is not a good model for the situation. People will remember that they do not like the ice cream, so what they buy each day is dependent on more than what they bought the day before.

7. **a.**

$$\begin{array}{c} \\ I \\ II \\ III \\ IV \end{array} \begin{array}{c} \begin{array}{cccc} I & II & III & IV \end{array} \\ \begin{pmatrix} 1 & 0 & 0 & 0 \\ 0 & 1 & 0 & 0 \\ 0 & 0.01994 & 0.98 & 0.00006 \\ 0.02 & 0.03 & 0 & 0.95 \end{pmatrix} \end{array}$$

   b.

$$\begin{array}{c} \\ I \\ II \\ III \\ IV \end{array} \begin{array}{c} \begin{array}{cccc} I & II & III & IV \end{array} \\ \begin{pmatrix} 1 & 0 & 0 & 0 \\ 0 & 1 & 0 & 0 \\ 0.0012 & 0.9988 & 0 & 0 \\ 0.4 & 0.6 & 0 & 0 \end{pmatrix} \end{array}$$

c. 0.0012

9. **a.**

$$\begin{array}{l} \\ Well \\ Ill \\ Immune \\ Dead \end{array} \begin{array}{c} \begin{array}{cccc} Well & Ill & Immune & Dead \end{array} \\ \begin{pmatrix} 0.80 & 0.20 & 0 & 0 \\ 0 & 0.55 & 0.40 & 0.05 \\ 0 & 0 & 1 & 0 \\ 0 & 0 & 0 & 1 \end{pmatrix} \end{array}$$

b. **i.** 0.7189
ii. 0.8028
iii. 0.7751
iv. 0.8866

c. The Markov chain does not have a stable state because of the absorbing states; however, the transition matrix raised to a large power does stabilize as

$$\begin{array}{l} \\ Well \\ Ill \\ Immune \\ Dead \end{array} \begin{array}{c} \begin{array}{cccc} Well & Ill & Immune & Dead \end{array} \\ \begin{pmatrix} 0 & 0 & 0.8889 & 0.1111 \\ 0 & 0 & 0.8889 & 0.1111 \\ 0 & 0 & 1 & 0 \\ 0 & 0 & 0 & 1 \end{pmatrix} \end{array}$$

INDEX